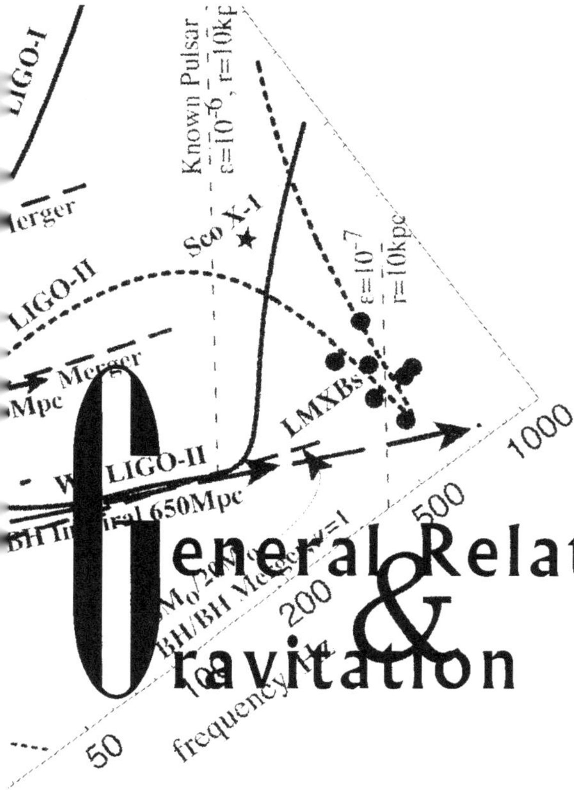

General Relativity & Gravitation

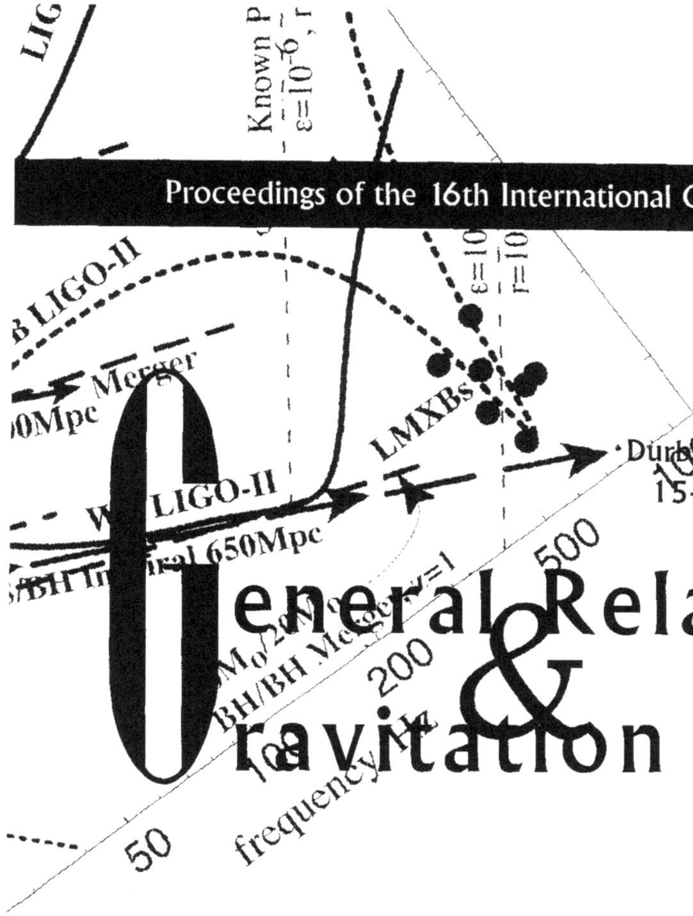

Proceedings of the 16th International Conference on

Durban, South Africa
15-21 July 2001

General Relativity & Gravitation

edited by

Nigel T. Bishop
Department of Mathematics,
Applied Mathematics and Astronomy,
University of South Africa, South Africa

Sunil D. Maharaj
School of Mathematical Sciences,
University of Natal, Durban, South Africa

World Scientific
New Jersey • London • Singapore • Hong Kong

Published by

World Scientific Publishing Co. Pte. Ltd.

5 Toh Tuck Link, Singapore 596224

USA office: 27 Warren Street, Suite 401-402, Hackensack, NJ 07601

UK office: 57 Shelton Street, Covent Garden, London WC2H 9HE

British Library Cataloguing-in-Publication Data
A catalogue record for this book is available from the British Library.

Proceedings of the 16th International Conference
GENERAL RELATIVITY AND GRAVITATION

ISBN-13 978-981-238-171-2
ISBN-10 981-238-171-6

PREFACE

GR16, the 16th conference of the International Society on General Relativity and Gravitation, was held at the International Convention Centre in Durban, South Africa, from 15 to 21 July 2001. The conference was attended by 450 delegates from all over the world. The scientific programme comprised

- 18 Plenary lectures
- 1 Public lecture
- 19 Workshops

which, excepting 3 plenary lectures, are presented in these Proceedings.

GR16 was the first major international conference on general relativity and gravitation to be held on the continent of Africa, and we are very grateful to the International Society for accepting our invitation to host the conference.

The successful running of a conference like GR16 requires input from many people, and we would like to thank the various organizing committees, which are listed below. In particular, the success of the conference owes much to the Scientific Organizing Committee chaired by Jorge Pullin, which put together an exciting academic programme. They chose plenary speakers who managed to present the latest results in a way that was understandable to the broad audience present. Malcolm MacCallum as secretary of the International Society was always available with advice and support. George Ellis and Dan Krige, as chairs of the national and local organising committees respectively, were very supportive before and during the conference.

We thank the plenary speakers and workshop chairpersons for providing us with material according to the specification and on time. A special word of thanks to the Deputy Minister of Education, Mosibudi Mangena, who officially opened the conference; Minister Mangena's presence had a particular resonance as he is a relativist by training. We are grateful to Monas Naidoo who assisted in the preparation of the proceedings and Kesh Govinder who helped with the figures. Robynne Savic and Jacqueline Ingram-Klopper of Conference InterAction, the official conference organisers, proved to be invaluable and provided a service beyond the call of duty. Lastly, we are grateful to World Scientific Publishers for agreeing to publish the Proceedings at a reasonable price; special thanks to Daniel Cartin for his assistance and patience during the preparation of this manuscript.

The conference received sponsorship amounting to R 414 000, and this enabled us to support, wholly or partially, some 50 persons - mainly from

Africa and other developing countries - who would not otherwise have been able to participate. We are grateful for this support.

Nigel Bishop & Sunil Maharaj, June 2002

ORGANIZING COMMITTEES

1 Scientific Organizing Committee

J. Pullin, USA (Chair)
B. Bertotti, Italy
N. Bishop, South Africa
S. Carroll, USA
M. Choptuik, Canada
N. Deruelle, France
G. Ellis, South Africa
D. Galtsov, Russia
C. Isham, UK
W. Israel, Canada
B. Iyer, UK
C. Kozameh, Argentina
R. Loll, Germany
M. MacCallum, UK
D. Matravers, UK
V. Moncrief, USA
R. Newman, USA
T. Newman, USA
N. Robertson, UK
M. Ryan, Mexico
M. Sasaki, Japan
B. Schmidt, Germany
S. Scott, Australia
E. Verdaguer, Spain
R. Weiss, USA

2 National Organizing Committee

G. Ellis (Chair)
A. Bawa
A. Beesham
P. Dunsby
T. Gebbie
C. Hellaby
A. Kgathi
W. Lesame
M. Maharaj
C. Pooe
S. Xulu

3 Local Organizing Committee

J. Krige (Chair)
S. Maharaj (Secretary)
N. Bishop
M. Govender
K. Govinder
S. Hansraj
P. Leach
D. Lortan
S. Moopanar
H. Murrell
J. Nevin

SPONSORSHIP

- Abdus Salaam International Centre for Theoretical Physics
- Alexander Forbes
- Anglo American Chairman's Fund
- Buffet Investments
- Department of Arts, Culture, Science and Technology, Republic of South Africa
- Gravity Research Foundation
- Hanno Rund Research Fund, School of Mathematical and Statistical Sciences, University of Natal
- International Society on General Relativity and Gravitation
- International Union of Pure and Applied Physics
- National Research Foundation of South Africa
- University of Natal

OPENING ADDRESS

MOSIBUDI MANGENA
Deputy Minister of Education, Private Bag X895, Pretoria 0001, South Africa

Director of Ceremonies, Ladies and Gentlemen:

It gives me great pleasure to be with you and participate in the official opening of the 16th International Conference of the Society on General Relativity and Gravitation. When I received the invitation from Professor Bishop earlier this year, he reminded me that we had met in Stockholm in 1986. How time flies. It was in 1986 that I last attended a conference of my past area of study. It is ironic but pleasant that I get an opportunity to again attend a conference of this stature, now, when I am so far removed from the Relativity and Gravitation environment. I am honoured to be your guest.

This is a landmark gathering in that a truly international conference is being held in Durban. This provides a forum for our local scientists, especially the younger scientists who would not get many opportunities of traveling, to hear and exchange ideas with their international counterparts. The organizing committee deserves much credit for mounting this event in Durban. I trust you will find the time between your deliberations to enjoy the hospitality of the city.

It can only be proper and fitting to open a conference of an organization dedicated to the pursuit of new knowledge by partly reflecting on the role of scientists in society. I am sure that scholars may differ in their opinions on what a scientist is or should be. I quote from Peter Medawar's Plato's Republic:

Scientists are people of very dissimilar temperaments, doing different things in very different ways. Among scientists are collectors, classifiers and compulsive tidiers. Many are detectives by temperament and many are explorers, some are artists and others artisans. There are poet scientists and philosopher scientists and even a few mystics.

I am hopeful that this quote will resonate with some of us in this hall today, and may elicit some debate. Whatever the outcome of that debate, I believe it is an appropriate way to highlight and acknowledge that scientists are ordinary people whose endeavours need to be celebrated.

Scientists also tend to share a set of values that are important for society. Bruce Alberts, the President of the US National Academy of Sciences cites three.

- **Honesty** The whole scientific enterprise depends on the honest reporting of the results of our experiments. The reputation of a scientist and his or her institution, as well as their past findings are at stake when integrity is lost.

- **Generosity** Many scientists spend an inordinate amount of time, giving selflessly to the community in voluntary roles for which they receive no monetary compensation. I am sure that the local organizing committee for this conference was not paid, because what they are doing is for the benefit of the society. Because the youth are the future, many of us give back what we gained by mentoring and nurturing the young scientists. Some of us are active in popularizing our disciplines and ensuring that the public not only understands, but also appreciates the role that science plays in our every day lives.

- The third is **the intrinsic worth of ideas and opinions**. One of the purposes of this gathering is to exchange ideas, to reflect on achievements and recent developments and to plan for the future. You have brought together to this conference, relativists, cosmologists, astrophysicists and mathematicians. Our society can only be enriched by your collective ideas.

These values should not just be restricted to the scientific community. They are values that we should all inculcate into our respective societies.

Your programme this year looks very appealing. If it were possible to extricate myself from my official duties, I would have enjoyed spending the week with you. Having been out of the system for so long, it would be an educational experience to be updated on *The state of cosmology* by Professor George Ellis and on *Quantum geometry and gravity* by Professor Ashtekhar, to cite but a few.

While the scope of the conference is vast and focuses on the technical issues of the discipline, I wonder if at your next conference you may wish to devote some space to the issue of the future of the profession. As you are acutely aware, fewer and fewer young people are coming into the profession. This is particularly worrisome in South Africa and the problem is further compounded by the legacy of our past. I am informed that South Africa, with a population of over 40 million does not have more than 20 black and practising mathematicians.

Worse still are the effects or potential effects of the HIV/AIDS pandemic on our youth. Perhaps in your future deliberations you may wish to build into some papers the strategies or constructive suggestions on how to deal with the effects of HIV/AIDS because the dangers posed by this malaise should be ever present in our minds and consciences.

I am sure you are also asking, what is government doing about these issues?

Figures relating to the participation and attainment of our learners in maths and science are both alarming and depressing. These figures show that participation rates in mathematics and science at Higher Grade, as reflected in the matric results for the years 1997 to 2000, is around 20%.

As a consequence the Minister of Education and myself, recently launched a major initiative to promote science, mathematics and technology education in our school systems. At the time of the launch, there was great public interest in the recent solar eclipse. Understanding the eclipse and being able to observe the phenomenon is something that we owe to scientific research done by many of the scientists gathered here today.

The strategy is multi-layered and comprises both short and medium-term thrusts that address teaching, learning and resources. A key principle in the strategy is to increase the numbers and attainment of learners in these subjects. The key proposals of the strategy include:

- Upgrading unqualified and under qualified mathematics and science teachers who are presently in service.
- Exploring ways of utilising retired or unemployed qualified mathematics and science teachers.
- Co-operating with other governments in training and upgrading our mathematics and science teachers.
- Recruiting and resourcing matriculants with good passes in mathematics and science and get them into the teaching profession.
- Identifying and resourcing, nationally 100 mathematics and science and technology education dedicated schools.

The revival of our schooling system in these critical subjects is paramount to the success of the higher education system and to your specific profession. I look forward to hearing from Professor Bishop and others, on how we can work together to ensure that these initiatives are successful.

On the 5th of March this year, government launched the National Plan for Higher Education. It is a plan that will shape the transformation of the South African system for decades to come.

The Plan is a living document. It provides the strategy for realising our policy goals for a new South African higher education system. Its implementation is well under way.

It is a product of engagement and consultation that has taken place over the past decade. The National Education Policy Investigation (NEPI) of the early 1990s kick started the debate by placing policy options on the table. The report of the National Commission on Higher Education (1996) laid the basis for the subsequent Green Paper, the Draft White Paper and Education White Paper 3 - A Programme for the Transformation of the Higher Education System (1997). In June last year, the Council on Higher Education (CHE) produced a report entitled *Towards a New Higher Education Landscape: Meeting the Equity, Quality and Social Development Imperatives of South Africa in the 21st Century.* The Plan marks the beginning of a new phase of delivery of a quality higher education system that we are confident will contribute to the social, economic and political challenges that face our country.

The plan establishes indicative targets for the size and shape of the higher education system, including overall growth and participation rates, institutional and programmes mixes, equity and efficiency goals. It provides a framework and outlines the process for the restructuring of the institutional landscape of the system. It also provides the signposts for the development of institutional plans.

The key proposals of the Plan include the following, that:

- The participation rate in higher education will be increased from 15% to 20% in the long-term, i.e. ten to fifteen years.
- There will be a shift in the balance of enrolments over the next five to ten years between the humanities, business and commerce and science, engineering and technology from the current ratio of 49: 26: 25 to 40: 30: 30 per cent, respectively.
- Institutions will be expected to establish student and staff equity targets, with student targets emphasised on the programmes in which blacks and women students are under-represented.
- Institutional diversity will be achieved through the approval of a distinct mission and academic programme profile for each university and technikon.
- The academic programme mix at each institution will be determined on the basis of its current programme profile, as well as the demonstrated capacity to add new programmes to the profile.
- The existing mission and programme differentiation between technikons and universities will be loosely maintained for at least the next five years.

- Redress for historically black institutions will be linked to agreed missions and programme profiles, including developmental strategies to build capacity.
- A single dedicated distance education institution will be established through the merger of the University of South Africa and Technikon South Africa and the incorporation of the distance education centre of Vista University into a single institution.

And probably what is of more interest to you is that:

- Research will be funded through a separate formula based on research outputs, including, a minimum for masters and doctoral graduates research publications.
- Earmarked funds will also be allocated to build research capacity, including scholarships to promote postgraduate enrolments.

As you can see, government has ambitious plans for the sector. My department will continue to work with higher education institutions, other national agencies such as the National Research Foundation and the Department of Arts Culture Science and Technology in the implementation of this plan. With your support we are confident of success.

In closing, I bid you a fruitful conference. Use it to recommit yourselves to be the best in your profession and to the pleasure and pain that go with it. Yours is a noble profession that is always in pursuit of the truth.

Chairperson, I now have the pleasure and honour of declaring the 16th Conference of the International Society on General Relativity and Gravitation open.

I thank you all.

CONTENTS

Plenary Lectures

SIMPLICIAL EUCLIDEAN AND LORENTZIAN QUANTUM GRAVITY

J. AMBJøRN
The Niels Bohr Institute,
Blegdamsvej 17, DK-2100 Copenhagen Ø, Denmark
email: ambjorn@nbi.dk

One can try to define the theory of quantum gravity as the sum over geometries. In two dimensions the sum over ***Euclidean*** geometries can be performed constructively by the method of ***dynamical triangulations***. One can define a ***proper-time*** propagator. This propagator can be used to calculate generalized Hartle-Hawking amplitudes and it can be used to understand the the fractal structure of ***quantum geometry***. In higher dimensions the philosophy of defining the quantum theory, starting from a sum over Euclidean geometries, regularized by a reparametrization invariant cut off which is taken to zero, seems not to lead to an interesting continuum theory. The reason for this is the dominance of singular Euclidean geometries. Lorentzian geometries with a global causal structure are less singular. Using the framework of dynamical triangulations it is possible to give a constructive definition of the sum over such geometries, In two dimensions the theory can be solved analytically. It differs from two-dimensional Euclidean quantum gravity, and the relation between the two theories can be understood. In three dimensions the theory avoids the pathologies of three-dimensional Euclidean quantum gravity. General properties of the four-dimensional discretized theory have been established, but a detailed study of the continuum limit in the spirit of the renormalization group and ***asymptotic safety*** is till awaiting.

1 Introduction

We are still searching for the theory of quantum gravity. While the four dimensional theory of quantum gravity may have to go beyond the conventional framework of quantum field theory it is possible to discuss two- and three-dimensional quantum gravity in terms of conventional field theoretical concepts. In this way one may hope to learn important lessons about the "real" theory of four-dimensional gravity, whether or not it can be defined as a non-perturbative quantum field theory.

Two-dimensional quantum gravity is an example of the subtlety present in a reparameterization invariant theory. In two dimensions the Einstein action is trivial as long as we do not sum over different topologies of *space-time* (which we are not attempting here). Thus we are left only with the cosmological term in the action and the classical theory is trivial. Diffeomorphism invariance ensures that no *field* degrees of freedom exist. Nevertheless, a finite number of quantum mechanical degrees of freedom survives, which still have a rich (quantum) geometrical description. Despite the fact that no field degrees of freedom exist, it still makes sense to talk about generalized Hartle-Hawking like amplitudes: the sum over all Euclidean two-geometries with boundaries of of lengths L_k, $k=1,\ldots,n$. Similarly one can define the concept of correlators, depending on geodesic distance and show that standard concepts, derived from Euclidean quantum field theory or the theory of critical phenomena, make sense even in a framework of fluctuating geometries.

Three-dimensional quantum gravity can be addressed in the same spirit. The new aspects of three-dimensional quantum gravity compared to two-dimensional quantum gravity are the following: the theory is a perturbative non-renormalizable theory in the gravitational coupling constant and the Euclidean action is unbounded from below. It is thus *a priori* unclear how to make sense of an Euclidean three-dimensional path integral. On the other hand we know that the conformal factor only appears as a constraint in a reduced phase-space quantization and thus should not really pose a problem in a Lorentzian quantization of three-dimensional quantum gravity. It is of interest to understand if it is possible at all to think about three-dimensional quantum gravity as a path integral involving the sum over a certain class of three-dimensional Euclidean geometries of a given topology.

Contrary to the situation in two dimensions this "purely" Euclidean approach has not been successful in three dimensions. This led to the concept of "Lorentzian" gravity, the idea being that only Lorentzian geometries with a global causal structure should be included in the path integral, and that a possible rotation to Euclidean geometries should respect this. Returning to two-dimensional quantum gravity where analytic solutions can be found, one can implement this program and one finds that Lorentzian quantum gravity *is* different from Euclidean quantum gravity. Using proper-time as an evolution parameter, two-dimension Euclidean quantum gravity can be understood as two-dimensional Lorentzian quantum gravity where it is possible to create at each space-time point a baby universe, which at later proper-time remains separated from the "parent" universe. In three-dimensions the path-integral over Lorentzian geometries can be studied by numerical as well as analytical methods, and it seems an interesting candidate for a theory of quantum gravity, defined as the sum over geometries, with relative weights dictated by the Einstein-Hilbert action.

2 Non-perturbative regularization

In a theory of quantum gravity it is not enough to study perturbation theory. The examples of the ϕ^4 field theory and QCD in four dimensions illustrates this. Both theories have well defined perturbative expansions, and observables can be calculated to any order in the coupling constant in both theories. Since both theories are renormalizable one can express the observables entirely in terms of the renormalized mass and coupling constant. However, it does not ensure that a genuine quantum field exists without a cut-off. In order to address this question one first has to define the theory outside perturbation theory and then show that observables calculated in the non-perturbatively defined theory become independent of a possible cut-off, used in the definition of the non-perturbative framework. Finally, one can then discuss to what extend the perturbation expansion, which is usually at most an asymptotic expansion, contains information about the non-perturbatively defined theory.

All the evidence suggests that QCD exists as a genuine quantum field theory in four-dimensional space-time. One needs a cut-off at an intermediate step when defining the theory, but this cut off will never show up in any continuum physical observables. The situation is opposite for the ϕ^4 theory. It seems impossible to define a non-trivial theory where the renormalized coupling constant is different from zero when the cut-off is removed. This situation is of course already hinted by low order perturbation theory where one observes a Landau pole at large energies.

In the case of a ϕ^4 theory, as it appears for instance in the Standard Model, one considers it as an effective low energy approximation to a more more elaborate, yet to be understood, theory at the GUT scale or the Planck scale. Clearly, if we want to apply conventional quantum field theoretical concepts to the theory of quantum gravity, such a situation is not good enough: it has to be a quantum theory independent of cut-off in the same way as QCD, since the quantum phenomena we want to consider take place precisely at the Planck scale. In order to understand whether or not the theory exists as a conventional quantum theory a non-perturbative framework is thus mandatory.

For non-Abelian gauge theories and the ϕ^4 scalar field theory the use of a space-time *lattice* has provided a useful regularization, in particularly when combined with the use of Monte Carlo simulations, taking advantage of the fast present days computers. It has allowed us to related Euclidean quantum field theory to *the statistical theory of critical phenomena*. Stated very shortly Euclidean quantum field theories can be extracted from second order phase transitions of generalized spin systems defined on the lattice. The way the continuum field theory is recovered from the lattice spin model is as follows: by fine-tuning the coupling constants of the model , generically denoted λ, one can approach the phase transition point λ_c. The spin-spin correlation length ξ, measured in lattice spacings, will diverge as

$$\xi(\lambda) \sim \frac{1}{|\lambda - \lambda_c|^\nu}. \tag{1}$$

The lattice spacing a serves as the cut-off. By Fourier transformation it is seen that the (lattice) momentum must be less that π/a. We introduce the physical correlation length L as $L = \xi(\lambda) \cdot a$. One now take the limit

$$L = \xi(\lambda) \cdot a \quad \text{fixed}, \quad \xi(\lambda) \to \infty, \quad a \to 0. \tag{2}$$

From eq. (1) we see that the interpretation of this continuum limit is that while the correlation length measured in lattice units diverges when $\lambda \to \lambda_c$, the lattice spacing (the cut-off) goes to zero as:

$$a(\lambda) \sim |\lambda - \lambda_c|^\nu \quad \text{for} \quad \lambda \to \lambda_c. \tag{3}$$

In this *scaling limit* the lattice becomes increasingly fine-grained compared to the physical correlation length L. The lattice structure becomes unimportant for the physics associated with these long range fluctuations and Euclidean invariance will be restored for the physics related to these fluctuations. Let $\phi(x_n)$ denote the spin at lattice point $x_n = n \cdot a$. Since the spin correlation functions will behaves

$$\langle \phi(x_n)\phi(0)\rangle \sim \mathrm{e}^{-n/\xi(\lambda)} \sim \mathrm{e}^{-x_n/L} \tag{4}$$

for large n, we see that the *renormalized physical mass* should be identified with $1/L$:

$$m_{ph} = 1/L. \tag{5}$$

The choice of physical correlation length in (2) is equivalent to a choice of renormalized mass in the continuum quantum field theory. We have used λ as a generic coupling constant and usually the "spin" system will have a number of coupling constants. Combined fine-tuning will allow to fix in addition the other renormalized coupling constants of the continuum theory. Finally the lattice "spin" $\phi(x_n)$ is assigned a dimension from the short distance behavior of the spin-spin correlation function. In eq. (4) it was assumed that $n \gg \xi(\lambda)$. In the opposite limit $1 \ll n \ll \xi(\lambda)$ one expects

$$\langle \phi(x_n)\phi(0)\rangle \sim n^{-d+2+\eta} \sim \frac{a^{d-2+\eta}}{x_n^{d-2+\eta}}, \quad a \ll x_n \ll L. \tag{6}$$

The factor $a^{d-2+\eta}$ reflects the scaling dimension of the *physical* field

$$\phi_{phys} = a^{-(d-2+\eta)/2}\phi_{lattice}, \tag{7}$$

and the ***anomalous scaling dimension*** η is related to wave function renormalization.

This way of defining a quantum field theory is closely related to use of the renormalization group in quantum field theory and it emphasizes the concept of *universality*: many different "spin" systems defined on lattices will lead to the same class of continuum field theories, characterized by *critical exponents* like ν and η. One can imagine an idealized spin system where all possible spin interactions are included in the spin Hamiltonian. The key assumption is then that in this infinite dimensional coupling constant space the critical surfaces where one can define continuum theories are of *finite co-dimensions*. Thus there is only a finite number of coupling constants which need to be fine-tuned in order that the system becomes critical. Translated to continuum physics only a finite number of coupling constants need renormalizations. In the context of a perturbative expansion around free field theory this leads to the usual concept of renormalizable theories. But the philosophy also applies to expansions about non-trivial critical points where there exist no free fields. Since four-dimensional quantum gravity is not renormalizable viewed as a perturbation theory around flat space, the simplest possible scenario still using the framework of conventional quantum field theory is that of a non-trivial fixed point in the sense described above. In the context of quantum gravity this was first emphasized by Weinberg, who called it "asymptotic safety" since only a finite number of coupling constants needed adjustment in order to recover continuum physics.

While the lattice formulation of quantum field theories has the virtues mentioned above it also has a number of drawbacks:

(1) The lattice formulation is usually not a convenient framework for analytic calculations.

(2) space-time symmetries are explicitly broken. In particular, the lattice seems not to be the best regularization for theories with local space-time symmetries.

It is no necessarily so. Two-dimensional Euclidean quantum gravity provides a counter example. There exists a simple lattice regularization of 2d Euclidean quantum gravity, called *dynamical triangulation*, which is

(1) convenient for analytic calculations,

(2) has no problems recovering the diffeomorphism invariant continuum limit,

(3) is defined directly on the space of geometries,

(4) serves as a textbook example of universality when viewed as a statistical field theory.

The use of dynamical triangulations [a] is an attempt to approximate *the space of geometries* by the class of piecewise linear geometries which can be constructed from *an elementary building block.* In two dimensions the building block is an equilateral triangle and the piecewise linear geometries are obtained by gluing together the triangles in all possible ways consistent with a given two-dimensional topology. If we consider the triangles as flat, the (piecewise linear) geometry is uniquely determined by the connectivity pattern of the triangulation. The length a of the links serves as the cut-off as for a regular lattice, and with the interpretation given here it is a diffeomorphism invariant cut-off, since we work directly with geometries. The hope is that when $a \to 0$ this set of geometries will be "dense" in the set of all geometries which enters into the path integral. Also, it is important to understand that in the spirit of universality there is nothing special about the use of equilateral triangles. For instance, one could have used squares as building blocks, and if we wanted a piecewise linear geometry associated with such generalized "quadrangulations" one could subdivide the square into two triangles at one of the diagonals, considering each triangle as "flat". The results we will mention in the following will indeed be independent of such details.

Using a lattice to define the path integral also includes defining the discretized action. In conventional lattice theories the action is usually chosen such that it for $a \to 0$ converges to the classical continuum action of the field theory one tries to quantize. There is a large freedom in choosing such discretized actions, all leading to the same continuum theory by the universality mentioned above. In the case of piecewise linear geometries the Einstein-Hilbert action can actually be implemented entirely in terms of the piecewise linear geometry, as first noted by Regge [5]. For a piecewise linear d-dimensional geometry the integrated curvature is

$$\int d^d\xi \sqrt{g}\, R = \sum_{\sigma_{d-2}} V(\sigma_{d-2})\, \varepsilon(\sigma_{d-2}), \tag{8}$$

where σ_{d-2} denotes a $(d-2)$-dimensional simplex in the d-dimensional triangulation, $V(\sigma_{d-2})$ the volume of the simplex and $\varepsilon(\sigma_{d-2})$ the deficit angle associated with the simplex. Eq. (8) is "exact" in the sense that it *is* the natural curvature one can associate to a piecewise linear geometry. As an example it leads to Eulers formula in the case of two-dimensional manifolds.

[a] The concept "Dynamical triangulations" was introduced in [1,2,3,4] mainly in an attempt to provide a non-perturbative definition of the bosonic string.

In our case it becomes even simpler because the piecewise linear geometries are constructed from a fundamental building block. The space-time volume is thus proportional to the total number of building blocks and the curvature associated with a (d–2)-simplex σ_{d-2} is related to the number $o(\sigma_{d-2})$ of d-simplices sharing it in the following way: call the angle in a d-simplex between two $(d-1)$-simplices sharing a $(d-2)$-simplex θ. Then the deficit angle associated with σ_{d-2} is

$$\varepsilon(\sigma_{d-2}) = 2\pi - \theta\, o(\sigma_{d-1}). \tag{9}$$

The Einstein-Hilbert action

$$S[g] = -\frac{1}{G}\int \mathrm{d}^d\xi\sqrt{g}R + \Lambda\int \mathrm{d}^d\xi\sqrt{g} \tag{10}$$

associated with this kind of piecewise linear geometries is thus very simple and can be expressed as

$$S_T = -k_{d-2}N_{d-2}(T) + k_d N_d, \tag{11}$$

where $N_{d-2}(T)$ denotes the number of $(d-2)$-simplices and $N_d(T)$ denotes the number of d-simplices of the triangulation T. The dimensionless coupling constants k_{d-2} and k_d are related to the coupling constants G and Λ as

$$k_{d-2} \sim \frac{a^{d-2}}{G}, \qquad k_d \sim \frac{a^{d-2}}{G} + a^d\Lambda. \tag{12}$$

This has the implication that in any dimension the calculation of the functional integral associated with the Einstein-Hilbert action is purely combinatorial:

$$Z(G,\Lambda) = \int \mathcal{D}[g_{\mu\nu}]\, \mathrm{e}^{-S[g]} \;\;\rightarrow\;\; Z(k_{d-2},k_d) = \sum_T \mathrm{e}^{-S_T}, \tag{13}$$

where the sum can be rewritten as

$$Z(x,y) = \sum_{N_d,N_{d-2}} x^{N_d}y^{N_{d-2}}\mathcal{N}(N_d,N_{d-2}). \tag{14}$$

In this formula $x = e^{k_{d-2}}$, $y = e^{-k_d}$ and $\mathcal{N}(N_d,N_{d-2})$ denotes the *number* of different piecewise linear geometries one can construct from N_d simplices, having N_{d-2} $(d-2)$-simplices. In this way the partition function $Z(x,y)$ becomes the generating function for the numbers $\mathcal{N}(N_d,N_{d-2})$ as it is used in standard combinatorial analysis. This kind of remarkable simplification is only possible because we have restricted the sum over geometries to the ones with can be constructed from the cut-off size building blocks and because the Einstein action by the Regge-prescription is very simple for piecewise linear geometries.

2.1 The continuum limit

The combinatorial aspect of the sum over geometries allows us to understand how to approach the continuum limit in the simplest situations: the number of geometries of a given space-time topology and space-time volume grows exponentially with space-time volume:

$$\sum_{N_{d-2}} \mathcal{N}(N_{d-2}, N_d) \sim e^{k_d^c N_d + o(N_d)}. \tag{15}$$

From this it follows that *for each value of k_{d-2} there exists a critical value* $k_d^c(k_{d-2})$ such that

(1) For $k_d > k_d^c(k_{d-2})$ the partition function $Z(k_{d-2}, k_d)$ is convergent, while it is divergent for $k_d < k_d^c(k_{d-2})$.

(2) The *infinite volume limit* is obtained by fine-tuning $k_d \to k_d^c(k_{d-2})$. This can be viewed as an additive renormalization of the bare "cosmological" constant:

$$k_d = k_d^c(k_{d-2}) + \Lambda a^d, \tag{16}$$

where Λ is viewed as a renormalized cosmological constant.

(3) It might require an additional fine-tuning of k_{d-2} to obtain a *continuum limit*, if it exists at all. Such a fine-tuning can be viewed as a renormalization of the gravitational coupling constant. (The analogy with the spin models might be helpful: the infinite volume limit means that the lattice is infinite, but it does not imply that we have a continuum field theory. That might require a fine-tuning of the coupling constants of the spin model).

If a continuum limit of the kind discussed above can be constructed, the corresponding quantum field theory might serve as a candidate for the *asymptotic safe* non-renormalizable theory in the sense of Weinberg.

3 Two-dimensional Euclidean quantum gravity

3.1 Hartle-Hawking wave functions

Two-dimensional gravity, defined via dynamical triangulations, serves as a test of some of the above mentioned ideas. The integral of the curvature term is a topological invariant in two dimensions. Thus, as long as we do not consider topology changes of space-time we can ignore the curvature term in two dimensions and we are left with the cosmological term. We write:

$$Z_a(k_2) = \sum_T \mathrm{e}^{-k_2 N_2(T)} = \sum_N \mathrm{e}^{-k_2 N} \mathcal{N}(N), \tag{17}$$

where $\mathcal{N}(N)$ denotes the number of different triangulations constructed from N equilateral triangles with link length a. Objects of interest in the context of quantum gravity are the (generalized) Hartle-Hawking wave functions where the geometry of spatial boundaries are kept fixed, and the amplitude for such a configuration is obtained by summing over all two-dimensional geometries which have the prescribed spatial boundary geometries. In two dimensions the geometry of the boundary is uniquely fixed by its length. In the regularization mentioned above a boundary will consist of l links of length a from the adjacent triangles. The length of the boundary is thus $L = l \cdot a$. The regularized Hartle-Hawking wave function can thus be written as

$$G_a(l_1, \ldots, l_n; k_2) = \sum_{T(l_1,\ldots,l_n)} \mathrm{e}^{-k_2 N_2(T)} = \sum_N \mathrm{e}^{-k_2 N} \mathcal{N}_{l_1,\ldots,l_n}(N), \tag{18}$$

where $\mathcal{N}_{l_1,\ldots,l_n}(N)$ denotes the number of different triangulations constructed from N triangles and with n boundaries consisting of $l_1, \ldots, l_n$ links, respectively.

It is seen that the calculation of $Z_a(k_2)$ and $G_a(l_1, \ldots, l_n; k_2)$ can be reduced to purely combinatorial problems: the counting of distinct triangulations of various kinds. One finds, using either combinatorial methods or so-called matrix models, that

$$\mathcal{N}(N) \sim \mathrm{e}^{k_2^c N} N^{\gamma - 3}, \tag{19}$$

where $\sim$ means the leading asymptotic behavior when N is large, and where γ depends on the topology of two-dimensional space-time. For the simplest topology of a sphere $\gamma = -1/2$. Thus we find

$$Z_a(k_2) \sim \sum_N \mathrm{e}^{-(k_2 - k_2^c)N} \; N^{\gamma - 3} \sim \frac{1}{(k_2 - k_2^c)^{\gamma - 2}}. \tag{20}$$

One observes a singular structure for $k_2 \to k_2^c$ which governs the large N behavior of the sum, and a continuum limit can be obtained by an additive renormalization of the cosmological constant:

$$k_2 = k_2^c + a^2 \Lambda. \tag{21}$$

An additive renormalization is expected since Λ has a positive mass dimension. The renormalization is "natural" in the sense that

$$(k_2 - k_2^c)N = (\Lambda a^2)N = \Lambda(a^2 N) = \Lambda\, V, \tag{22}$$

where V denotes the continuum space-time volume. The partition function becomes

$$Z_a(k_2) \sim (a^2\Lambda)^{2-\gamma} \sim a^{2-\gamma} \cdot Z(\Lambda), \qquad Z(\Lambda) = \Lambda^{2-\gamma}. \tag{23}$$

The divergent a-factor can be viewed as the kind of wave function renormalization which is always present in the path integral version of quantum field theory.

The counting can also be performed in the case of triangulations with boundaries. The result is as follows:

$$\mathcal{N}_{l_1,\ldots,l_n}(N) \sim e^{k_2 N} e^{k_b(l_1+\cdots+l_n)} N^{\gamma_n} (l_1 \cdots l_n)^{\gamma_b} F(l_i/\sqrt{N}). \tag{24}$$

The exponential growth with N is counteracted by a renormalization of the cosmological constant as above while the exponential growth with respect to the boundary length can be controlled by adding a boundary cosmological constant to the action. This should anyway be done when we have space-like boundaries (see [6] for details). In this way we can take a continuum limit by scaling $l_i \sim \sqrt{N}$, which is what one expects from the canonical dimensions of boundaries and bulk in two-dimensional quantum gravity. The resulting continuum Hartle-Hawking wave functions can be found (see [6,7]). We list the results in the case three boundaries (which is particularly simple):

$$G(L_1, L_2, L_3, \Lambda) = \Lambda^{-1/2} \sqrt{L_1 L_2 L_3}\, e^{-\sqrt{L}(L_1+L_2+L_3)}. \tag{25}$$

3.2 Universality

It is important to understand that the results mentioned above are *universal*. The precise nature of the short distance regularization used is not important. We could have constructed our two-dimensional complexes from any combination of triangles, squares, pentagons etc with relative weights $g_3, g_4, g_5, \ldots$ and we would have obtained the same continuum Hartle-Hawking wave function as long as the relative weights of the various kinds of polygons are non-negative. If some of the weights are taken negative the system can no longer be given a simple interpretation as two-dimensional quantum gravity. However, it can be shown that they can be given the interpretation as certain matter fields coupled to two-dimensional random lattices of the kind used above in constructing the theory of two-dimensional quantum gravity. The fine-tuning of the coupling constants give us new continuum systems which can be viewed as certain conformal field theories coupled to two-dimensional quantum gravity [6,8,9].

Thus we can view these generalized systems as a school book example of critical systems: they are defined in an infinite dimensional coupling constant space $g_3, g_4, \ldots$. Fine-tuning the coupling constants bring us to a critical hyper-surface of co-dimension one, describing two-dimensional quantum gravity, and further fine-tuning leads to the critical behavior of certain conformal field theories coupled to two-dimensional quantum gravity. The critical exponents of the conformal field theories are changed (relative to their value in flat two-dimensional space) due to the influence of the fluctuating geometry, and likewise the behavior of two-dimensional quantum gravity is changed by the back-reaction of matter (see [6] for details).

3.3 Quantum geodesic distance

The critical behavior of the conformal field theories coupled to quantum gravity can be calculated by looking at globally defined matter correlators. Let $\phi(x)$ be a scalar matter field. Then

$$\int \mathcal{D}[g(x)]\mathcal{D}\phi(x)\; \mathrm{e}^{-S[g,\phi]} \int \mathrm{d}^2x\sqrt{g(x)} \int \mathrm{d}^2y\sqrt{g(y)}\; \phi(x)\phi(y) \tag{26}$$

is reparameterization invariant and the scaling behavior of this correlator with respect to the cosmological constant Λ determines the scaling dimension of the fields and the critical exponents. The reason for this is that the average volume of space-time is monitored by the cosmological constant,

$$\langle V\rangle \sim \Lambda^{-1}, \tag{27}$$

and thus the finite-size scaling relations of the system are determined as functions of the cosmological constant.

However, from the theory of critical phenomena we know that the underlying universality of large scale fluctuations originating from quite different microscopic interactions is due to a *divergent* correlation length $\xi(g_3, g_4, \ldots)$. How do we define such a length in quantum gravity? An object like $\langle\phi(x)\phi(y)\rangle$ has clearly no reparameterization invariant meaning if we view x and y as coordinates. Further, we have to integrate over all geometries and the (geodesic) distance between x and y will vary. One can define the concept of an invariance correlation function as follows [10,?]:

$$\begin{aligned}\langle\phi(D)\phi(0)\rangle \equiv \int \mathcal{D}[g(z)]\mathcal{D}\phi(z)\; \mathrm{e}^{-S[g,\phi]} & \qquad (28)\\ \int \mathrm{d}^2x\sqrt{g(x)} \int \mathrm{d}^2y\sqrt{g(y)}\; \delta\Big(D_g(x,y) - D\Big)\, \phi(x)\phi(y). &\end{aligned}$$

In (28) $D_g(x,y)$ denotes the *geodesic distance* between the space-time points labeled x and y. Let us consider the simplest case where there are no matter Lagrangian at all and $\phi(x) = 1$ in (28). In this case (28) can be viewed as the partition function $G(D)$ for universes where two marked points are separated a given geodesic distance D. Again the calculation of this partition function reduces to the combinatorial problem of counting a certain class of two-geometries. And again one can solve the combinatorial problem. In the regularized theory one obtains for k_2 close to the critical point [10]:

$$G_a(d;k_2) = (k_2 - k_2^c)^{3/4} \frac{\cosh d\sqrt[4]{k_2 - k_2^c}\, d}{\sinh^3 d\sqrt[4]{k_2 - k_2^c}}. \tag{29}$$

Here d denotes a suitable discretized geodesic distance [b]

If we take the continuum limit for the above "two-point" function it is seen that we are forced to scale geodesic distance *anomalously*:

$$D = d\sqrt{a}. \tag{30}$$

With this scaling we obtain

$$G(D;\Lambda) = \Lambda^{3/4} \frac{\cosh \sqrt[4]{\Lambda} D}{\sinh^3 \sqrt[4]{\Lambda} D}. \tag{31}$$

This function behaves as two-point function: it falls of exponentially for large distances and it behaves power-like for distances smaller that $1/\sqrt[4]{\Lambda}$. The anomalous dimension of the geodesic distance is also reflected in the average space-time volume V enclosed in disc of (geodesic) radius D. One finds

$$\langle V \rangle_D \sim R^4, \qquad R \ll 1/\sqrt[4]{\Lambda}. \tag{32}$$

Thus the Hausdorff dimension of a typical two-dimensional geometry is four, as first realized in [12]. This is a genuine quantum phenomenon.

Until now it has not been possible to calculate analytically the two-point function (28) when matter fields are present. However, it can be studied numerically and it can be verified that for the Ising model and the three-state Potts model coupled to two-dimensional quantum gravity one really obtains a *divergent* spin-spin correlation length when the spin couplings approach their critical values (see [13,14] for details).

---

[b]One expects "universality" in such choice. For instance one can define a distance between two vertices in the triangulation as the smallest number of links connecting them. Other definitions lead to identical results (see [11] for details).

3.4 The proper-time propagator and quantum geometry

An important tool used in the calculation of $G(D;\Lambda)$ is the proper-time propagator. In Euclidean geometry proper-time T is equivalent to geodesic distance, which we have denoted D. Let B_1 be a boundary. The distance of a point P to B_1 is defined as the minimal geodesic distance of P to the points of B_1. We define another boundary B_2 to be separated from it by a geodesic distance D, or proper-time T, if the distance of *each* point on B_2 to B_1 is D (note that the definition is asymmetric with respect to B_1 and B_2). The proper-time propagator $G_\Lambda(B_1, B_2; T)$ is defined by summing over all two-geometries with boundaries B_1 and B_2, such that B_2 has geodesic distance T to B_1, the weight of each configuration given as usually by e^{-S}. It was first calculated in [12], and in [15] it was shown that the equation which determines it can be obtained by very simple "quantum geometric" reasoning. Consider Fig. 1. It shows a disc amplitude with one marked interior point P (the marked point on the boundary is irrelevant for the considerations to follow) Since P can be anywhere, marking a point corresponds to multiplying with the space-time volume V or (equivalently) differentiating after the cosmological constant. However, one can also make the following decomposition: P has a geodesic distance T to the boundary "1". Form the connected loop through P which has distance T to the boundary. One obtains all geometries with one marked point by integrating over T and the length of the loop through P. We thus have the following consistency equation involving the Hartle-Hawking wave function and the proper-time propagator:

$$\frac{\partial G(L_1;\Lambda)}{\partial \Lambda} = \int \mathrm{d}T\mathrm{d}L_2\; G_\Lambda(L_1, L_2; T)G(L_2;\Lambda) \tag{33}$$

Under reasonable scaling assumptions one can show that eq. (33) determines the behavior of both $G(L;\Lambda)$ and $G_\Lambda(L_1, L_2; T)$. Quite surprisingly one finds *two* different solutions [15]. One scaling behavior leads to the Hartle-Hawking wave function already encountered. The other consistent scaling behavior leads to a different two-dimensional theory of quantum gravity which we call "Lorentzian" quantum gravity for reasons which will become clear in Sec. 5. At the moment we concentrate on the "Euclidean" quantum gravity theory discussed until now. Rather than discussing the explicit form of the proper-time propagator, let us give a further example of simple relations derived using "quantum geometry" (for more examples and a description of "operator product expansions" in two-dimensional quantum gravity see [16]). The example will illustrate that the proper-time propagator contains *all* information about quantum gravity. We have already mentioned that the usual Hartle-Hawking wave function $G(L;\Lambda)$ can be derived from properties of $G_\Lambda(L_1, L_2; T)$. Let

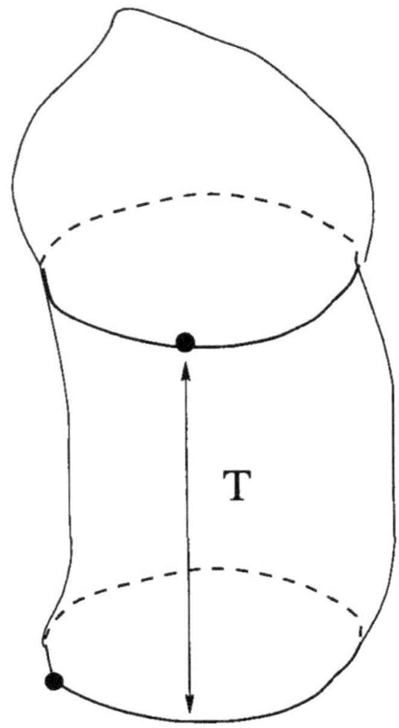

Figure 1. A geometry contributing to the punctured disc amplitude (the punctured Hartle-Hawking wave function), decomposed into geometries contributing to the proper-time propagator and the disc amplitude.

us consider instead the two-loop function $G(L_1, L_2; \Lambda)$. It is not simply obtained from by integrating $G_\Lambda(L_1, L_2; T)$ over T since boundary "2" of the space-times included in $G_\Lambda(L_1, L_2; T)$ has a well-defined distance to boundary "1". No such relation exists between the boundaries in $G(L_1, L_2; \Lambda)$. However, starting at boundary "1" and moving in successive proper-time steps one will end in the situation shown in Fig. 2. The "figure 8" amplitude [17] with a boundary of length L is related to the disc amplitude by

$$G_{fig-8}(L; \Lambda) \sim LG(L; \Lambda), \tag{34}$$

simply because it is obtained by pinching the boundary of the disc amplitude. Finally we can write:

$$G(L_1, L_2; \Lambda) = \int \mathrm{d}T \mathrm{d}L_3 \; G_\Lambda(L_1, L_3; T)\,(L_2 + L_3) G(L_2 + L_3; \Lambda). \tag{35}$$

Like the expression for the disc amplitude, which could be derived from general properties of the proper-time propagator, (35) is also an example of

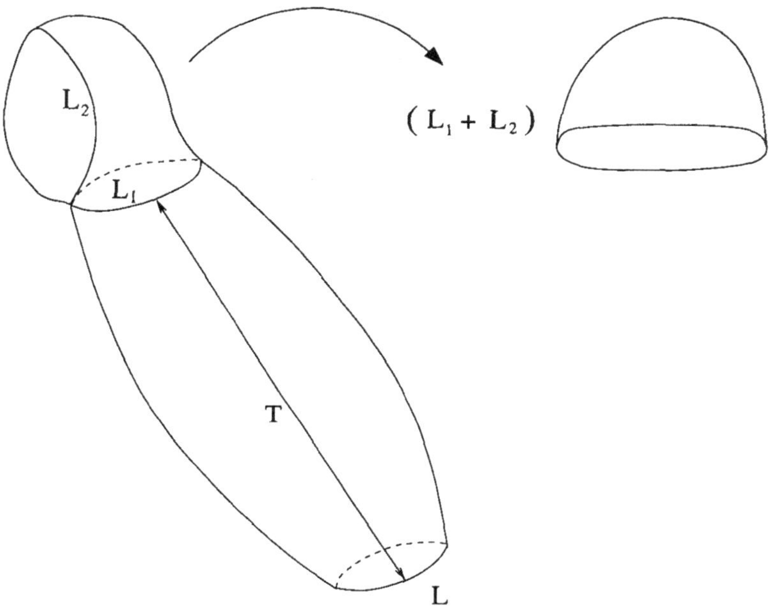

Figure 2. A geometry contributing to $G(L, L_2; \Lambda)$ decomposed into geometries contributing to $G_\Lambda(L, L_1; T)$ and $G_{fig\text{-}8}(L_1 + L_2; \Lambda)$. The latter is related to the disc amplitude (the Hartle-Hawking wave function) $G(L_1 + L_2; \Lambda)$ as shown.

"quantum geometric" considerations, which in essence are just combinatorial identities.

3.5 Summary

Euclidean two-dimensional quantum gravity as described above as the continuum limit of a regularized theory of two-dimensional geometries is still awaiting a precise mathematical description. What is meant by this is the following: Consider the random walk process in R^d. One can take the continuum limit in much the same way as we did in the construction of $G(L_1, \ldots, L_n; \Lambda)$. The number of random walks between point x and y in R^d grows exponentially with the number of steps in the random walk process and by fine-tuning the fugacity or stopping probability of the random walk process one can obtain the relativistic scalar particle propagator from point x to point y. This is in fact the typical "first quantized" path integral derivation of the free propagator. In this case we know the correct mathematical measure on the space of continuous path from x to y: it is the Wiener measure [c]. We would like to construct the corresponding measure on the space of two-dimensional continuous geometries. It should exist! Let us list the analogies between the the random walk and the "random geometries": Their numbers grow exponentially with the number of "elements" (steps in the random walk, triangles etc in the case of random geometries). The constants in the exponents are non-universal, but the leading corrections to the exponential growth *are* universal. A "typical" path or random geometry is fractal, in fact in almost the same way. The length of a "typical" path between x and y grows like $|x-y|^2$ rather than like $|x-y|$. In the same way the area of a random two-geometry of (geodesic) diameter D grows as D^4 rather than D^2. It thus seems that the class of continuous two-dimensional geometries constitutes a well-defined class of geometries, not much "wilder" than the class of continuous path [d]. Our results suggest that it should be possible to define a measure on this set, since we can constructively define the path integral over the set of two-geometries.

Some of the results mentioned above can be "guessed" from the continuum formulation of two-dimensional quantum gravity known as "Liouville quantum gravity". Let us consider a fixed topology, for instance that of the sphere, which we represent as the complex plane with ∞ identified as the north-pole. Any metric can be decomposed as

$$g_{\mu\nu} = \mathrm{e}^{\phi}\delta_{\mu\nu}, \tag{36}$$

and has to satisfy certain regularity conditions at ∞. The corresponding classical action is simply

$$S = \Lambda \int \mathrm{d}^2 x \sqrt{g} = \Lambda \int \mathrm{d}^2 x \; \mathrm{e}^{\phi}. \tag{37}$$

However, $\phi(x)$ is not a genuine dynamical field and the kinetic term comes entirely from the gauge fixing in the path integral:

$$Z(\Lambda) = \int \mathcal{D}\phi \; \mathrm{e}^{-\int \mathrm{d}^2 x (\Lambda \mathrm{e}^{\phi} + \alpha(\partial\phi)^2)}. \tag{38}$$

---

[c] A subtlety should be mentioned here: The usual Wiener measure is defined on the class of *parameterized* path. We need it on the class of *unparameterized* path. See [6] for a detailed discussion.

[d] A more precise description of the geometries will be given below when Lorentzian quantum gravity is discussed

There are two problems associated with this partition function. The first is that a cut-off a should be invariant under reparameterization, i.e.

$$a^2 = \mathrm{e}^{\phi(x)}(dx_1^2 + dx_2^2), \tag{39}$$

and the cut-off in parameter space R^2 is thus field dependent. The next problem is the geometric observables we have discussed so far become non-local and complicated when formulated in terms of $\phi(x)$. For instant the concept of geodesic distance as used above in the combinatorial approach is difficult to treat in terms of $\phi(x)$. Nevertheless, a number of the results from the combinatorial approach has been verified in quantum Liouville theory via the study of vertex operators of the theory in the following sense: the study of the scaling properties of vertex operators allows one to make reasonable guesses of the scaling behavior of the "observables" $G(L_1, \ldots, L_n; \Lambda)$ and in this way recover the results obtained constructively by combinatorial methods (see for instance [18]). It is worth emphasizing this, because it shows that there *is* an underlying continuum quantum theory at the critical point of the discretized theory, described by a Lagrangian and some standard rules of quantization, which however are technically difficult to implement because of the reparameterization invariance of the theory.

4 Higher dimensions

Encouraged by the success of two-dimensional Euclidean quantum gravity one can try to follow the program outlined in the introduction and look for a non-trivial fixed point of four-dimensional quantum gravity in the spirit of *asymptotic safety* as defined by Weinberg. It *is* indeed possible to find a candidate fixed point using the simplest action mentioned in Sec. 2 [19,20]. This fixed point seemed originally to relate in very interesting ways to various other proposals for a theory quantum gravity [21,22], but extensive computer simulations eventually established that the phase transition was a (weakly) *first order* transition [23] and thus could not serve as the desired fixed point of second or higher order [e]. In terms of the properties of "typical" configura-

[e] The situation is slightly more complicated: There is still an infinite volume limit for *all* values of the bare gravitational coupling constant by an appropriate fine-tuning of the cosmological coupling constant, as described in Sec. 2. However, unlike in two dimensions this infinite volume limit seems to be uninteresting from a physical point of view. We expect in four-dimensional quantum gravity to have genuine dynamical propagating degrees of freedom and the situation is thus more like two-dimensional quantum gravity coupled to for instance an Ising model. While we can always take the infinite volume limit, there will only be one value of β (of k_2) where the Ising spin correlation length diverges (where the graviton is massless).

tions of four-dimensional Euclidean geometries there is a sharp contrast with the two-dimensional case: a typical geometry, constructed by gluing together equilateral four-simplices with weight "one" is singular: the four-dimensional space-time volume grows at least exponentially with geodesic radius of the universe: Thus one cannot talk about a definite fractal dimension of a "typical" geometry. The first order phase transition observed in the numerical simulations is a transition from such "crumpled" geometries of no extension to another phase where the geometries are "maximally elongated" and fractal, of fractal dimension two, namely so-called branched polymers [24]. Neither set of geometries can serve as an underlying set of geometries defining an interesting theory of four-dimensional quantum gravity.

The original hope was that *at* the transition point one could define a set of geometries of a well defined fractal dimension larger than or equal to four, which could serve as the underlying set of geometries used in the definition of the continuum path integral. However, the first order transition rules out this possibility: there is no smooth interpolation between to two extreme classes of geometries mentioned above.

No simple modification of the Einstein-Hilbert action seems to change the order of the transition [25]. The conclusion is that a continuum theory of Euclidean four-dimensional quantum gravity does not exist as a limit of the simplicial formulation given here.

5 Lorentzian quantum gravity

The relation between the theory of critical phenomena and Euclidean quantum field theory is well established. However, if the field theory in question is "quantum gravity" the situation is unclear. The results for Euclidean gravity mentioned in the former Section indicate that in higher than two dimensions one should not sum over Euclidean geometries if one wants a viable theory of quantum gravity. Twenty years ago Teitelboim suggested [26] that one should only sum over causal space-time histories in the (Minkowskian) path integral if one wanted to maintain causality in the quantum theory of gravity [f]. This idea was made concrete in simplicial quantum gravity in [15,27] and was denoted *Lorentzian simplicial quantum gravity.* More precisely, consider the proper-time propagator already defined above in the context of Euclidean quantum gravity. Let the allowed geometries entering in the sum between two spatial boundaries, separated by proper time T be all non-degenerate causal geome-

[f] Related ideas have been discussed in [28,29].

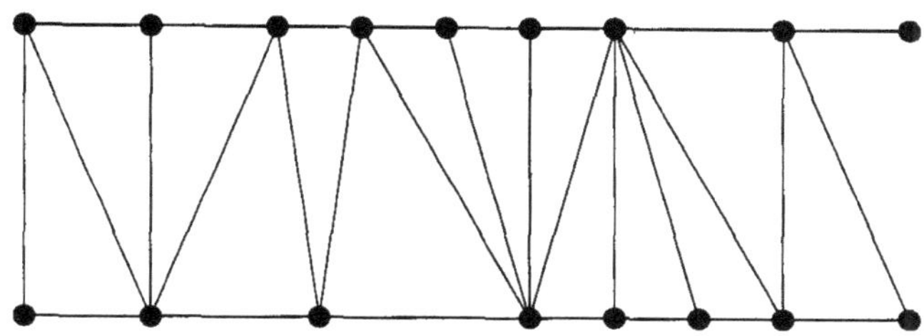

Figure 3. Two successive spatial slices connected by fundamental building blocks. The spatial link of a triangle has $ds^2 = a^2$ while the (two) time-like links have $ds^2 = -a^2$ before rotating to Euclidean space-time.

tries which allow a proper time slicing [9]. In the context of simplicial gravity this can be made precise (see [15] for details): starting with a given spatial geometry one defines constructively the class of space-time geometries with a boundary separated from initial spatial surface by proper-time a, a been the lattice spacing, i.e. the cut-off. This is illustrated in the simplest case of two-dimensional Lorentzian gravity in Fig. 3, but can be generalized to three– and four-dimensional cases [27]. Proceeding this way one defines the simplicial space-time geometries with two spatial boundaries separated a proper-time $T = a \cdot n$ and which allow a proper-time slicing.

The class of simplicial geometries defined this way has the virtue that each geometry allows a natural rotation to Euclidean signature. Also the action is rotated as follows: $iS_M \to -S_E$, as in ordinary quantum field theory. This opens for the possibility to follow the standard procedure and first rotate to Euclidean signature, then perform the summation over space-time histories and finally to rotate proper-time back to Minkowskian signature. The difference from the Euclidean path integral considered in Sec. 4 is the requirement that each Euclidean geometry entering in the sum over space-time histories comes from a Minkowskian non-degenerate geometry which allows a proper-time slicing. In particular, this implies that the spatial geometry obtained at a given proper-time is always connected. This is quite natural from the point of view of canonical quantization. The possibility of splitting space in several components is not natural in the framework of canonical quantization. However, if we return to two-dimensional Euclidean quantum gravity it is seen that the proper-time slicing does not respect this. Starting out

[9] It is often stated that proper-time is a bad "time" choice since it can become singular in the context of initial value problems of General Relativity. However, in the path integral such singular behavior is most likely unimportant and of zero measure.

with a connected boundary of a "typical" two-dimensional Euclidean geometry and implementing a proper-time slicing, the spatial slices will in general consist of many disconnected components. In fact, the main difference between the class of of geometries used in Euclidean two-dimensional quantum gravity and Lorentzian two-dimensional quantum gravity can be understood by looking at the proper-time slicing [30]. In Lorentzian quantum gravity the typical fluctuating geometry is genuining *two-dimensional* [h]. One obtains a typical Euclidean geometry by allowing the outgrowth of a baby universe at each space-time point. In this way the four-dimensional fractal structure of a typical quantum configuration can be understood as a kind of "product" of ordinary two-dimensional space-times.

In view of the universality mentioned in connection with two-dimensional Euclidean quantum gravity it is surprising that there exists another fixed point belonging to a different universality class and corresponding to Lorentzian two-dimensional quantum gravity, but this universality class is now well understood in the context of critical phenomena [31]. To make this possibility of two different types of critical behavior more clear let us consider a model which allows for the creation of baby universes along with the propagation of the two-dimensional Lorentzian space-time. Such a possibility is shown in Fig. 4. Compared to Fig. 3 one allows for the possibility of a baby universe to be created at any point at the spatial slice at proper time t and then to develop independently of the "parent universe". Let $G_\Lambda(L_1, L_2; T)$ be the proper-time propagator mentioned above, T being the proper-time and let $G_\Lambda(X, Y; T)$ be the Laplace transform of $G_\Lambda(L_1, L_2; \Lambda)$ (and similarly $G(X; \Lambda)$ the Laplace transform of the Hartle-Hawking wave function $G(L; \Lambda)$). Again it has a combinatorial interpretation and from this one can derive the following equation (see [15] for details):

$$a^\varepsilon \frac{\partial}{\partial T} G_\Lambda(X, Y; T) = -a \frac{\partial}{\partial X}\Big([X^2 - \Lambda + \kappa a^{\eta-2} G(X; \Lambda)] G_\Lambda(X, Y; T)\Big) \qquad (40)$$

Here κ is the coupling constant for creating a baby universe at a space-time point. The exponents ε and η are related to the dimension of proper-time (or geodesic distance) and the Hartle-Hawking wave function: $T \sim a^\varepsilon$ and $G(X, \Lambda) \sim a^{-\eta}$, where $\sim$ just indicates the dimension. One can show, using eq. 33 (i.e. Fig. 1) that there are only two consistent choices: (1) $\kappa = 0$, $\varepsilon = 1$, and $\kappa > 0$, $\varepsilon = 1/2$ and $\eta = 3/2$ (see again [15] for details). The first case

[h] The fact that a typical Lorentzian geometry is not *fractal* does not mean that it is not fluctuating. In fact, in the case of two-dimensional Lorentzian geometries, the fluctuation of the spatial volume (length) $L(t)$, where t denotes proper time, is maximal: $\langle L^2\rangle - \langle L\rangle^2 \sim \langle L\rangle^2$.

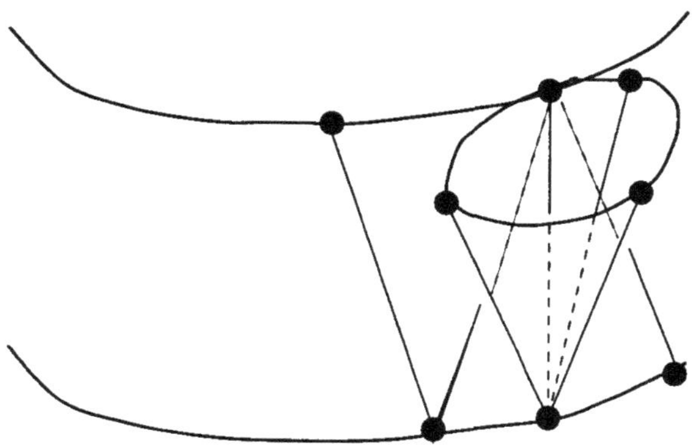

Figure 4. The creation of a baby universe in the time-evolution of two-dimensional space-times indicated in Fig. 3. At successive time-steps the baby universe should be considered as detached from the "parent universe".

corresponds to two-dimensional Lorentzian quantum gravity where proper-time scales canonically in accordance with the fact that the fractal dimension of a typical geometry is two. The second case corresponds to two-dimensional Euclidean quantum gravity where the geodesic distance scales anomalously and the fractal dimension of a typical geometry is four.

The coupling of matter to two-dimensional Lorentzian gravity is also (partially) understood. The Ising model maintains its flat space critical exponents [32], and the interaction between matter fields and quantum gravity is weak [i] as long as the central change c of the matter field is less than one [j]. This is in contrast to Euclidean quantum gravity where the critical exponents of matter as well as the critical exponents of quantum gravity itself are changed due to the coupling.

5.1 Higher dimensional Lorentzian quantum gravity

The theory of Lorentzian simplicial quantum gravity has been formulated in two, three and four dimensions. It has been solved analytically in two dimensions, as described above and its relation to two-dimensional Euclidean quan-

---

[i] Again it should be emphasized that a "weak" interaction does not imply that there is no interaction, only that it does not change general properties of the class of geometries used to define the continuum path integral, i.e. the fractal dimension of these geometries.

[j] When $c > 1$ computer simulations have shown a strong back-reaction of matter on geometry [33].

tum gravity is understood. The main question is of course if it provides us with a viable four dimensional theory of quantum gravity in the spirit of asymptotic safety. So far only the three-dimensional theory has been investigated in some detail, this made possible because it can be mapped on a matrix model which allows us to do perform some analytical calculations [34] and because it is possible to perform numerical simulations [35]. Three-dimensional quantum gravity is not without interest. Although it contains no propagating gravitons, i.e. no field-theoretical degrees of freedom, it is formally non-renormalizable when expanded around a fixed background geometry. The results obtained so far are quite encouraging. They indicate that one can take a continuum limit which describe quantum fluctuations around a background geometry created by the presence of a positive cosmological constant. However, more work is needed to substantiate this interpretation. If this works out and one can establish a firm connection to canonically quantized three-dimensional quantum gravity, Lorentzian simplicial quantum gravity will be a most interesting candidate for a regularized, background independent theory of four-dimensional quantum gravity, formulated entirely in terms of fluctuating geometries.

Acknowledgments

It is a pleasure to thank K. Anagnostopoulos, J. Jurkiewics, R. Loll and Y. Makeenko collaborations associated with the work reported here. This work was supported by "MaPhySto", the Center of Mathematical Physics and Stochastics, financed by the National Danish Research Foundation as well as by the EU network on "Discrete Random Geometry", grant HPRN-CT-1999-00161, and by the ESF network no.82 on "Geometry and Disorder".

References

1. V. A. Kazakov, A. A. Migdal and I. K. Kostov, *Critical Properties Of Randomly Triangulated Planar Random Surfaces*, Phys. Lett. B 157 (1985) 295.
2. F. David, *A Model Of Random Surfaces With Nontrivial Critical Behavior*, Nucl. Phys. B 257 (1985) 543.
3. J. Ambjørn, B. Durhuus and J. Frohlich, *Diseases Of Triangulated Random Surface Models, And Possible Cures*, Nucl. Phys. B 257 (1985) 433; *The Appearance Of Critical Dimensions In Regulated String Theories, II.*, Nucl. Phys. B 275 (1986) 161.
4. J. Ambjørn, B. Durhuus, J. Frohlich and P. Orland, *The Appearance Of Critical Dimensions In Regulated String Theories*, Nucl. Phys. B 270

(1986) 457.
5. T. Regge: *General relativity without coordinates*, Nuovo Cim. A 19 (1961) 558-571. [hep-th/9807216].
6. J. Ambjørn, B. Durhuus and T. Jonsson, *Quantum Geometry. A Statistical Field Theory Approach*, Cambridge Univ. Press (1997) 363 p.
7. J. Ambjørn, J. Jurkiewicz and Y. M. Makeenko, *Multiloop Correlators For Two-Dimensional Quantum Gravity*, Phys. Lett. B 251 (1990) 517.
8. V. A. Kazakov, *The Appearance Of Matter Fields From Quantum Fluctuations Of 2-D Gravity*, Mod. Phys. Lett. A 4 (1989) 2125.
9. M. Staudacher, *The Yang-Lee Edge Singularity On A Dynamical Planar Random Surface*, Nucl. Phys. B 336 (1990) 349.
10. J. Ambjørn and Y. Watabiki: *Scaling in quantum gravity*, Nucl. Phys. B 445 (1995) 129-144 [hep-th/9501049].
11. J. Ambjørn, J. Jurkiewicz and Y. Watabiki: *On the fractal structure of two-dimensional quantum gravity*, Nucl. Phys. B 454 (1995) 313-342 [hep-lat/9507014].
12. H. Kawai, N. Kawamoto, T. Mogami and Y. Watabiki: *Transfer matrix formalism for two-dimensional quantum gravity and fractal structures of space-time*, Phys. Lett. B 306 (1993) 19-26 [hep-th/9302133].
13. J. Ambjørn, K. N. Anagnostopoulos, U. Magnea and G. Thorleifsson, Phys. Lett. B 388 (1996) 713 [hep-lat/9606012].
14. J. Ambjørn and K. N. Anagnostopoulos, Nucl. Phys. B 497, 445 (1997) [hep-lat/9701006].
15. J. Ambjørn and R. Loll: *Non-perturbative Lorentzian quantum gravity, causality and topology change*, Nucl. Phys. B 536 (1998) 407-434 [hep-th/9805108].
16. H. Aoki, H. Kawai, J. Nishimura and A. Tsuchiya: *Operator product expansion in two-dimensional quantum gravity*, Nucl. Phys. B 474 (1996) 512-528 [hep-th/9511117].
17. E. Adi and S. Solomon, *The Solution to Wheeler-DeWitt is eight*, Phys. Lett. B 336 (1994) 152 [hep-th/9404079].
18. G. W. Moore, N. Seiberg and M. Staudacher, *From loops to states in 2-D quantum gravity*, Nucl. Phys. B 362 (1991) 665.
19. J. Ambjørn and J. Jurkiewicz: *Four-dimensional simplicial quantum gravity*, Phys. Lett. B 278 (1992) 42-50.
20. M.E. Agishtein and A.A. Migdal: *Simulations of four-dimensional simplicial quantum gravity*, Mod. Phys. Lett. A 7 (1992) 1039-1061.
21. I. Antoniadis, P. O. Mazur and E. Mottola, *Scaling behavior of quantum four - geometries*, Phys. Lett. B 323 (1994) 284 [hep-th/9301002]. *Criticality and scaling in 4D quantum gravity*,' Phys. Lett. B 394 (1997)

49 [hep-th/9611145].

22. H. Kawai and M. Ninomiya, *Renormalization Group And Quantum Gravity*, Nucl. Phys. B 336 (1990) 115.
H. Kawai, Y. Kitazawa and M. Ninomiya, *Ultraviolet stable fixed point and scaling relations in (2+epsilon)-dimensional quantum gravity*, Nucl. Phys. B 404 (1993) 684 [hep-th/9303123].
23. P. Bialas, Z. Burda, A. Krzywicki and B. Petersson: *Focusing on the fixed point of 4-d simplicial gravity.*, Nucl. Phys. B 472 (1996) 293-308 [hep-lat/9601024].
24. J. Ambjørn and J. Jurkiewicz, *Scaling in four-dimensional quantum gravity*, Nucl. Phys. B 451 (1995) 643 [hep-th/9503006].
25. J. Ambjørn, K. N. Anagnostopoulos and J. Jurkiewicz, *Abelian gauge fields coupled to simplicial quantum gravity*, JHEP 9908 (1999) 016 [hep-lat/9907027].
26. C. Teitelboim, *Causality Versus Gauge Invariance In Quantum Gravity And Supergravity*, Phys. Rev. Lett. 50 (1983) 705.
27. J. Ambjørn, J. Jurkiewicz and R. Loll: *A nonperturbative Lorentzian path integral for gravity*, Phys. Rev. Lett. 85 (2000) 924-927 [hep-th/0002050].
Dynamically triangulating Lorentzian quantum gravity, Nucl. Phys. B 610 (2001) 347 [hep-th/0105267].
28. F. Markopoulou and L. Smolin, *Causal evolution of spin networks*, Nucl. Phys. B 508 (1997) 409-430, [gr-qc/9702025;
Quantum geometry with intrinsic local causality, Phys. Rev. D 58 (1998) 084032, [gr.qc/9712067].
29. L. Bombelli, J. Lee, D. Meyer and R. Sorkin, *Space-time as a causal set*, Phys. Rev. Lett. 59 (1987) 521-524.
D.P. Rideout and R.D. Sorkin, *Classical sequential growth dynamics for causal sets*, Phys. Rev. D 61 (2000) 024002 [gr-qc/9904062].
30. J. Ambjørn, J. Correia, C. Kristjansen and R. Loll: *The relation between Euclidean and Lorentzian 2D quantum gravity*, Phys. Lett. B 475 (2000) 24-32 [hep-th/9912267].
31. P. Di Francesco, E. Guitter and C. Kristjansen: *Integrable 2-d Lorentzian gravity and random walks*, Nucl. Phys. B 567 (2000) 515-553 [hep-th/9907084];
Generalized Lorentzian gravity in (1+1)D and the Calogero Hamiltonian, Nucl. Phys. B 608 (2001) 485 [hep-th/0010259].
32. J. Ambjørn, K.N. Anagnostopoulos and R. Loll: *A new perspective on matter coupling in 2d quantum gravity*, Phys. Rev. D 60 (1999) 104035 [hep-th/9904012].
33. J. Ambjørn, K.N. Anagnostopoulos and R. Loll: *Crossing the c=1 barrier*

in 2d Lorentzian quantum gravity, Phys. Rev. D 61 (2000) 044010 [hep-lat/9909129].
34. J. Ambjørn, J. Jurkiewicz, R. Loll and G. Vernizzi, *Lorentzian 3d gravity with wormholes via matrix models*, JHEP 0109 (2001) 022 [hep-th/0106082].
35. J. Ambjørn, J. Jurkiewicz and R. Loll, *Non-perturbative 3d Lorentzian quantum gravity*, Phys. Rev. D 64 (2001) 044011 [hep-th/0011276].

QUANTUM GEOMETRY AND GRAVITY: RECENT ADVANCES

ABHAY ASHTEKAR

Center for Gravitational Physics and Geometry,
Physics Department, Penn State, University Park, PA 16802, USA

Over the last three years, a number of fundamental physical issues were addressed in loop quantum gravity. These include: A statistical mechanical derivation of the horizon entropy, encompassing astrophysically interesting black holes as well as cosmological horizons; a natural resolution of the big-bang singularity; the development of spin-foam models which provide background independent path integral formulations of quantum gravity and 'finiteness proofs' of some of these models; and, the introduction of semi-classical techniques to make contact between the background independent, non-perturbative theory and the perturbative, low energy physics in Minkowski space. These developments spring from a detailed quantum theory of geometry that was systematically developed in the mid-nineties and have added a great deal of optimism and intellectual excitement to the field. The goal of this article is to communicate these advances in general physical terms, accessible to researchers in all areas of gravitational physics represented in this conference.

1 Introduction

Let us begin by recalling some of the central conceptual and physical questions of quantum gravity.

• *Big-Bang and other singularities:* It is widely believed that the prediction of a singularity, such as the big-bang of classical general relativity, is primarily a signal that the theory has been pushed beyond the domain of its validity. A key question to any quantum gravity theory, then, is: What replaces the big-bang? Qualitatively, classical geometry may be a mean field like 'magnetization', which provides an excellent macroscopic description of a ferromagnet. However, at the Curie temperature, magnetization goes to zero and susceptibility diverges. But there is no physical infinity; we simply have to turn to the correct *microscopic* description in terms of spin-systems to describe physics. Does something similar happen at the big-bang and other singularities? Is there a mathematically consistent description of the quantum state of the universe which replaces the classical big-bang? What is the analog of the microscopic spin-system that underlies magnetism? What can we say about the 'initial conditions', i.e., the quantum state of geometry and matter that correctly describes the big-bang? If they have to be imposed externally, is there a *physical* guiding principle?

• *Black holes:* In the early seventies, using imaginative thought experiments, Bekenstein argued that black holes must carry an entropy proportional to their area. About the same time, Bardeen, Carter and Hawking (BCH) showed that black holes in equilibrium obey two basic laws, which have the same form as the zeroth and the first laws of thermodynamics, provided one equates the black hole surface gravity κ to some multiple of the temperature T in thermodynamics and the horizon area a_{hor} to a corresponding multiple of the entropy S. However, at first this similarity was thought to be only a formal analogy because the BCH analysis was based on *classical* general relativity and simple dimensional considerations show that the proportionality factors must involve Planck's constant $\hbar$. Two years later, using quantum field theory on a black hole background space-time, Hawking showed that black holes in fact radiate quantum mechanically as though they are black bodies at temperature $T = \hbar\kappa/2\pi$. Using the analogy with the first law, one can then conclude that the black hole entropy should be given by $S_{\rm BH} = a_{\rm hor}/4G\hbar$. This conclusion is striking and deep because it brings together the three pillars of fundamental physics —general relativity, quantum theory and statistical mechanics. However, the argument itself is a rather hodge-podge mixture of classical and semi-classical ideas, reminiscent of the Bohr theory of atom. A natural question then is: what is the analog of the more fundamental, Pauli-Schrödinger theory of the Hydrogen atom? More precisely, what is the statistical mechanical origin of black hole entropy? What is the nature of a quantum black hole and what is the interplay between the quantum degrees of freedom responsible for entropy and the exterior curved geometry? Can one derive the Hawking effect from first principles of quantum gravity? Is there an imprint of the classical singularity on the final quantum description, e.g., through 'information loss' ?

• *Planck scale physics and the low energy world:* Perhaps the central lesson of general relativity is that *gravity is geometry*. There is no longer a background metric, no inert stage on which dynamics unfolds. Geometry itself is dynamical. Therefore, one expects that a fully satisfactory quantum gravity theory would also be free of a background space-time geometry. However, of necessity, a background independent description must use physical concepts and mathematical tools that are quite different from those of the familiar, low energy physics. A major challenge then is to show that this low energy description does arise from the pristine, Planckian world in an appropriate sense. In this 'top-down' approach, does the fundamental theory admit a "sufficient number" of semi-classical states? Do these semi-classical sectors provide enough of a background geometry to anchor low energy physics? can one recover the familiar description? Furthermore, can one pin point why the standard 'bottom-up' perturbative approach fail? That is, what is the essential feature which makes the fundamental description mathematically coherent but is absent in the standard perturbative quantum gravity?

Of course, this is by no means a complete list of challenges. There are many others: the issue of time, of measurement theory and the associated questions of interpretation of the quantum framework, the issue of diffeomorphism invariant observables and practical methods of computing their properties, practical methods of computing time evolution and S-matrices, exploration of the role of topology and topology change, ... The purpose of this report is to summarize recent advances in the non-perturbative approach based on quantum geometry which has led to illuminating answers to many of these questions and opened-up avenues to address others. The plenary session of this conference also covered dynamical triangulations and string theory which, in a certain sense, complement our approach. And there are other approaches as well, ranging from twistors and causal sets to Euclidean path integrals. Unfortunately, due to space limitation, I will not be able to discuss these; indeed, even within the approach I focus on, I can discuss only a few illustrative examples. I apologize in advance to the authors whose very interesting contributions could not be referred to in this brief report.

2 A bird's eye view of loop quantum gravity

In this section, I will briefly summarize the salient features and current status of loop quantum gravity. The emphasis is on structural and conceptual issues; details can be found in references [1-9].

2.1 Viewpoint

In this approach, one takes the central lesson of general relativity seriously: gravity *is* geometry whence, in a fundamental theory, there should be no background metric. In quantum gravity, geometry and matter should *both* be 'born quantum mechanically'. Thus, in contrast to approaches developed by particle physicists, one does not begin with quantum matter on a background geometry and use perturbation theory to incorporate quantum effects of gravity. There *is* a manifold but no metric, or indeed any other physical fields, in the background.[a]

[a]In 2+1 dimensions, although one begins in a completely analogous fashion, in the final picture one can get rid of the background manifold as well. Thus, the fundamental theory can be formulated combinatorially [3,1]. To achieve this goal in 3+1 dimensions, one needs a much better understanding of the theory of (intersecting) knots in 3 dimensions.

In the classical gravity, Riemannian geometry provides the appropriate mathematical language to formulate the physical, kinematical notions as well as the final dynamical equations. This role is now taken by *quantum* Riemannian geometry, discussed below. In the classical domain, general relativity stands out as the best available theory of gravity, some of whose predictions have been tested to an amazing accuracy, surpassing even the legendary tests of quantum electrodynamics. Therefore, it is natural to ask: *Does quantum general relativity, coupled to suitable matter* (or supergravity, its supersymmetric generalizations) *exist as consistent theories non-perturbatively ?* There is no a priori implication that such a theory would be the final, complete description of Nature. Nonetheless, this is a fascinating open question, at least at the level of mathematical physics.

In the particle physics circles, the answer is often assumed to be in the negative, not because there is concrete evidence against non-perturbative quantum gravity, but because of an analogy to the theory of weak interactions. There, one first had a 4-point interaction model due to Fermi which works quite well at low energies but which fails to be renormalizable. Progress occurred not by looking for non-perturbative formulations of the Fermi model but by replacing the model by the Glashow-Salam-Weinberg renormalizable theory of electro-weak interactions, in which the 4-point interaction is replaced by $W^{\pm}$ and Z propagators. Therefore, it is often assumed that perturbative non-renormalizability of quantum general relativity points in a similar direction. However this argument overlooks the crucial fact that, in the case of general relativity, there is a qualitatively new element. Perturbative treatments pre-suppose that the space-time can be assumed to be a continuum *at all scales* of interest to physics under consideration. In the gravitational case, the scale of interest is given by the Planck length $\ell_{\rm Pl}$ and there is no physical basis to pre-suppose that the continuum picture should be valid down to that scale. The failure of the standard perturbative treatments may simply be due to this grossly incorrect assumption and a non-perturbative treatment which correctly incorporates the physical micro-structure of geometry may well be free of these inconsistencies.

As indicated above, even if quantum general relativity did exist as a mathematically consistent theory, there is no a priori reason to assume that it would be the 'final' theory of all known physics. In particular, as is the case with classical general relativity, while requirements of background independence and general covariance do restrict the form of interactions between gravity and matter fields and among matter fields themselves, the theory would not have a built-in principle which *determines* these interactions. Put differently, such a theory would not be a satisfactory candidate for unification of all known forces. However, just as general relativity has had powerful implications in spite of this limitation in the classical domain, quantum general relativity should have qualitatively new predictions, pushing further the existing frontiers of physics. Indeed, unification does not appear to be an essential criterion for usefulness of a theory even in other interactions. QCD, for example, is a powerful theory even though it does not unify strong interactions with electro-weak ones. Furthermore, the fact that we do not yet have a viable candidate for the grand unified theory does not make QCD any less useful.

2.2 Quantum Geometry

Although there is no natural unification of dynamics of all interactions in loop quantum gravity, it does provide a kinematical unification. More precisely, in this approach one begins by formulating general relativity in the mathematical language of connections, the basic variables of gauge theories of electro-weak and strong interactions. Thus, now the configuration variables are not metrics as in Wheeler's geometrodynamics, but certain spin connections; the emphasis is shifted from distances and geodesics to holonomies and Wilson loops [1]. Consequently, the basic kinematical structures are the same as those used in gauge theories. A key difference, however, is that while a background space-time metric is available and crucially used in gauge theories, there are no background fields whatsoever now. This absence is forced on us by the requirement of diffeomorphism invariance (or 'general covariance').

This is a key difference and it causes a host of conceptual as well as technical difficulties in the passage to quantum theory. For, most of the techniques used in the familiar, Minkowskian quantum theories are deeply rooted in the availability of a flat back-ground metric. It is this structure that enables one to single out the vacuum state, perform Fourier transforms to decompose fields canonically in to creation and annihilation parts, define masses and spins of particles and carry out regularizations of products of operators. Already when one passes to quantum field theory in curved space-times, extra work is needed to construct mathematical structures that are adequate for physics. In our case, the situation is much more drastic: there is no background metric what so ever. Therefore new physical ideas and mathematical tools are now necessary. Fortunately, they were constructed by a number of researchers in the mid-nineties and have given rise to a detailed quantum theory of geometry [4,5,6,8].

Because the situation is conceptually so novel and because there are no direct experiments to guide us, reliable results require a high degree of mathematical precision to ensure that there are no hidden infinities. Achieving this precision has been a high priority in the program. Thus, while one is inevitably motivated by heuristic, physical ideas and formal manipulations, the final results are mathematically rigorous. In particular, due care is taken in constructing function spaces, defining measures and functional integrals, regularizing products of field operator, and calculating eigenvectors and eigenvalues of geometric operators. The final results are all free of divergences, well-defined, and respect the background independence and diffeomorphism invariance.

Let me now turn to specifics. Our basic configuration variable is an SU(2)-connection, A_a^i on a 3-manifold Σ representing 'space' and, as in gauge theories, the momenta are the 'electric fields' E_i^a. However, in the present gravitational context, they acquire a geometrical significance: they can be naturally interpreted as orthonormal triads (with density weight 1) and determine the dynamical, Riemannian geometry of Σ. Thus, in contrast to Wheeler's geometrodynamics, the Riemannian structures, including the positive-definite metric on Σ, is now built from *momentum* variables. The basic kinematic objects are holonomies of A_a^i, which dictate how spinors are parallel transported along curves, and the triads E_i^a, which determine the geometry of Σ. (Matter couplings to gravity have also been studied extensively [2,1].)

In the quantum theory, the fundamental excitations of geometry are most conveniently expressed in terms of holonomies [3,4]. They are thus one-dimensional, polymer-like and, in analogy with gauge theories, can be thought of as 'flux lines of the electric field'. More precisely, they turn out to be flux lines of areas: an elementary flux line deposits a quantum of area on any 2-surface S it intersects. Thus, if quantum geometry were to be excited along just a few flux lines, most surfaces would have zero area and the quantum state would not at all resemble a classical geometry. This state would be analogous, in Maxwell theory, to a 'genuinely quantum mechanical state' with just a few photons. In the Maxwell case, one must superpose photons coherently to obtain a semi-classical state that can be approximated by a classical electromagnetic field. Similarly, here, semi-classical geometries can result only if a huge number of these elementary excitations are superposed in suitable dense configurations [13,14]. The state of quantum geometry around you, for example, must have so many elementary excitations that $\sim 10^{68}$ of them intersect the sheet of paper you are reading. Even in such states, the geometry is still distributional, concentrated on the underlying elementary flux lines; but if suitably coarse-grained, it can be approximated by a smooth metric. Thus, the continuum picture is only an approximation that arises from coarse graining of semi-classical states.

These quantum states span a specific Hilbert space $\mathcal{H}$ consisting of functions on the space of connections which are square integrable with respect to a natural, diffeomorphism invariant (regular, Borel) measure [4]. This space is very large. However, it can be conveniently decomposed in to a family of orthonormal, *finite* dimensional sub-spaces $\mathcal{H} = \oplus_{\gamma,\vec{j}}\mathcal{H}_{\gamma,\vec{j}}$, labelled by graphs γ each edge of which itself is labelled by a spin (i.e., half-integer) j [5]. One can think of γ as a 'floating lattice' in Σ —'floating' because its edges are arbitrary, rather than 'rectangular'. (Indeed, since there is no background metric on Σ, a rectangular lattice has no invariant meaning.) Mathematically, $\mathcal{H}_{\gamma,\vec{j}}$ can be regarded as the Hilbert space of a spin-system. These spaces are extremely simple to work with; this is why very explicit calculations are feasible. Elements of $\mathcal{H}_{\gamma,\vec{j}}$ are referred to as *spin-network states* [5].

As one would expect from the structure of the classical theory, the basic quantum operators are the holonomies $\hat{h}_p$ along paths p in Σ and the triads $\hat{E}^a_i$ [6]. (Both sets are densely defined and self-adjoint on $\mathcal{H}$. Furthermore, a striking result is that *all eigenvalues of the triad operators are discrete.* This key property is, in essence, the origin of the fundamental discreteness of quantum geometry. For, just as the classical Riemannian geometry of Σ is determined by the triads E^a_i, all Riemannian geometry operators —such as the area operator $\hat{A}_S$ associated with a 2-surface S or the volume operator $\hat{V}_R$ associated with a region R— are constructed from $\hat{E}^a_i$. However, since even the classical quantities A_S and V_R are non-polynomial functionals of the triads, the construction of the corresponding $\hat{A}_S$ and $\hat{V}_R$ is quite subtle and requires a great deal of care. But their final expressions are rather simple [6].

In this regularization, the underlying background independence turns out to be a blessing. For, diffeomorphism invariance constrains the possible forms of the final expressions *severely* and the detailed calculations then serve essentially to fix numerical coefficients and other details. Let us illustrate this point with the example of the area operators $\hat{A}_S$. Since they are associated with 2-surfaces S while the states are 1-dimensional excitations, the diffeomorphism covariance requires that the action of $\hat{A}_S$ on a state $\Psi_{\gamma,\vec{j}}$ must be concentrated at the intersections of S with γ. The detailed expression bears out this fact: the action of $\hat{A}_S$ on $\Psi_{\gamma,\vec{j}}$ is dictated simply by the spin labels j_I attached to those edges of γ which intersect S. For all surfaces S and 3-dimensional regions R in Σ, $\hat{A}_S$ and $\hat{V}_R$ are densely defined, self-adjoint operators. *All their eigenvalues are discrete* [6]. Naively, one might expect that the eigenvalues would be uniformly spaced, given by, e.g., integral multiples of the Planck area or volume. This turns out *not* to be the case; the distribution of eigenvalues is quite subtle. In particular, the eigenvalues crowd rapidly as areas and volumes increase. In the case of area operators, the complete spectrum is known in a closed form, and the first several hundred eigenvalues have been explicitly computed numerically. For a large eigenvalue a_n, the separation $\Delta a_n = a_{n+1} - a_n$ between consecutive eigenvalues decreases exponentially: $\Delta a_n \leq \ell_{\rm Pl}^2 \exp -(\sqrt{a_n}/\ell_{\rm Pl})$! Because of such strong crowding, the continuum approximation becomes excellent quite rapidly just a few orders of magnitude above the Planck scale. At the Planck scale, however, there is a precise and very specific replacement. This is the arena of quantum geometry. The premise is that the standard perturbation theory fails because it ignores this fundamental discreteness.

There is however a key mathematical subtlety [2,7]. This non-perturbative quantization has a one parameter family of ambiguities labelled by $\gamma > 0$. This γ is called the Barbero-Immirzi parameter and is rather similar to the well-known θ-parameter of QCD. In QCD, a single classical theory gives rise to inequivalent sectors of quantum theory, labelled by θ. Similarly, γ is classically irrelevant but different values of γ correspond to unitarily inequivalent representations of the algebra of geometric operators. The overall mathematical structure of all these sectors is very similar; the only difference is that the eigenvalues of all geometric operators scale with γ. For example, the simplest eigenvalues of the area operator $\hat{A}_S$ in the γ quantum sector is given by [b]

$$a_{\{j\}} = 8\pi\gamma\ell_{\rm Pl}^2 \sum_I \sqrt{j_I(j_I+1)} \tag{1}$$

where $I = 1, \ldots N$ for some N and each j_I is a half-integer. Since the representations are unitarily inequivalent, as usual, one must rely on Nature to resolve this ambiguity: Just as Nature must select a specific value of θ on QCD, it must select a specific value of γ in loop quantum gravity. With one judicious experiment —e.g., measurement of the lowest eigenvalue of the area operator $\hat{A}_S$ for a 2-surface S of any given topology— we could determine the value of γ and fix the theory. Unfortunately, such experiments are hard to perform! However, we will see in Section 3.2 that the Bekenstein-Hawking formula of black hole entropy provides an indirect measurement of this lowest eigenvalue of area for the 2-sphere topology and can therefore be used to fix

[b]In particular, the lowest non-zero eigenvalue of area operators is proportional to γ. This fact has led to a misunderstanding: in circles outside loop quantum gravity, γ is sometimes thought of as a regulator responsible for discreteness of quantum geometry. As explained above, this is *not* the case; γ is analogous to the QCD θ and quantum geometry is discrete in *every* permissible γ-sector. Note also that, at the classical level, the theory is equivalent to general relativity only if γ is *positive*; if one sets $\gamma = 0$ by hand, one can not recover *even the kinematics of general relativity. Similarly, at the quantum level, setting* $\gamma = 0$ would lead to a meaningless theory in which *all* eigenvalues of geometric operators vanish identically.

the value of γ.

2.3 Quantum dynamics

Quantum geometry described above provides a mathematical arena to formulate non-perturbative dynamics of candidate quantum theories of gravity, without any reference to a background classical geometry. In the case of general relativity, it provides the tools to write down quantum Einstein's equations in the Hamiltonian approach and transition amplitudes in the path integral approach. Until recently, effort was focussed primarily on the Hamiltonian approach. However, over the last three years, path integrals —called *spin foams*— have drawn a great deal of attention. This work has led to fascinating results suggesting S that, thanks to the fundamental discreteness of quantum geometry, path integrals defining quantum general relativity may be finite. These developments will be discussed in some detail in Section 4.1. In this Section, I will summarize the status of the Hamiltonian approach. For brevity, I will focus on source-free general relativity, although there has been considerable work on matter couplings.

For simplicity, let me suppose that the 'spatial' 3-manifold Σ is spatially compact. Then, in any theory without background fields, Hamiltonian dynamics is governed by constraints. Roughly, this is because, in these theories, diffeomorphisms correspond to gauge in the sense of Dirac. Recall that, on the Maxwell phase space, gauge transformations are generated by the functional $\mathcal{D}_a E^a$ which is constrained to vanish on physical states due to Gauss law. Similarly, on phase spaces of background independent theories, diffeomorphisms are generated by Hamiltonians which are constrained to vanish on physical states. In the case of general relativity, there are three sets of constraints. The first set consists of the three Gauss equations

$$\mathcal{G}_i := \mathcal{D}_a\, E^a_i = 0,$$

which, as in Yang-Mills theories, generates internal SU(2) rotations on the connection and the triad fields. The second set consists of a co-vector (or diffeomorphism) constraint

$$\mathcal{C}_b := E^a F_{ab} = 0,$$

which generates spatial diffeomorphism on Σ (modulo internal rotations generated by $\mathcal{G}_i$). Finally, there is the key scalar (or Hamiltonian) constraint

$$\mathcal{S} := \epsilon^{ijk} E^a_i E^b_j F_{ab}k + \ldots = 0$$

which generates time-evolutions. (The ... represent extrinsic curvature terms whose form is not important here.) Our task in quantum theory is three-folds:

i) Elevate these constraints (or their 'exponentiated versions') to well-defined operators on the kinematic Hilbert space $\mathcal{H}$; ii) Select physical states by asking that they be annihilated by these constraints; iii) introduce an inner-product and interesting observables, and develop approximation schemes, truncations, etc to explore physical consequences.

Step i) has been completed. Since the action of the Gauss and the co-vector constraints have a simple geometrical meaning, completion of i) in these cases is fairly straightforward. For the scalar constraint, on the other hand, there are no such guiding principles whence the procedure is subtle. In particular, specific regularization choices have to be made. Consequently, the answer is not unique. At the present stage of the program, such ambiguities are inevitable; one has to consider all viable candidates and analyze if they lead to sensible theories. The availability of well-defined Hamiltonian constraint operators is a notable technical success. For example, the analogous problem in quantum geometrodynamics —a satisfactory regularization of the Wheeler-DeWitt equation— is still open although the formal equation was written down some thirty five years ago. To be specific, I will focus on the procedure developed by Rovelli, Smolin, Lewandowski and others which culminated in a specific construction due to Thiemann [9].

Step ii) has been completed for the Gauss and the co-vector constraints [8]. The mathematical implementation required a very substantial extension [8,1] of the algebraic quantization program initiated by Dirac, and the use of the spin-network machinery [5] of quantum geometry. Again, the detailed implementation is a non-trivial technical success and the analogous task has not been completed in geometrodynamics because of difficulties associated with infinite dimensional spaces. The quantum scalar constraint is defined *on the space of solutions to the Gauss and co-vector constraints* and subsequent work (also by others) has led to the completion of step ii) for this constraint as well. More precisely, the problem of finding a general solution has been systematically reduced to that of finding *elementary* solutions, a task that requires only analysis of linear operators on certain finite dimensional spaces. While this is a striking result, it is still unclear whether the resulting theory is physically satisfactory; at this stage, it is in principle possible that it captures only an 'exotic' sector of quantum gravity. A key open problem in loop quantum gravity is to show that the scalar/Hamiltonian constraint —either Thiemann's or alternatives such as the one of Gambini and Pullin [9]— admit a 'sufficient number' of semi-classical states. Progress on this problem has been slow because, as explained in Section 1, the general issue of semi-classical limits is itself difficult in background independent approaches. However, as discussed in Section 4.2 below, a systematic understanding has now begun to emerge and is providing the 'infra-structure' needed to analyze the key problem mentioned above. More generally, while there are promising ideas to complete step iii), substantial further work is necessary to solve this problem. Recent advance in quantum cosmology, described in Section 3.1, is an example of the progress in this direction and it provides a strong support for the Thiemann scheme, but of course only within the limited context of mini-superspaces.

Remark: There has been another concern about this class of regularizations which, however, is less specific. It stems from the structure of the constraint algebra. To analyze this issue, the domain of definition of the scalar constraint had to be extended to *certain* states which are not diffeomorphism invariant, so that the commutators could be meaningfully calculated. It was then found that the commutator between any two Hamiltonian constraints vanishes identically [9], while in the classical theory, the corresponding Poisson brackets vanishes only on solutions to the diffeomorphism constraint. However, it was also shown that the operator representing the right side of the classical Poisson bracket *also vanishes* on all the quantum states considered, including the ones which are not diffeomorphism invariant. Therefore, while the vanishing of the commutator of the Hamiltonian constraint was unexpected, this analysis does not reveal a clear-cut problem with these regularizations.

Furthermore, one can follow this scheme step by step in 2+1 gravity where one knows what the result should be. One can obtain the 'elementary solutions' mentioned above and show that all the standard quantum states —including the semi-classical ones— can be recovered as linear combinations of these elementary ones. As is almost always the case with constrained systems, there are *many more solutions* which can be eliminated only by the requirement that their physical inner-product should be finite. In 2+1 gravity, the connection formulation used here naturally leads to a complete set of Dirac observables and the inner-product can be essentially fixed by the requirement that they be self-adjoint. In 3+1 gravity, by contrast, we do not have this luxury and the problem of constructing the physical inner-product is therefore much more difficult. However, the concern here is that of weeding out unwanted solutions rather than having a 'sufficient number' of semi-classical ones, which is a much less serious issue at this stage.

3 Applications of quantum geometry

In this section, I will summarize two recent developments that answer several of the questions raised under first two bullets in the Introduction.

3.1 Big bang

Over the last three years, quantum geometry has led to some striking results of direct physical interest. The first of these concerns the fate of the big-bang singularity.

Traditionally, in quantum cosmology one has proceeded by first imposing spatial symmetries —such as homogeneity and isotropy— to freeze out all but a finite number of degrees of freedom *already at the classical level* and then quantizing the reduced system. In the simplest case, the basic variables of the reduced classical system are the scale factor a and matter fields ϕ. The symmetries imply that space-time curvature goes as $\sim 1/a^2$ and Einstein's equations predict a big-bang, where the scale factor goes to zero and the curvature blows up. As indicated in Section 1, this is reminiscent of what happens to ferro-magnets at the Curie temperature: magnetization goes to zero and the susceptibility diverges. By analogy, the key question is: Do these 'pathologies' disappear if we re-examine the situation in the context of an appropriate quantum theory? In traditional quantum cosmologies, without an additional input, they do not. That is, typically, to resolve the singularity one either has to introduce matter with unphysical properties or introduce boundary conditions, e.g., by invoking new principles.

In a series of seminal papers [10], Bojowald has shown that the situation in loop quantum cosmology is quite different: the underlying quantum geometry makes a *qualitative* difference very near the big-bang. In the standard procedure summarized above, the reduction is carried out at the classical level and this removes all traces of the fundamental discreteness. Therefore, the key idea in Bojowald's analysis is to retain the essential features of quantum geometry by first quantizing the kinematics of the *full theory* as in Section 2.2 and then restricting oneself to *quantum* states which are spatially homogeneous and isotropic. As a result, the scale factor operator $\hat{a}$ has *discrete eigenvalues.* The continuum limit is reached rapidly. For example, the gap between an eigenvalue of $\hat{a}$ of $\sim$ 1cm and the next one is less than $\sim 10^{-30}\,\ell_{\rm Pl}$! Nonetheless, near $a \sim \ell_{\rm Pl}$ there are surprises and predictions of loop quantum cosmology are very different from those of traditional quantum cosmology.

The first surprise occurs already at the kinematical level. Recall that, in the classical theory curvature is essentially given by $1/a^2$, and blows up at the big-bang. What is the situation in quantum theory? Denote the Hilbert space of spatially homogeneous, isotropic kinematical quantum states by $\mathcal{H}_{\rm HI}$. A self-adjoint operator $\widehat{\rm curv}$ corresponding to curvature can be constructed on $\mathcal{H}_{\rm HI}$ and turns out to be *bounded from above.* This is very surprising because $\mathcal{H}_{\rm HI}$ admits an eigenstate of the scale factor operator $\hat{a}$ with a discrete, zero eigenvalue! At first, it may appear that this could happen only by an artificial trick in the construction of $\widehat{\text{curv}}$ and that this quantization can not possibly be right because it seems to represent a huge departure from the classical relation $(\text{curv})\, a^2 = 1$. However, these concerns turn out to be misplaced. The procedure for constructing $\widehat{\text{curv}}$ is natural and, furthermore, descends from the full theory; Bojowald essentially repeats a key step in Thiemann's procedure of defining the quantum scalar/Hamiltonian constraint in the full theory. Let us examine the properties of $\widehat{\text{curv}}$. Its upper bound u_{curv} is finite but absolutely huge:

$$u_{\text{curv}} \sim \frac{256}{81}\frac{1}{\ell_{\text{Pl}}^2} \equiv \frac{256}{81}\frac{1}{G\hbar} \tag{2}$$

or, about 10^{77} times the curvature at the horizon of a solar mass black hole. The functional form of the upper bound is also illuminating. Recall that, in the case of an hydrogen atom, energy is unbounded from below classically but, thank to $\hbar$, we obtain a finite value, $E_0 = -\,(me^4/n^2)\,(\frac{1}{\hbar^2})$, in quantum theory. Similarly, u_{curv} is finite because $\hbar$ is non-zero and tends to the classical answer as $\hbar$ tends to zero. At curvatures as large as u_{curv}, it is natural to expect large departures from classical relations such as $(\text{curv})\, a^2 = 1$. But is this relation recovered in the semi-classical regime? The answer is in the affirmative. In fact it is somewhat surprising how quickly this happens. As one would expect, one can simultaneously diagonalize $\hat{a}$ and $\widehat{\text{curv}}$. If we denote their eigenvalues by a_n and b_n respectively, then $a_n \cdot b_n - 1$ is of the order 10^{-4} at $n = 100$ and decreases rapidly as n increases. These properties show that, in spite of the initial surprise, the quantization procedure is viable. Furthermore, one can apply it also to more familiar systems such as a particle moving on a circle and obtain results which at first seem surprising but are in complete agreement with the standard quantum theory of these systems.

Since the curvature is bounded above in the entire Hilbert space, one might hope that the quantum evolution may be well-defined right through the big-bang singularity. Is this in fact the case? The second surprise is that although the quantum evolution is close to that of the Wheeler-DeWitt equation of standard quantum cosmology for large a, there are dramatic differences near the big-bang which makes it well defined even *at* the big-bang, without any additional input. As one might expect, the 'evolution' is dictated by the quantum scalar constraint operator. To obtain this operator, Bojowald again follows, step by step, the procedure used by Thiemann in the full theory. Let us expand out the full quantum state as $\mid \Psi >= \sum_n \psi_n(\phi) \mid n>$ where $\mid n>$ are the eigenstates of the scale factor operator and ϕ denotes matter fields.

Then, the scalar constraint takes the form:

$$c_n\psi_{n+8}(\phi)+d_n\psi_{n+4}(\phi)+e_n\psi_n(\phi)+f_n\psi_{n-4}(\phi)+g_n\psi_{n-8}(\phi) = \gamma\ell_{\rm Pl}^2\ \hat{H}_\phi\psi_n(\phi) \tag{3}$$

where $c_n, \ldots g_n$ are fixed numerical coefficients, γ the Barbero-Immirzi parameter and $\hat{H}_\phi$ is the matter Hamiltonian. (Again, using the Thiemann regularization, one can show that the matter Hamiltonian is a well-defined operator.) Primarily, being a constraint equation, (3) constrains the physically permissible $\psi_n(\phi)$. However, *if* we choose to interpret the scale factor (more precisely, the square of the scale factor times the determinant of the triad) as a time variable, (3) can be interpreted as an 'evolution equation' which evolves the state through discrete time steps. In a (large) neighborhood of the big-bang singularity, this 'deparametrization' is viable. For the choice of factor ordering used in the Thiemann regularization, one can evolve in the past through $n = 0$, i.e. right through the classical singularity. Thus, the infinities predicted by the classical theory at the big-bang are artifacts of assuming that the classical, continuum space-time approximation is valid right up to the big-bang. In the quantum theory, the state can be evolved through the big-bang without any difficulty. However, the classical space-time description fails near the big-bang; quantum evolution is well-defined but the classical space-time 'dissolves'.

The 'evolution' equation (3) has other interesting features. To begin with, the space of solutions is 16 dimensional. Can we single out a preferred solution by imposing a ***physical*** condition? One possibility is to impose ***pre-classicality***, i.e., to require that the quantum state not oscillate rapidly from one step to the next at *late* times when we know our universe behaves classically. Although this is an extra input, it is not a theoretical prejudice about what should happen at (or near) the big-bang but an observationally motivated condition that is clearly satisfied by our universe. The coefficients $c_n, \ldots g_n$ of (3) are such that this condition singles out a solution uniquely. One can ask what this state does at negative times, i.e., before the big-bang. (Time becomes negative because triads flip orientation on the 'other side'.) Preliminary indications are that the state does not become pre-classical there. If this is borne out by detailed calculations, then the qualitative analogy with ferro-magnets would become closer, our side of the big-bang being analogous to the ferro-magnetic phase in which classical geometry (the analog of magnetization) is both meaningful and useful and the 'other' side being analogous to the para-magnetic phase where it is not. Another interesting feature is that the standard Wheeler-DeWitt equation is recovered if we take the limit $\gamma \to 0$ and $n \to \infty$ such that the eigenvalues of $\hat{a}$ take on continuous values. This is completely parallel to the limit we often take to coarse grain the quantum description of a rotor to 'wash out' discreteness in angular momentum eigenvalues and arrive at the classical description. From this perspective, then, one is led to say that the most striking of the consequences of loop quantum gravity are not seen in standard quantum cosmology because it 'washes out' the fundamental discreteness of quantum geometry.

Finally, the detailed calculations have revealed another surprising feature. The fact that the quantum effects become prominent near the big bang, completely invalidating the classical predictions, is pleasing but not unexpected. However, prior to these calculations, it was not clear how soon after the big-bang one can start trusting semi-classical notions and calculations. It would not have been surprising if we had to wait till the radius of the universe became, say, a few million times the Planck length. These calculations strongly suggest that a few tens of Planck lengths should suffice. This is fortunate because it is now feasible to develop quantum numerical relativity; with computational resources commonly available, grids with $(10^6)^3$ points are hopelessly large but one with $(100)^3$ points are readily available.

3.2 Black-holes

Loop quantum cosmology illuminates dynamical ramifications of quantum geometry but within the context of mini-superspaces where all but a finite number of degrees of freedom are frozen. In this sub-section, I will discuss a complementary application where one considers the full theory but probes consequences of quantum geometry which are not sensitive to full quantum dynamics —the application of the framework to the problem of black hole entropy. This discussion is based on joint work with Baez, Corichi and Krasnov [11] which itself was motivated by earlier work of Krasnov, Rovelli and others.

As explained in the Introduction, since mid-seventies, a key question in the subject has been: What is the statistical mechanical origin of the black hole entropy $S_{\rm BH} = \frac{a_{\rm hor}}{4\ell_{\rm Pl}^2}$? What are the microscopic degrees of freedom that account for this entropy? This relation implies that a solar mass black hole must have $(\exp 10^{77})$ quantum states, a number that is *huge* even by the standards of statistical mechanics. Where do all these states reside? To answer these questions, in the early nineties Wheeler had suggested the following heuristic picture, which he christened 'It from Bit'. Divide the black hole horizon in to elementary cells, each with one Planck unit of area, $\ell_{\rm Pl}^2$ and assign to each cell two microstates, or one 'bit'. Then the total number of states $\mathcal{N}$ is given by $\mathcal{N} = 2^n$ where $n = (a_{\rm hor}/\ell_{\rm Pl}^2)$ is the number of elementary cells, whence entropy is given by $S = \ln \mathcal{N} \sim a_{\rm hor}$. Thus, apart from a numerical coefficient, the entropy ('It') is accounted for by assigning two states ('Bit') to each elementary cell. This qualitative picture is simple and attractive. But the key open issue was: can these heuristic ideas be supported by a systematic analysis from first principles? Quantum geometry has supplied the required analysis.[c]

A systematic approach requires that we first specify the class of black holes of interest. Since the entropy formula is expected to hold unambiguously for black holes in equilibrium, most analyses were confined to *stationary*, eternal black holes (i.e., in 4-dimensional general relativity, to the Kerr-Newman family). From a physical viewpoint however, this assumption seems overly restrictive. After all, in statistical mechanical calculations of entropy of ordinary systems, one only has to assume that the given system is in equilibrium, not the whole world. Therefore, it should suffice for us to assume that the black hole itself is in equilibrium; the exterior geometry should not be forced to be time-independent. Furthermore, the analysis should also account for entropy of black holes which may be distorted or carry (Yang-Mills and other) hair. Finally, it has been known since the mid-seventies that the thermodynamical considerations apply not only to black holes but also to cosmological horizons. A natural question is: Can these diverse situations be treated in a single stroke? Within the quantum geometry approach, the answer is in the affirmative. The entropy calculations have been carried out in the recently developed framework of 'isolated horizons' which encompasses all these situations. Isolated horizons serve as 'internal boundaries' whose intrinsic geometries (and matter fields) are time-independent, although space-time geometry as well as matter fields in the external space-time region can be fully dynamical. The zeroth and first laws of black hole mechanics have been extended to isolated horizons. Entropy associated with an isolated horizon refers to the family of observers in the exterior for whom the isolated horizon is a physical boundary that separates the region which is accessible to them from the one which is not. This point is especially important for cosmological horizons where, without reference to observers, one can not even define horizons. States which contribute to this entropy are the ones which can interact with the states in the exterior; in this sense, they 'reside' on the horizon.

In the detailed analysis, one considers space-times admitting an isolated horizon as inner boundary and carries out a systematic quantization. The quantum geometry framework can be naturally extended to this case. The isolated horizon boundary conditions imply that the intrinsic geometry of the

[c]However, I should add that this account does not follow chronology. Black hole entropy was computed in quantum geometry quite independently and the realization that the 'It from Bit' picture works so well was somewhat of a surprise.

quantum horizon is described by the so called U(1) Chern-Simons theory on the horizon. This is a well-developed, topological field theory. A deeply satisfying feature of the analysis is that there is a seamless matching of three otherwise independent structures: the isolated horizon boundary conditions, the quantum geometry in the bulk, and the Chern-Simons theory on the horizon. In particular, one can calculate eigenvalues of certain physically interesting operators using purely bulk quantum geometry without any knowledge of the Chern-Simons theory, or using the Chern-Simons theory without any knowledge of the bulk quantum geometry. The two theories have never heard of each other. Yet, thanks to the isolated horizon boundary conditions, the two infinite sets of numbers match exactly, providing a coherent description of the quantum horizon.

In this description, the polymer excitations of the bulk geometry, each labelled by a spin j_I, pierce the horizon, endowing it an elementary area a_{j_I} given by (1). The sum $\sum_I a_{j_I}$ adds up to the total horizon area $a_{\rm hor}$. The intrinsic geometry of the horizon is flat except at these puncture, but at each puncture there is a *quantized* deficit angle. These add up to endow the horizon with a 2-sphere topology. For a solar mass black hole, a typical horizon state would have 10^{77} punctures, each contributing a tiny deficit angle. So, although the quantum geometry *is* distributional, it can be well approximated by a smooth metric.

The counting of states can be carried out as follows. First one constructs a micro-canonical ensemble by restricting oneself only to those states for which the total area, angular momentum, and charges lie in small intervals around fixed values $a_{\rm hor}, J_{\rm hor}, Q^i_{\rm hor}$. (As is usual in statistical mechanics, the leading contribution to the entropy is independent of the precise choice of these small intervals.) For each set of punctures, one can compute the dimension of the surface Hilbert space, consisting of Chern-Simons states compatible with that set. One allows all possible sets of punctures (by varying both the spin labels and the number of punctures), subject to the constraint that the total area $a_{\rm hor}$ be fixed, and adds up the dimensions of the corresponding surface Hilbert spaces to obtain the number $\mathcal{N}$ of permissible surface states. One finds that the horizon entropy $S_{\rm hor}$ is given by

$$S_{\rm hor} := \ln \mathcal{N} = \frac{\gamma_o}{\gamma}\frac{a_{\rm hor}}{\ell_{\rm Pl}^2} + \mathcal{O}(\frac{\ell_{\rm Pl}^2}{a_{\rm hor}}), \quad \text{where} \quad \gamma_o = \frac{\ln 2}{\sqrt{3}\pi} \tag{4}$$

Thus, for large black holes, entropy is indeed proportional to the horizon area. This is a non-trivial result; for examples, early calculations often led to proportionality to the square-root of the area. However, even for large black holes, one obtains agreement with the Hawking-Bekenstein formula only in the sector of quantum geometry in which the Barbero-Immirzi parameter γ takes the value $\gamma = \gamma_o$. Thus, while all γ sectors are equivalent classically, the standard quantum field theory in curved space-times is recovered in the semi-classical theory only in the γ_o sector of quantum geometry. It is quite remarkable that thermodynamic considerations involving *large* black holes can be used to fix the quantization ambiguity which dictates such Planck scale properties as eigenvalues of geometric operators. Note however that the value of γ can be fixed by demanding agreement with the semi-classical result just in one case —e.g., a spherical horizon with zero charge, or a cosmological horizon in the de Sitter space-time, or, ... Once the value of γ is fixed, the theory is completely fixed and we can ask: Does this theory yield the Hawking-Bekenstein value of entropy of *all* isolated horizons, irrespective of the values of charges, angular momentum, and cosmological constant, the amount of distortion, or hair. The answer is in the affirmative. Thus, the agreement with quantum field theory in curved space-times holds in *all* these diverse cases.

Why does γ_o not depend on other quantities such as charges? This important property can be traced back to a key consequence of the isolated horizon boundary conditions: detailed calculations show that only the gravitational part of the symplectic structure has a surface term at the horizon; the matter symplectic structures have only volume terms. (Furthermore, the gravitational surface term is insensitive to the value of the cosmological constant.) Consequently, there are no independent surface quantum states associated with matter. This provides a natural explanation of the fact that the Hawking-Bekenstein entropy depends only on the horizon geometry and is independent of electro-magnetic (or other) charges.

Finally, let us return to Wheeler's 'It from Bit'. One can ask: what are the states that dominate the counting? Perhaps not surprisingly, they turn out to be the ones which assign to each puncture the smallest quantum of area (i.e., spin value $j = \frac{1}{2}$), thereby maximizing the number of punctures. In these states, each puncture defines Wheeler's 'elementary cell' and his two states correspond to whether the deficit angle is positive or negative.

To summarize, quantum geometry naturally provides the micro-states responsible for the huge entropy associated with horizons. In this analysis, all black holes and cosmological horizons are treated in an unified fashion; there is no restriction, e.g., to near-extremal black holes. The sub-leading term has also been calculated and shown to be proportional to $\ln a_{\rm hor}$. Finally, in this analysis quantum Einstein's equations *are* used. In particular, had we not imposed the quantum Gauss and co-vector/diffeomorphism constraints on surface states, the spurious gauge degrees of freedom would have given an infinite entropy. However, because of the isolated horizon boundary conditions, the scalar/Hamiltonian constraint has to be imposed just in the bulk. Since in the entropy calculation one traces over bulk states, the final result is insensitive to the details of how this (or any other bulk) equation is imposed. Thus, as in other approaches to black hole entropy, the calculation does not require a complete knowledge of quantum dynamics.

4 Current work

Work is now in progress along many directions, ranging from the fate of the 'final' black hole singularity in quantum geometry and the associated issue of 'information loss', to predictions of quantum cosmology for structure formation, to possible experimental tests of quantum geometry. To illustrate these developments, in this section I will discuss two main thrusts.

4.1 Spin foams

Spin foams can be thought of as histories traced out by 'time evolution' of spin networks and provide a path integral approach to quantum dynamics. I will focus on the fascinating 'finiteness' results [12] obtained by Crane, Rovelli and especially Perez based on earlier work also by Baez, Barrett, De Pietri, Freidel, Krasnov, Mikovič, Reisenberger and others.

In the gravitational context, the path integral can play two roles. First, as in standard quantum field theories, it can be used to compute 'transitions amplitudes'. However outside, say, perturbation theory about a background space-time, there still remain unresolved conceptual questions about the physical meaning of such amplitudes. The second role is 'cleaner': as in the Euclidean approach of Hawking and others, it can be considered as a device to extract physical states, i.e. solutions to all the quantum constraint equations. In this role as an *extractor*, it can shed new light on the difficult issue of regularization of the Hamiltonian constraint discussed in Section 2.3.

The well-defined quantum kinematics of Section 2.2 has motivated specific proposals for the definition of path integrals, often called 'state sum models'. Perhaps the most successful of these is the Barrett-Crane model (modified slightly to handle a technical issue). At the classical level, one regards general relativity as a topological field theory, *called the* BF *theory*, supplemented with an algebraic constraint. The BF theory is itself a generalization of the Chern-Simons theory mentioned in Section 3.2 and has been investigated in detail in the mathematical physics literature. However, the role of the additional constraint is very important. Indeed, the BF theory has no local degrees of freedom; it is the extra constraint that reduces the huge gauge freedom of the BF theory, thereby recovering the local degrees of freedom of general relativity. The crux of the problem in quantum gravity is the appropriate incorporation of this constraint. At the classical level, the constrained B-F theory is equivalent to general relativity. To obtain Euclidean general relativity, one has to start with the BF theory associated with $SO(4)$ while the Lorentzian theory results if one uses $SO(3,1)$ instead. The Barrett-Crane model is a specific proposal to define path integrals for the constrained BF theory in either case.

Fix a 4-manifold M bounded by two 3-manifolds Σ_1 and Σ_2. Spin-network states on the two boundaries can be regarded as 'initial' and 'final' quantum geometries. One can then consider histories, i.e., quantum 4-geometries, joining them. Each history is a spin-foam. Each vertex of the initial spin-network on Σ_1 'evolves' to give a 1-dimensional edge in the spin-foam and each edge, to give a 2-dimensional face. Consequently, each face carries a spin label j. However, in the course of 'evolution' *new vertices* can appear, making the dynamics non-trivial and yielding a non-trivial amplitude for an 'initial' spin-network with n_1 vertices to evolve in to a 'final' spin-network with n_2 vertices. For mathematical clarity as well as physical intuition, it is convenient to group spin-foams associated with the same 4-dimensional graph but differing from one another in the labels, such as the spins j carried by faces. Each group is said to provide a *discretization* of M. Physically, a discretization has essentially just the topological information. The geometrical information —such as the area associated with each face— resides in the labels. This is a key distinction from lattice gauge theories with a background metric, where a discretization itself determines, e.g., the edge lengths and hence how refined the lattice is.

A key recent development is the discovery that the non-perturbative path integral, defined by the (modified) Barrett-Crane model is, in a certain precise sense, equivalent to a manageable *group field theory* (GFT) in the sense specified below. The GFT is a rather simple quantum field theory, defined on four copies of the underlying group —$\mathrm{SL}(2,C)$ in the case of Lorentzian gravity and Spin(4) in the case of Euclidean. (Note that these are just double covers of the Lorentz group and the rotation group in Euclidean 4-space.) Thus GFTs live in high dimensions. The action has a 'free part' and an interaction term with a coupling constant λ. But the free part is non-standard and does not have the familiar kinetic term whence the usual non-renormalizability arguments for higher dimensional, interacting theories do not apply. In fact, the first key recent result is that *this GFT is finite order by order in Feynman perturbation expansion.* The second key result is $A_{\mathrm{BC}}(n) = A_{\mathrm{GFT}}(n)$, where $A_{\mathrm{BC}}(n)$

is the Barret-crane amplitude obtained by summing over all geometries (i.e., spin labels j) for a fixed discretization and $A_{\mathrm{GFT}}(n)$ is the coefficient of λ^n in the Feynman expansion of the GFT. Together, the two results imply that, in this approach to quantum gravity, *sum over geometries for a fixed discrete topology is finite.* This is a highly non-trivial result because, on each face, the sum over js ranges from zero to infinity; there is no cut-off.

These results do not show that the full path integral is finite because the discrete topology is kept fixed in the sum. Work is in progress on removing this dependence. But even as they stand, the results have a striking, qualitative similarity with the finiteness results that have been recently obtained in other approaches. The order by order finiteness is reminiscent of the order by order ultraviolet finiteness of string perturbation theory. In the present case, the situation is somewhat better: there is both ultra-violet *and* infra-red finiteness. Qualitatively, the first is ensured by the discreteness of underlying geometry while the second by the fact that arbitrarily large values of js are allowed in the sum. The second result —equivalence of the GFT with a certain definition of quantum gravity— is reminiscent of the conjectures discussed by Maldacena in this conference. In both cases, there is a mathematical equivalence between quantum gravity and certain field theories which have no knowledge of the physical space-time. On the one hand, results are stronger in the present case: both sides of the equality are separately defined and the thrust is on proving the equality, not just on constructing evidence in support of it. However, in contrast to Maldacena's bold conjecture, here the equality in question is only order by order. Finally, recently Luscher and Reuter have found evidence for nonperturbative renormalizability of 4-dimensional Euclidean quantum general relativity (stemming from the existence of a nontrivial fixed point) [12]. It is natural to ask: Is there a relation to Perez's finiteness results in the Euclidean signature?

4.2 Relation to low energy physics

The basic mathematical structures underlying loop quantum gravity are very different from those used in the text-book treatments of low energy physics. For example, in quantum geometry the fundamental excitations are one dimensional, polymer-like; a convenient basis of states is provided by spin-networks; and, eigenvalues of the basic geometric operators are discrete. By contrast, in the Fock framework of low energy physics the fundamental excitations are 3-dimensional, wavy; the convenient basis is labelled by the number of particles, their momenta and helicities; and, all geometric operators have continuous spectra. The challenge is to bridge the gap between these apparently disparate frameworks. These differences are inevitable, given that the standard quantum field theory is constructed in Minkowski space, while loop quantum gravity does not refer to any background geometry. Nonetheless, if loop quantum gravity is correct, it should be possible to recover from it the standard Fock framework in a suitable semi-classical approximation. Over the last decade, semi-classical states approximating classical geometries have been constructed with various degrees of precision [13]. However, the relation to Fock quantization has been explored only recently [14]. My summary of this development will be even briefer than that of other advances because this topic was discussed in some detail in Rovelli's workshop D1i.

The recent developments are based on a key mathematical insight due to Varadarajan. Fock states do not belong to the Hilbert space $\mathcal{H}$ of polymer excitations. However, neither do the physical states, i.e., the solutions to quantum constraints. This is not a peculiarity of general relativity. Even for quantum mechanical systems such as a free particle in Minkowski space, the physical states belong not to the kinematical Hilbert space, but to a natural extension of it, constructed from the refinement of the Dirac quantization program for constrained systems, mentioned in Section 2.3. Varadarajan showed that Fock states do belong to this extension $\mathcal{E}$ of $\mathcal{H}$. Thus, the natural habitat $\mathcal{E}$ for the physical states in loop quantum gravity also accommodates the standard Fock states of low energy physics.

At first, this result seems very surprising. For example, in the polymer description of a quantum Maxwell field, fluxes of electric field are quantized. On the Fock space, on the other hand, these flux operators are not even defined unless surfaces are thickened and, when they are, the quantization is lost. This phenomenon succinctly captures the tension between the polymer and Fock excitations. It turns out that the flux operators are indeed well-defined on $\mathcal{E}$ but they do not map the Fock space to itself; flux quantization is lost because the Fock space is 'too small' and fails to accommodate any of the eigenvectors of these operators with discrete eigenvalues. The situation with quantum geometry is completely analogous: The discreteness of quantum geometry is lost if one insists that *all* quantum gravity states must reside in the Fock space of gravitons. To see the discreteness, one has to allow states which are 'purely quantum mechanical' and can not be regarded as excitations on *any* classical background geometry.

The interplay between the Fock and polymer excitations can now be studied in detail, thanks to a key property: each Fock state casts a 'shadow' in each finite dimensional space $\mathcal{H}_{\gamma,\vec{j}}$ of $\mathcal{H}$ and can be fully recovered from the collection of these shadows. Using these shadows one can analyze a number of features of the Fock framework, such as construction of coherent states, expectations values and fluctuations of fields, needed in the analysis of semi-classical issues. Thus, one can now begin to analyze two key questions: Can the background-independent, non-perturbative theory reproduce the familiar low-energy physics on suitable coarse graining? and, Can one pin-point where and why the standard perturbation theory fails?

5 Conclusion

In the last three sections, I have summarized recent advances which have answered some of the long standing questions of quantum gravity raised in the Introduction. Personally, I find it very satisfying that a number of the key ideas came from younger researchers —from Bojowald in quantum cosmology, from Krasnov in the understanding of quantum black holes, Perez in spin foams and Varadarajan in the relation to low energy physics.

Throughout the development of loop quantum gravity, unforeseen simplifications have arisen regularly, leading to surprising solutions to seemingly impossible difficulties. Progress could occur because some of the obstinate problems which had slowed developments in background independent approaches, sometimes for decades, evaporated when 'right' perspectives were found. I will conclude with a few examples.

• Up until the early nineties, it was widely believed that spaces of connections do not admit non-trivial diffeomorphism invariant measures. This would have made it impossible to develop a background independent approach. Quite surprisingly, such a measure could be found by looking at connections in a slightly more general perspective. It is simple, natural, and has just the right structure to support quantum geometry. This geometry, in turn, supplied some missing links, e.g., by providing just the right expressions that Ponzano-Regge had to postulate without justification in their celebrated, early work.

• Fundamental discreteness first appeared in a startling fashion in the construction of the so-called weave states, which approximate a classical 3-geometry. In this construction, the polymer excitations were introduced as a starting point with the goal of taking the standard continuum limit. It came as a major surprise that, if one wants to recover a given classical geometry *on large scales*, one can not take this limit, i.e., one can not pack the polymer excitations arbitrarily close together; there is an in-built discreteness.

• At a heuristic level, it was found that the Wilson loop functionals of a suitably defined connection around a smooth loop solve the notoriously difficult quantum scalar constraint automatically. No one expected to find such simple and natural solutions even heuristically. This calculation suggested that the action of the constraint operator is concentrated at 'nodes', i.e., intersections, which in turn led to strategies for its regularization.

• As I indicated in some detail, unforeseen insights arose in the well-studied subject of quantum cosmology essentially by taking an adequate account of the quantum nature of geometry, i.e., by respecting the fundamental discreteness of the eigenvalues of the scale factor operator. Similarly, in the case of black holes, three quite distinct structures —the isolated horizon boundary conditions, the bulk quantum geometry and the surface Chern-Simons theory— blended together unexpectedly to provide a coherent theory of quantum horizons.

Repeated occurrence of such 'unreasonable' simplifications suggest that the ideas underlying loop quantum gravity may have captured an essential germ of truth.

Acknowledgements I would like to thank Martin Bjowald, Jerzy Lewandowski and Alejandro Perez for their comments on the manuscript and Carlo Rovelli for correspondence. This work was supported in part by the NSF grant PHY-0090091 and the Eberly research funds of Penn State.

References

1. *Books:*
   A. Ashtekar, *Lectures on non-perturbative canonical gravity,* (World Scientific, Singapore, 1991)
   R. Gambini and J. Pullin, *Loops, knots, gauge theories and quantum gravity,* (CUP, Cambridge, 1996)
2. *Classical theory:*
   A. Ashtekar, Phys. Rev. Lett. **57**, 2244-2247, (1986); Phys. Rev. **D36**, 1587-1602 (1987); A. Ashtekar, J.D. Romano and R. S. Tate, **D40**, 2572-2587 (1989); F. Barbero, Phys. Rev. **D51**, 5507-5510 (1996)
3. *Connctions and loops:*
   C. Rovelli and L. Smolin, Nucl. Phys. **B331** 80-152 (1990); A. Ashtekar, V. Husain, C. Rovelli, J. Samuel and L. Smolin, Class. Quant. Grav. **6**, L185-L193 (1989); A. Ashtekar, C. Rovelli and L. Smolin, Phys. Rev. **D44**, 1740-1755 (1991).
4. *Quantum Geometry: Basics*
   A. Ashtekar and C. J. Isham, Class. Quant. Grav. **9**, 1433-1467 (1992); A. Ashtekar and J. Lewandowski, Representation theory of analytic holonomy algebras, in *Knots and Quantum Gravity*, edited by J. C. Baez, Oxford U. Press, Oxford, (1994); J. C. Baez, Lett. Math. Phys.

31, 213-223 (1994); D. Marolf and J. Mourão, Commun. Math. Phys. **170**, 583-606 (1995); A. Ashtekar and J. Lewandowski, Jour. Math. Phys. **36**, 2170-2191 (1995); J. C. Baez and S. Jour. Funct. Analysis **150**, 1-27 (1997)

5. *spin networks:*
R. Penrose, Angular momentum: an approach to combinatorial spacetime, in *Quantum Theory and Beyond*, edited by Ted Bastin, (Cambridge University Press, 1971); C. Rovelli and L. Smolin, Phys. Rev. **D52**, 5743-5759 (1995); J. C. Baez, Spin networks in non-perturbative quantum gravity, in *The Interface of Knots and Physics,* edited by L. Kauffman, American Mathematical Society, Providence, 1996, pp. 167-203; J. C. Baez, Spin networks in gauge theory, Adv. Math. **117**, 253-272 (1996); J. C. Baez and S. Sawin, Diffeomorphism-invariant spin network states, Jour. Funct. Analysis **158**, 253-266 (1998)
6. *Geometric operators and their properties*
C. Rovelli and L. Smolin, Nucl. Phys. **B442**, 593-622 (1995); Erratum: Nucl. Phys. **B456**, 753 (1995); R. Loll, Phys. Rev. Lett. **75** 3048-3051 (1995); A. Ashtekar and J. Lewandowski, Jour. Geo. & Phys. **17**, 191-230 (1995); A. Ashtekar and J. Lewandowski, Class. Quant. Grav. **14**, A55-A81 (1997); Adv. Theo. Math. Phys. **1**, 388-429 (1997); R. Loll, Class. Quant. Grav. **14** 1725-1741 (1997); T. Thiemann, Jour. Math. Phys. **39**, 3372-3392 (1998)
7. *Barbero-Immirzi ambiguity*
G. Immirzi, Nucl. Phys. Proc. Suppll. **57**, 65-72 (1997); C. Rovelli and T. Thiemann, Phys. Rev. **D57** 1009-1014 (1998); R. Gambini O. Obregon and J. Pullin, Phys. Rev. **D59** 047505 (1999)
8. *Gauss and difeomorphism constraints*
D. Marolf, `gr-qc/9508015`; A. Ashtekar, J. Lewandowski, D. Marolf, J. Mourão, and T. Thiemann, Jour. Math. Phys. **36**, 6456-6493 (1995); J. Lewandowski and T. Thiemann, Class. Quant. Grav. **16**, 2299-2322 (1999)

9. *Hamiltonian constraint*
C. Rovelli and L. Smolin, Phys. Rev. Lett. **72**, 446-449 (1994); T. Thiemann, Phys. Lett. **B380** 257-264 (1996); Class. Quant. Grav. **15** 839-873 (1998); Class. Quant. Grav. **15** 1207-1247 (1998); Class. Quant. Grav. **15** 1281-1314 (1998); R. Gambini, J. Lewandowski, D. Marolf and J. Pullin, Int. J. Mod. Phys. **D7**, 97-109 (1998); J. Lewandowski and D. Marolf, Int. J. Mod. Phys. **D7**, 299-330 (1998); C. Di Bartolo, R. Gambini, J.

Griego, J. Pullin, Phys. Rev. Lett. **84**, 2314-2317 (2000)

10. *Big-bang:*
M. Bojowald, Phys. Rev. Lett. **86**, 5227-5230 (2001); Phys. Rev. Lett. **87**, 121301 (2001); Phys. Rev. **D64** 084018 (2001); Class. Quant. Grav. **18** 1055–1070 (2001); Class. Quant. Grav. **18** 1071–1088 (2001); *Quantum Geometry and Symmetry*, (Shaker-Verlag, Aachen, 2001)
11. *Black holes:*
A. Ashtekar, J. Baez, A. Corichi, K. Krasnov, Phys. Rev. Lett. **80** 904-907 (1998); A. Ashtekar, A. Corichi and K. Krasnov, Adv. Theor. Math. Phys. **3**, 418-471 (1999); A. Ashtekar, J. Baez, K. Krasnov, Adv. Theo. Math. Phys., **4**, 1-95 (2000); R. K. Kaul and P. Majumdar, Phys. Rev. Lett. **84**, 5255-5257 (2000)
12. *Spin foams and finiteness:*
A. Perez, Nucl. Phys. **B599**, 427 (2001); L. Crane, A. Perez, and C. Rovelli, Phys. Rev. Lett. **87** 181301, (2001); A. Perez and C. Rovelli, Phys. Rev. **D63**, 041501 (2001); J. W. Barrett, L. Crane, J. Math. Phys. **39**, 3296 (1998); J. Baez, Class. Quant. Grav. **15**, 1827-1858 (1998); J. W. Barrett, L. Crane, Class. Quant. Grav. **17** 3101-3118 (2000); O. Lauscher and M. Reuter, hep-th/0110021
13. *Semi-classical states:* A. Ashtekar, C. Rovelli and L. Smolin, Phy. Rev. Lett. **69**, 237-240 (1992); M. Arnsdorf, S. Gupta, Nucl. Phys. **B577** 529-546 (2000); T. Thiemann, Class. Quant. Grav. **18** 2025-2064 (2001); H. Sahlmann, T. Thiemann and O. Winkler, gr-qc/0102038
14. *Fock states in the polymer picture:*
M. Varadarajan, Phys. Rev. **D61**, 104001 (2000); Phys. Rev. **D 64**, 104003 (2001); A. Ashtekar and J. Lewandowski, Class. Quant. Grav. **18**, L117-L127 (2001)
L. Bombelli, gr-qc/0101080.

GRAVITATIONAL WAVES AND DYNAMICS OF COMPACT BINARY SYSTEMS

LUC BLANCHET

Institut d'Astrophysique de Paris (C.N.R.S.),
98bis boulevard Arago, 75014 Paris, France
E-mail: blanchet@iap.fr

Part A of this article is devoted to the general investigation of the gravitational-wave emission by post-Newtonian sources. We show how the radiation field far from the source, as well as its near-zone inner gravitational field, can (in principle) be calculated in terms of the matter stress-energy tensor up to any order in the post-Newtonian expansion. Part B presents some recent applications to the problems of the dynamics and gravitational-wave flux of compact binary systems. The precision reached in these developments corresponds to the third post-Newtonian approximation.

1 Introduction

The theory of gravitational radiation from isolated sources, in the context of general relativity, takes on importance because of the possibility of comparing the theory with contemporary astrophysical observations. We have in mind the large-scale optical interferometers LIGO, VIRGO, GEO and TAMA, which should routinely observe the gravitational waves produced by massive and rapidly evolving systems such as inspiralling compact binaries. To prepare these experiments the required theoretical work consists of (i) carrying out a sufficiently general solution of the Einstein field equations, valid for a large class of matter systems, (ii) applying the latter solution to specific systems like binaries.

For general isolated sources the basic problem is to relate the asymptotic gravitational-wave form, at the location of a detector in the source's wave zone, to the stress-energy tensor $T^{\alpha\beta}$ of the matter fields. In this article we shall explore this problem using the post-Newtonian expansion, when the speed of light $c \to +\infty$, as the basic approximation. This approximation has proved in the past to constitute an invaluable tool for understanding many aspects of the problems of radiation and motion in general relativity (for reviews see Refs. [1,2,3,4,5,6]).

Concerning the application to compact binaries, the point is that a model made of two structureless point-particles, characterized by two mass parameters m_1 and m_2 (and possibly two spins), should be sufficient to describe the "inspiral" phase, driven by gravitational radiation reaction, which precedes the final binary coalescence. Most of the non-gravitational effects, usually plaguing the dynamics of binary star systems, are dominated by (post-Newtonian) gravitational effects. However the real justification of the model of point particles is that the effects due to the finite size of the compact bodies are small [a]. Inspiralling compact binaries are ideally suited for application of a post-Newtonian wave generation formalism, because it has been shown [7,8] that the post-Newtonian precision required to implement successively the optimal filtering technique in the LIGO/VIRGO detectors corresponds grossly, in the case of neutron-star binaries, to the 3PN approximation, namely $1/c^6$ beyond the quadrupole moment formalism.

In this article we review (Part A) and apply (Part B) a wave-generation formalism based on systematic multipole decompositions of Einstein's theory, and initiated by Blanchet and Damour [9,10], following preceding work by Bonnor [11] and Thorne [12]. The basic multipole moments in this approach are some appropriately defined *source* moments, which are some functions of time, supposed to describe the actual physical source. They are iterated by means of a post-Minkowskian expansion of the vacuum field equations (valid in the source's exterior). Technically, the post-Minkowskian approximation scheme is greatly simplified by the assumption of multipolar expansion, because one can consider separately the iteration of the different multipole pieces composing the exterior field. In our formalism, the radiation field at future null infinity is described by the standard *radiative* multipole moments [12], which are obtained as some explicit non-linear functionals of the more basic source moments [13,14,15]. In the current situation, the *closed-form* expressions of the source multipole moments have been established in the case where the source is post-Newtonian [16,17] (existence of a small parameter which can be regarded as a slow motion estimate $\sim v/c$). For post-Newtonian sources the domain of validity of the post-Newtonian expansion (*viz* the near zone) overlaps the exterior weak-field region, so that there exists an intermediate zone in which the post-Newtonian and multipolar expansions can be matched together [b]. Furthermore, the complete expression of the gravitational field inside the source, valid formally up to any post-Newtonian order, has also been determined by the latter matching [20].

The post-Newtonian wave-generation formalism permitted to compute the total energy flux, $\mathcal{L}$, emitted by inspiralling compact binaries. $\mathcal{L}$ was

[a] For instance the influence of the Newtonian quadrupole moments, Q_1 and Q_2, induced by tidal interaction between two compact objects (for which $a_1 \simeq \frac{Gm_1}{c^2}$ and $a_2 \simeq \frac{Gm_2}{c^2}$), is, by a simple Newtonian argument, comparable (formally) with gravitational effects at the 5PN order.

[b] This is an application of the method of matched asymptotic expansions [18,19].

completed first to the 2PN order by Blanchet, Damour and Iyer [21], and, independently, by Will and Wiseman using their own formalism [22,23]. Higher-order tail effects at the 2.5PN and 3.5PN orders, as well as a crucial contribution of tails generated by the tails themselves (the so-called "tails of tails") arising at the 3PN order, were subsequently obtained [15]. Notice that unlike the 1.5PN, 2.5PN and 3.5PN orders which are entirely composed of tail terms, the 3PN order involves also, besides the tails of tails, many non-tail contributions coming from the relativistic corrections in the binary's multipole moments. These have been recently computed by Blanchet, Iyer and Joguet [24].

The dynamics of compact binaries, i.e. their equations of motion, is also needed in this theoretical work. The 1PN approximation in the equations of motion is standard [25]. The 2PN terms have been derived by Damour and Deruelle [26,27], Kopeikin and Grishchuk [28,29], Blanchet, Faye and Ponsot [30], Itoh *et al* [31], Pati and Will [32]. The 3PN order has been obtained independently by two groups : Jaranowski and Schäfer [33,34], Damour, Jaranowski and Schäfer [35,36] employ the ADM-Hamiltonian formalism of general relativity; Blanchet and Faye [37,38,39,40], de Andrade, Blanchet and Faye [41] compute directly the equations of motion (instead of a Hamiltonian) in harmonic coordinates. To the 3PN order the end results have been shown to be physically equivalent in the sense that there exists a unique "contact" transformation of the binary's dynamical variables, that changes the harmonic-coordinates Lagrangian of Ref. [41] into a new Lagrangian whose Legendre transform coincides exactly with the Hamiltonian given in Ref. [35]. This equivalence is of course quite satisfying.

PART A : POST-NEWTONIAN SOURCES

We describe the general solution of the field equations, valid for isolated post-Newtonian sources, in a series of mathematical statements, presented here without proofs. This presentation has the advantage of emphasizing some key results and outlining the basic "skeleton" of the formalism. We define $h^{\alpha\beta} = \sqrt{-g}\, g^{\alpha\beta} - \eta^{\alpha\beta}$, where $g^{\alpha\beta}$ is the metric, $\eta^{\alpha\beta}$ is the Minkowski metric, and $g = \det(g_{\alpha\beta})$. The field equations in harmonic coordinates, $\partial_\beta h^{\alpha\beta} = 0$, read

$$\Box h^{\alpha\beta} = \frac{16\pi G}{c^4}\tau^{\alpha\beta} , \tag{1}$$

$$\tau^{\alpha\beta} \equiv |g|T^{\alpha\beta} + \frac{c^4}{16\pi G}\Lambda^{\alpha\beta}(h, \partial h, \partial^2 h) , \tag{2}$$

where $\Box$ is the flat d'Alembertian operator, $T^{\alpha\beta}$ is the matter tensor and $\Lambda^{\alpha\beta}$ denotes the non-linear gravitational source term.

2 Linearized vacuum equations : general solution

The first result concerns the solution of the linearized field equations in the vacuum region exterior to the isolated source.

Theorem 1 [12,9] *The most general solution of the linearized vacuum field equations, $\Box h_1^{\alpha\beta} = 0 = \partial_\beta h_1^{\alpha\beta}$, outside some time-like world tube enclosing the source, and stationary in the past (when $t \leq -\mathcal{T}$), reads*

$$h_1^{\alpha\beta} = k_1^{\alpha\beta} + \partial^\alpha \varphi_1^\beta + \partial^\beta \varphi_1^\alpha - \eta^{\alpha\beta}\partial_\gamma \varphi_1^\gamma \,. \tag{3}$$

The vector φ_1^α parametrizes a linearized gauge transformation. The first term is given by [c]

$$k_1^{00} = -\frac{4}{c^2}\sum_{l\geq 0}\frac{(-)^l}{l!}\partial_L\left(\frac{1}{r}I_L(u)\right) \,, \tag{4}$$

$$k_1^{0i} = \frac{4}{c^3}\sum_{l\geq 1}\frac{(-)^l}{l!}\left\{\partial_{L-1}\left(\frac{1}{r}I^{(1)}_{iL-1}(u)\right) + \frac{l}{l+1}\varepsilon_{iab}\partial_{aL-1}\left(\frac{1}{r}J_{bL-1}(u)\right)\right\} \,, \tag{5}$$

$$k_1^{ij} = -\frac{4}{c^4}\sum_{l\geq 2}\frac{(-)^l}{l!}\left\{\partial_{L-2}\left(\frac{1}{r}I^{(2)}_{ijL-2}(u)\right) + \frac{2l}{l+1}\partial_{aL-2}\left(\frac{1}{r}\varepsilon_{ab(i}J^{(1)}_{j)bL-2}(u)\right)\right\} \,, \tag{6}$$

where I_L and J_L are two sets of STF*-tensorial multipole moments, functions of the retarded time $u = t - r/c$, and satisfying the conservation laws : $I =$* const, $I_i =$ const, $J_i =$ const.

3 Exterior vacuum solution : post-Minkowskian iteration

Consider the so-called post-Minkowskian expansion, or formal series in powers of Newton's constant G,

---

[c] Multi-indices are denoted by $L = i_1 i_2 \cdots i_l$; the derivative operator ∂_L is a short-hand for $\partial_{i_1}\cdots\partial_{i_l}$; a tensor I_L is STF if and only if for any pair of indices i_p, $i_q \in L$, we have $I_{\cdots i_p \cdots i_q \cdots} = I_{\cdots i_q \cdots i_p \cdots}$ and $\delta_{i_p i_q} I_{\cdots i_p \cdots i_q \cdots} = 0$. The superscript (p) refers to the time derivation.

$$h_{\rm ext}^{\alpha\beta} = \sum_{n=1}^{+\infty} G^n h_n^{\alpha\beta} \ , \tag{7}$$

that is based on the general linearized exterior solution $h_1^{\alpha\beta}[I_L, J_L]$ of Theorem 1. We obtain a particular solution, at any post-Minkowskian order n, in the form [9]

$$h_n^{\alpha\beta} = \mathcal{FP}\, \Box_{\rm ret}^{-1}[\Lambda_n^{\alpha\beta}] + v_n^{\alpha\beta} \ , \tag{8}$$

where $\Box_{\rm ret}^{-1}$ is the usual retarded integral and $\mathcal{FP}$ a specific operation (defined by complex analytic continuation) of taking the finite part. The finite part is necessary because we are looking for the exterior solution, in the form of a multipole expansion which is singular at the origin $r = 0$, supposed to be located inside the compact-support source. The term $v_n^{\alpha\beta}$ is a particular source-free retarded solution constructed in such a way that the harmonic coordinate condition ($\partial_\beta h_n^{\alpha\beta} = 0$) is satisfied.

Theorem 2 [9] *The most general solution of the Einstein field equations, in the vacuum region outside the source, admitting post-Minkowskian and multipolar expansions (and stationary in the past), is given by the construction (7)-(8). It depends on the source multipole moments $I_L(u)$ and $J_L(u)$, and on four supplementary functions parametrizing the gauge vector in Eq. (3).*

4 General structure of the exterior solution

The solution is valid in all the (weak-field) exterior region of the source. We investigate its expansion when $r \to 0$ and $r \to +\infty$.

Theorem 3 [9] *The structure of the expansion of the post-Minkowskian exterior metric in the near-zone (when $r \to 0$) is of the type* [d] *: $\forall N \in \mathbb{N}$,*

$$h_n^{\alpha\beta}(\mathbf{x}, t) = \sum \hat{n}_L r^m (\ln r)^p F_{L,m,p,n}^{\alpha\beta}(t) + o(r^N) \ , \tag{9}$$

where $m \in \mathbb{Z}$, with $m_0 \le m \le N$ (and m_0 becoming more and more negative when n grows), $p \in \mathbb{N}$ with $p \le n-1$. The functions $F_{L,m,p,n}^{\alpha\beta}$ are multilinear functionals of the source multipole moments.

---

[d] We denote by $\hat{n}_L$ (or sometimes $n_{<L>}$) the STF projection of the product of unit vectors $n_L = n_{i_1} \cdots n_{i_l}$; an expansion into STF tensors $\hat{n}_L = \hat{n}_L(\theta, \phi)$ is equivalent to the usual expansion in spherical harmonics $Y_{lm}(\theta, \phi)$. Similarly we pose $x_L = x_{i_1} \cdots x_{i_l} = r^l n_L$ and $\hat{x}_L = r^l \hat{n}_L$.

This theorem shows in particular that the general form of the post-Newtonian expansion (i.e. when $c \to +\infty$) is necessarily of the type $\sum (\ln c)^p/c^m$.

Theorem 4 [10] *The most general multipolar and post-Minkowskian solution, stationary in the past, (i) admits radiative coordinates* $(T, \mathbf{X})$, *for which the expansion at infinity,* $R \to +\infty$ *with* $U \equiv T - R/c = \text{const}$, *takes the form :*

$$H_n^{\alpha\beta}(\mathbf{X}, T) = \sum \frac{\hat{N}_L}{R^k} K_{L,k,n}^{\alpha\beta}(U) + \mathcal{O}\left(\frac{1}{R^N}\right) , \tag{10}$$

where the functions $K_{L,k,n}^{\alpha\beta}$ *are computable functionals of the source multipole moments, (ii) is asymptotically simple in the sense of Penrose* [42], *perturbatively to any post-Minkowskian order.*

In radiative coordinates the retarded time U is a null coordinate at future null infinity. Theorem 4 permits to connect our work with the study by Bondi *et al* [43] of the asymptotic structure of radiative gravitational fields. Most importantly, it permits to calculate the relations between the source multipole moments I_L and J_L and the radiative ones, say U_L and V_L, which parametrize the asymptotic wave form. These relations involve many non-linear multipole interactions associated with tails and related effects [13,14,15].

5 Post-Newtonian wave generation

Suppose that there exists a solution $h^{\alpha\beta}$, valid everywhere inside and outside a post-Newtonian source, whose multipole expansion coincides with the previously constructed exterior field : $\mathcal{M}(h^{\alpha\beta}) \equiv h_{\text{ext}}^{\alpha\beta}$. Our basic assumption is that of the matching between the near-zone expansion, $r \to 0$, of the exterior multipole expansion : i.e. $\overline{\mathcal{M}(h^{\alpha\beta})}$, and the far-zone expansion, $r \to +\infty$, of the inner post-Newtonian expansion : i.e. $\mathcal{M}(\overline{h^{\alpha\beta}})$. The notation is $\mathcal{M}$ for the operation of taking the multipole expansion, and an overline for the post-Newtonian (re-)expansion. As shown by the matching, the near-zone expansion of the outer field is actually a post-Newtonian expansion, and the far-zone expansion of the inner field is identical to the multipole decomposition. The matching equation reads

$$\overline{\mathcal{M}(h^{\alpha\beta})} = \sum \hat{n}_L r^m (\ln r)^p F_{L,m,p}^{\alpha\beta}(t) = \mathcal{M}(\overline{h^{\alpha\beta}}) , \tag{11}$$

where we have given the common structure of both types of asymptotic expansions, which is given, after formal post-Minkowskian re-summation, by Theorem 3.

Theorem 5 [16,17] *Under the hypothesis of matching, Eq. (11), the multipole expansion of the field exterior to a post-Newtonian source reads*

$$\mathcal{M}(h^{\alpha\beta}) = \mathcal{FP}\,\Box^{-1}_{\rm ret}[\mathcal{M}(\Lambda^{\alpha\beta})] - \frac{4G}{c^4}\sum_{l=0}^{+\infty}\frac{(-)^l}{l!}\partial_L\left\{\frac{1}{r}\mathcal{F}^{\alpha\beta}_L(t-r/c)\right\}\,, \qquad (12)$$

where the "multipole moments" are given by

$$\mathcal{F}^{\alpha\beta}_L(u) = \mathcal{FP}\int d^3\mathbf{y}\,\hat{y}_L\int_{-1}^{1} dz\,\delta_l(z)\,\overline{\tau}^{\alpha\beta}(\mathbf{y},u-z|\mathbf{y}|/c)\,. \qquad (13)$$

Here, $\overline{\tau}^{\alpha\beta}$ denotes the post-Newtonian expansion of the stress-energy pseudo-tensor defined by Eq. (2).

The function $\delta_l(z)$ entering the moments is [e]

$$\delta_l(z) = \frac{(2l+1)!!}{2^{l+1}l!}(1-z^2)^l\,. \qquad (14)$$

A different formalism, alternative to Theorem 5, has been developed by Will and Wiseman [22], and proved [5,6] to be strictly equivalent to ours.

With Theorem 5 in hands, one can obtain the expressions of the source multipole moments I_L and J_L parametrizing the exterior field of Sections 2-3. These are the ones to be inserted into the linearized metric (3)-(6) and iterated by means of the post-Minkowskian algorithm (7)-(8).

$$I_L(u) = \mathcal{FP}\int d^3\mathbf{y}\int_{-1}^{1} dz\left\{\delta_l\hat{y}_L\Sigma - \frac{4(2l+1)}{c^2(l+1)(2l+3)}\delta_{l+1}\hat{y}_{iL}\Sigma^{(1)}_i\right.$$
$$\left. + \frac{2(2l+1)}{c^4(l+1)(l+2)(2l+5)}\delta_{l+2}\hat{y}_{ijL}\Sigma^{(2)}_{ij}\right\}(\mathbf{y},u-z|\mathbf{y}|/c)\,, \qquad (15)$$

$$J_L(u) = \mathcal{FP}\int d^3\mathbf{y}\int_{-1}^{1} dz\,\varepsilon_{ab<i_l}\left\{\delta_l\hat{y}_{L-1>a}\Sigma_b\right.$$
$$\left. - \frac{2l+1}{c^2(l+2)(2l+3)}\delta_{l+1}\hat{y}_{L-1>ac}\Sigma^{(1)}_{bc}\right\}(\mathbf{y},u-z|\mathbf{y}|/c)\,, \qquad (16)$$

where we have posed

[e] This function approaches the Dirac delta-function in the limit of large l : $\lim_{\,l\to+\infty}\delta_l(z) = \delta(z)$, and its integral is normalized to one : $\int_{-1}^{1} dz\,\delta_l(z) = 1$.

$$\Sigma = \frac{\overline{\tau}^{00} + \overline{\tau}^{ii}}{c^2}\,, \quad \Sigma_i = \frac{\overline{\tau}^{0i}}{c}\,, \quad \Sigma_{ij} = \overline{\tau}^{ij}\,. \tag{17}$$

6 Post-Newtonian field in the near zone

The post-Newtonian metric can be systematically computed by means of a specific variant of the operator of the "symmetric" potentials,

$$\mathcal{I}^{-1} = \frac{1}{\Delta - \frac{1}{c^2}\partial_t^2} = \sum_{k=0}^{+\infty} \frac{1}{c^{2k}} \partial_t^{2k} \Delta^{-k-1}\,. \tag{18}$$

Theorem 6 [20] *The expression of the post-Newtonian field inside the source, satisfying correct boundary conditions at infinity (no incoming radiation), follows from the matching equation (11) as*

$$\begin{aligned}\overline{h}^{\alpha\beta} &= \frac{16\pi G}{c^4} \mathcal{FP}\, \mathcal{I}^{-1} \left[\overline{\tau}^{\alpha\beta}\right] \\ &- \frac{4G}{c^4} \sum_{l=0}^{+\infty} \frac{(-)^l}{l!} \partial_L \left\{ \frac{\mathcal{A}_L^{\alpha\beta}(t - r/c) - \mathcal{A}_L^{\alpha\beta}(t + r/c)}{2r} \right\}\,.\end{aligned} \tag{19}$$

The "anti-symmetric" waves are parametrized by $\mathcal{A}_L^{\alpha\beta}(u) = \mathcal{F}_L^{\alpha\beta}(u) + \mathcal{R}_L^{\alpha\beta}(u)$, *where* $\mathcal{F}_L^{\alpha\beta}$ *is given by Eq. (13), and where* [f]

$$\mathcal{R}_L^{\alpha\beta}(u) = \mathcal{FP} \int d^3\mathbf{y}\, \hat{y}_L \int_1^{+\infty} dz\, \gamma_l(z)\, \mathcal{M}(\tau^{\alpha\beta})(\mathbf{y}, u - z|\mathbf{y}|/c)\,. \tag{20}$$

The anti-symmetric waves are associated with radiation-reaction effects. For instance the $\mathcal{F}_L^{\alpha\beta}$-contribution yields the dominant radiation reaction force at the 2.5PN order [18,19], while the $\mathcal{R}_L^{\alpha\beta}$-contribution contains the gravitational-wave tails in the radiation reaction, which arise at the 4PN order [44,45].

PART B : COMPACT BINARIES

We recall that the forthcoming detection and analysis of the gravitational waves emitted by inspiralling compact binaries should necessitate the prior knowledge of the equations of motion and gravitational radiation up to the high 3PN order [7,8].

---

[f] We denote $\gamma_l(z) = -2\delta_l(z)$. The normalization of $\gamma_l(z)$: $\int_1^{+\infty} dz\, \gamma_l(z) = 1$ (where the integral is computed by analytic continuation in $l \in \mathbb{C}$), is consistent with that of $\delta_l(z)$.

7 The problem of self-field regularization

It makes sense in post-Newtonian approximations to model the compact objects by point-like particles, but we then face the problem of the regularization of the infinite self-field of the particles. The Hadamard regularization has been extensively used in this field, but it turned out, as we now review, to yield an unknow coefficient at the 3PN order (and not before that order).

Jaranowski and Schäfer [33,34] noticed that, in their computation of the equations of motion within the ADM-Hamiltonian formulation of general relativity, the "standard" Hadamard regularization leads to some ambiguous results for the computation of certain integrals at the 3PN order. They showed that there are *two and only two* types of ambiguous terms in the 3PN Hamiltonian, which they parametrized by two unknown numerical coefficients ω_{static} and ω_{kinetic}.

Motivated by this result, Blanchet and Faye [39,40] introduced an "improved" Hadamard regularization, based on a theory of pseudo-functions and generalized distributional derivatives, which is free of ambiguities, and leads to unique results for the computation of all the integrals occuring at the 3PN order. Unfortunately, this regularization turned out to be in a sense incomplete, because it was found [37,38] that the 3PN equations of motion involve *one and only one* unknown numerical constant, called λ, which cannot be determined within the method. The comparison of this result with the work of Jaranowski and Schäfer [33,34], on the basis of the computation of the invariant energy of binaries moving on circular orbits, revealed [37] that

$$\omega_{\text{kinetic}} = \frac{41}{24}\,, \tag{21}$$

$$\omega_{\text{static}} = -\frac{11}{3}\lambda - \frac{1987}{840}\,. \tag{22}$$

The ambiguity ω_{kinetic} is therefore fixed, while λ is equivalent to the other ambiguity ω_{static}.

Damour, Jaranowski and Schäfer [35] recovered the value of ω_{kinetic} given in Eq. (21) by proving that this value is the unique one for which the global Poincaré invariance of their formalism is verified. Since the coordinate conditions associated with the ADM approach do not manifestly respect the Poincaré symmetry, they had to prove that the Hamiltonian is compatible with the existence of generators for the Poincaré algebra. By contrast, the harmonic-coordinates condition preserves the Poincaré invariance, and therefore the associated equations of motion should be Lorentz-invariant, as was indeed found to be the case by Blanchet and Faye [37,38], thanks to their use of a Lorentz-invariant regularization [40] (hence their determination of ω_{kinetic}).

More recently, the other parameter ω_{static} was computed by Damour, Jaranowski and Schäfer [36] by means of a dimensional regularization, instead of an Hadamard-type one, within the ADM-Hamiltonian formalism. The result, which in principle fixes λ according to Eq. (22), is

$$\omega_{\text{static}} = 0 \iff \lambda = -\frac{1987}{3080} . \tag{23}$$

As Damour *et al* [36] argue, clearing up the ambiguity is made possible by the fact that the dimensional regularization, contrary to the Hadamard one, respects all the basic properties of the algebraic and differential calculus of ordinary functions [g].

Blanchet, Iyer and Joguet [24], in their computation of the 3PN radiation field of two point particles, used the (standard) Hadamard regularization and found necessary to introduce three additional regularization constants ξ, κ and ζ, playing a role analogous to the equation-of-motion constant λ. Such unknowns come from the computation of the 3PN binary's quadrupole moment I_{ij}. It was found [24] that the total gravitational-wave flux, in the case of circular orbits, depends in fact only on a single combination of these three constants,

$$\theta = \xi + 2\kappa + \zeta . \tag{24}$$

8 The equations of binary motion

The equations are expressed in Newtonian-like form, by means of the coordinate positions y_1^i, y_2^i, and velocities $v_1^i = dy_1^i/dt$, $v_2^i = dy_2^i/dt$, of the bodies in the harmonic coordinate system. The 3PN acceleration of one of the bodies, say 1, reads

$$\frac{dv_1^i}{dt} = A_{\text{N}}^i + \frac{1}{c^2}A_{1\text{PN}}^i + \frac{1}{c^4}A_{2\text{PN}}^i + \frac{1}{c^5}A_{2.5\text{PN}}^i + \frac{1}{c^6}A_{3\text{PN}}^i + \mathcal{O}\left(\frac{1}{c^7}\right) , \tag{25}$$

The first term is the famous Newtonian law,

$$A_{\text{N}}^i = -\frac{Gm_2}{r_{12}^2} n_{12}^i , \tag{26}$$

[g] For instance, the Hadamard regularization unavoidably violates at some stage the Leibniz rule for the differentiation of a product.

(where $r_{12} = |\mathbf{y}_1 - \mathbf{y}_2|$ and $\mathbf{n}_{12} = (\mathbf{y}_1 - \mathbf{y}_2)/r_{12}$). All the other terms represent the successive post-Newtonian corrections. The 3PN equations of motion, that we show below, possess the correct perturbative limit given by the geodesics of the Schwarzschild metric, stay invariant when we perform a global Lorentz transformation, and are derivable from a Lagrangian or Hamiltonian formulation (when the radiation-reaction term is neglected). The 1PN term is the Einstein-Infeld-Hoffmann [25] acceleration :

$$A^i_{1\mathrm{PN}} = \left[5\frac{G^2 m_1 m_2}{r_{12}^3} + 4\frac{G^2 m_2^2}{r_{12}^3} + \frac{G m_2}{r_{12}^2}\left(\frac{3}{2}(n_{12}v_2)^2 - v_1^2 + 4(v_1 v_2) - 2v_2^2\right)\right] n_{12}^i + \frac{G m_2}{r_{12}^2}\Big[4(n_{12}v_1) - 3(n_{12}v_2)\Big] v_{12}^i \,. \tag{27}$$

The parenthesis indicate the usual Euclidean scalar product, e.g. $(n_{12}v_1) = \mathbf{n}_{12}.\mathbf{v}_1$, and $v_{12}^i \equiv v_1^i - v_2^i$. At the 2PN order, the result is [26,27,28,29,30,31,32]

$$\begin{aligned}
A^i_{2\mathrm{PN}} = \Bigg[&-\frac{57}{4}\frac{G^3 m_1^2 m_2}{r_{12}^4} - \frac{69}{2}\frac{G^3 m_1 m_2^2}{r_{12}^4} - 9\frac{G^3 m_2^3}{r_{12}^4} \\
&+ \frac{G m_2}{r_{12}^2}\bigg(-\frac{15}{8}(n_{12}v_2)^4 + \frac{3}{2}(n_{12}v_2)^2 v_1^2 \\
&- 6(n_{12}v_2)^2(v_1 v_2) - 2(v_1 v_2)^2 + \frac{9}{2}(n_{12}v_2)^2 v_2^2 + 4(v_1 v_2)v_2^2 - 2v_2^4 \bigg) \\
&+ \frac{G^2 m_1 m_2}{r_{12}^3}\bigg(\frac{39}{2}(n_{12}v_1)^2 - 39(n_{12}v_1)(n_{12}v_2) \\
&+ \frac{17}{2}(n_{12}v_2)^2 - \frac{15}{4}v_1^2 - \frac{5}{2}(v_1 v_2) + \frac{5}{4}v_2^2 \bigg) \\
&+ \frac{G^2 m_2^2}{r_{12}^3}\bigg(2(n_{12}v_1)^2 - 4(n_{12}v_1)(n_{12}v_2) - 6(n_{12}v_2)^2 - 8(v_1 v_2) + 4v_2^2 \bigg) \Bigg] n_{12}^i \\
&+ \Bigg[\frac{G^2 m_2^2}{r_{12}^3}\bigg(-2(n_{12}v_1) - 2(n_{12}v_2) \bigg) \\
&+ \frac{G^2 m_1 m_2}{r_{12}^3}\bigg(-\frac{63}{4}(n_{12}v_1) + \frac{55}{4}(n_{12}v_2) \bigg) \\
&+ \frac{G m_2}{r_{12}^2}\bigg(-6(n_{12}v_1)(n_{12}v_2)^2 + \frac{9}{2}(n_{12}v_2)^3 + (n_{12}v_2)v_1^2 - 4(n_{12}v_1)(v_1 v_2) \\
&+ 4(n_{12}v_2)(v_1 v_2) + 4(n_{12}v_1)v_2^2 - 5(n_{12}v_2)v_2^2 \bigg) \Bigg] v_{12}^i \,.
\end{aligned} \tag{28}$$

The 2.5PN term – an "odd" term $\sim 1/c^5$ – corresponds to the damping effect of radiation reaction :

$$A^i_{2.5\text{PN}} = \frac{4}{5}\frac{G^2 m_1 m_2}{r_{12}^3}\left[-6\frac{Gm_1}{r_{12}} + \frac{52}{3}\frac{Gm_2}{r_{12}} + 3v_{12}^2\right](n_{12}v_{12})n_{12}^i + \frac{4}{5}\frac{G^2 m_1 m_2}{r_{12}^3}\left[2\frac{Gm_1}{r_{12}} - 8\frac{Gm_2}{r_{12}} - v_{12}^2\right]v_{12}^i \,. \tag{29}$$

Finally, the very lengthy 3PN coefficient is given by [37,38]

$$\begin{aligned}
A^i_{3\text{PN}} = \Bigg[&\frac{Gm_2}{r_{12}^2}\bigg(\frac{35}{16}(n_{12}v_2)^6 - \frac{15}{8}(n_{12}v_2)^4 v_1^2 + \frac{15}{2}(n_{12}v_2)^4(v_1v_2) \\
&+ 3(n_{12}v_2)^2(v_1v_2)^2 - \frac{15}{2}(n_{12}v_2)^4 v_2^2 + \frac{3}{2}(n_{12}v_2)^2 v_1^2 v_2^2 \\
&- 12(n_{12}v_2)^2(v_1v_2)v_2^2 - 2(v_1v_2)^2 v_2^2 \\
&+ \frac{15}{2}(n_{12}v_2)^2 v_2^4 + 4(v_1v_2)v_2^4 - 2v_2^6\bigg) + \frac{G^2 m_1 m_2}{r_{12}^3}\bigg(-\frac{171}{8}(n_{12}v_1)^4 \\
&+ \frac{171}{2}(n_{12}v_1)^3(n_{12}v_2) - \frac{723}{4}(n_{12}v_1)^2(n_{12}v_2)^2 + \frac{383}{2}(n_{12}v_1)(n_{12}v_2)^3 \\
&- \frac{455}{8}(n_{12}v_2)^4 + \frac{229}{4}(n_{12}v_1)^2 v_1^2 - \frac{205}{2}(n_{12}v_1)(n_{12}v_2)v_1^2 \\
&+ \frac{191}{4}(n_{12}v_2)^2 v_1^2 - \frac{91}{8}v_1^4 - \frac{229}{2}(n_{12}v_1)^2(v_1v_2) + 244(n_{12}v_1)(n_{12}v_2)(v_1v_2) \\
&- \frac{225}{2}(n_{12}v_2)^2(v_1v_2) + \frac{91}{2}v_1^2(v_1v_2) - \frac{177}{4}(v_1v_2)^2 + \frac{229}{4}(n_{12}v_1)^2 v_2^2 \\
&- \frac{283}{2}(n_{12}v_1)(n_{12}v_2)v_2^2 + \frac{259}{4}(n_{12}v_2)^2 v_2^2 - \frac{91}{4}v_1^2 v_2^2 + 43(v_1v_2)v_2^2 - \frac{81}{8}v_2^4\bigg) \\
&+ \frac{G^2 m_2^2}{r_{12}^3}\bigg(-6(n_{12}v_1)^2(n_{12}v_2)^2 + 12(n_{12}v_1)(n_{12}v_2)^3 + 6(n_{12}v_2)^4 \\
&+ 4(n_{12}v_1)(n_{12}v_2)(v_1v_2) + 12(n_{12}v_2)^2(v_1v_2) + 4(v_1v_2)^2 \\
&- 4(n_{12}v_1)(n_{12}v_2)v_2^2 - 12(n_{12}v_2)^2 v_2^2 - 8(v_1v_2)v_2^2 + 4v_2^4\bigg) \\
&+ \frac{G^3 m_2^3}{r_{12}^4}\bigg(-(n_{12}v_1)^2 + 2(n_{12}v_1)(n_{12}v_2) + \frac{43}{2}(n_{12}v_2)^2 + 18(v_1v_2) - 9v_2^2\bigg) \\
&+ \frac{G^3 m_1 m_2^2}{r_{12}^4}\bigg(\frac{415}{8}(n_{12}v_1)^2 - \frac{375}{4}(n_{12}v_1)(n_{12}v_2) + \frac{1113}{8}(n_{12}v_2)^2
\end{aligned}$$

$$- \frac{615}{64}(n_{12}v_{12})^2\pi^2 + \frac{123}{64}v_{12}^2\pi^2 + 18v_1^2 + 33(v_1v_2) - \frac{33}{2}v_2^2\Big)$$

$$+ \frac{G^3m_1^2m_2}{r_{12}^4}\Big(- \frac{45887}{168}(n_{12}v_1)^2 + \frac{24025}{42}(n_{12}v_1)(n_{12}v_2)$$

$$- \frac{10469}{42}(n_{12}v_2)^2 + \frac{48197}{840}v_1^2 - \frac{36227}{420}(v_1v_2) + \frac{36227}{840}v_2^2$$

$$+ 110(n_{12}v_{12})^2 \ln\left(\frac{r_{12}}{r'_1}\right) - 22v_{12}^2 \ln\left(\frac{r_{12}}{r'_1}\right)\Big)$$

$$+ \frac{G^4m_1^3m_2}{r_{12}^5}\left(- \frac{3187}{1260} + \frac{44}{3}\ln\left(\frac{r_{12}}{r'_1}\right)\right)$$

$$+ \frac{G^4m_1^2m_2^2}{r_{12}^5}\left(\frac{34763}{210} - \frac{44}{3}\lambda - \frac{41}{16}\pi^2\right) + 16\frac{G^4m_2^4}{r_{12}^5}$$

$$+ \frac{G^4m_1m_2^3}{r_{12}^5}\left(\frac{10478}{63} - \frac{44}{3}\lambda - \frac{41}{16}\pi^2 - \frac{44}{3}\ln\left(\frac{r_{12}}{r'_2}\right)\right)\Big]n_{12}^i$$

$$+ \Big[\frac{Gm_2}{r_{12}^2}\Big(\frac{15}{2}(n_{12}v_1)(n_{12}v_2)^4 - \frac{45}{8}(n_{12}v_2)^5 - \frac{3}{2}(n_{12}v_2)^3v_1^2$$

$$+ 6(n_{12}v_1)(n_{12}v_2)^2(v_1v_2) - 6(n_{12}v_2)^3(v_1v_2) - 2(n_{12}v_2)(v_1v_2)^2$$

$$- 12(n_{12}v_1)(n_{12}v_2)^2v_2^2 + 12(n_{12}v_2)^3v_2^2 + (n_{12}v_2)v_1^2v_2^2$$

$$- 4(n_{12}v_1)(v_1v_2)v_2^2 + 8(n_{12}v_2)(v_1v_2)v_2^2$$

$$+ 4(n_{12}v_1)v_2^4 - 7(n_{12}v_2)v_2^4\Big)$$

$$+ \frac{G^2m_2^2}{r_{12}^3}\Big(- 2(n_{12}v_1)^2(n_{12}v_2) + 8(n_{12}v_1)(n_{12}v_2)^2 + 2(n_{12}v_2)^3$$

$$+ 2(n_{12}v_1)(v_1v_2) + 4(n_{12}v_2)(v_1v_2) - 2(n_{12}v_1)v_2^2 - 4(n_{12}v_2)v_2^2\Big)$$

$$+ \frac{G^2m_1m_2}{r_{12}^3}\Big(- \frac{243}{4}(n_{12}v_1)^3 + \frac{565}{4}(n_{12}v_1)^2(n_{12}v_2)$$

$$- \frac{269}{4}(n_{12}v_1)(n_{12}v_2)^2 - \frac{95}{12}(n_{12}v_2)^3 + \frac{207}{8}(n_{12}v_1)v_1^2$$

$$- \frac{137}{8}(n_{12}v_2)v_1^2 - 36(n_{12}v_1)(v_1v_2)$$

$$+ \frac{27}{4}(n_{12}v_2)(v_1v_2) + \frac{81}{8}(n_{12}v_1)v_2^2 + \frac{83}{8}(n_{12}v_2)v_2^2\Big)$$

$$+ \frac{G^3m_2^3}{r_{12}^4}\Big(4(n_{12}v_1) + 5(n_{12}v_2)\Big)$$

$$+ \frac{G^3 m_1 m_2^2}{r_{12}^4}\left(- \frac{307}{8}(n_{12}v_1) + \frac{479}{8}(n_{12}v_2) + \frac{123}{32}(n_{12}v_{12})\pi^2\right)$$
$$+ \frac{G^3 m_1^2 m_2}{r_{12}^4}\left(\frac{31397}{420}(n_{12}v_1) - \frac{36227}{420}(n_{12}v_2)\right.$$
$$\left.\left. - 44(n_{12}v_{12}) \ln \left(\frac{r_{12}}{r'_1}\right)\right)\right] v_{12}^i \,. \qquad (30)$$

The latter harmonic-coordinates equations depend at the 3PN order on two arbitrary constants r'_1 and r'_2 parametrizing some logarithmic terms, however these constants are not physical in the sense that they can be removed by a coordinate transformation. By contrast, the equations of motion in ADM-coordinates [33,34,35] do not involve logarithms and are free of such type of constants.

9 Energy of circular-orbit binaries

Most inspiralling compact binaries will have been circularized at the time when they become visible by the LIGO and VIRGO detectors. In the case of circular orbits the equations (25)-(30) simplify drastically. Also we can translate the origin of coordinates to the binary's center of mass by imposing that the mass-dipole be $I_i = 0$ (notation of Part A). Mass parameters are the total mass $m = m_1 + m_2$, the reduced mass $\mu = m_1 m_2 / m$, and the useful symmetric mass ratio

$$\nu \equiv \frac{\mu}{m} \equiv \frac{m_1 m_2}{(m_1 + m_2)^2} \,, \qquad (31)$$

such that $0 < \nu \leq \frac{1}{4}$, with $\nu = \frac{1}{4}$ in the case of equal masses, and $\nu \to 0$ in the "test-mass" limit for one of the bodies. The energy E of the binary in the center-of-mass frame is deduced from the 3PN harmonic-coordinates Lagrangian [41] and expressed in invariant form (the same in different coordinate systems). This is achieved with the help of the angular orbital frequency ω of the circular orbit, or, more conveniently, the parameter defined by

$$x \equiv \left(\frac{G m \omega}{c^3}\right)^{2/3} \,. \qquad (32)$$

After replacement we discover that the two gauge constants r'_1 and r'_2 cancel out. The end result is

$$E = -\frac{\mu c^2 x}{2}\left\{1 + \left(-\frac{3}{4} - \frac{1}{12}\nu\right)x + \left(-\frac{27}{8} + \frac{19}{8}\nu - \frac{1}{24}\nu^2\right)x^2 + \left(-\frac{675}{64} + \left[\frac{209323}{4032} - \frac{205}{96}\pi^2 - \frac{110}{9}\lambda\right]\nu - \frac{155}{96}\nu^2 - \frac{35}{5184}\nu^3\right)x^3\right\}. \quad (33)$$

As an application of Eq. (33) let us determine the location of the minimum of the energy E as a function of the orbital frequency for circular orbits [46], say $E_{\rm min} = E(\omega_{\rm min})$.

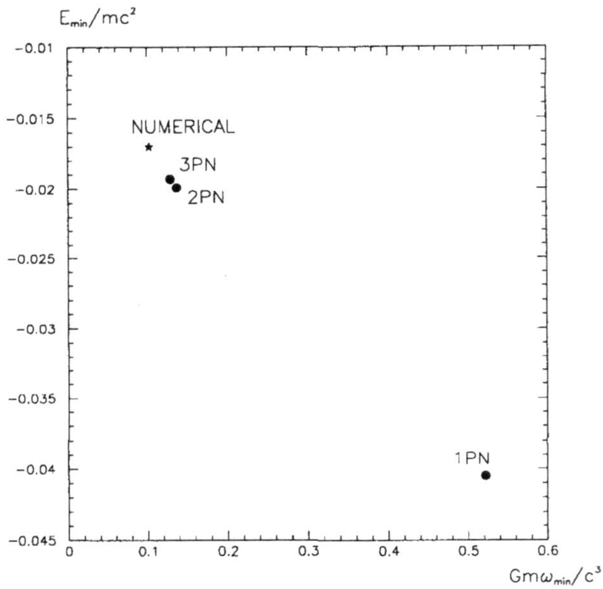

Figure 1. Results for $E_{\rm min}$ in terms of $\omega_{\rm min}$ for equal-mass binaries ($\nu = \frac{1}{4}$). The asterisk marks the result calculated by numerical relativity. The point 3PN corresponds to the value for the ambiguity parameter given by Eq. (23).

In Fig. 1 we display the results given by the successive 1PN, 2PN and 3PN orders [the 1PN and 2PN orders are defined by the obvious truncation of Eq. (33)], and compare them with the prediction recently calculated numerically by Gourgoulhon, Grandclément and Bonazzola [47,48] under the assumption of exactly circular orbits (helicoidal symmetry). As we see the post-Newtonian and numerical predictions agree reasonably well. The post-Newtonian approximation seems to converge well toward the "exact" solution, even at this very relativistic stage where the orbital velocity reaches $v/c \sim (Gm\omega/c^3)^{1/3} \sim 0.5$. However, the 1PN order is clearly not precise enough at this stage.

10 Flux of circular-orbit binaries

The calculation of the flux necessitates, following the formalism described in Part A, the expressions of the source multipole moments I_L and J_L. The crucial moment to be calculated is the mass-quadrupole, which is to be determined with the maximal post-Newtonian precision. It has been obtained by Blanchet, Iyer and Joguet [24], at the 3PN order in the case of circular orbits [h]. We compute the time-derivatives of the moments with the help of the 3PN equations of motion (25)-(30), replace them into the total energy flux developed at the 3PN order, and add all the non-linear tail and related effects found up to the 3.5PN order in Ref. [15]. We must also express the result in terms of the frequency-dependent parameter (32). Finally we get

$$\begin{aligned}
\mathcal{L} = \frac{32c^5}{5G}\nu^2 x^5 \Bigg\{ & 1 + \left(-\frac{1247}{336} - \frac{35}{12}\nu\right) x + 4\pi x^{3/2} \\
& + \left(-\frac{44711}{9072} + \frac{9271}{504}\nu + \frac{65}{18}\nu^2\right) x^2 + \left(-\frac{8191}{672} - \frac{535}{24}\nu\right)\pi x^{5/2} \\
& + \left(\frac{6643739519}{69854400} + \frac{16}{3}\pi^2 - \frac{1712}{105}C - \frac{856}{105}\ln(16\,x)\right. \\
& \left. + \left[-\frac{11497453}{272160} + \frac{41}{48}\pi^2 + \frac{176}{9}\lambda - \frac{88}{3}\theta\right]\nu - \frac{94403}{3024}\nu^2 - \frac{775}{324}\nu^3\right) x^3 \\
& + \left(-\frac{16285}{504} + \frac{176419}{1512}\nu + \frac{19897}{378}\nu^2\right)\pi x^{7/2} \Bigg\} . \qquad (34)
\end{aligned}$$

We have $\theta = \xi + 2\kappa + \zeta$ (and $C \simeq 0.577$ is the Euler constant). The flux (34) agrees perfectly, in the test-mass limit $\nu \to 0$, with the result of linear black-holes perturbations calculated by Tagoshi and Sasaki [49].

As an application of Eqs. (34) and (33) one can obtain the orbital phase $\phi = \int \omega dt$ of an inspiralling compact binary at the 3.5PN order simply by

---

[h] The 3PN quadrupole is found to involve three ambiguity constants ξ, κ and ζ [see Eq. (24)], as well as some r_0 coming from the definition of the finite part process $\mathcal{FP}$ employed in Part A, and some r'_0 which is in fact the "logarithmic barycenter" of the constants r'_1 and r'_2 present in the equations of motion (30). Both r_0 and r'_0 disappear from the result at the end, in agreement with general expectations.

using the energy balance equation

$$\frac{dE}{dt} = -\mathcal{L} \implies \phi = -\int \frac{\omega dE}{\mathcal{L}} . \tag{35}$$

The orbital phase is the crucial observable to be implemented in the construction of accurate templates for the detection and analysis of inspiralling-binary signals in the LIGO and VIRGO experiments.

References

1. T. Damour, in *300 years of Gravitation*, S. W. Hawking and W. Israel (eds.), Cambridge U. Press (1987).
2. K. S. Thorne, in *300 years of Gravitation*, S. W. Hawking and W. Israel (eds.), Cambridge U. Press (1987).
3. C.M. Will, in Proc. of the 8th Nishinomiya-Yukawa Symposium on Relativistic Cosmology, M. Sasaki (ed.), Universal Acad. Press, Japan (1994).
4. L. Blanchet, in *Relativistic gravitation and gravitational radiation*, J.A. Marck and J.P. Lasota (eds.), Cambridge U. Press (1996).
5. L. Blanchet, in *Einstein's field equations and their physical implications*, B.G. Schmidt (ed.), Springer (2000).
6. L. Blanchet, submitted to *Living Reviews in Relativity*, Albert Einstein Institute, Potsdam (2002).
7. C. Cutler, T.A. Apostolatos, L. Bildsten, L.S. Finn, E.E. Flanagan, D. Kennefick, D.M. Markovic, A. Ori, E. Poisson, G.J. Sussman and K.S. Thorne, Phys. Rev. Lett. **70**, 2984 (1993).
8. C. Cutler, L.S. Finn, E. Poisson and G.J. Sussman, Phys. Rev. D**47**, 1511 (1993).
9. L. Blanchet and T. Damour, Philos. Trans. R. Soc. London A **320**, 379 (1986).
10. L. Blanchet, Proc. R. Soc. Lond. A **409**, 383 (1987).
11. W.B. Bonnor, Philos. Trans. R. Soc. London A 251, 233 (1959).
12. K.S. Thorne, Rev. Mod. Phys. **52**, 299 (1980).
13. L. Blanchet and T. Damour, Phys. Rev. D **46**, 4304 (1992).
14. L. Blanchet, Class. Quantum. Grav. **15**, 89 (1998).
15. L. Blanchet, Class. Quantum Grav. **15**, 113 (1998).
16. L. Blanchet, Phys. Rev. D **51**, 2559 (1995).
17. L. Blanchet, Class. Quantum. Grav. **15**, 1971 (1998).
18. W.L. Burke and K.S. Thorne, in Relativity, M. Carmeli et al. (eds) (Plenum Press, New York, 1970) p. 208-209.

19. W.L. Burke (1971) J. Math. Phys. 12, 401.
20. O. Poujade and L. Blanchet, submitted to Phys. Rev. D, gr-qc 0112057.
21. L. Blanchet, T. Damour and B.R. Iyer, Phys. Rev. D **51**, 5360 (1995).
22. C.M. Will and A.G. Wiseman, Phys. Rev. D **54**, 4813 (1996).
23. M.E. Pati and C.M. Will, Phys. Rev. D **62**, 124015 (2000).
24. L. Blanchet, B.R. Iyer and B. Joguet, to appear in Phys. Rev. D, gr-qc 0105098.
25. A. Einstein, L. Infeld and B. Hoffmann, Ann. Math. **39**, 65 (1938).
26. T. Damour and N. Deruelle, Phys. Lett. **87A**, 81 (1981).
27. T. Damour, in *Gravitational Radiation*, N. Deruelle and T. Piran (eds.), North-Holland Company (1983).
28. S.M. Kopejkin, Astron. Zh. **62**, 889 (1985).
29. L.P. Grishchuk and S.M. Kopejkin, in *Relativity in Celestial Mechanics and Astrometry*, J. Kovalevsky and V.A. Brumberg (eds.), Reidel, Dordrecht (1986).
30. L. Blanchet, G. Faye and B. Ponsot, Phys. Rev. D**58**, 124002 (1998).
31. Y. Itoh, T. Futamase and H. Asada, Phys. Rev. D**63**, 064038 (2001).
32. M.E. Pati and C.M. Will, gr-qc 0201001.
33. P. Jaranowski and G. Schäfer, Phys. Rev. D**57**, 7274 (1998).
34. P. Jaranowski and G. Schäfer, Phys. Rev. D**60**, 124003 (1999).
35. T. Damour, P. Jaranowski and G. Schäfer, Phys. Rev. D**62**, 021501R (2000); Erratum Phys. Rev. D**63**, 029903 (2001).
36. T. Damour, P. Jaranowski and G. Schäfer, Phys. Lett. B **513**, 147 (2001).
37. L. Blanchet and G. Faye, Phys. Lett. A **271**, 58 (2000).
38. L. Blanchet and G. Faye, Phys. Rev. D**63**, 062005 (2001).
39. L. Blanchet and G. Faye, J. Math. Phys. **41**, 7675 (2000).
40. L. Blanchet and G. Faye, J. Math. Phys. **42**, 4391 (2001).
41. V.C. de Andrade, L. Blanchet and G. Faye, Class. Quantum Grav. **18**, 753 (2001).
42. R. Penrose, Proc. R. Soc. Lond. A **284**, 159 (1965).
43. H. Bondi, M.G.J. van der Burg and A.W.K. Metzner, Proc. R. Soc. Lond. A**269**, 21 (1962).
44. L. Blanchet and T. Damour, Phys. Rev. D**37**, 1411 (1988).
45. L. Blanchet, Phys. Rev. D**47**, 4392 (1993).
46. L. Blanchet, submitted to Phys. Rev. D, gr-qc 0112056.
47. E. Gourgoulhon, P. Grandclément and S. Bonazzola, gr-qc 0106015.
48. P. Grandclément, E. Gourgoulhon and S. Bonazzola, gr-qc 0106016.
49. H. Tagoshi and M. Sasaki, Prog. Theor. Phys. **92**, 745 (1994).

AN OVERVIEW OF GRAVITATIONAL-WAVE SOURCES

CURT CUTLER

Max Planck Institute for Gravitational Physics, 14476 Golm, Germany

KIP S. THORNE

Theoretical Astrophysics, California Institute of Technology, Pasadena, CA 91125

in consultation with

LARS BILDSTEN, ALESSANDRA BUONANNO, CRAIG HOGAN, VASSILIKI KALOGERA, BENJAMIN J. OWEN, E. STERL PHINNEY, THOMAS A. PRINCE, FREDERIC A. RASIO, STUART L. SHAPIRO, KENNETH A. STRAIN, GREG USHOMIRSKY AND ROBERT V. WAGONER.

We review current best estimates of the strength and detectability of the gravitational waves from a variety of sources, for both ground-based and space-based detectors, and we describe the information carried by the waves.

1 Preamble

Major relativity conferences such as GR16 traditionally include an overview talk on gravitational wave (GW) sources. Some excellent recent ones include those by Flanagan [1], Finn [2], and Bender et al. [3]. Such talks are always by theorists, and can be described as basically informed speculation, since until now detectors sufficiently sensitive to detect GW's have not existed. But km-scale laser interferometers (IFO's) are now coming on-line, and it seems very likely these will detect mergers of compact binaries within the next seven years, and possibly much sooner. There are several other classes of sources that the ground-based IFO's might well detect: massive star collapse (supernovae and hypernovae), rapidly rotating neutron stars, and possibly a GW stochastic background created in the early universe. A space-based interferometer, LISA, is also planned (though not yet fully funded), and could fly in $\sim$ 2011. There is one type of source — short-period galactic binaries — that LISA is guaranteed to observe (at its planned sensitivity), plus a list of very promising candidates: the inspiral and merger of supermassive black holes (SMBH's), the inspiral and capture of compact objects by SMBH's, and sources in the very early universe.

In this article we review the various GW sources that have been studied, for both ground and space-based detectors, summarizing the best available estimates for rates and source strengths and describing some of the information that may be extracted from the waves. We will try to emphasize what is new, but much here is not, and indeed many sections of this paper have been adapted, with minor updating, from an earlier review by Thorne [4] or from a publicly available LIGO-Project document by Thorne [5] (based on extensive input from the colleagues listed as consultants at the beginning of this article).

Gravitational-wave detection efforts focus on four frequency bands: (i) The extremely low frequency (ELF) band, 10^{-15} to 10^{-18} Hz, in which waves will be sought via their imprint on the polarization of the cosmic microwave background (CMB) radiation [6], and the only expected source is primordial gravitational fluctuations amplified by inflation of the universe. (ii) The very low frequency (VLF) band, 10^{-7} to 10^{-9} Hz, in which the waves are sought using high-stability pulsar timing [7], and the expected sources are processes in the very early universe and extremely massive binary black hole systems. (iii) The low frequency (LF) band, 10^{-4} to 1 Hz, in which LISA will operate, and (iv) the high frequency (HF) band, 1 to 10^4 Hz, in which earth-based detectors (interferometers and resonant-mass detectors) operate.

In this article we shall confine attention to the LF and HF bands, since that is where the richest variety of sources reside, that is where the vast majority of the current experimental activity concentrates, and that is where the first firm detections are likely to be made. We shall not say much about details or current status or plans for gravitational wave detection, as there are many other review articles devoted to this, e.g. those by Kawamura [8], Schutz [9], Robertson [10], and Barish and Weiss [11].

In Sec. 2 we shall discuss high-frequency sources being sought by earth-based detectors, and in Sec. 3, low-frequency sources that will be sought by LISA (and some of which are already being sought by Doppler tracking of spacecraft [12]).

2 High-Frequency Sources Sought by Earth-Based Detectors

2.1 Fiducial Noise Curves

In discussing gravitational-wave sources, it is important to put them in the context of detector sensitivities, both near-term and planned. As is discussed in Kawamura's chapter in this volume [8], the first-generation kilometer-scale interferometric detectors (LIGO, VIRGO, GEO600, TAMA300) were being assembled at the time of GR16 and will begin their initial GW searches about the time that our review is published. In about 2007 these first-generation interferometers (IFOs) will be upgraded to produce second-generation IFOs, with noise curves that can be adjusted by adjusting the position and reflectivity of one mirror. In parallel with this IFO experimental effort, a network of narrow-band resonant-mass detectors is operating and being improved [13].

In this review we shall find it convenient to evaluate prospects for detection with respect to three representative, fiducial noise curves: (i) that for the first-generation LIGO IFOs (the *LIGO-I* noise curve); (ii) that for the second-generation LIGO IFOs (sometimes called **advanced** LIGO IFOs and sometimes called **LIGO-II** IFOs), when they are operated in a "wide-band" mode (the *WB LIGO-II* noise curve); and (iii) that for second-generation IFOs operated in a "narrow-band" mode that is optimized for sources near 600 Hz frequency (the *NB LIGO-II* noise curve). Though focusing on LIGO smacks of American egocentrism, the reason is pragmatic: the first-generation VIRGO noise curve would largely lie between ithe initial LIGO (LIGO-I) curve and the advanced LIGO (LIGO-II) curves anyway, VIRGO does not publish an official second-generation noise curve, and GEO600 and TAMA300 are purely first-generation instruments.

Our three fiducial noise curves are shown in Fig. 1 along with the wave strengths for a number of HF sources to be discussed below. The horizontal axis in the figure is GW frequency f; the vertical axis is the square root of the spectral density of an IFO's arm-length difference, i.e. the IFO's "strain per root Hertz" $\tilde{h}(f) \equiv \sqrt{S_{\Delta L/L}(f)}$. The (dimensionless) rms noise in a bandwidth Δf around frequency f is given by $h_{\rm rms} = \tilde{h}(f)\sqrt{\Delta f}$.

Notice that, compared to LIGO I, the WB LIGO-II IFOs (i) will lower the amplitude noise by a factor ~ 15 at the frequencies of best sensitivity $f \sim 100$–200 Hz, and (ii) will widen the band of high sensitivity at both low frequencies (pushing it down to ~ 20 Hz) and high frequencies (pushing it up to ~ 1000 Hz). The lowered WB noise at optimal frequencies will increase the event rate for distant, extragalactic sources by a factor $\sim 15^3 \simeq 3000$. Opening up lower and higher frequency bands will bring us into the domains of new sources: colliding, massive black holes and stochastic background at low frequencies; low-mass X-ray binaries, fast pulsars and tidal disruption of neutron stars by black holes at high frequencies. Noise-curve reshaping, as illustrated by the NB LIGO-II IFO, will reduce the noise by a factor $\sim$3 to 5 relative to the WB IFO within some chosen narrow frequency band $\Delta f/f \sim 0.2$ in which targeted periodic sources (e.g. low-mass X-ray binaries) are expected to lie. We shall refer to this as "narrow-band tuning" of the IFO.

The expected LIGO II noise curves depend somewhat on whether the mirrors' substrate is made from sapphire (the preferred material) or from fused silica (the backup). We assume sapphire in this article. If problems with sapphire force fused silica to be used, the WB IFOs will likely be a few tens of percent more noisy than Fig. 1 near the noise-curve minimum, and

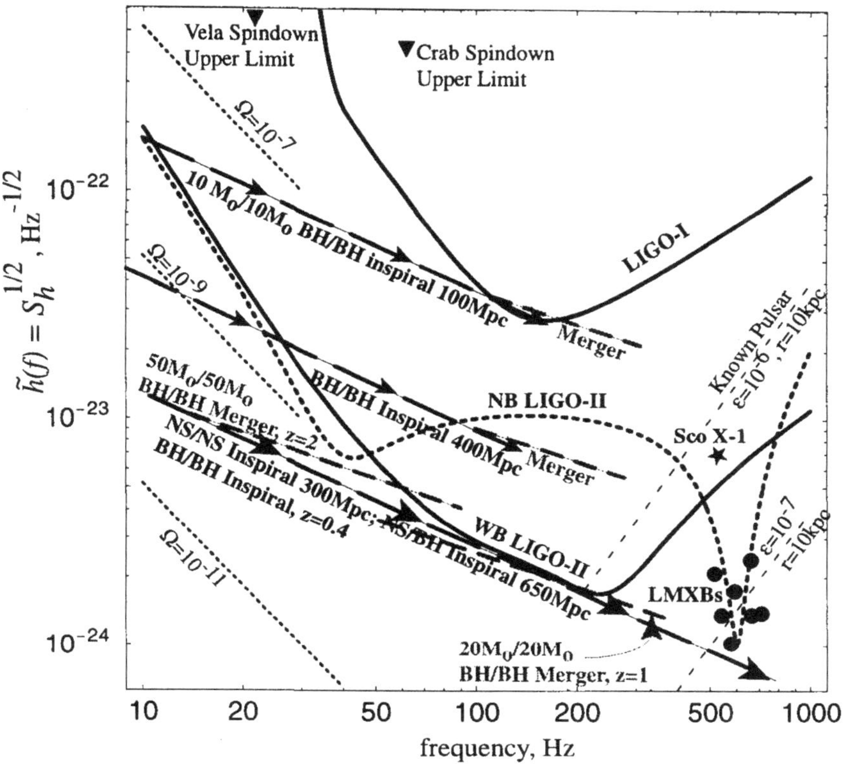

Figure 1. The noise $\tilde{h}(f)$ in several planned LIGO interferometers plotted as a function of gravity-wave frequency f, and compared with the estimated signal strengths $\tilde{h}_s(f)$ from various sources. The signal strength $\tilde{h}_s(f)$ is defined in such a way that, *wherever a signal point or curve lies above the interferometer's noise curve, the signal, coming from a random direction on the sky and with a random orientation, is detectable with a false alarm probability of less than one per cent;* see the text for greater detail and discussion.

the overall GW event rate may be worse by about a factor 2.

Figure 1 shows, along with the noise curves, the estimated signal strengths $\tilde{h}_s(f)$ for various sources. These signal strengths are defined in such a way [14] that the ratio $\tilde{h}_s(f)/\tilde{h}(f)$ is equal to the ratio of signal S to noise threshold T, rms averaged over source directions and orientations, $\tilde{h}_s(f)/\tilde{h}(f) = \langle S^2/T^2\rangle^{1/2}$, with the threshold being that at which the false alarm probability is one per cent when using the best currently known, practical data analysis algorithm. (For broad-band sources, the algorithm is assumed to integrate over a bandwidth equal to frequency and to use the output from LIGO's two 4km IFOs and one 2km IFO [15], thereby removing all non-Gaussian noise. For periodic sources such as spinning neutron stars, the algorithm uses data from only one 4 km IFO, usually narrow banded, the noise again is assumed Gaussian, and the signal is integrated for 10^7 sec, except in cases such as Low Mass X-Ray Binaries where there is little gain from integrating so long.) This definition of $\tilde{h}_s(f)$ means that, *wherever a signal point or signal curve lies above the IFO noise curve, the signal, coming from a random direction on the sky and with a random orientation, is detectable with a false alarm probability of less than one per cent.*

2.2 Overview of High-Frequency Sources

In this section, we shall discuss the high-frequency sources briefly, in turn, and then in subsequent sections we shall discuss them in greater detail, focusing on the likelihood of detection and the science we expect to extract from detected waves.

The three arrowed, long-dashed lines in Fig. 1 represent the signal $\tilde{h}_s(f)$ from **neutron-star (NS) and black-hole (BH) binaries in the last few minutes of their inspiral**, assuming masses $M = 1.4M_\odot$ for each NS and $M = 10M_\odot$ for each BH. These sources are best searched for by the method of matched filters [16]. Using matched filters, LIGO's initial IFOs can detect NS/NS inspirals (with a 1 per cent false alarm probability) out to a distance of 20 Mpc [top arrowed line]; the wide-band advanced (LIGO-II) IFOs can do so 15 times farther, out to 300 Mpc for NS/NS and out to 650 Mpc for NS/BH [bottom arrowed line]. For BH/BH inspiral, the wide-band IFOs can see so far that cosmological effects are important. For definiteness, throughout this proposal we assume a Hubble expansion rate $H_o = 65$ km/s/Mpc, a cold-matter density 0.4 of that required to close the universe $\Omega_M = 0.4$, and a vacuum energy density (cosmological constant) 0.6 of closure $\Omega_\Lambda = 0.6$. Then the wide-band IFOs can see $(10M_\odot/10M_\odot)$ BH/BH inspirals out to a cosmological redshift $z = 0.4$. The binary inspiral rates at these advanced IFO distances are likely to be many per year; see Box 1 at the end of this section. The middle arrowed line is the signal from BH/BH inspiral at 400Mpc, where the geometric mean of the event-rate estimates for BH/BH field binaries is three per year (third column of table in Box 1).

The **tidal disruption of a NS by its BH companion** at the endpoint of NS/BH inspiral should produce gravitational waves that carry detailed information about the NS structure and equation of state. The advanced IFOs may detect these waves and extract their information; see Sec. 2.4. This tidal disruption is a promising candidate for the **trigger of some gamma-ray bursts**, as is the final merger of the two NS's in a NS/NS binary. A gamma-burst / gravitational-wave coincidence would be of great value in revealing the nature of gamma-burst sources; see Sec. 2.8

For BH/BH binaries much heavier than $\sim 10M_{\odot}/10M_{\odot}$, most of the gravitational signal is likely to come from the **black holes' merger and the vibrational ringdown of the final black hole**, rather than from the inspiral. Rough estimates discussed in Sec. 2.5 suggest that, if the holes are rapidly spinning (within a few per cent of the fastest spin allowed, i.e. $a/m \gtrsim 0.98$), then the wide-band IFOs can see the merger waves from two $20M_{\odot}$ holes out to $z = 1$ and two $50M_{\odot}$ holes out to $z = 2$; see the down-sloping, non-arrowed, dashed lines in Fig. 1. The event rates at these distances could well be many per year, and the waves from such mergers will carry rich physical and astrophysical information; see Sec. 2.5.

The triangles, star, large dots, and up-sloaping short-dashed lines in Fig. 1 represent signals from **slightly deformed, spinning neutron stars**. The most interesting of these is a class of objects called **low-mass X-ray binaries (LMXB's)**. These are neutron stars that are being torqued by accretion from a companion, but that seem to be locked into spin periods in the range $\sim 300 - 600$ revolutions per second. The most plausible explanation for this apparent locking is that the accretion is producing an asymmetry that radiates gravitational waves, which torque the star's spin down at the same rate as accretion torques it up [17,18]. Assuming this to be true, one can deduce an LMXB's wave strength $\tilde{h}_s$ from its measured X-ray flux and its spin frequency [with the frequency inferred, sometimes to within a few Hz but not better, from nearly coherent oscillations (NCOs) in type-I X-ray bursts, or less reliably from frequency splittings of quasiperiodic (QPO) X-ray oscillations [19].] The spin frequency, and thence the gravitational-wave frequency, will wander somewhat due to fluctuations in accretion (which can be estimated by monitoring the X-ray flux) and due to poorly known orbital parameters. As a result, in searching for an LMXB's waves one can only perform coherent integrations for about 20 days; thereafter, one must stack the signals incoherently, allowing for unknown shifts of the wave frequency [20,21]. When one uses this "stack-slide" method of data analysis, the resulting signal strengths may improve only slightly for integration times longer than 20 days [20]. (The most sensitive LMXB searches will probably be with NB IFO's, with the optimum sensitivity band re-adjusted every ~ 20 days to cover the whole range $\sim 400 - 1000$ Hz.)

Assuming 20 days of integration using a single 4-km IFO, the estimated signal strengths and frequencies for the strongest known LMXB's are shown by the big dots and the star in Fig. 1. The estimated strengths assume a steady-state balance of accretion torque by gravitational-wave torque, and that the GW's are generated by density inhomogneneities (as opposed to NS wobble or unstable modes), so f_{GW} is twice the spin frequency [17]. For every LMXB in Fig. 1 except the weakest one, the estimated rotation frequency is based on QPO splittings rather than NCOs, which means that about half of the frequencies might be double these estimates: $\sim$ 1200 Hz for density inhomogeneities and $\sim$ 800 Hz for r-mode sloshing, rather than $\sim$ 600 Hz. Doubling the frequency above 600 Hz reduces the emission amplitude by a factor 2 (at fixed X-ray flux assuming a steady-state torque balance), and increases the amplitude noise $S_h^{1/2}$ in an advanced NB IFO by a factor $\simeq \sqrt{2}$. As a result, at $\sim$ 1200 Hz frequency, Sco X-1 would still be readily detectable in a NB IFO but not in a WB IFO, and the strongest of the other LMXB's would be marginal. By contrast, if the frequencies are $\sim$ 600 Hz or $\sim$ 400 Hz and the waves are in a steady state, then Sco X-1 should be very easily detected by a WB IFO, and several LMXB's should be detectable by narrow banding (NB).

The advanced IFOs can perform interesting searches for waves from **known radio pulsars**, such as the **Crab and Vela** for which the current upper limits (based on the pulsars' observed spindown rates) are shown as triangles. A Crab search, using coherent integrations based on the star's observed (slightly wandering) rotation period, could improve the limit on the Crab's wave amplitude by a factor 100 and constrain the star's gravitational ellipticity to $\epsilon \lesssim 7 \times 10^{-6}$ — which is approaching the realm of physically plausible ellipticities, $\epsilon \lesssim 10^{-6}$. (We shall discuss the physical origins of neutron-star ellipticities and their plausible ranges in Sec. 2.7.)

More interesting will be searches for waves from **known, fast pulsars**, since the signal strength scales as $\tilde{h}_s \propto \epsilon f^2/r$ (where r is distance to the source). The up-sloping short-dashed lines in Fig. 1 show some examples of signal strengths. With a narrow-band IFO tuned to the vicinity of such a fast pulsar, the waves would be detectable when $\epsilon \gtrsim 2\times10^{-6}(100\text{Hz}/f)^2(r/10\text{kpc})$, which is in the realm of plausible ellipticities for pulsars throughout our galaxy so long f exceeds 200 Hz, i.e. the spin frequency exceeds 100 Hz.

Also of great interest will be searches for **previously unknown spinning neutron stars**, for which the signal strengths $\tilde{h}_s$ will be reduced by a factor of a few to $\sim$ 15 by the lack of prior information about the frequency and its evolution and the direction to the source (which determines the time-evolving doppler shift produced by the earth's motion) [22,20]. A tunable, narrow-band IFO will be crucial to such searches. One can search more deeply by a factor of several, using a narrow-banded IFO that dwells on a given frequency and its neighborhood for a few days or weeks and then moves on to another frequency, than using a broad-interferometer that collects signal at all frequencies simultaneously for a year. Such searches will be in the band of physically plausible ellipticities, for stars throughout our galaxy, if $f \gtrsim 400$ Hz (spin frequency above ~ 200 revolutions per second).

One can search for a **stochastic background** of gravitational waves by cross correlating the outputs of LIGO's two 4 km IFOs [23]. For such a search the signal strengths $\tilde{h}_s(f)$ are shown in Fig. 1 as downward-sloping dotted lines, assuming a cross-correlation of 4 months of (not necessarily contiguous) data, and isotropic waves. The lines are labeled by the waves' energy density $\Omega(f)$ in a bandwidth equal to frequency and in units of the density to close the universe, $\Omega(f) = (f dE_{\rm GW}/df)/\rho_{\rm closure}$. Unfortunately, the frequency of optimal *a priori* sensitivity, $f \sim 70$ Hz (where the noise curve is parallel to the dotted Ω lines), is near the center of a dead band for LIGO. This dead band arises from the fact that $1/70\text{Hz} \simeq 14$ ms is about the round-trip gravity-wave travel time between the two LIGO sites [24]. The result is a net debilitation of the stochastic background sensitivity by a factor of a few: the initial IFOs can detect an isotropic background with Ω down to $\sim 10^{-5}$, while advanced, wide-band IFOs can reach down to $\Omega \sim 5 \times 10^{-9}$, and a factor ~ 2 lower than this if it is reoptimized for low frequencies. These are interesting sensitivities, able to test a wide range of speculations about the physics of the early universe and perhaps even detect waves — but *not* able to reach the level predicted by standard inflation, $\Omega \lesssim 10^{-15}$. Such a detection could have profound implications for physics and cosmology; see Sec. 2.9.

Other waves that advanced IFOs in LIGO will seek and may detect are those from **the stellar core collapse that triggers supernovae** and the **boiling of the nascent neutron star** at the endpoint of that collapse (Sec. 2.6), **accretion induced collapse of white dwarfs** (Sec. 2.6), and **totally unknown sources**.

We now give a more detailed discussion of each of the high-frequency sources, focusing especially on event rate estimates and the information that the waves may bring. Box 1 gives a brief summary of the various sources and their detectability by LIGO I and II (WB and NB).

2.3 Inspiraling NS and BH Binaries with $M_{BH} \lesssim 10 M_\odot$

As we have discussed, wide-band advanced-LIGO (LIGO-II) IFOs can detect the waves from NS/NS inspirals out to 300Mpc, NS/BH out to 650 Mpc, and BH/BH out to $z = 0.4$. The event rates out to these distances can be esti-

Box 1

Brief Summary of Detection Capabilities of Advanced LIGO (LIGO-II) Interferometers

• **Inspiral of NS/NS, NS/BH and BH/BH Binaries:** The table below [25,15] shows estimated rates $\mathcal{R}_{\rm gal}$ in our galaxy (with masses $\sim 1.4 M_\odot$ for NS and $\sim 10 M_\odot$ for BH), the distances $\mathcal{D}_{\rm I}$ and $\mathcal{D}_{\rm WB}$ to which initial IFOs and advanced WB IFOs can detect them, and corresponding estimates of detection rates $\mathcal{R}_{\rm I}$ and $\mathcal{R}_{\rm WB}$; Sec. 2.3.

	NS/NS	NS/BH	BH/BH in field	BH/BH in clusters
$\mathcal{R}_{\rm gal}$, yr^{-1}	10^{-6}–5×10^{-4}	$\lesssim 10^{-7}$–10^{-4}	$\lesssim 10^{-7}$–10^{-5}	$\sim 10^{-6}$–10^{-5}
$\mathcal{D}_{\rm I}$	20 Mpc	43 Mpc	100	100
$\mathcal{R}_{\rm I}$, yr^{-1}	$3\times10^{-4}-0.3$	$\lesssim 4\times10^{-4}-0.6$	$\lesssim 4\times10^{-3}-0.6$	$\sim 0.04-0.6$
$\mathcal{D}_{\rm WB}$	300 Mpc	650 Mpc	$z=0.4$	$z=0.4$
$\mathcal{R}_{\rm WB}$, yr^{-1}	$1-800$	$\lesssim 1-1500$	$\lesssim 30-4000$	$\sim 300-4000$

• **Tidal disruption of NS by BH in NS/BH binaries:** First crude estimates suggest WB IFOs can measure onset of disruption at 140Mpc well enough to deduce the NS radius to 15% accuracy (compared to current uncertainties of a factor ~ 2); see table above for rates; Sec. 2.4.

• **BH/BH merger and ringdown:** Rough estimates suggest detectability, by WB IFOs out to the cosmological distances shown in Fig. 2(b); rates for BH/BH total mass $\sim 20 M_\odot$ are in table above; rates for much larger masses are unknown; Sec. 2.5.

• **Low-Mass X-Ray Binaries:** If accretion's spin-up torque on NS due is counterbalanced by gravitational-wave-emission torque, then WB IFOs can detect Sco X-1, and NB IFOs can detect ~ 6 other known LMXB's; Sec. 2.7.

• **Fast, Known Spinning NS's (Pulsars with pulse frequency above 100 Hz):** Detectable by a advanced NB IFO in 3 months' integration time, if NS ellipticity is $\epsilon \gtrsim 2\times10^{-8}(1000{\rm Hz}/f)^2(r/10{\rm kpc})$, where f is gravity wave frequency (twice the pulsar frequency) and r is distance; actual ellipticities are unknown, but plausible range is $\epsilon \lesssim 10^{-6}$; Sec. 2.7.

• **Fast, Unknown Spinning NS's:** Unknown frequency wandering and doppler shifts degrade the detectable ellipticity ϵ by a factor of a few to ~ 15, so detection with a NB IFO requires $\epsilon \gtrsim (0.6$ to $3)\times10^{-5}(100{\rm Hz}/f)^2(r/10{\rm kpc})$; Secs. 2.2 and 2.7.

• **Centrifugally Hung-Up Proto Neutron Stars in White-Dwarf Accretion-Induced Collapse and in Supernovae:** Dynamics of star very poorly understood; if instability deforms star into tumbling bar, may be detectable by WB IFOs to ~ 20 Mpc (the Virgo Cluster), and possibly farther; event rates uncertain but could be enough for detection; Sec. 2.6.

• **Convection of Supernova Core:** May be detectable by WB IFOs, via correlations with neutrinos, for supernovae in our Galaxy and possibly Magellanic Clouds; Sec. 2.6.

• **Gamma Ray Bursts:** If triggered by NS/BH mergers, a few per year could be detectable by WB IFOs; if none are seen individually, statistical studies could nevertheless confirm gravity-wave emission by the gamma-burst triggers; Sec. 2.8.

• **Stochastic Background:** Detectable by cross correlating Hanford and Livingston 4km detector outputs, if $\Omega =$ (gravitational-wave energy in $\Delta f \sim f \sim 40$ Hz) $\gtrsim 5\times10^{-9}$; there are many possible sources of such waves in very early universe, all very speculative; Sec. 2.9.

mated from observational data in our own galaxy [25], and an extrapolation out through the universe based on the density of massive stars (which can be deduced by several different methods) [25,26,27]. The resulting rates (as compiled and evaluated by V. Kalogera [25]) are quite uncertain but very promising; see Box 1.

For NS/NS binaries, the event rate in our galaxy is constrained by the results of radio astronomers' searches for binary pulsars that will merge, due to gravitational radiation emission, in less than the age of the universe, and by other aspects of pulsar searches [25]. The resulting constraints, 10^{-6}/yr $\lesssim \mathcal{R}_{\rm gal} \lesssim 5 \times 10^{-4}$/yr, extrapolate to a NS/NS event rate for advanced IFOs between 1/yr and 800/yr. Searches in our galaxy for NS/BH binaries in which the NS is a pulsar have failed to find any as yet, so we must turn to much less reliable estimates based on "population synthesis" (simulations of the evolution of a population of progenitor binary systems to determine the number that make NS/BH binaries compact enough to merge in less than the universe's age). Population synthesis gives a NS/BH event rate in our galaxy in the range $\sim 10^{-7}$/yr $\lesssim \mathcal{R}_{\rm gal} \lesssim 10^{-4}$/yr [28], though it is possible the rate could be even less than this (hence the $\lesssim 10^{-7}$/yr in Box 1). Extrapolating out into the universe, Kalogera finds a NS/BH rate in WB advanced IFOs between $\lesssim$ 1/yr and about 1500/yr. A similar analysis for BH/BH binaries based on population synthesis (third column of the table in Box 1) gives a rate between $\lesssim$ 30/yr and $\sim$ 4000/yr [29]. That population synthesis estimate ignores the likely role of globular clusters and other types of dense star clusters as "machines" for making BH/BH binaries [30]: single black holes, being heavier than most stars in a globular, sink to the center via tidal friction, find each other, and make binaries; then the BH/BH binaries get "hardened" (made more compact) by interaction with other black holes, reaching sizes where gravitational-wave emission will cause them to merge in less than the age of the universe; and further interactions will often eject the BH/BH binaries from the globular, to interstellar space where they merge. Simulations by McMillan and Portegies Zwart [30] suggest that each dense cluster will make a number of such BH/BH binaries, and extrapolations into the universe predict an event rate in WB advanced IFOs between $\sim$ 300 and $\sim$ 4000/yr — though the uncertainties are probably larger than suggested by these numbers from the literature [30,25].

These event rates are very encouraging. They make it seem quite likely that the advanced IFOs will observe tens to thousands of BH and NS inspirals per year, while the initial IFOs will be lucky to observe $\sim$ 1 per year.

The observed inspiral waves will last for between $\sim$ 1000 and 10,000 cycles depending on the binary's masses, and will carry detailed information about the binary and about general relativistic deviations from Newtonian gravity. This information can be extracted with good precision using the method of matched filters. Specifically (denoting by $M = M_1 + M_2$ the binary's total mass and $\mu = M_1 M_2 / M$ its reduced mass): [31,32]

(i) The binary's chirp mass $M_c \equiv \mu^{3/5} M^{2/5}$ will typically be measured, from the Newtonian part of the signal's upward frequency sweep, to $\sim 0.04\%$ for a NS/NS binary and $\sim 0.3\%$ for a system containing at least one BH. (ii) *If* we are confident (e.g., on a statistical basis from measurements of many previous binaries) that the binary's spins are a few percent or less of the maximum physically allowed, then the reduced mass μ will be measured to $\sim 0.5\%$ for NS/NS and NS/BH binaries, and $\sim 2\%$ for BH/BH binaries. (iii) Because the frequency dependences of the (relativistic) μ effects and spin effects are not sufficiently different to give a clean separation between μ and the spins, if we have no prior knowledge of the spins, then the spin/μ correlation will worsen the typical accuracy of μ by a large factor, to $\sim 50\%$ for NS/NS, $\sim 90\%$ for NS/BH, and a factor ~ 1.5 for BH/BH. These worsened accuracies should be improved significantly (though we do not yet know how much) by waveform modulations due to spin-induced precession of the orbit [33,34,35], and even without modulational information, a certain combination of μ and the spins will be determined to a few per cent. (iv) The distance to the binary ("luminosity distance" at cosmological distances) can be inferred, from the observed waveforms, to a precision $\sim 3/\rho \lesssim 30\%$, where $\rho \equiv S/N$ is the amplitude signal-to-noise ratio in the total LIGO network (which must exceed about 8 in order that the false alarm rate be less than the threshold for detection). (v) With the aid of VIRGO, the location of the binary on the sky can be inferred, by time of flight between the detector sites, to a precision of order five degrees [36]. (This angular resolution is rather worse than one might expect from a simple estimate based on time-of-flight alone. The reason is that because two LIGO detectors are nearly parallel, the addition of VIRGO to the network still leaves a near-degeneracy between polarization and time-of-flight information. The addition of a fourth detector with comparable sensitivity in Australia would decrease the error box on the sky $\Delta\Omega$ by a factor ~ 4 [36].)

Advanced LIGO will likely produce a catalog of hundreds or thousands of binary inspirals and their inferred parameters; this catalog will be a valuable data base for observational astronomy and cosmology.

Important examples of the general relativistic effects that can be detected and measured with precision, in the inspiral waves, are these: (i) As the waves emerge from the binary, some of them get backscattered one or more times off the binary's spacetime curvature, producing wave *tails*. These tails act back on the binary, modifying its radiation reaction force and thence its inspiral rate in a measurable way. [37] (ii) If the orbital plane is inclined to one or both of the binary's spins, then the spins drag inertial frames in the binary's vicinity (the "Lense-Thirring effect"), this frame dragging causes the orbit to precess, and the precession modulates the waveforms [33]. This precession and modulation should be very strong in a significant fraction of NS/BH binaries [28].

2.4 Tidal Disruption of a NS by a BH: Measuring the Nuclear Equation of State

As a NS/BH binary spirals inward, its NS experiences ever increasing tidal forces from the BH's gravitational field (its spacetime curvature). In many cases these tidal forces may tear the NS apart before it begins its final, quick plunge into the hole's horizon. The gravitational waves from this tidal disruption and from the termination of inspiral should carry detailed information about the NS's equation of state (the equation of state of bulk nuclear matter at $\sim$ 10 times the density of an atomic nucleus). The disruption waves lie largely in the frequency band $\sim$ 300Hz $\lesssim f \lesssim$ 1000 Hz, where the wide-band, advanced IFOs have good sensitivity, and where narrow-band or some other optimized IFO can do even better. This suggests that the advanced IFOs may be able to extract new information about the nuclear equation of state from the tidal-disruption waves. A first, crude estimate [38] suggests, for example, that for NS/BH binaries at 140 Mpc distance (where the event rate could be a few per year; see the table in Box 1), tidal-disruption observations may enable the NS radius R to be measured to a precision $\sim$ 15%, by contrast with its present uncertainty (for fixed NS mass) of about a factor 2. From the measured radii would follow the desired equation-of-state information. Detailed numerical-relativity simulations will be required to firm up this estimate, and will be essential as a foundation for interpreting any tidal-disruption waves that are observed.

The merger waves from NS/NS binaries, by contrast with NS/BH, are likely to lie outside the band of good advanced-IFO sensitivity — at frequencies $f \gtrsim$ 1500 Hz. However the onset of NS/NS merger, triggered by a plunge of the two NS's toward each other, may produce a strong "cliff" in the waves' spectrum, in a range $f \sim 400 - 1000$ Hz of good sensitivity, and by measuring the cliff frequency we may strongly constrain the NS radii (assuming the NS masses have been determined accurately from the earlier portion of the waveform), and hence also learn about the nuclear equation of state [39].

2.5 BH/BH Mergers and Ringdown: Observing the Nonlinear Dynamics of Spacetime Curvature

A BH/BH binary evolves through three epochs: a gradual, adiabatic inspiral driven by gravitational radiation reaction (discussed in Sec. 2.3 above); a merger epoch, during which the holes' spacetime geometry undergoes violent, highly nonlinear oscillations; and a ringdown epoch, during which the merged black hole undergoes damped pulsations as it settles down into its final, Kerr-metric, quiescent state. The inspiral waves are rather well understood (Sec. 2.3), as are the ringdown waves; but the merger dynamics and waves are hardly understood at all.

It is likely that in many cases the holes are spinning rapidly and the system's three angular momentum vectors (two spins and one orbital) are substantially misaligned. In such cases, the merger dynamics might be quite rich, with spin-spin and spin-orbit coupling perhaps strong enough to produce spin flips and complex contortions and flows of the spacetime geometry as the two holes convert themselves into one.

Flanagan and Hughes [40,41] have made plausible estimates about the unknown aspects of the waves from each epoch — inspiral, merger and ringdown — and have used these estimates to evaluate the signal-to-noise ratios in LIGO and LISA for equal-mass, fast-spinning BH binaries. Figures 2 and 3 show their results for the initial (LIGO-I) and advanced (LIGO-II) IFOs, and Fig. 5 (Sec. 3.3 below) shows them for LISA. [The LIGO-II results in Fig. 3 are updates [5] of the original Flanagan-Hughes results, taking account of the currently planned noise curve for LIGO-II, Fig. 1.]

Figures 2 and 3 suggest that, for $M \lesssim 24M_\odot$ (two $12M_\odot$ holes), the inspiral waves produce most of the signal in LIGO; for $24M_\odot \lesssim M \lesssim 200M_\odot$ the merger waves dominate; and for $200M_\odot \lesssim M \lesssim 1000M_\odot$, the merger and ringdown waves have roughly equal strength. (We say "suggest" because the plausible assumptions made by Flanagan and Hughes are far from secure; numerical-relativity simulations and/or observations are needed to firm them up.) For $M = 20M_\odot$ (two $10M_\odot$ holes), LIGO's initial and advanced IFOs can detect the waves out to 100Mpc and $z = 0.4$ respectively, at which distance population synthesis calculations suggest event rates $\lesssim 1$/year and ~ 30 to ~ 4000/year respectively (Box 1 and Sec. 2.3 above). Black-hole binaries more massive than $M \sim 50M_\odot$ (two $25M_\odot$ holes) are unlikely to form from main-sequence progenitors, but could well form in dense star clusters. The event rates at the distances seen by the initial IFOs (~ 200 to 800 Mpc for masses up to $\sim 1000M_\odot$) and by the advanced IFOs (redshifts $z \sim 1$ to 2 for masses up to $\sim 2000M_\odot$) are exceedingly uncertain; but for the advanced

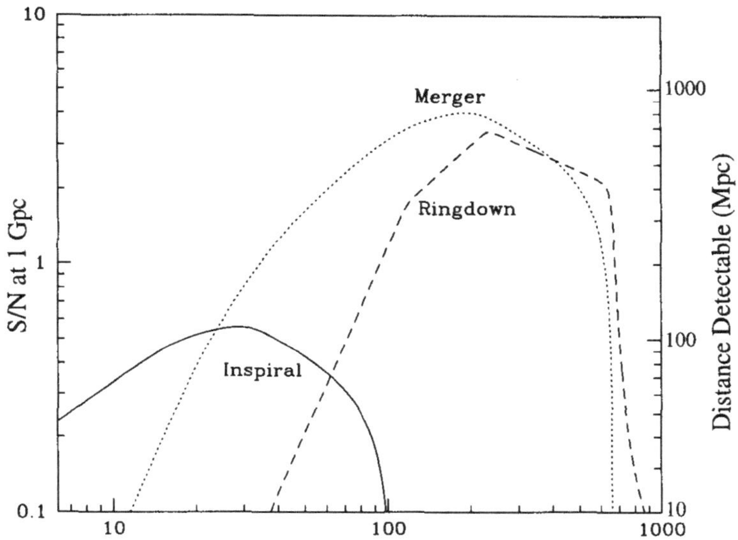

Figure 2. The inspiral, merger, and ringdown waves from equal-mass black-hole binaries as observed by LIGO's initial interferometers: The distance to which the waves are detectable (right axis) and the signal-to-noise ratio for a binary at 1Gpc (left axis), as functions of the binary's total mass (bottom axis). (Figure adapted from Flanagan and Hughes [40].)

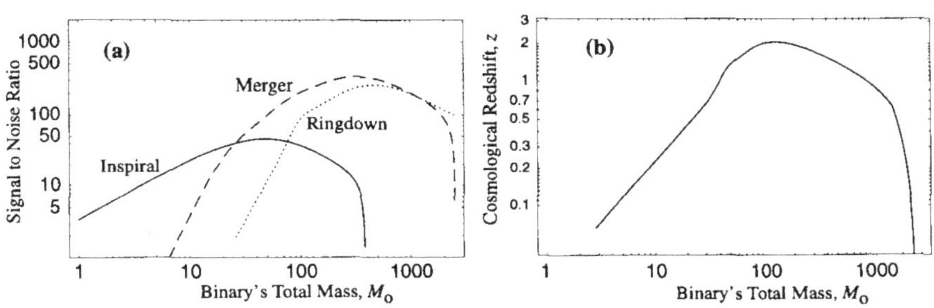

Figure 3. (a) The signal-to-noise ratio for the inspiral, merger and ringdown waves from fast-spinning, equal-mass black-hole binaries at a fiducial distance of 1Gpc, as observed by advanced (LIGO-II) interferometers. (b) The cosmological redshift out to which these waves can be seen, assuming a cosmology with Hubble expansion rate $H_o = 65$ km/s/Mpc, cold-matter density 0.4 of that required to close the universe $\Omega_M = 0.4$, and vacuum energy density (cosmological constant) 0.6 of closure, $\Omega_\Lambda = 0.6$. (Figure from Ref. [5], based on the LIGO-II wide-band noise curve in Fig. 1 and the analysis of Flanagan and Hughes [40].)

IFOs they could be large.

2.6 Supernovae and Accretion-Induced Collapse of White Dwarfs

Type-II supernovae are triggered by the violent collapse of a stellar core to form a NS or BH. Despite decades of extensive modeling, we still understand only poorly the details of the collapse and how it triggers the explosion, and the birth throes of the NS or BH. The optical display brings little information, as the stellar material is so optically thick that the display is produced hours after the collapse and from radii $\sim 10,000$ times larger than the NS or BH. Neutrinos bring information about the core's temperature during the first few seconds after the collapse, and gravitational waves could reveal the nonspherical dynamics of the core's mass distribution on millisecond timescales and longer, and the nature of the core's final state: BH or NS.

Numerical models of supernovae suggest that, even if it is slowly rotating, a newborn proto-neutron-star will be convectively unstable, and that the gravitational waves from the convective overturn in the first ~one second of the star's life may be detectable throughout our galaxy and its orbiting companions, the Magellanic Clouds [42]. Although the supernova rate is low in our galaxy and its companions ($\lesssim 1/30$yrs), one observed event could be very valuable scientifically: The bulk of the supernova's neutrinos are thought to come from the same convecting material as produces the gravitational waves, so there should be correlations between the neutrinos and the waves, which could teach us much about the proto-neutron-star's dynamics.

Observations show that many type-II supernovae produce neutron stars and give them kicks of magnitude as large as ~ 1000 km/s. Indeed, about half of all radio pulsars are born with a kick larger than 500 km/s [43]. The observed kicks suggest that at least some newborn proto-neutron-stars may be strongly asymmetric, perhaps due to fast rotation, and therefore could produce significant gravitational radiation. Rapidly rotating proto-neutron-stars may also be produced by the accretion-induced collapse of white dwarf stars (AIC) in some cases, depending on the white-dwarf composition, central density and accretion rate [44].

If the newborn proto-neutron-star in an AIC or supernova is spinning so fast that it hangs up centrifugally at a radius large compared to that of the final NS, then it may be dynamically (or at least secularly) unstable to deforming into a bar-shaped object that tumbles end-over-end, emitting gravitational waves in LIGO's band of good sensitivity [45]. Recent simulations suggest that the bar will be long-lived, rather than just wrapping itself up into an axisymmetric shape and disappearing [46,47]. The simulations show that the waves from such a bar may sweep upward in frequency, due to a gradual shrinkage of the proto-neutron-star, or may sweep downward due to development of "Dedekind-like" internal circulation [48]. The discovery of the waves from such a proto-neutron-star and observations of their frequency evolutions (which should mirror the bar's evolution) would teach us much. Though the strengths of the waves and the best signal processing techniques and thresholds are all ill-understood, a rough estimate [49] suggests that the advanced IFOs' range for detection might be the distance of the VIRGO cluster of galaxies (about 15 Mpc) and conceivably significantly larger, suggesting event rates that could be some per year but might be far less.

If the advanced WB IFOs detect no gravitational waves from a supernova at distance r, during a time T preceeding the beginning of the optical outburst or over a time T during a neutrino outburst, then one can thereby place a limit on the emitted gravitational-wave energy. The limit (as devised by Finn[50]) is based on the statistics of a cross correlation of the outputs of the two 4km LIGO interferometers, over the chosen time T and over some frequency band near the minimum of the advanced LIGO noise curve. In the band from ~ 100 to 300 Hz this limit is $\Delta E_{\rm GW} \lesssim 0.05(r/15{\rm Mpc})^2(T/1{\rm h})^{1/2}M_\odot$. In the case of a neutrino-emitting supernova in our own galaxy, with $T \sim 1$ sec, this limit is very impressive: $\sim 10^{-9}M_\odot$.

2.7 Spinning Neutron Stars

As a newborn neutron star settles down into its final state, its crust solidifies (crystalizes). The solid crust will assume nearly the oblate axisymmetric shape that centrifugal forces are trying to maintain. Thereafter the crust has a preferred shape (oblate about some preferred axis). One can define the NS's *deformation ellipticity* ϵ_d as the residual ellipticity the NS would then maintain (due to induced crustal shear stresses) if the NS were spun down to zero frequency without the crust breaking or otherwise relaxing [51,52]:

$$\epsilon_d \sim 7 \times 10^{-8}\left(\frac{\nu_{\rm rel}}{\rm kHz}\right)^2 \tag{1}$$

where $\nu_{\rm rel}$ is the spin frequency for which the crust is (most) relaxed. Again, this is the level of ellipticity that is "frozen into" the crust, as opposed to the ellipticity that arises from centrifugal forces and so "follows" the star's spin . If the star's angular momentum $\vec{J}$ deviates from the crust's preferred symmetry axis, the NS will wobble as it spins, with small "wobble angle" θ_w. The NS will emit GW's with $h \propto \theta_w \epsilon_d$ at frequency $f_{GW} = f_{\rm rot} + f_{\rm prec}$, where the *body frame precession frequency* $f_{\rm prec}$ $[= \epsilon_d f_{\rm rot}(I_{\rm crust}/I)]$ is orders of magnitude smaller than $f_{\rm rot}$; thus f_{GW} is just slightly higher than $f_{\rm rot}$ [53,51]. The NS may also deviate slightly from axisymmetry about its principal axis; i.e., it may have a slight ellipticity $\epsilon_e \equiv (I_1 - I_2)/I_3$ in its equatorial plane. It will then emit GW's with $h \propto \epsilon_e$ at twice the rotation frequency, $f_{GW} = 2f_{\rm rot}$ [53]. The gravitational ellipticity ϵ referred to in Fig. 1 and in Sec. 2.2 is $\epsilon = \epsilon_e$ for deformations around the principal axis, and $\epsilon = 2\theta_w \epsilon_d$ for deformatins in which $\vec{J}$ is inclined to the preferred symmetry axis, producing wobble.

How large will be ϵ_e and $\theta_w \epsilon_d$? Ushomirsky et al. [18] have shown that, for typical NS parameters, $\epsilon_e < 5 \times 10^{-7}(\sigma_{max}/10^{-2})$, where σ_{max} is the (highly uncertain) crustal yield strain, estimated to be somewhere in the range $10^{-5} < \sigma_{max} < 10^{-1}$. Thus rough upper limits are $\epsilon_e < 10^{-5}$ and [from Eq. 1] $\theta_w \epsilon_d < 10^{-7}$. These upper limits are for ellipticities that are basically supported by crustal shear stresses, but there are other possible sources of ellipticity. For example, inside the superconducting cores of many neutron stars there may be trapped magnetic fields with mean strength $B_{\rm core} \sim 10^{13}$G or even 10^{15}G. Because such a field is actually concentrated in flux tubes with $B = B_{\rm crit} \sim 6 \times 10^{14}$G surrounded by field-free superconductor, its mean pressure is $p_B = B_{\rm core} B_{\rm crit}/8\pi$. This pressure could produce ellipticity $\epsilon_{\rm e} \sim \epsilon_d \sim p_B/p \sim 10^{-6} B_{\rm core}/10^{15}$G (where p is the core's material pressure). We are extremely ignorant, and correspondingly there is much to be learned from searches for gravitational waves from spinning neutron stars.

From this discussion we see that the range of plausible gravitational ellipticities for a spinning neutron star is $\epsilon \lesssim 10^{-6}$, though they may get as large as 10^{-5}. The strengths of the waves produced by such ellipticities are shown in Fig. 1 and the waves' detectability is discussed in Sec. 2.2.

Neutron stars gradually spin down, due in part to gravitational-wave emission but perhaps more strongly due to electromagnetic torques associated with their spinning magnetic fields and pulsar emission. This spin-down reduces the strength of centrifugal forces, and thereby causes the star's oblateness to decrease, with an accompanying breakage and resolidification of its crust's crystal structure (a "starquake") [54]. In each starquake, θ_w, ϵ_e, and ϵ_d will all change suddenly, thereby changing the amplitudes and frequencies of the star's two gravitational "spectral lines" $f = 2f_{\rm rot}$ and $f = f_{\rm rot} + f_{\rm prec}$. After each quake, there should be a healing period in which the star's fluid core and solid crust, now rotating at different speeds, gradually regain synchronism. By monitoring the amplitudes, frequencies, and phases of the two gravitational-wave spectral lines, and by comparing with timing of the electromagnetic pulsar emission, one might learn much about the physics of the neutron-star interior.

A spinning NS can radiate not only due to built-in asymmetries, as discussed above, but also due to a gravitational-radiation-reaction-driven instability first discovered by Chandrasekhar [55] and elucidated in greater detail by Friedman and Schutz [56]. In this "CFS instability", density waves travel around the star in the opposite direction to its rotation, but are dragged forward by the rotation. These density waves produce gravitational waves that carry positive energy as seen by observers far from the star, but negative energy from the star's viewpoint; and because the star thinks it is losing negative energy, its density waves get amplified. This intriguing mechanism is similar to that by which spiral density waves are produced in galaxies. Although the CFS instability was once thought ubiquitous for spinning stars [56,57], we now know that neutron-star viscosity will kill it, stabilizing the star and turning off the waves, when the star's temperature is above some limit $\sim 10^{10}$K [58] and below some limit $\sim 10^9$K [59]; and correspondingly, the instability should operate only during the first few years of a neutron star's life, when $10^9\text{K} \lesssim T \lesssim 10^{10}\text{K}$, and only for NS spins greater than $\sim$ 90% of the upper limit set by mass-shedding at the equator [60].

Shortly after GR15 there was a surge of renewed interest in the CFS instability when Andersson [61] and Friedman and Morsink [62] showed that r-modes (driven by Coriolis forces) are unstable at all spin values (again, in the absence of viscosity), and then Lindblom et al. [63] and Andersson et al. [64] estimated that in a newborn NS, the instability would *not* be damped by shear or bulk viscosity, but instead would rapidly spin down the NS to $\nu \lesssim 100$ Hz. However, since then, a host of other damping mechanisms have been suggested, including: (i) wind-up of magnetic fields [65], (ii) dissipation at the crust-core bondary [66], (iii) bulk viscous damping from weak interactions involving hyperons in the inner core, which is predicted to kill the instability unless the star is less masive than about $1.2M_\odot$ [67,68]; and (iv) nonlinear mode-mode coupling, specifically a 3-mode interaction, which is predicted to damp out r-modes at a maximum dimensionless amplitude $\sim 10^{-3}(\text{spin frequency}/1000\text{Hz})^{5/2}$ [69]. It therefore appears that r-modes will not be significant GW sources, after all.

2.8 Gamma Ray Bursts

Cosmic gamma-ray bursts, observed $\sim$ once per day by detectors on spacecraft, are thought to be produced by extremely hot shocks in a relativistic fireball, which is ignited by rapid inflow of gas onto a newborn black hole [70]. The fireball is so dense that electromagnetic observations cannot probe the BH's vicinity; the observed gamma rays are produced at radii $\sim 10^{13}$ to 10^{15} cm — millions to billions of times larger than the hole itself. The only hope for probing the fireball's trigger is via gravitational waves: The BH and its accretion flow are thought to form violently, by the collapse of a massive star (perhaps initiated by merger with a companion) ["a hypernova"], or by the merger of a binary made of compact objects: a NS/NS binary, a BH/NS binary, a BH/white-dwarf or a BH/He-core binary [70]. Each of these gamma-burst triggers should emit strong gravitational waves that carry detailed information about its source.

The distribution of γ-ray bursts is bimodal: there are *long bursts* (duration $\gtrsim$ 2 s) and *short bursts* ($\lesssim$ 2 s). Beppo-Sax has detected afterglows from several long bursts, allowing their redshifts to be measured. No redshifts are known for short bursts, but their distance can be estimated from the $\langle V/V_{max}\rangle$ statistic [71] (assuming a standard profile, and that their event rate tracks the star formation rate). The best event rate estimates (by M. Schmidt [71]) currently are: 1/yr to $D = 170(\theta_0/0.1)^{2/3}$ Mpc for long bursts, and 1/yr to D = $250(\theta_0/0.1)^{2/3}$ Mpc for short bursts. Here the gamma-rays are assumed beamed into solid angle $2\pi\theta_0^2$. Clearly, these rates (for $\theta_0 = 0.1$) are quite consistent with the estimated NS-NS merger rates given in Box 1, and marginally consistent with the BH-NS rate.

If BH-NS mergers are indeed responsible for either long or short bursts, the (WB) LIGO II detection rate would be $\sim 15 - 50$/yr. Of course, these closest sources will generallly not be seen in γ-rays, since they are typically beamed away from us; the closest burst for which we see the γ-rays (in 1 yr) is at $D = 1.0(1.5)$ Gpc for long (short) bursts. Could these be detected in coincidence with GW inspiral events? Box 1 states that LIGO II (WB) could detect BH-NS (NS-NS) inspirals to 0.65 (0.30) Gpc. But those distances assume a random orientation of the source, and assume that $S/N > 8$ is required for confident detection. If the γ-rays are beamed perpendicular to the orbital plane, then in cases where we detect the burst, we are also at optimal "viewing angle" for the GW's. The amplitude h of the GW's beamed "up the axis" are $\sqrt{\frac{5}{2}} \sim 1.58$ larger than the rms value (averaged over the sky, as seen from the source). Also, the near-coincident detection of a γ-ray burst in time (within several seconds of each other) and on a consistent area of the sky, effectively reduces by another factor ~ 1.5 the GW S/N required for detection with 99% confidence [72]. Thus the distances out to which LIGO I and II could detect binary mergers associated with γ-ray bursts are ~ 2.4 times farther than the distances given in Box 1; for LIGO II: 720 Mpc for NS/NS and 1.55 Gpc for NS/BH mergers. Thus, if NS/BH mergers are responsible for long and/or short bursts, with LIGO II one might expect ~ 2/yr simultaneous detections in both GW's and γ-rays. If NS/NS mergers are responsible, one might expect to wait ~ 5 yr.

M. van Putten[73] points out that in the above γ-ray burst scenarios, the accretion torus itself might develop large non-axisymmetries and so be a strong source of GW's. He develops a picture of long bursts as due to "suspended" accretion around a rapidly rotating BH, with most of the BH's angular momentum getting transferred to the torus via Maxwell stresses (similar to the Blandford-Znajek effect). If the torus develops "lumps", van Putten estimates that $\sim$ 10% of the BH's spin energy could get radiated away in GW's–so $E_{GW} \sim 0.3 M_{\odot}(M_{BH}/10 M_{\odot})$. He predicts a sinusoidal waveform with frequency $f \sim$ 1–2 kHz that wanders slowly in time — which is simple enough to allow near-optimal data analysis. Would this be detectable? The rms S/N would then be (taking $z \ll 1$ and approximating f and $S_h(f)$ as constants over the signal band, for simplicity),

$$\frac{S}{N} = \frac{\sqrt{2E_{GW}}}{\pi D f^{1/2} h_n(f)}.$$

yielding $S/N = 1.4$ for $E_{GW} = 0.3 M_{\odot}$, $D = 150$ Mpc, $f = 10^3$Hz, and $h_n(\equiv \sqrt{5 f S_h(f)}) = 8 \times 10^{-22}$ (LIGO II WB value at 1 kHz). This does not seem especially promising, but given the uncertainty in E_{GW} and D (for the closest source in one year), it seems worth looking for. Moreover, one could improve the sensitivity in LIGO-II by a factor of a few by reshaping the noise curve so it is lower at $f \sim 1$ kHz, at the price of higher noise at lower frequencies (cf. the difference between the NB and WB LIGO-II curves in Fig. 1).

If gravitational waves are detected from one or more gamma-burst triggers, the waves will almost certainly reveal the physical nature of the trigger. Moreover, by comparing the arrival times of the gravitational waves and the earliest gamma rays, it should be possible to measure the relative propagation speeds of light and gravitational waves to an accuracy ~ 1 sec$/10^{10}$ yr $\sim 10^{-17}$.

If no gravitational waves are detected from any individual gamma burst, then (as Mohanty, Finn and Romano [74] have shown) the correlation between gamma bursts and gravitational waves might nevertheless be established by statistical studies of the advanced IFOs' gravitational-wave data in narrow time windows preceeding the gamma bursts.

2.9 Sources of Stochastic Background

Searches for a GW stochastic background with Earth-based detectors begin with the assumptions that 1) the GW background will certainly be smaller than the instrumental noise at all frequencies, and 2) the instrumental noise level will not be known sufficiently well *a* priori that one can simply search for "excess noise" in each instrument. Thus, stochastic background searches with Earth-based detectors are based on cross-correlating the outputs of two highly independent detectors. (The two 4 km LIGO detectors at Hanford, Washington and Livingston, Louisiana give the best sensitivity).

The most plausible sources of a stochastic gravitational-wave background in LIGO's frequency band are processes in the very early universe. The current best limit on the strength of such waves is $\Omega \lesssim 6 \times 10^{-6}$, where $\Omega \equiv \int \Omega(f) d(\ln f)$ is the integrated GW energy density from all frequency bands. A wave energy larger than this would have caused the universe to expand too rapidly through the era of primordial nucleosynthesis (universe age $\sim$ a few minutes), thereby distorting the universal abundances of light elements away from their observed values. Other known bounds on $\Omega(f)$ are: 1) $h_{100}^2 \Omega(f \sim 10^{-8}\text{Hz}) < 10^{-8}$ from pulsar timing; and 2)$h_{100}^2 \Omega(f) < 7 \times 10^{-11}(f/H_0)^{-2}$ for $3 \times 10^{-18} h_{100}\text{Hz} < f < \times 10^{-16} h_{100}\text{Hz}$, from microwave background anisotropy measurements. Here h_{100} is the Hubble constant in units of 100 km/s/Mpc.

The LIGO II interferometers would improve the current limit at $f \sim 40$ Hz by a factor $\sim 10^3$, to $\Omega(f \sim 40\text{Hz}) \simeq 5 \times 10^{-9}$ — an improvement enabling LIGO to test a number of current speculations about the very early universe. Inflationary models of the early universe predict that vacuum fluctuations, created in the Planck era when the universe was being born, should have been parametrically amplified, during the first $\sim 10^{-25}$ sec of the universe's life, to produce a stochastic gravitational wave background in the LIGO band. Unfortunately, if "standard" inflation theory is correct, then the amplified waves are much too weak for LIGO to detect, $\Omega \lesssim 10^{-15}$ [75]. The most plausible modifications of standard inflation push Ω downward from this [75], but some less plausible modifications push it upward, to the point of detectability [76].

The first, tentative efforts to combine superstring theory with inflationary ideas have produced a new description of the very early universe called the "pre-big-bang model", in which string effects cause the gravitational-wave spectrum to rise steeply at high frequencies — most likely at frequencies above LIGO's band, but quite possibly in or below that band [77]. The result could be waves strong enough for LIGO to detect. A non-detection would significantly constrain the pre-big-bang model.

There are a wide variety of postulated mechanisms that could have produced strong gravitational waves, with wavelengths of order the horizon size, at various epochs in the very early universe. Those waves (if any) produced at (universe age)$\sim 10^{-25}$ sec, corresponding to (universe temperature T) $\sim 10^9$ GeV, would have been redshifted into the LIGO band today and might be detectable. The temperature (energy) region $\sim 10^9$ GeV is *tera incognita*;

LIGO's advanced detectors will provide our first opportunity for a serious experimental exploration of it. Among the speculated wave-generating mechanisms that could operate there, and that LIGO could constrain (or discover!), are these:

- A first-order phase transition in the states of quantum fields at $T \sim 10^9$ K. Such a phase transition would nucleate bubbles of the new phase that expand at near the speed of light and collide to produce gravitational waves; and their collisions would also generate turbulence that radiates waves. If the transition is strongly first order, the waves would be strong enough for LIGO's advanced IFOs to detect [78].
- Goldstone modes (coherent, classical excitations) of scalar fields that arise in supersymmetric and string theories. If strongly excited, these modes will entail coherent flows of energy that radiate gravitational waves strong enough for detection [79].
- Coherent excitations of our 3+1 dimensional universe, regarded as a "brane" (defect surface) in a higher dimensional universe. The excitations could be of a "radion" field that controls the size or curvature of the additional dimensions, or they could be of the location and shape of our universe's brane in the higher dimensions; in either case, if there is an equipartition of energy between these excitations, in the very early universe, and other forms of energy, then the excitations will produce gravitational waves easily strong enough for detection by LIGO's advanced IFOs. LIGO would thereby probe one or two additional dimensions of size or curvature length $\sim 10^{-10}$ - 10^{-13} mm; by contrast, LISA's lower-frequency observations would probe lengths ~ 1 - 10^{-5} mm. If the number of extra dimensions is larger than 2, the probes reach to much smaller scales [80,81]

Cosmic strings (not to be confused with superstrings), produced in the early universe, were once regarded as candidates for seeding galaxy formation, but recent cosmological observations have ruled them out as seeds. Nevertheless, it remains possible that a network of vibrating cosmic strings too weak to seed galaxy formation was formed in the early universe. LIGO can search for the presence of such a network in two ways: (i) Via the stochastic background of gravitational waves that the strings' vibrations produce; this background would be strong enough for the advanced IFOs to detect if the strings' mass per unit length is $\gtrsim 10^{-8}$. [82] (ii) Via occasional, individually detectable, strong non-Gaussian bursts ("spikes") of gravitational waves produced by kinks or cusps along the string [83].

Cusps are points on the string that, at one instant, move at the speed of light; kinks typically arise when two strings cross each other and interconnect. We now briefly discuss the associated GW spikes as analyzed by Damour and Vilenkin [83]. These signals have no free parameters, except the overall amplitude and the arrival time t_c: $h(t) \propto |t - t_c|^{1/3}$ for cusps and $h(t) \propto |t - t_c|^{2/3}$ for kinks. The time spent in the detector band is $\sim 1/f_c$ where f_c is the center of the band. Matched filter searches should therefore be trivial, with coincidence between detectors (and other vetoes) used to remove instrumental bursts. Cusp bursts are more highly beamed than kink bursts, have higher amplitude, and are more detectable (the higher amplitude more than compensates for the decreased probability that any given cusp burst is beamed in our direction). Cusp bursts should be detectable for a large range of string tension μ not otherwise constrained: LIGO II should detect cusp bursts for $\mu \gtrsim 10^{-11}$, while LISA could detect $\mu \gtrsim 10^{-13}$. Burst searches thus provide a stronger limit on cosmic strings than searches for a stochastic, Gaussian background.

Each of these cosmological speculations is plausible, though not highly likely. (For further explication of most of them, we highly recommend the review by Allen [84].) Perhaps their greatest value is to remind us of how terribly ignorant we are of physics and astrophysics in the domain that LIGO's advanced IFOs will probe. Our ignorance may well be even greater than that of the pioneering radio astronomers of the 1930s and X-ray astronomers of the 1960s; and as there, so also here, the first waves to be discovered may well be from sources that were previously unknown. LIGO II and its partners could bring us a revolution of insights into the universe comparable to the revolutions wrought by radio and X-ray astronomy.

3 Low-Frequency Sources Sought by LISA

3.1 LISA's Noise Curve and Conventions for Wave Strengths

In discussing low-frequency sources, we shall place them in the context of the sensitivity of LISA, the *Laser Interferometer Space Antenna* [85]. As is discussed in Schutz's chapter in this volume [9], LISA consists of three drag-free spacecraft at the corners of a 5-million-kilometer triangle. The spacecraft track the distances to each other using laser beams, thereby searching for gravitational waves in the 10^{-4}–1 Hz low-frequency band. LISA is tentatively scheduled for launch in 2011 as a joint European-American mission with a nominal lifetime of 5 years, though it could well continue to collect GW data much longer than that. By the time LISA begins its GW searches, the second-

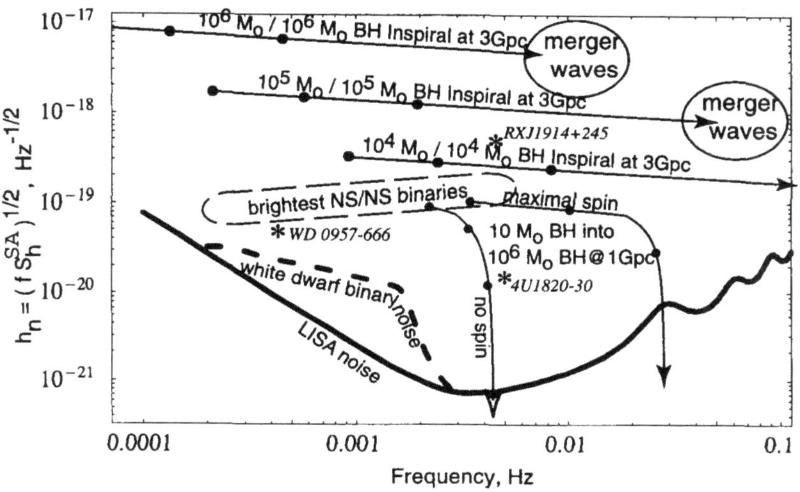

Figure 4. LISA's sky- and polarization-averaged rms noise $h_n(f)$ in a bandwidth equal to frequency, [86] compared with the strengths of the waves from several low-frequency sources. The wave strengths are plotted so the height above the noise curve is the S/N for a wave search using optimal signal processing; see text for further detail. For frequency-sweeping waves (arrowed curves) the dots are, from left to right, the signal strength and frequency 1 year, 1 month, and one day before the end. The thick-dashed curve is a background of waves from WD-WD binaries that are so numerous they cannot be removed from the data, and so constitute a noise when searching for other waves.

generation earth-based detectors (e.g. the advanced LIGO-II IFOs) described in the last section could be nearing the end of their observations, and a transition to a third generation might be near at hand.

Figure 4 depicts LISA's noise and compares it with the wave strengths for various low-frequency sources, in much the same manner as Fig. 1 depicted LIGO's noise and compared it with high-frequency wave strengths. However, there are important differences in conventions between the LISA Fig. 4 and the LIGO Fig. 1:

First: **For LIGO** we plotted vertically the square root of the spectral density of the fractional arm-length difference, $\tilde{h}(f) \equiv \sqrt{S_{\Delta L/L}(f)}$ (noise per root Hz) for a single 4 km IF0. **For LISA**, by contrast, we plot the rms

GW strain noise in a bandwidth equal to frequency $\Delta f = f$, averaged over the sky and over GW polarizations, for the full 3-spacecraft LISA system: $h_n = \sqrt{f S_h^{\rm SA}(f)}$, where $S_h^{\rm SA}(f)$ is the noise spectral density inverse-averaged over the sky and polarizations [87]. (For LIGO the analogous quantity is $h_n = \sqrt{5f}\tilde{h}$, where the $\sqrt{5}$ accounts for the average over the sky and polarizations.) In many discussions of LISA a different quantity is used to characterize the noise: the amplitude sensitivity to periodic sources for a one-year integration time and a signal-to-noise ratio of 5, averaged over the sky and over source polarizations. That quantity, $h^{\rm SA}_{\rm SN5,1yr}$, is related to our rms noise h_n by

$$h_n = \frac{\sqrt{f \times 1{\rm yr}}}{5} h^{\rm SA}_{\rm SN5,1yr} \,. \tag{2}$$

Second: Our wave-strength convention **for LISA** (Fig. 4) is also different from that for LIGO (Fig. 1): The height of a source point or curve above the LISA noise curve is equal to the signal-to-noise ratio S/N in a search for the source's waves using optimal signal processing. By contrast, **for LIGO** (Fig. 1), the height above the noise curve is the signal-to-threshold ratio S/T, with the threshold being that at which the false alarm probability is one per cent when using the best currently known, practical data analysis algorithm (typically $T \gtrsim 5N$). The difference in convention is motivated by the fact that we know little as yet about practical LISA data analysis algorithms and thus don't know with much confidence how to set thresholds, whereas LIGO data analysis is fairly mature.

For LISA, when evaluating the S/N's that determine the height of the source points in Fig. 4, we assume for periodic sources (e.g. white dwarf binaries) an integration time of five years (the nominal mission lifetime), and for sources with a large frequency sweep, we assume at a given frequency f an integration time that is the shorter of the time until the signal stops, and the time to sweep through a frequency band $\Delta f = f$.

Notice that LISA's minimum noise $h_n \simeq 10^{-21}$ is nearly the same as that of LIGO's first interferometers (Fig. 1 with $h_n = \sqrt{5f}\tilde{h}$), but at 100,000 times lower frequency f. Since the waves' energy flux scales as $f^2 h^2$, this corresponds to 10^{10} better energy sensitivity than LIGO-I.

LISA can detect and study, simultaneously, a wide variety of different sources scattered over all directions on the sky. The key to distinguishing the different sources is the different time evolution of their waveforms. The key to determining each source's direction, and confirming that it is real and not just noise, is the manner in which its waves' amplitude and frequency are modulated by LISA's complicated orbital motion—a motion in which the interferometer triangle rotates around its center and around the normal to the ecliptic plane once per year as the spacecraft orbit the sun; see the discussion by Schutz [9] elsewhere in this volume. Most sources will be observed for of order a year or longer, thereby making full use of these modulations.

Our understanding of sources and data analysis for LISA has evolved considerably since GR15. The principal sources are (i) short-period stellar-mass binaries — both galactic and extragalactic, (ii) the inspiral and merger of massive ($\sim 10^6 M_\odot$) BH binaries, (iii) the inspiral and capture of stellar-mass compact objects (white dwarfs, NS's and BH's) into massive BH's, (iv) the gravitational collape of supermassive stars, and (v) a stochastic GW background from the early universe. We discuss these in turn in the following sections.

3.2 Short-Period Binaries

Unlike the case for ground-based detectors, there are known GW sources that LISA is certain to detect at its planned sensitivity: short-period binary stars in our galaxy. There are currently about a dozen galactic binaries with GW frequencies above 0.1 mHz that LISA would be able to detect and study in detail. These include: [88] (i) White-Dwarf / white-dwarf (WD/WD) binaries; the most detectable known example is WD 0957-666 (shown in Fig. 4), with masses 0.37 and $0.32 M_\odot$, distance from Earth 100 pc, and time to merger 2×10^8 years. (ii) Am CVn stars — a white dwarf (WD) that accretes from a low-mass helium-star companion; the most detectable known example is RXJ1914+245 (shown in Fig. 4), a $0.07 M_\odot$ helium star orbiting a $0.6 M_\odot$ WD at 100pc distance from Earth. (iii) Low-mass X-ray binaries — a NS that accretes from a low-mass companion; the most detectable known example is 4U1820-30 (shown in Fig. 4), a $< 0.1 M_\odot$ star orbiting a $1.4 M_\odot$ NS at 8100 pc from Earth. These sources will provide an important test that LISA is functioning as expected (as well as a test of the general relativistic theory of GW emission, in the weak-gravity limit).

These optically-known binaries are just the tip of the iceberg; there are an estimated 10^{8-9} galactic binaries with GW frequencies $f > 0.1$ mHz (most are WD/WD). At frequencies below ~ 2 mHz, there are many binaries per resolvable frequency bin $\Delta f = 1/T_{obs} \sim 10^{-8}$Hz — too many to be fit and removed from LISA's data, so they constitute a source of "confusion noise" that will dominate over the instrumental noise at these frequencies. Thus LISA truly suffers from an "embarrassment of riches." But there is also some science to be extracted from this confusion noise: because the WD/WD background is anisotropic, the confusion noise level is modulated by LISA's time-varying antenna pattern on the sky. From this one can extract with few-percent accuracy the $m = 1$ and $m = 2$ moments (in spherical coordinates centered on the Sun) of the WD/WD distribution [89].

Above $\sim$ 2 mHz the density of sources thins to less than $\sim$ 0.5 per frequency bin, and one expects to be able to measure individually all these binaries, about 10^4 of them, and remove them from the data. The confusion noise therefore drops dramatically above $\sim$ 2 mHz — to a level set by the extragalactic binary background (about 10 times smaller than the galactic background) — and crosses below the instrumental noise curve; see Fig. 4.

Although it should be possible to remove the binaries above 2 mHz from LISA's data so as to search there for other waves (e.g. from inspiral of small mass BH's into supermassive BH's; cf. Fig. 4), no effort has yet been put into the task of developing practical data analysis algorithms to perform the removal. Designing and optimizing such algorithms is an important challenge for the next few years.

For the $\sim 10^3$ highest-frequency binaries, one should be able to measure the binary's rate of change of frequency $\dot{f}$ (due to inspiral driven by GW emission). From the measured $\dot{f}$ and f one can infer the binary's chirp mass M_c, and from M_c and the measured amplitudes of the binary's two polarizations once can infer its inclination to the line of sight and its distance. For binaries with $f \sim 10^{-2}$ Hz, one expects to measure the angular location to $\sim 1°$ [90] and the distance to $\sim 2\%$ [91]. Thereby one can obtain a 3-D map of the galaxy with distances unsullied by problems of dust, etc.

3.3 Mergers of Supermassive Black Holes

The merger of two supermassive black holes (SMBHs; $10^4 - 10^7 M_\odot$) at $z = 1$ could be detected by LISA with S/N of thousands; see Fig. 5. This S/N level is at least 10 times greater than we can expect with ground-based detectors (Figs. 2 and 3) — so high that the inspiral portion of the waveform could typically yield the binary parameters to the following accuracies: individual masses to $\Delta M_i/M_i \sim 10^{-4}$, the so-called spin-orbit parameter β to $\Delta\beta \sim 10^{-3}$, the distance to $\Delta D/D \sim 10^{-3} - 10^{-2}$, and the position of the source on the sky to $\sim 10^{-4}$ sr [92]. Knowing the binary's initial conditions with such high accuracy, it should be possible to make high-precision comparisons of numerical relativity simulations with the observed merger waveforms — thereby extending LIGO's exploration of the nonlinear dynamics of curved spacetime (Sec. 2.5) into a domain of far higher accuracy. For example, as the gravitational waves from a binary's merger depart from their source, the waves' energy should create (via the nonlinearity of Einstein's field equations) a secondary wave called the "Christodoulou memory" [93,94,95], which LISA

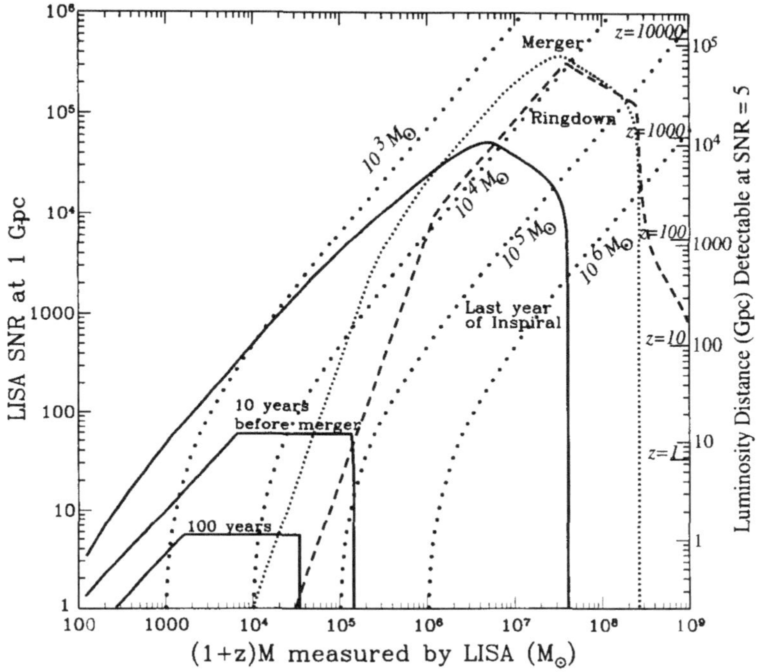

Figure 5. The waves from equal-mass, supermassive black-hole binaries as observed by LISA in one year of integration time. Plotted horizontally is the binary's total mass M multiplied by 1 plus the binary's cosmological redshift z. The solid, dashed, and tight-dotted curves refer to the left axis, which shows the estimated S/N for LISA's observations when the binary is 1 Gpc from earth (redshift $z \simeq 0.2$). The wide-spaced dots are curves of constant binary mass M, for use with the right axis, which shows the luminosity distance and redshift to which the binary can be detected, with S/N = 5 (look at the intersection of a wide-dotted curve with a solid, dashed, or tight-dotted curve and then project horizontally onto the right axis). (The assumed cosmology is H_o = 65 km/s/Mpc, $\Omega_\Lambda = 0.6$, $\Omega_M = 0.4$.) The bottom-most curves are the signal strengths after one year of signal integration, for BH/BH binaries 10 years and 100 years before their merger. (Figure adapted from Flanagan and Hughes [40].)

should easily be able to measure, while LIGO might not quite see it.

Unfortunately, it is far from obvious whether the event rate for such SMBH mergers will be interestingly high: By contrast with advanced LIGO IFOs, where Kalogera and others estimate tens to thousands of BH/BH mergers per year (Box 1), Haehnelt [96] and others estimate for LISA's SMBH mergers out to $z = 1$ a rate in the range $10^{-1} - 10^{2}$ per year. Even if the coalescence rate is only 0.1/yr, then LISA should still see ~ 3 BH/BH binaries with $3000 M_{\odot} \lesssim M \lesssim 10^{5} M_{\odot}$ that are ~ 30 years away from their final merger. These slowly inspiraling binaries should be visible, with one year of integration, out to a redshift $z \sim 1$ (bottom part of Fig. 5).

Because the SMBH-merger S/N is so high at $z = 1$, LISA could detect such mergers to redshifts $z \gg 1$. SMBH's are not expected to form in the universe at redshifts any larger than $z \sim 30$, and perhaps somewhat less than this. Figure 5 shows that the merger waves can be detected to distances well beyond $z \sim 30$ for masses in the range $M \sim 10^{4}$ to $\sim 10^{7} M_{\odot}$. (Below a few times $10^{5} M_{\odot}$ a crucial role in making this possible is played by the redshift itself, which lowers the waves' frequency, bringing them into LISA's band of good sensitivity; see Fig. 4.)

In the standard picture of bottom-up structure formation, large galaxies are formed by the merger of protogalaxies (which are themselves formed from the mergers of smaller protogalaxies). If the the merging protogalaxies contain $10^{5} M_{\odot}$ BH's at their centers, dynamical friction will bring the BH's to the common center, plausibly (though not assuredly) to the point where gravitational radiation takes over and drives the BH's to merger in less than a Hubble time. Based on this picture, Phinney estimates that the merger rate for $10^{5} M_{\odot}$ BH's would be ~ 1500/yr at $z \sim 7$, ~ 500/yr at $z \sim 10$, and ~ 1/yr at $z \sim 20$ [88,97] — rates comparable to those of advanced LIGO IFOs (Box 1). Thus LISA has the possibility to reach back and shed light on the earliest epochs of galaxy and SMBH formation

3.4 Inspirals of $\sim 1 - 1000 M_{\odot}$ Compact Objects into Massive BH's

LISA has the potential to observe the final inspiral waves from compact objects (WD's, NS's and BH's) of any mass $\mu \gtrsim 1 M_{\odot}$ spiraling into central bodies of mass $3 \times 10^{5} M_{\odot} \lesssim M \lesssim 3 \times 10^{7} M_{\odot}$ out to 3Gpc. Figure 4 shows the example of a $10 M_{\odot}$ black hole spiraling into a $10^{6} M_{\odot}$ SMBH at 1 Gpc distance. The inspiral orbit and waves are strongly influenced by the SMBH spin. Two cases are shown [87]: an inspiraling circular orbit around a non-spinning hole, and a prograde, circular, equatorial orbit around a near maximally spinning hole ($a/M = 0.999$ in Kerr-metric language). For each case three dots are shown on the signal curve — from left to right the signal strength and frequency one year, one month, and one day before the small BH plunges into the horizon of the SMBH. Table 1 shows, at each of these times, the ratio of the orbital circumference to the SMBH horizon circumference, C/C_{H}, the number of GW cycles left until plunge, $N_{\rm cyc}$, and the signal to noise in a

Table 1. The waves from a $10M_\odot$ in circular equatorial orbit around a $10^6 M_\odot$ SMBH.

SMBH Spin	Time to Plunge	$\mathcal{C}/\mathcal{C}_H$	N_{cyc}	S/N
0.999	1 yr	6.80	185,000	200
0.999	1 mo	3.05	40,900	60
0.999	1 day	1.30	2,320	5
0	1 yr	9.46	85,000	60
0	1 mo	7.14	9,650	70
0	1 day	6.22	366	20

bandwidth equal to frequency (or, if that band extends beyond plunge, then the total signal to noise in the remainder of the waves), S/N.

The numbers in Table 1 are remarkable: The small BH spends its last year in the immediate vicinity of the SMBH horizon, emitting $\sim 100,000$ cycles of waves. These waves should carry, encoded in themselves, high-accuracy information about the SMBH [98]:

Ryan [99] has shown that, when any small-mass, compact object spirals into a much more massive, compact central body, the inspiral waves will carry a "map" of the massive body's external spacetime geometry. Since that geometry is uniquely characterized by the values of the body's multiple moments, we can say equivalently that the inspiral waves carry, encoded in themselves, the values of all the body's multipole moments. By measuring the inspiral waveforms and extracting their map (i.e., measuring the lowest few multipole moments), we can determine whether the massive central body is a black hole or some other kind of exotic compact object [99]. The huge number of wave cycles, in the example in Table 1, implies the possibility of a high-precision map.

While the S/N values in Table 1 are large enough (up to 200) to seem encouraging for detecting these waves and extracting the map that they carry, there are serious worries. These S/N's assume optimal signal processing (via matched filters), but it is highly unlikely that optimal processing can be achieved. The reasons are two:

First: Models of galactic nuclei, where massive holes (or other massive central bodies) reside, suggest that inspiraling stars and small holes typically will be in rather eccentric orbits [100,101]. This is because they get injected into such orbits via gravitational deflections off other stars, and by the time gravitational radiation reaction becomes the dominant orbital driving force, there is not enough inspiral left to strongly circularize their orbits. Such orbital eccentricity will complicate the waveforms and complicate the extraction of information from them. In fact, the waveforms are likely to be so complex that the number of discrete templates needed in a in one-year coherent-integration search for these waves may be enormously larger than modern computers can handle. As a result, the signal processing may have to be done coherently over short stretches of time with incoherent techniques used to combine successive stretches, leading to a significant loss of S/N. Efforts to generate waveforms from generic orbits and then scope out this problem are barely getting started, as of winter 2001–02. (The waveforms are fairly well understood for generic orbits in the SMBH equatorial plane [102] and for circular orbits inclined to the equator, [103] but inclined generic orbits have not yet been studied, and the counting of templates even for equatorial orbits and inclined circular orbits has barely begun [104]. Although the influence of radiation reaction on the orbital evolution is not yet fully understood [105], it is probably known well enough to scope out the data analysis problem.)

Second: There may be a significant loss of S/N due to "source confusion" — i.e. due to the problem of separating out individual inspiral waveforms from the background due to other, weaker inspiral sources. At high frequencies this confusion noise could perhaps dominate all other noise sources, just as WD/WD confusion noise dominates at $10^{-4} - 10^{-3}$ Hz (Fig. 4). Efforts to scope this out are also barely getting started.

The event rates for inspirals into SMBH's are not well known. Phinney [97] (relying in part on simulations of star-cluster evolution around a SMBH by Freitag [106]) has shown that the detection rate is likely to be dominated by $\sim 10M_\odot$ BH's spiraling into $\sim 10^6 M_\odot$ SMBH's. (Phinney shows that $\sim 10^6 M_\odot$ SMBH's are likely to have larger relaxed stellar cusps around them than either smaller or bigger BH's; coincidentally $10^6 M_\odot$ is the optimum SMBH mass for LISA. And while inspirals of $\sim 10M_\odot$ BH's probably have a smaller space density than inspirals of $\sim 1.4M_\odot$ NS's, because of their larger S/N, they can be detected throughout a volume that is $(\sqrt{10/1.4})^3 \sim 20$ times larger.)

The estimated inspiral rate for $\sim 10M_\odot$ BH's is $\sim 1 - 10$/yr out to 1 Gpc [97,101], where the S/N's would be ~ 60 to ~ 200 *if optimal signal processing were possible.* If the loss of S/N due to signal-processing problems is much larger than an order of magnitude (which it might be), then there could be difficulty detecting these inspirals. This motivates exploring a change of LISA's design so its noise curve is lowered by a factor 4. Since such a change might significantly increase LISA's cost, it is greatly hoped that a scoping of the data analysis problem and a firming of the event rate will reveal that no design change is needed!

3.5 Collapse of Supermassive Stars

New and Shapiro [107] have investigated the possibility of substantial GW emission from the collapse of supermassive stars (SMS's) to form SMBH's in galactic nuclei. The formation mechanism for SMBH's is still highly uncertain; one path could lead through rotating SMS's. While non-rotating SMS's suffer from a well-known radial collapse instability (arising from GR effects), even a small amount of rotation can stabilize the star [108]. As the SMS cools and contracts, the ratio of $\beta = T/|W|$ (ratio of kinetic to gravitational binding energy) grows, and (for sufficiently large initial β) the star eventually becomes unstable either to mass-shedding at the equator or to a dynamical, $m = 2$ (bar-mode) instability. New and Shapiro show that if viscosity and magnetic fields are insufficient to enforce uniform rotation (so that angular momentum is conserved on cylinders during the contraction), then the bar-mode instability is likely to occur first.

If the bar-mode perturbation grows to order unity and is long-lived, a substantial fraction of the star's rest mass-energy could be radiated in GW's. These waves end up in the LISA band for a range of plausible initial parameters. For example, for SMS mass $M \sim 10^6 M_\odot$, initial (pre-collapse) radius $R \sim 10^{17}$cm and initial $\beta \sim 10^{-6}$, New and Shapiro [107] estimate the SMS could radiate away $\sim 10^3 M_\odot$ in GW's at $f \sim 3 \times 10^{-4}$Hz, in a "burst" lasting $\hat{\tau} \sim 1$yr, which is $n = f\hat{\tau} \sim 10^4$ cycles. *If* optimal signal processing could be used to search for these waves, the signal to noise ratio would be $S/N \simeq h_c/h_n$ where $h_c \equiv h\sqrt{n} \sim 10^{-18}$ at 1Gpc distance; here h is the wave amplitude as estimated by New and Shapiro (corrected by them from a different value given in their paper), and h_c is called the waves' *characteristic amplitude* [109]. Comparing with LISA's noise curve (Fig. 4) we see that $S/N \sim 40$ for optimal signal processing. With sufficiently accurate templates from numerical simulations of the collapse, it might be possible to achieve a search within a factor 4 of optimal, so $S/N \sim 10$, which would be adequate for confident detection; but as yet it is far from clear one can do so well — especially as the search must be performed in the thick of the WD-WD background. To get a clearer handle on the prospects will require considerable analysis — both source simulations and data-analysis explorations.

3.6 LISA and the GW Stochastic Background

While ground-based detectors can search for a stochastic background by cross-correlating their outputs, this method is not available to LISA, since only one space-based detector is currently planned.

However, it is possible to combine the signals from the 3 spacecraft to construct a "symmetrized Sagnac observable", which is practically insensitive to both laser phase noise and gravitational waves [86]. With the aid of this observable, one can directly measure (symmetrized averages of) the proof-mass noise and optical-path noise, and thereby determine the instrumental noise power in the (sum of the) 3 Michelson channels used to detect GW's. Any excess "noise" above this power is due to GW's. This method is most effective in frequency bands where either the proof-mass noise or optical-path noise greatly dominates over the other. With this approach, in a one-year integration LISA could detect $h_{100}^2\Omega \gtrsim 10^{-11}$ in either the $10^{-4.5} < f < 10^{-3.5}$Hz or $10^{-2.8} < f < 10^{-1.6}$Hz bands [110]. Unfortunately, LISA's sensitivity to a GW background of *early-universe* origin is about an order-of-magnitude worse than this, since any primordial source with $\Omega(f) \lesssim 10^{-10}$ will be "covered up" by (the isotropic portion of) the galactic and extragalactic WD/WD binary backgrounds. LISA's low-frequency (LF) sensitivity of $\Omega \sim 10^{-10}$ is about two orders of magnitude better than the adjacent-band sensitivities of advanced LIGO IFOs ($\sim 5 \times 10^{-9}$ in the HF band) and current pulsar timing ($\sim 10^{-8}$ in the VLF band); Sec. 2.9.

In Sec. 2.9 we discussed a number of speculative early-universe sources that could produce GWs in earth-based detectors' HF band. Variants of all these sources could also radiate in LISA's LF band: (i) Universal inflation (with the standard model predicting waves much too weak for LISA to see, $\Omega \lesssim 10^{-16}$–10^{-15}, while some less plausible models such as the pre-big-bang suggest stronger waves); (ii) Phase transitions of quantum fields (LISA's frequency band is ideal for probing the electroweak phase transition, which should have occurred at universe age $\sim 10^{-12}$ sec and if strongly first order could have produced waves strong enough for LISA to see [111]); (iii) Goldstone modes of scalar fields that arise in supersymmetric and string theories; (iv) excitations of our universe as a brane in higher dimensions (with LISA probing excitation lengthscales ~ 1–10^{-5} mm if there are one or two relevant higher dimensions); and (v) Cosmic strings (with LISA's searches for bursts from string cusps and kinks able to constrain string tensions to $\mu \lesssim 10^{-13}$). For further discussion and references, see Sec. 2.9.

There is some hope that a follow-on mission to LISA might reach the sensitivity required for probing standard inflation, $\Omega(f) \sim 10^{-16} - 10^{-15}$. To do so will require that the mission's detectors operate outside the regime dominated by close binaries. A "short-LISA" with ~ 100 times shorter arms would have good sensitivity at $f \sim 1$Hz, where the WD binaries disappear and the binary background is dominated by NS binaries; a constellation of 3 short-LISA's spaced evenly around the Sun (along Earth's orbit) could have angular resolution $\Delta\theta \sim 10^{-3}$ rad, allowing (probably) the NS binaries to be detected individually and subtracted out. At present, such a constellation of $\sim 9-12$ satellites seems the best possibility for probing the inflationary GW background from space.

4 Conclusions

It is now about 40 years since Joseph Weber initiated his pioneering development of gravitational-wave detectors [112], and 30 years since Robert Forward [113] and Rainer Weiss [114] initiated work on interferometric detectors. Since then, hundreds of talented experimental physicists have struggled to improve the sensitivities of gravitational-wave detectors, and hundreds of theorists have explored general relativity's predictions.

These two parallel efforts are now intimately intertwined and are pushing toward an era in the not distant future, when measured gravitational waveforms will be compared with theoretical predictions to learn how many and what kinds of relativistic objects *really* populate our Universe, and how these relativistic objects *really* are structured and *really* behave when quiescent, when vibrating, and when colliding. By about the time of GR-18, if not sooner, we should have some answers.

Acknowledgments

Much of this manuscript's Section 2 (high-frequency waves) was written in close consultation with Lars Bildsten, Alessandra Buonanno, Craig Hogan, Vassiliki Kalogera, Benjamin J. Owen, E. Sterl Phinney, Thomas A. Prince, Frederic A. Rasio, Stuart L. Shapiro, Kenneth A. Strain, Greg Ushomirsky, and Robert V. Wagoner. We warmly thank them for their significant contributions to this review. We also thank Sam Finn for especially valuable input into our sections on supernovae, pulsars, and gamma ray bursts. Thorne's contributions to this review were supported in part by NSF grant PHY-0099568 and NASA grant NAG5-10707. Cutler's work was partly supported by NASA grant NAG5-4093.

References

1. E. E. Flanagan. In Naresh Dadhich and Jayant Narlikar, editors, *Gravitation and Relativity: At the turn of the Millennium, 15th International Conference on General Relativity and Gravitation.* Inter-University Centre for Astronomy and Astrophysics, 1998.

2. Lee Samuel Finn. In Joan M. Centrella, editor, *Astrophysical Sources for Ground-based Gravitational Wave Detectors*, volume 575 of *AIP Conference Proceedings*, Melville, New York, 2001. American Institute of Physics. gr-qc/0104042.
3. S. Hughes, S. Marka, P.L. Bender, and C.J. Hogan. In *Proceedings of the 2001 Snowmass Meeting*. astro-ph/0110349.
4. K.S. Thorne. In R.M. Wald, editor, *Black Holes and Relativistic Stars*, page 41. University of Chicago Press, 1998.
5. K.S. Thorne. The scientific case for advanced ligo interferometers. 2001. LIGO Document Number P-000024-00-R; available from LIGO Document Control Center, http://admdbsrv.ligo.caltech.edu/dcc/.
6. M. Kamionkowski and A. Jaffe. *Int. J. Mod. Phys. A*, 16:116, 2001.
7. A.N. Lommen and D.C. Backer. *Astrophys. J.*, 2002. in press; astro-ph/0107470.
8. S. Kawamura. 2002. article in this volume.
9. B.F. Schutz. 2002. article in this volume.
10. N.A. Robertson. *Class. Quant. Grav.*, 17:R19, 2000.
11. B. Barish and R. Weiss. *Physics Today*, 52(10):44, 2001.
12. M. Tinto and J.W. Armstrong. *Phys. Rev. D*, 58:042002, 1998.
13. See http://igec.lnl.infn.it/.
14. This $\tilde{h}_s(f)$ is related to the *characteristic amplitude* $h_c(f)$ widely used in the literature in the following way: $h_c = \sqrt{5f}(T/N)\tilde{h}_s$, where the $\sqrt{5}$ comes from averaging over the sky, the $\sqrt{f}$ is due to h_c being the signal strength in a bandwidth equal to frequency, and T/N is the ratio of the threshold to the rms noise at the endpoint of signal processing.
15. If, in its advanced ("LIGO-II") incarnation, the 2km interferometer is replaced by a narrow-banded 4km interferometer, then for the most important wide-band sources, inspiraling binaries, the 3-IFO threshold will be lowered (improved) slightly and correspondingly the distances to which the sources can be seen will be increased slightly beyond those in Box 1. If the 2km interferometer is replaced by a wide-band 4km interferometer, then the threshold will be lowered by a factor 1.16, the binaries' observable distances will be increased correspondingly by 1.16, and the predicted event rates will go up by $1.16^3 \simeq 1.5$.
16. See, e.g., B.J. Owen and B.S. Sathyaprakash. *Phys. Rev. D*, 60:022002, 1999. and references therein.
17. L. Bildsten. *Astrophys. J. Lett.*, 501:L89, 1998.
18. G. Ushomirsky, C. Cutler, and L. Bildsten. *Mon. Not. Roy. Astron. Soc.*, 319:902, 2000.
19. M. van der Klis. *Ann. Rev. Astron. Astrophys.*, 38:717, 2000.

20. P.R. Brady and T. Creighton. *Phys. Rev. D*, 61:082001, 2000.
21. M.A. Papa, B.F. Schutz, and A.M. Sintes. In *Proceedings of ICTP Conf. on Gravitational waves: a challenge to theoretical astrophysics.* gr-qc/0011034.
22. P.R. Brady, T. Creighton, C. Cutler, and B.F. Schutz. *Phys. Rev. D*, 57:2101, 1998.
23. B. Allen and J.D. Romano. *Phys. Rev. D*, 59:102001, 1999.
24. See the discussion of the overlap reduction function in Sec. III.B of Allen and Romano (1999), especially Fig. 2.
25. The NS/NS numbers in this table are taken from V. Kalogera, R. Narayan, D. Spergel, and J.H. Taylor, Astrophys. J., 556:340, 2001 — as augmented slightly by a new channel for NS/NS formation analyzed by K. Belczynski and V. Kalogera, Astrophys. J. Lett., 550:L183, 2001 — and with adjustment of $\mathcal{D}_{\rm WB}$ from 350 Mpc (appropriate for three 4 km interferometers) to 300 Mpc (for one 2 km and two 4 km interferometers [15]). The NS/BH and BH/BH numbers are based on Kalogera's survey of all estimates of $\mathcal{R}_{\rm gal}$ [V. Kalogera, private communication winter 2000-01; V. Kalogera in *Gravitational Waves, Proceedings of the Third Edoardo Amaldi Conference*, AIP Conference Proceedings Vol. 523, ed. Sydney Meshkov (American Institute of Physics, 2000), astro-ph/9911532; and V. Kalogera in *Astrophysical Sources of Gravitational Radiation for Ground-Based Detectors*, ed. J.M. Centrella (AIP, Woodbury NY), astro-ph/0101047] — extrapolated into the distant universe by the B-band method of Phinney [26] as updated by Kalogera et. al. The last column is based on the estimates by McMillan and Portegies Zwart [30]. The numbers preceded by "$\lesssim$" represent Kalogera's personal best estimate of the lower limit with the "$<$" being a warning that some researchers have argued for lower numbers.
26. E.S. Phinney. *Astrophys. J.*, 380:L17, 1991.
27. P. Madau. *Physica Scripta*, T85:156, 2000. astro-ph/9902228.
28. V. Kalogera. *Astrophys. J.*, 541:319, 2000.
29. When extrapolating out to cosmological distances, $z \gtrsim 0.1$, we must take account of cosmological effects. We assume a cosmological model with $H_o = 65$ km/s-Mpc, $\Omega_M = 0.4$, and we take account of the evolution of the numbers of massive stars and thence BH/BH binaries using Eq. (4) of Madau, Ref. [27]; the result is an event rate, inside redshift z, that scales as z^3 for $0 < z < 4$, aside from an enhancement by a factor ~ 2 near $z = 1$ and a decrement by a factor ~ 2 at $z \sim 2.5 - 4$.
30. S. McMillan and S. Portegies Zwart. *Astrophys. J. Lett.*, 528:L17, 2000.
31. C. Cutler and E. E. Flanagan. *Phys. Rev. D*, 49:2658, 1994.

32. E. Poisson and C. M. Will. *Phys. Rev. D*, 52:848, 1995.
33. T. A. Apostolatos, C. Cutler, G. J. Sussman, and K. S. Thorne. *Phys. Rev. D*, 49:6274, 1994.
34. L. E. Kidder. *Phys. Rev. D*, 52:821, 1995.
35. T. A. Apostolatos. *Phys. Rev. D*, 52:605, 1995.
36. P. Jaranowski, K.D. Kokkotas, A. Krolak, and G. Tsegas. *Class. Quant. Grav.*, 13:1279, 1996.
37. E. Poisson. *Phys. Rev. D*, 47:1497, 1993.
38. M. Vallisneri. *Phys. Rev. Lett.*, 84:3519, 2000.
39. J.A. Faber and F.A. Rasio. *Phys. Rev. D.* in press; gr-qc/0201040.
40. E. E. Flanagan and S. A. Hughes. *Phys. Rev. D*, 57:4535, 1998.
41. E. E. Flanagan and S. A. Hughes. *Phys. Rev. D*, 57:4566, 1998.
42. A. Burrows, J. Hayes, and B. Fryxell. *Astrophys. J.*, 450:830, 1995.
43. http://online.itp.ucsb.edu/online/neustars00/chernoff/, especially slide 34.
44. K. Nomoto, in Proc. 13th Texas Symposium on Relativistic Astrophysics, ed. M. Ulmer (World Scientific: Singapore); K. Nomoto and Y. Kondo, Astrophys. J. Lett., 367:L19, 1991.
45. See, e.g., recent simulations of dynamical bar formation by M. Shibata, T.W. Baumgarte and S.L. Shapiro, Astrophys. J. 542:453, 2000, and M. Saijo, J. Shibata, T.W. Baumgarte and S.L. Shapiro, Astrophys. J., 548:919, 2001, and references therein. These simulations show that general relativity enhances the instability to bar formation.
46. K.C.B. New, J.M. Centrella, and J.E. Tohline. *Phys. Rev. D*, 62:064019, 2000.
47. D. Brown. *Phys. Rev. D*, 62:084024, 2000.
48. D. Lai and S.L. Shapiro. *Astrophys. J.*, 442:259, 1995.
49. In order to shrink to normal neutron-star size, the proto-neutron-star must get rid of its excess angular momentum. In the idealized case that (i) the angular momentum is lost solely to gravitational waves and not to hydrodynamic or magnetic processes, and (ii) the LIGO data are analyzed by optimal signal processing, the integrated signal strength depends primarily on the excess angular momentum and not on the ellipticity of the bar, and the distance to which the signal can be seen is of order 150Mpc (about half that for a NS/NS binary whose mass is twice as large). A factor ten reduction of amplitude signal to noise (factor 100 reduction of power signal to noise), due to non-optimal signal processing and to loss of angular momentum via non-gravitational-wave channels, would still leave the source detectable at $\sim$ 15 Mpc, the distance of Virgo.
50. L.S. Finn. Section titled "Core-Collapse Supernovae" in Ref. [2], and pri-

vate communication about further details.
51. C. Cutler and D.I. Jones. *Phys. Rev. D*, 63:024002, 2001.
52. C. Cutler, G. Ushomirsky, and B. Link. in preparation.
53. M. Zimmermann and E. Szedenits. *Phys. Rev. D*, 20:351, 1979.
54. S. L. Shapiro and S. A. Teukolsky. Wiley: Interscience, 1983. Section 10.10 and references cited therein.
55. S. Chandrasekhar. *Phys. Rev. Lett.*, 24:611, 1970.
56. J. L. Friedman and B. F. Schutz. *Astrophys. J.*, 222:281, 1978.
57. R. V. Wagoner. *Astrophys. J.*, 278:345, 1984.
58. L. Lindblom. *Astrophys. J.*, 438:265, 1995.
59. L. Lindblom and G. Mendell. *Astrophys. J.*, 444:804, 1995.
60. L. Lindblom. *Astrophys. J.*, 438:265, 1995.
61. N. Andersson. *Astrophys. J.*, 502:708, 1998.
62. J. Friedman and S. Morsink. *Astrophys. J.*, 502:714, 1998.
63. L. Lindblom, B.J. Owen, and S. Morsink. *Phys. Rev. Lett.*, 80:4843, 1998.
64. N. Andersson, K. Kokkotas, and B. Schutz. *Astrophys. J.*, 510:846, 1999.
65. L. Rezzolla, F.L. Lamb, D. Markovic, and S.L. Shapiro. *Phys. Rev. D*, 64:104014, 2001.
66. L. Bildsten and G. Ushomirsky. *Astrophys. J.*, 529:L33, 2000.
67. P.B. Jones. *Phys. Rev. D*, 64:084003, 2001.
68. L. Lindblom and B.J. Owen. Submitted; astro-ph/0110558.
69. P. Arras, E.E. Flanagan, S.M. Morsink, A.K. Schenk, S.A. Teukolsky, and I. Wasserman. *Astrophys. J.*., 2002. Submitted; astro-ph/0202345.
70. See, e.g., P. Meszaros, M.J. Rees and R.A.M.J. Wijers, New Astron., 4:303, 1999.
71. M. Schmidt. *Astrophys. J.*, 559:L79, 2001.
72. C.S. Kochanek and T. Piran. *Astrophys. J.*, 417:L17, 1993.
73. M.H.P.M. van Putten. *Physics Reports*, 345:1, 2001.
74. L.S. Finn, S.D. Mohanty, and J. Romano. *Phys. Rev. D*, 60:121101, 1999.
75. M.S. Turner. *Phys. Rev. D*, 55:435, 1997.
76. L. P. Grishchuk. *Soviet Physics–JETP*, 40:409, 1974.
77. R. Brustein, M. Gasperini, M. Giovannini, and G. Veneziano. *Phys. Lett. B*, 361:45, 1995. For a review of more recent literature, see M. Maggiore. Phys. Rep., 331:283, 2000.
78. M. Kamionkowski, A. Kosowsky, and M.S. Turner. *Phys. Rev. D*, 49:2837, 1994.
79. See, e.g., C.J. Hogan, Phys. Rev. Lett **74**, 3105 (1995), and C.J. Hogan, astro-ph/9809364.

80. C.J. Hogan. *Phys. Rev. Lett.*, 85:2044, 2000.
81. C.J. Hogan. *Phys. Rev. D*, 62:121302, 2000.
82. B. Allen and J.D. Romano. *Phys. Rev. D*, 59:102001, 1999.
83. T. Damour and A. Vilenkin. *Phys. Rev. D.*, 64:064008, 2001.
84. B. Allen. In J. A. Miralles, J. A. Morales, and D. Saez, editors, *Proceedings of the Spanish Relativity Meeting, '96*, page 3. 1996. gr-qc/9604033.
85. P. Bender et. al. *LISA, Laser interferometer space antenna for the detection and observation of gravitational waves: Pre-Phase A Report.* Max-Planck-Institut für Quantenoptik, MPQ 208, December 1995.
86. J.W. Armstrong, F.B. Estabrook, and M. Tinto. *Class. Quant. Grav.*, 18:4059, 2001.
87. L. S. Finn and K. S. Thorne. *Phys. Rev. D*, 62:124021, 2000.
88. E.S. Phinney. Lisa science requirements, 2001. Presentation to LISA International Science Team, 12 December 2001; available at LIST Working Group 1 Website, http://www.tapir.caltech.edu/listwg1/.
89. C. Ungarelli and A. Vecchio. *Phys. Rev. D*, 64:121501, 2001.
90. C. Cutler and A. Vecchio. In W.M. Folkner, editor, *Proceedings of the 2nd International LISA Symposium*, page 95. American Institute of Physics, 1998.
91. S. Hughes, private communication.
92. A. Vecchio. In S. Meshkov, editor, *Proceedings of the Third Amaldi Conference on Gravitation al Waves*, page 238. American Institute of Physics, 2000. gr-qc/9912099.
93. D. Christodoulou. *Phys. Rev. Lett.*, 67:1486, 1991.
94. K. S. Thorne. *Phys. Rev. D*, 45:520–524, 1992.
95. A. G. Wiseman and C. M. Will. *Phys. Rev. D*, 44:R2945, 1991.
96. M. G. Haehnelt. *Mon. Not. Roy. Astron. Soc.*, 269:199, 1994.
97. E.S. Phinney, private communication.
98. F. D. Ryan. *Phys. Rev. D*, 56:1845, 1997.
99. F. D. Ryan. *Phys. Rev. D*, 52:5707, 1995.
100. D. Hils and P. Bender. *Astrophys. J. Lett*, 445:L7, 1995.
101. S. Sigurdsson and M. J. Rees. *Mon. Not. Roy. Astron. Soc.*, 284:318, 1997.
102. C. Glampedakis and D. Kennefick. 2002. In preparation.
103. S.A. Hughes. *Phys. Rev. D*, 64:064004, 2001. And references therein.
104. S. Hughes. LIST Working Group 1 Website, http://www.tapir.caltech.edu/listw g1/.
105. L. Barack, Y. Mino, H. Nakano, A. Ori, and M. Sasaki. *Phys. Rev. Lett.*, 88:091101, 2002. And references therein.
106. M. Freitag. *Class. Quant. Grav.*, 18:4033, 2001.

107. K.C.B. New and S.L. Shapiro. *Astrophys. J.*, 548:439, 2001. And private communication.
108. W.A. Fowler. *Astrophys. J.*, 184:180, 1966.
109. K. S. Thorne. In S. W. Hawking and W. Israel, editors, *Three Hundred Years of Gravitation*, pages 330–458. Cambridge University Press, 1987.
110. C.J. Hogan and P.L. Bender. *Phys. Rev. D*, 64:062002, 2001.
111. A. Kosowsky and M. S. Turner. *Phys. Rev. D*, 49:2837, 1994.
112. J. Weber. *Phys. Rev.*, 117:306, 1960.
113. G. E. Moss, L. R. Miller, and R. L. Forward. *Applied Optics*, 10:2495, 1971.
114. R. Weiss. *Quarterly Progress Report of RLE, MIT*, 105:54, 1972.

THE STATE OF COSMOLOGY 2001: TWO VIEWS AND A MIDDLE WAY

GEORGE F R ELLIS

Cosmology Group, Department of Mathematics and Applied Mathematics University of Cape Town, Rondebosch 7701, South Africa

E–mail: ellis@maths.uct.ac.za

The first part of this article gives a survey of cosmology to the present day, and then puts two radically diverging views on what has been achieved. The second part explores a middle way between these divergent views: a positive but critical approach, looking in turn at issues arising in observations, geometry, physics, and finally at the foundational (philosophical) issues that necessarily arise. For a related review, see [26].

1 Development of cosmology

1: Observations

Observational cosmology started seriously in the 1920's through optical observations of galaxies, leading to an understanding of the nature of faint 'nebulae' as galactic systems equivalent to our own Milky Way, and discovery of the vast size of the Universe. Hubble established the cosmological distance scale (later revised significantly) and the linear relation between redshift and distance, and used number counts to establish rough homogeneity of the Universe [46].

2: Geometry/relativity

The static cosmologies of Einstein and de Sitter (1917) were replaced, after some initial resistance, by the expanding universes of Friedmann and Lemaître ('FL'). These universes, surveyed in Robertson's splendid 1933 review paper [73], are a family of exact solutions of the Einstein field equations ('EFE')', with Robertson–Walker ('RW') geometry (they are spatially homegenous and isotropic). Imposing the EFE, the Friedmann equation

$$\frac{\dot{S}^2}{S^2} = \frac{\kappa\mu}{3} + \frac{\Lambda}{3} - \frac{k}{S^2} \tag{1}$$

relates the scale factor $S(t)$, energy density $\mu(t)$, curvature constant k (normalized to ± 1 if it is non-zero), and cosmological constant Λ. Using units such that $c = 1$ (here and throughout), the energy density obeys the conservation equation

$$\dot{\mu} + 3(\mu + p)\frac{\dot{S}}{S} = 0 \qquad (2)$$

which becomes determinate when a suitable equation of state relates the pressure $p(t)$ to the energy density $\mu(t)$. The Raychaudhuri equation for this context is

$$3\frac{\ddot{S}}{S} = -\frac{\kappa}{2}(\mu + 3p) + \Lambda \qquad (3)$$

which is a consequence of the other two when $\dot{S} \neq 0$.

Solutions of these equations for the case of pressure-free matter are characterised by three observational parameters: the Hubble constant $H_0 = \left[\frac{\dot{S}}{S}\right]_0$, matter density μ_0, and deceleration parameter $q_0 \equiv -\left[\frac{\ddot{S}}{S}\right]_0 \frac{1}{H_0^2}$. They are related to the cosmological constant Λ by equation (3) in the form

$$6{H_0}^2 q_0 = \kappa\mu_0 - 2\Lambda, \qquad (4)$$

when the pressure of the matter is negligible, as is presumably true at recent times.

For ordinary matter and $\Lambda = 0$, q_0 is negative (gravity slows down the expansion of the Universe), and a minimum radius ('bounce') is possible only if $k = +1$. Further work by Robertson and Walker clarified geometrical aspects of these spatially homogeneous and isotropic Universe models, while the link to galactic observations was established by various workers, culminating in Sandage's 1961 paper [75] on the (m, z)– and (N, m)–relations in these models. However, source evolution prevented determination of q_0 from these observations, as had been hoped.

3: Astrophysical cosmology

Astrophysical cosmology [69] centred initially on the properties and evolution of galaxies, clusters of galaxies, and the intergalactic medium, and later also properties of radio sources, quasi-stellar objects, and X-ray sources. This provides a crucial link to cosmological observations, for this information plays a key role in interpreting those observations. A major discovery is that these sources evolve with time, confirming the view of an evolving and changing Universe (radio source counts are inconsistent with a steady state universe). Key questions are: What is the origin of structures on various scales?, of rotation?, of magnetic fields? This led in particular to Lifshitz's pioneering 1946 studies [58] of perturbed Friedmann–Lemaître Universe models and Peebles' analysis [68] of the matter covariance function (the probability of finding another object within a given distance of a given object). A further key question is: What is the density of various forms of matter?; with analysis of observed velocities leading to the discovery of dark matter that dominates astrophysical dynamics on large scales. Indeed, it appears [15] that the dimensionless density parameter Ω_{m0} for matter has a value

$$\Omega_{m0} \equiv \frac{\kappa\mu_0}{3{H_0}^2} \approx 0.3, \tag{5}$$

substantially less than the critical value of unity.

4: Nuclear cosmology and the Hot Big Bang

The further issue: What is the origin of the elements?, is of a different nature than the others, because it relates to nuclear physics during the Hot Big Bang ('HBB') era in the early expansion of the Universe. As was worked out in the period 1948–1965, the unbounded temperature in this era implies matter–radiation equilibrium and an epoch of nucleosynthesis when the light elements form in the early Universe at a temperature of about 10^9 K, with heavy elements forming much later in stars [67,83]. The fundamentally important further point was the understanding that cosmic blackbody radiation[a] ('CBR') would be left over as a remnant of the HBB era, with experimental discovery of the CBR in 1965 at a temperature of 2.75 K vindicating the application of standard physics to the early Universe. This value agreed well with nucleosynthesis theory (based on nuclear physics together with the Friedmann equation) and element abundance observations, provided the baryon density in the early Universe is low,

$$\Omega_{b0} \approx 0.01;$$

and hence, together with the density estimate from astrophysical cosmology, provides evidence for the important conclusion that there is much more non-baryonic dark matter than baryonic matter in the Universe [15]. Furthermore, this concordance required there be no more than three neutrino species, a conclusion confirmed later by experiments at CERN [52].

5: Particle cosmology and inflation

In the very early Universe particle physics plays a major role. Because quantum field theory allows a violation of the standard energy condition

---

[a]Often called the cosmic microwave background radiation ('CMBR'). However, it is only microwave radiation at the present time, so CBR is better — it applies from decoupling to the present day. Note also that there are many other components (in particular, radio and X-ray) of the cosmic background radiation spectrum.

$(\mu+3p) > 0$, at very early times there can be a period of accelerated expansion driven by scalar fields, leading to 'inflation' when such expansion rapidly takes place through many e-foldings before a subsequent HBB era begins (when the inflationary field has decayed to ordinary matter and radiation). Inflation can lead to a vast increase in the size of each comoving region in the Universe. The power of this concept was demonstrated in 1980 by Guth [41], who showed it could in principle give an explanation of the smoothness, flatness, and horizon problems,[b] and later it became clear it could also provide an origin for structure formation through quantum fluctuations being expanded to galactic scales [39,56]. This has led to a vast burst of activity examining the resulting links of the very small (particle physics effects) to the very large (cosmological structure). In particular, structure formation theories based on inflation predict a scale-free power spectrum for the matter distribution and a series of peaks in the power spectrum of CBR anisotropies [66,56]. It is greatly to the theory's credit that existence of the first three such peaks has been confirmed by balloon measurements. The position and heights of these peaks give estimates of the cosmological parameters that are consistent with the other data available. It should be made clear however that inflation is not a single theory, but rather a family of theories, with no particular version unambiguously indicated as the correct one. A popular version is 'chaotic inflation', in which even today pockets of inflation surround ordinary expanding Universe regions which occur on many different scales, resulting in an overall fractal-like structure [59,42].

6: Observational transformation

Since the 1960's, observational cosmology has gone through a major transformation, with the discovery of quasi-stellar objects, X-ray sources, gravitational lenses, and so on. This has been made possible by a revolution in observational technology based on the one hand on new telescopes, both ground-based (e.g. Keck) and borne aloft in balloons and satellites (e.g. IRAS, the HUBBLE SPACE TELESCOPE), operating at all wavelengths from radio to γ-ray, and often with features such as multiple-mirrors and adaptive optics, and on the other hand on new detectors, particularly photo-multipliers, CCD's, and interferometers [54,10]. Use of optical fibres has allowed a vast increase in measured redshifts and so determination of 3-dimensional structures, automated measuring machines have resulted in major new surveys and better source statistics, and improved detectors have allowed much deeper observa-

[b] An important point is that although inflation implies that the Universe is likely to be nearly flat today, it does *not* imply it is *exactly* flat (i.e. $k = 0$).

tions and number counts than before.

Three particularly important consequences for cosmology have been, firstly, the systematic study of gravitational lensing and its use for studying the distribution of dark matter in galaxies and clusters [76]; second, the systematic study of decay of supernovae in distant galaxies [71], and consequent first reliable detection of the deceleration parameter q_0, showing there is presently an effective positive cosmological constant, with

$$\Omega_\Lambda \approx 0.7; \tag{6}$$

and thirdly, measurements of the spectrum and anisotropy of the CBR by the COBE satellite and BOOMERANG and MAXIMA balloons, with reliable detection of the quadrupole anisotropy and angular power spectrum peaks [5]. Together with the supernova data this suggests that

$$\Omega_0 = \Omega_{m0} + \Omega_\Lambda \approx 1.0 \tag{7}$$

(the Universe has nearly flat space sections, as expected on the basis of inflationary theory) [1,82]. In parallel to this, there has been major development on the one hand in the theory of observations, relating to measurement theory, selection effects, and how these are affected by astrophysical source evolution, and on the other hand in data reduction, relating to Fourier analysis, statistical methods, and large volume data-handling.

7: Geometry/relativity transformation

In the same period, the geometric theory has also gone through a major transformation, perhaps not recognised as much as the others mentioned above. This has centred on the one hand on a growing understanding of generic geometrical and dynamical behaviour of cosmological models, and on the other on developing specific families of anisotropic and inhomogeneous models. In particular:

a: *Exact 1+3 covariant and tetrad equations* have been developed [21,53,22,31] showing the generic relation between kinematic variables, dynamic variables, and Weyl curvature tensor components (the 'free gravitational field'), for both the fluid and kinetic theory cases, including in particular the energy and momentum conservation equations and the Raychaudhuri equation (the basic equation of gravitational attraction), namely

$$\dot{\Theta} + \frac{1}{3}\Theta^2 - a^b{}_{;b} + 2\sigma^2 - 2\omega^2 + \frac{\kappa}{2}(\mu + 3p) - \Lambda = 0, \tag{8}$$

where Θ is the fluid expansion, and a^b, ω and σ are its acceleration, vorticity scalar, and shear scalar, respectively.[c] This shows that shear and a positive active gravitational density $\mu_{grav} \equiv (\mu + 3p)$ cause gravitational attraction,[d] whereas vorticity and a positive cosmological constant Λ cause gravitational repulsion. Corresponding equations for massless particles determine the geometry of observations in cosmological space-times, in particular implying null-ray focusing and distortion (leading to weak gravitational lensing) and the formation of caustics (leading to strong gravitational lensing).

b: *Exact theorems* have been developed that hold when there are no symmetries present, in particular [44]:

(i) theorems relating causal properties to existence of geodesics and spacelike surfaces,

(ii) theorems proving existence of singularities under various plausible circumstances, for example in the case of irrotational pressure-free matter on the one hand, when the result follows directly from (8), and when closed trapped surfaces exist on the other, when the result is given by the Hawking–Penrose singularity theorems [45]; the latter result implies that existence of singularities at the start of the Universe is strongly indicated by existence of the CBR (which implies the existence of past-directed close trapped surfaces associated with the refocusing of our past light cone),

(iii) existence and uniqueness theorems for the EFE (the dual of the singularity theorems) under various circumstances.

c: *Families of cosmological exact solutions* with specific symmetries have been examined in detail, in particular [31] Kantowski–Sachs and Bianchi (spatially homogeneous) and Lemaître–Tolman–Bondi (spherically symmetric) models, together with properties of associated dynamical systems, self-similar solutions, and phase planes. The latter identify attractors, limit points, and saddle points, and so characterise typical early, late, and intermediate behaviour, as well as relations between the dynamical development of different families of models (with self-similar space-times acting as attractors in many cases) [80].

d: *Approximate cosmological solutions* determine dynamical behaviour in an open neighbourhood of exact solutions, but raise subtle gauge issues

[c]This gives the form (3) when $\sigma = \omega = a^a = 0$,and $\Theta = 3\dot{S}/S$.
[d]Inflation occurs when $\mu_{grav} < 0$.

(resulting from there being no fixed background space-time) which are best solved by developing gauge invariant perturbation variables and equations for fluids [2,62,30,11,31] and kinetic theory [33,13]. These provide the link to realistic observations, in particular relating structure formation theories to anisotropies of the CBR [13,56].

8: Quantum cosmology

Before the inflationary era, some kind of quantum gravity effects will dominate the dynamics of the Universe and provide the initial conditions for inflation. Many brave attempts have been made to determine an effective cosmological theory in these circumstances even though no consistent theory of quantum gravity is known. These attempts include examination of the wave function of the Universe, either through the Wheeler–DeWitt equation and mini-superspace or via path integral methods [43], 'pre-big bang' theory based on an attempt to extend some of the dualities of M-theory [12], and various string cosmology approaches involving higher-dimensional theories, including the brane cosmology approach of Randall and Sundrum where our Universe lives on a 4-dimensional 'brane' imbedded in a 5-dimensional spacetime [57] and the ekpyrotic universe where two such branes collide [50]. In all cases the problem is making a solid link to observational tests, because the proposed particle interactions and/or extension of classical gravitational theory is not directly testable; however, these theories are indirectly testable via their effects on the inflationary perturbation spectrum, and so on structure formation.

9: Probability and ensembles

A final theme that has been coming to the fore in many recent analyses has been that of an ensemble of cosmologies. This has been proposed to occur in various ways: as related to the Everett many-universes idea of quantum theory, to the continual production of 'big bangs' in Linde's chaotic inflationary scenario [59], as related to multiple vacuum energy minima, or as completely unconnected universes [61]. The merit of this approach is that it potentially brings cosmology within the realm of statistical analysis, and so for example provides the only scientific basis for attempts to use the anthropic principle to determine the values of parameters such as the cosmological constant [4], for which we otherwise have no plausible explanation.

2 Different views

The developments summarised in the previous section are very impressive. As a result, we have a series of models of decreasing certainty:

A: Standard Hot Big Bang model

We have a very well established and commonly agreed standard model of a HBB era followed by the observed Hubble expansion, including a solidly based theory for nucleosynthesis and the origin of the CBR as remnant radiation of the early equilibrium era. Typical concordance model parameters, fitting all the current cosmological observations (particularly the supernova data and the BOOMERANG and MAXIMA CBR anisotropy observations), are [1,82]:

(i) *Hubble constant* $H_0 = 100h\,\text{km}/(\text{sec Mpc})$, $h = 0.57 \pm 0.11$, *age* $t_0 = 14.6 \pm 2.0\,\text{Gyr}$.

(ii) *Matter density parameters*: *baryons* $\Omega_b h^2 = 0.017 \pm 0.003$, *cold dark matter* $\Omega_{CDM} h^2 = 0.17 \pm 0.02$.

(iii) *Cosmological constant* $\Omega_\Lambda = 0.66 \pm 0.06$, *spatial curvature* $\Omega_k = -0.08 \pm 0.06$.

(iv) *Spectral index* $n = 1.03 \pm 0.06$, where the power spectrum is $P(k) \propto k^{n-1}$(k is the wave number).

B: Extended standard model (inflation, baryosynthesis)

Here we suppose additionally a period of inflation preceding the HBB epoch, and followed by baryosynthesis. While broad outlines of many possible mechanisms are available for both inflation and baryosynthesis, in neither case is there a unique physical proposal for which specific theory actually holds, so our understanding of these epochs is necessarily more speculative than that for the HBB era. One should note here that although structure formation models and the link to CBR anisotropies were first calculated in the context of inflationary models, the actual fit to observations — as represented in the concordance HBB model mentioned above — is independent of the details of inflation (if this was not so, some specific inflationary model would be indicated), and so is in fact a post-inflationary model fit, based on the physics of the HBB era (the matter–radiation interaction during this epoch determines the accoustic wave properties and the Jeans length).

C: Speculative models (quantum cosmology)

Here we consider the various quantum cosmology options for producing the initial conditions for inflation: the wave function of the Universe, brane universes, etc. The physics is highly speculative and its major features are contested, including whether it can in fact produce the required almost smooth conditions required for inflation to start (if there is some kind of space-time foam then, this could be problematic). There is no consensus on which model, if any, is most plausible.

2.1 Evaluation:

How to evaluate them? Here are two extant views (perhaps slightly parodied, but not by much):

View 1: *Cosmology is solved!* — the era of precision cosmology has arrived, only details remain to be sorted out. The extended standard model gives us a solid understanding of cosmology, our main concern is to get ever more accurate values for the various parameters involved. Querying the inflationary 'standard model' or searching for alternatives is a waste of time, and quantum cosmology does not have to be solved because its details get ironed out by inflation. The remaining relevant uncertainties will vanish with improved observations, which will enable us to refine the theory as needed.

View 2: *Cosmology is a disaster!* — there are many observational problems, and we don't understand any of its main dynamical features.[e] While we have indeed identified these dynamical features, simply giving them a name ('inflaton', 'dark matter', 'quintessence') does not mean we understand them. There is no clarity concerning the quantum gravity era which sets the stage for inflation, we don't have any clear idea what specific physical feature underlies the supposed era of inflation, and we don't understand the dominant dynamical features of the present era because on the one hand we don't know what the dark matter is that is dynamically dominant on galactic to cluster scales, and on the other hand we have no idea why there is a small non-zero cosmological constant — or perhaps non-constant 'quintessence' field — dominating the largest observable scales. Indeed, quantum field theory gives an answer for this constant that is wrong by 120 orders of magnitude. This has been characterised as 'arguably the greatest quantitative discrepancy between theory and observation in the history of science' [60], with no clear resolution in sight, but inflationary physics depends crucially on this faulty theory (which also diverges at second order).

I suggest that both these views are extreme, and the best approach is a third one:

---

[e]See Disney [17] for a strong statement along these lines.

View 3: *A middle way: a positive but critical approach* — the standard model above, including an inflationary epoch, is the best we have, but with significant problems. We can cautiously adopt this standard model provided we do so in a critical way, using all available observations to delimit its nature, carrying out all possible consistency tests, acknowledging the observational limits and resulting uncertainty, and looking for all viable alternatives. We can additionally tentatively explore the various quantum cosmology options, but in so doing clearly acknowledge the uncertainties and foundational problems involved.

In the remainder of this paper I develop some issues that are important in such a cautious approach, looking sequentially at observations, geometry, physics, and unavoidable philosophical issues. A major problem with the triumphalist approach (View 1 above) is that it denies there are any such philosophical problems intruding on the physics. I suggest that view is seriously misleading.

3 Real observations

There is a beautiful simple theory of cosmological observations applicable in any cosmological space-time [53,22], but practice is much more complex, and introduces many problems. The basic observational relations for discrete sources in a FL cosmological model are straightforward, relating redshift, magnitude, angular size, and numbers in a given volume [75]. However there are no ideal standard candles, and so in practice we have to deal also with source statistical properties, particularly the luminosity function (the number of sources in a given region characterised by apparent luminosity) and power spectra (the number of sources in a given region in each physical size range). And that is just the start of the complexity of real observations.

1: Auxiliary functions

Each basic theoretical concept ('cosmological redshift', 'source luminosity', 'angular size', 'source numbers', etc.) is related to actual observations by complex procedures that introduce new relations and associated concepts. Thus, as emphasized by Imre Lakatos, the hard core of a theory (here, the FL model with specific equations of state) is related to data by a belt of auxiliary hypotheses; when there is a conflict of theoretical predictions with observations, one usually resolves this by altering the auxiliary hypotheses, while leaving the basic theory unchanged. In the case of cosmology, one fits theory to data by specifying a variety of auxiliary functions and parameters, e.g. absorbtion, source evolution, bias, tilt, related to the astrophysics of the problem. One can't fit the data without them, but their existence then loosens the link between theory and observations because you can fit almost any observations by varying these auxiliary functions. One can systematise this statement by deriving corresponding theorems, for example any spherically symmetric geometry can be fitted to any galaxy observations by appropriate choice of source evolution functions [63,31], often treated as arbitrarily adjustable (and these functions must be included in the analysis, because otherwise a FL model is incompatible with number count observations). Thus the basic theory is invulnerable to test, unless one can pin down the auxiliary functions by adequate astrophysical understanding. The problem then is that any astrophysical theory to deal with this issue (e.g. by predicting source evolution) usually itself introduces new auxiliary functions or parameters, which also have to be determined. The problem recurs.

2: Selection functions

Of particular importance in resolving the statistics of observations are the detection and selection functions relating the number of source images detected and identified to the number of sources present. Indicative of the problems here are that any attempt to characterise selection effects in terms of a galaxy distribution function depending on only one parameter (e.g. a Press–Schechter luminosity function) is basically flawed, for these effects necessarily depend on two parameters [34]. However, a distribution function depending on two galaxy parameters has only just been obtained [47], and has not so far been used in examining detection effects. This kind of issue is crucial in estimating the number of undetected objects in the Universe, and so determining the amount of dark matter in the Universe.

3: Inhomogeneity effects

Most of the Universe is empty, with matter concentrated in isolated clumps separated by vast regions of empty space; the homogeneity of the FL models is only a realistic representation when we average on fairly large averaging scales. The optical scalars determining the focusing of null geodesics, and hence the area distance in cosmological models, obey the equations [44]

$$\frac{d\hat{\theta}}{d\lambda} + \frac{1}{2}\hat{\theta}^2 + \hat{\sigma}^2 = -\,R_{ab}k^a k^b$$

$$\frac{d\hat{\sigma}_{ab}}{d\lambda} + \hat{\theta}\hat{\sigma}_{ab} = -\,C_{acbd}k^c k^d$$

along null geodesics $x^a(\lambda)$ with tangent vector $k^a = dx^a(\lambda)/d\lambda$, where $\hat{\theta}$ is their convergence and $\hat{\sigma}_{ab}$ their shear. The light rays by which we observe in fact mainly move in the empty space between galaxies, where the Ricci curvature term $R_{ab}k^a k^b$ (determined algebraically by the matter along the path) is zero but the Weyl curvature term $C_{acbd}k^c k^d$ (determined non-locally by the matter concentrations around that path) is non-zero. However, we usually calculate the observational relations using the FL smoothed out approximation, where exactly the opposite is true: the Ricci curvature term is non-zero but the Weyl curvature term is zero.[f] This description is likely to be misleading along the narrow bundles of rays whereby we observe distant supernovae (even though we see them in different directions in the sky, these observations are not sky-averaged by including over-dense as well as under-dense regions). This suggests we should use different area-distance formulae to analyse these observations, for example the Dyer–Roeder distance [19,76] which depends on the degree of clumping of matter. This can make a difference to the value we assign to the cosmological constant through analysis of the supernovae observations [49].

4: Horizons

It is crucial that — unless we live in a 'small universe' where we are able to see many times round the Universe [23], which does not seem to be the case [18] — there are essential limits to what we can observe because of the existence of particle horizons (the limit of causal contact) [72,44,79] and visual horizons (the limit of visual contact) [36]. We can have no direct knowledge whatever of what lies beyond the particle horizon, even if we develop ideal gravitational wave and neutrino telescopes, for we can never receive any radiation whatever from these regions in the time available since the beginning of the Universe, and we can obtain no detailed knowledge of what lies behind the visual horizon (the furthermost matter we can observe by particles travelling at the speed of light since decoupling of matter and radiation), because we can never observe these regions by electromagnetic radiation. Thus any theoretical predictions of what lies at greater distances (i.e. many Hubble radii) away from us are completely unverifiable. Obvious problems arise in making any scientific claims that any large-scale cosmological model is the correct model of the Universe. In particular we will never be able to observationally test the idea of chaotic inflation, where there are many different expanding regimes (with different geometric and physical parameters) beyond the particle horizon [59,42].

5: Statistics

Perhaps the most fundamental theoretical problems underlying our anal-

---

[f] This is essentially what underlies the Mattig formulae quoted by Sandage [75].

yses of observations are those arising from relating observations of unique observed objects to theories that make statistical predictions. We can compare simulations with observational data in a statistical sense, and the approach works well when there are observations of a statistically representative sample of objects, but fails when observing rare objects or specific configurations and we then end up with the strange concept of "cosmic variance"[48] — an expression of dissatisfaction with the real Universe because it is not a typical enough representative of a theoretical ensemble of universes we have derived! To get round this one has to then use Bayesian statistics, which has its own pitfalls — in that the results depend completely on the priors introduced[g] — or constrained simulations [51]. This issue is seldom considered in depth but is one of the most profound facing observational cosmology, where we are observing the one unique Universe so frequentist statistical techniques cannot properly be deployed (we can only observe one member of any supposed ensemble of similar objects). This is one of the limits in analysing the CBR observations at large angular scales, for example, but comes up in many other contexts, for example in analysing large-scale motions and the nature of the associated 'great attractor'.

4 Real geometry

The key issue here is that one needs to examine inhomogeneous and anisotropic models in order to adequately model the real Universe and to examine stability and probability of cosmological solutions, in particular determining generic behaviour through phase planes and exact theorems, and observational properties through examining null geodesics in these models. One cannot examine these concepts adequately by exclusively using FL models based on the evolving RW geometries, let alone models with even higher symmetry (such as de Sitter spacetimes) — but that is what is frequently attempted. These models are of measure zero in the space of all models, and there is no reason to believe they demonstrate generic behaviour — indeed, through a variety of counter-examples, we know that they do not do so. Hence, use of these models alone to understand cosmology could be misleading because almost all universes — and in particular realistic Universe models representing the inhomogeneity of the real Universe — would not necessarily behave in this way.

[g] It is to the credit of this approach that this dependence is made explicit. It is still there in other approaches, but is hidden.

1: Present geometry

The first question is how we can observationally prove that the observable region of the Universe is almost FL — that is, how can we show that it is almost spatially homogeneous on some suitable averaging scale, and hence, because we observe almost-isotropy around ourselves, is also almost isotropic everywhere. Isotropy is easy to measure, but homogeneity is not, both because of projection effects, making distance estimates crucial, and the finite speed of light, making evolution effects important.

The best observationally-based argument for spatial homogeneity is using the *almost-EGS theorem* [77]: namely, *if a family of geodesically moving observers in a space-time region $\mathcal{U}$ all measure almost-isotropy of the CBR, then the Universe is almost FL in $\mathcal{U}$.* This provides the desired proof of spatial homogeneity if we adopt a weak Copernican assumption, namely, that the CBR is measured to be almost isotropic everywhere in the observable region of the Universe (thus extending our own observations throughout that region, on the basis of the assumption that we are not exceptional). This assumption of CBR almost-isotropy at distant events can in principle be weakly tested via accurate measurements of the CBR spectrum in directions where the Sunyaev–Zel'dovich effect is significant [40].

While this gives a reasonably convincing argument for spatial homogeneity, it still deserves further exploration because the detailed statement of the theorem depends not just on the pointwise isotropy as measured by CBR multipole moments, but also on their space and time derivatives; and there are examples showing that this dependence is essential — if rapid oscillations in the anisotropy of the Universe occur (as is possible for example in many Bianchi universes [80]) then the CBR can be almost isotropic even if the universe is not [64]. Even though very contrived, this analysis is important because it motivates checks on whether the usual, taken for granted geometric assumptions are indeed correct. The almost-EGS analysis is likely to be robust, but limits on the CBR multipoles do not easily translate into direct statements characterising the averaging scale at which the Universe is well-described by an almost-FL model [77]. It should be emphasized here that the usual CBR anisotropy analyses [66,56h] do not *prove* almost-homogeneity, they *assume it.* An alternative approach of some interest is that of uniform thermal histories [9], which also needs further elucidation.

2: Past and future geometry

Almost-isotropy today and in the entire observable region of the Universe

[h] All based on the classic paper by Sachs and Wolfe [74].

does not imply almost-isotropy in the distant past or in the distant future. Indeed phase-plane analyses of Bianchi universes show that models which are very nearly isotropic today can become very anisotropic in both directions of time. The converse result is interesting: we can have models that start out very unlike a FL model but experience ***intermediate isotropisation***, where for a very long period of time their geometry is very close to a FL geometry [80,81,31]. Thus for example anisotropic Bianchi II models can develop flatness and almost-isotropy in this way; similarly G_2 inhomogeneous models [80,37] can develop homogeneity from a very inhomogeneous starting state. Thus we can geometrically obtain many of the results inflation aims for, in terms of developing an almost homogeneous and isotropic Universe with flat spatial sections, without an inflationary phase. It is of course also possible to have an inflationary phase in some anisotropic and inhomogeneous models, but they have been relatively little studied, even though one cannot assess the success of inflation in dealing with inhomogeneity and anisotropy without such studies.

3: Problem of initial conditions and probability

Whatever dynamics precedes the HBB era, whether or not inflation took place, one eventually runs into the problem of establishing initial conditions for those dynamics. The usual aim is to demonstrate either that these dynamics are probable, or that their probability is irrelevant. High probability is difficult to establish without the proper setting of a well-determined probability measure on the space of generic cosmological models, which is needed to establish stability and fragility of results obtained [78]. For example, the often used idea of 'fine tuning' implies showing a model is improbable, which implies consideration of the family of all plausible cosmological models, including all geometrical possibilities. However, we have no well-established measure of probability on the family of cosmological models. Showing the irrelevance of probability is the aim of a varied family of models, including inflation. It is claimed [42] that no matter how small the probability is of inflation starting in this generic geometric setting, once it has started it will come to dominate because of the explosive nature of inflationary expansion. However, it is difficult to see how this can be turned from a reasonably plausible idea into a precise provable statement. The challenge is to devise and prove some theorems that show this idea is in fact correct. A key point here is that we don't know that the postulated physics, which is required to make this work, is in fact correct (see below).

4: Distant regions and global inhomogeneity

If the chaotic inflationary idea is right, then the Universe beyond the visual horizon consists of numerous FL-like patches, some inflating and some not, with different dynamics and expansion parameters. There then have to be inhomogeneous regions between them, where physics and geometry have drastic variation, and perhaps even discontinuities (if these domains are confined to thin-wall regions). Almost nothing has been written about how these inhomogeneous regions can consistently bridge between the supposed FL-like domains, or how the physical and gravitational interactions that may be expected to take place there can be confined to these regions, rather than spreading out through the FL-like neighbouring domains, thereby disrupting their smoothness.

5: Averaging problem

It is known that averaging a given geometry on different length-scales does not commute with calculating the EFE for a specific geometry. This non-commutativity will have an effect both on observations, as discussed in the previous section, and on dynamics [24,8]. It is commonly supposed that the resultant effect, often classified as 'back-reaction' (for example, when we discuss the effect of gravitational radiation on the large-scale space-time geometry), will be negligible in most cosmological settings. However, as cosmological predictions become more precise this will probably not remain true, and indeed there are varied investigations of the effect underway from a variety of viewpoints. One problem is that it is difficult, perhaps even impossible, to examine this effect satisfactorily in a covariant way; but that does not mean it will be physically insignificant in all circumstances.

6: Consistency tests

Given all the uncertainties emphasized above, there is one positive statement one can make: it is important to carry out all consistency tests one can of the geometry and physics of the standard models of cosmology. Some claim that this is unnecessary and indeed is a waste of telescope time, but I suggest that this is far from the case. Because of all the underlying theoretical and observational problems facing us in understanding the unique existing Universe, it is crucial we test our models in all ways we can. Specific examples of important consistency tests are:

(i) determining the CBR temperature as function of z, and showing it varies as $(1+z)$ [55],

(ii) determining element abundances as a function of z, and so verifying the homogeneity of the Universe [65],

(iii) verifying the existence of a dipole anisotropy in number counts of cosmological sources, occurring in the same direction as the CBR dipole anisotropy [29,7].

Insofar as these have been carried out, they confirm our standard model. This is an important observational confirmation of our overall present cosmological understanding.

5 Real physics

The issue here is relating dynamics and observations to real physics, as opposed to potential physics. For example, just writing down a Lagrangian or a potential for a scalar field and then deriving the consequences does not necessarily mean one is relating to real physics. That depends on whether such Lagrangians or potentials mirror the behaviour of matter or fields that actually exist at some time in the real Universe; and whether or not that is so is ultimately decided by experiment.

1: Einstein field equations and matter

A fundamental issue is in what sense one is solving the field equations in most applications in present day cosmology. There are two basic options. One can solve the EFE[i]

$$R_{ab} - \frac{1}{2}Rg_{ab} = \kappa T_{ab} \tag{9}$$

from right to left ('matter implies geometry'), which is the standard approach where, given some matter description and a family of geometrical models, one essentially tries to solve these equations regarded as second order non-linear partial differential equations for the metric tensor components implicit in the Ricci tensor components.[j] Alternatively, one can solve from left to right ('geometry implies matter'). In this case, one is using the *Synge trick*: one can choose any geometry one wants, differentiate the metric tensor to determine its curvature tensor, and then substitute into the left-hand side of the field equations. Algebraically solving for the stress tensor on the right then gives an exact solution of these equations. No differential equations have to be solved. This can be done for any geometry whatever, determining the

---

[i]We will from now on regard the cosmological constant as a matter component ('the vacuum') with pressure $p = -\mu$; consequently it is not explicitly represented in these equations.
[j]Or an equivalent first-order set of equations obtained by using tetrad components and rotation coefficients as variables, see Ellis and van Elst [31] for a summary.

hypothesized 'matter' required to give the desired geometrical behaviour. The problem is that almost always, the resulting 'matter' does not correspond to anything physical — it does not satisfy any suitable equations of state.[k] It gives an exact mathematical solution of the equations, but only rarely is it of physical significance.[l]

The main point to make, then, is that *important parts of what is done in present day theoretical cosmology are essentially this Synge trick, carried out in the context of RW geometries. Arbitrary time evolutions are related in this way either to a hypothetical fluid equation of state, or to a potential for a hypothetical scalar field.* One can do this for any evolution of a RW geometry, and also to some degree for perturbations around this metric. Thus the theory has no predictive content: it can explain anything.

Indeed, one can systematize this in a set of precise statements:

- Note first that the geometry of a RW universe is characterised by only one free function $S(t)$ and one constant k.

- Note second that whatever the underlying dynamics, be it string theory, pre-big bang theory, a variable gravitational constant, scalar field, or whatever, the effective dynamics is characterised by only two free functions $\mu(t)$ and $p(t)$ (because of the RW symmetry); and irrespective of what fundamental theory underlies the dynamics, an effective Einstein theory can be obtained by moving all non-Einsteinian terms to the right hand side of (9) and regarding them as contributions to $\mu(t)$ and $p(t)$. That is, one can always write the field equations of a FL type model in the form (1)–(3) as an effective fluid theory, if we define $\mu(t)$ and $p(t)$ suitably.

- Consequently, one can then at least locally[m] fit any evolution $S(t)$ with suitable choice of the function $p(\mu)$. Specifically, for any $S(t)$ and any k

[k]The corresponding procedure in electromagnetism is to choose an arbitrary potential A_a, determine the resulting electromagnetic field tensor F_{ab}, and then substitute into the 'divergence' Maxwell's equation to determine the current needed to produce this arbitrary field. Again no differential equations are solved — one simply differentiates to 'solve' the equations.

[l]This problem does not arise if one adopts the Einstein–Wheeler position (labelled 'geometrodynamics' by Wheeler) that only vacuum solutions of the field equations should be countenanced [84].

[m]Constraints arise if one of the concerned functions does not evolve monotonically with time. However, this does not seem to be the case for the evolution of the actual physical Universe at least from the start of inflation to the present time.

we define $\mu(t)$ and $p(t)$ by

$$\kappa\mu(t) = 3\left(\frac{\dot{S}^2(t)}{S^2(t)} + \frac{k}{S^2(t)}\right), \tag{10}$$

$$\kappa p(t) = \left(\frac{\dot{S}^2(t)}{S^2(t)} + \frac{k}{S^2(t)}\right) - 2\frac{\ddot{S}(t)}{S(t)}, \tag{11}$$

then (1)–(3) will be satisfied, and we have 'solved' the field equations for whatever evolution $S(t)$ we care to imagine. If we can observationally determine the form of $S(t)$, for example from (m, z)–curves associated with supernovae data, this is essentially how we can then determine that some kind of 'dark energy' or 'quintessence' is required to give that evolution and we can find the equation of state implied by eliminating t between these two equations. This is, however, not a *physical* explanation until we have either in some independent experimental test demonstrated that matter of this form exists, or have theoretically shown why this matter or field has the form it does in some more fundamental terms than simply a phenomenological fit.

– One way to propose such an interpretation is to consider there may be a scalar field $\phi(t)$ with dynamics governed by a potential function $V(\phi)$ in the usual way via the Klein–Gordon equation. This arbitrary function allows us then to 'explain' any desired dynamics $S(t)$. This is made explicit by a theorem [32] showing how to obtain such solutions by first solving the equations for $V(t)$ and $\phi(t)$, and then eliminating t to get $V(\phi)$.[n] However, this 'solution' is not related to real physics unless $V(\phi)$ is shown to exist and have that form by independent means — by some other physical test. Unless this is done, we have a hypothetical scalar potential, not linked to actual physics (it may be related in important ways to invariance principles and proposed physical symmetries, but its effective existence for the relevant energy range then has to be demonstrated by observations). If *all* that is explained by this potential is the desired evolution $S(t)$, for example in an inflationary era in the early

[n] The key issue is to show that *all* the EFE, conservation equations, and equations of motion are satisfied. This is indeed the case, essentially because the Friedmann equation is a first integral of the Raychaudhuri and conservation equations, and the conservation equation is a consequence of the Klein–Gordon equation. Thus, in the end we only have to satisfy two non-trivial equations. We can do so because all the arbitrariness in $S(t)$ and k can be absorbed in $V(\phi)$ and $\phi(t)$.

Universe[o] or in determining the nature of 'quintessence', it is not really a physical explanation at all; rather it is a rewriting of the proposed dynamical evolution in one of many possible alternative forms. It becomes real physics when some other experiment or observation shows that some scalar field exists in nature with the desired properties. In principle the observations of the matter and CBR perturbation spectra might provide the extra constraint needed to test the form of the potential, but in practice they only determine the derivatives of the potential at a fixed epoch, and so can be used to constrain the needed form of the potential to get any desired perturbation spectrum [16], but do not provide an independent test of the proposed potential over the same range of values as the basic inflationary expansion. Its existence remains hypothetical physics rather than tested physics.[p]

To be fair, important properties of inflationary models are relatively independent of the nature of the potential, provided a slow roll approximation applies, so — within broad limits — the specific nature of the potential is not critical to the inflationary paradigm as a whole. Nevertheless, without a specified and verified potential, it is not a tried and tested physical theory in the usual sense.

2: Limits of testing — the physics horizon

We have a hope of finding these fields and testing their properties when they are dominant at late times, as for example in the case of the non-baryonic 'dark matter' that is being searched for in various laboratory tests. However, we cannot do so when the relevant fields are important at very early times, with the relevant energies way beyond what can be achieved in accelerators on Earth. This is for example probably the case for the hypothetical fields suggested to be responsible for chaotic inflation. This is an attractive theory, but only becomes a fully defined physical explanation when the field is firstly uniquely defined, and secondly then shown in some other way to exist with the desired properties. At present these fields are postulated in a generic way, and are not proven to exist with the desired properties, apart from showing that particular forms would solve specific cosmological issues. The hoped for link with particle physics is hypothetical rather than actual.

This is part of the major problem of *theories and extrapolation* facing us in

[o]Note that this direct procedure is not often used; rather an indirect trial-and-error process of searching for a potential that will give the desired effect is employed — with ultimately the same physics implications. The process stated here is a more efficient way of determining the desired potential.

[p]A problem shared by much of present-day theoretical physics.

any theories of the very early Universe. We can test for proposed particles and fields in the local context of the solar system and the Earth at present times, but not in the conditions that held in the very early Universe, i.e. significantly before the quark-hadron phase transition. Thus we face the existence of a *physics horizon* which is just as important for our testing of cosmological theories as the particle horizon and visual horizon. This horizon is defined as the energy range separating those physical properties and theories which can be tested by experiments in the solar system from those that cannot. This applies both to gravitational theory and to particle physics. We can test and verify the physics or theories relating to events governed by physics this side of the physics horizon (e.g. nucleosynthesis) but not that for events the other side (e.g. the Wheeler–DeWitt equation for quantum cosmology). Indeed this is why we are often told that the early Universe is the only laboratory where we can test many aspects of physics [85]: these are domains beyond the reach of physical experimentation on Earth.

The implication is that we have to extrapolate physics from the testable to the untestable in order to determine the laws that apply in the very early Universe. The key issue then is the *extrapolation problem*: *which aspects of known and tested physics are chosen as the features to extrapolate to this unknown domain?* We have to use our physical understanding and intuition to decide which aspects of known physics are the guides to the true underlying nature of physics, and so will remain true even under the most extreme conditions in the early Universe, and which are misleading and will lead us up false trails. It is the different choices made here that lead to the various different approaches to the very early Universe and to quantum cosmology. We have to make these choices on philosophical grounds ("simplicity", "beauty", "naturalness", "unification", etc.) because we can't test them independently in the relevant conditions.

In particular this applies firstly to *quantum field theory,* which is routinely extrapolated a huge amount from the laboratory domain where it is tested, for example in calculations for inflation, despite all the problems with this theory even in the range where it is testable: in particular, the zero order disaster of the cosmological constant and the second order disaster of divergences. There are major questions surrounding this massive extrapolation — how do we know it leads to a true theory? And what is most remarkable is how this extrapolation is not usually commented on.

Secondly, this applies to gravitational theories in the early Universe, applied in a dynamic range which has never been tested in the quasi-static situation of the solar system.

Thirdly, it applies to any quantum gravity proposal whatever. Should we extrapolate Lorentz invariance, Lagrangians, Hamiltonians, path integrals, symmetry groups? Or should we rather extrapolate the features that underlie general relativity such as general covariance and the equivalence principle (leading to the loop integral approach)? How do we handle the problem of interpretation and observation (collapse of the wave function) in these extreme conditions? If we adopt a *wave function of the Universe approach*, under what circumstances would we regard problems with the Wheeler–DeWitt equation and the question of time so formidable that we would give up on this approach, despite its apparent naturalness in other ways? Even in considering inflation we run into *transplanckian issues* [20], where the outcome of the structure formation process depends on the initial conditions for inflation which are determined by the *quantum gravity era* and in particular by the *smoothness and causality assumptions* we make about that era. These affect the resultant theory and its outcome by setting initial conditions for inflation, and raise potential problems for that theory [70]. The advent of promising ideas such as M-Theory, brane cosmology, and the ekpyrotic universe have not resolved the issues for the simple reason that there is more than one theory on the market; they are all basically untestable except in terms of their residual effects on low energy physics, and none is a fully developed and self-consistent whole that is clearly superior to all the others. We have to choose between them on philosophical grounds in these circumstances — and this is where the choice of what aspect of tested physics to extrapolate to these extreme domains becomes crucial.

3: The problem of initial conditions

The fourth domain where the issue of testability becomes paramount is how to handle within physical theory the question of the origin of the Universe — in essence, the setting of initial conditions which determine the outcome of the quantum gravity era, and hence in turn the outcome of inflation.[q] Here we run full tilt into the fundamental feature of the uniqueness of the Universe. It is not possible to test any such theory even in principle, because there is only one Universe and we can't re-run it with the same or altered conditions. Insofar as physics involves determining the effects of tried and tested laws on the nature of physical reality, we are outside its domain. Thus the question of laws for creation of the Universe lies outside physics proper (indeed it precedes the existence of physics), and is outside the domain of scientific testability. Proposals for such laws can be tested by their cosmological implications, but

[q]There are various proposals which try to sidestep this issue by claiming that in one way or another there is no beginning to the Universe. These proposals do not avoid the issue of its unique form: why it has one specific nature and not another [27].

not independently of them. The outcome is unique only if there is a unique theory that gives the desired result; but this is not the case. We end up facing squarely the uniqueness of the Universe and our inability to test proposed theories for its creation [28].

6 Philosophical issues

Because of these considerations, any fundamental approach to cosmology must necessarily face up to philosophical issues that give the essential framework of understanding cosmology. The major issues arising are the uniqueness of the Universe combined with observational and testing uncertainties [27,28,25]. I will conclude with just a few points in this regard.

1: The scope of cosmology

What is regarded as a satisfactory cosmological theory depends on what issues of explanation and cause we choose to address by that theory. For example, do these include issues of the existence of life and questions of meaning, or do they solely relate to physical cosmology issues? Do they extend to questions of ultimate causation, or only to physical effects, with the nature of the laws of physics taken for granted? Depending on the answer given here, the resulting theory will be more or less dependent on the underlying philosophical stances adopted. But, of course, choice of this answer is itself a philosophical stance.

2: Causation and local physics

One of the fundamental issues one can either emphasize or ignore is the relation of local physics to cosmology. The question here is to what degree there is top-down causation in cosmology (with the whole influencing the nature and behaviour of the parts) rather than just the bottom-up action which is the focus of standard cosmological theory (where local physical laws together with boundary or initial conditions determine the large-scale structure of the Universe). Examples of the former are Olbers' paradox, Mach's principle, and the issue of the arrow of time [35] — the latter two being difficult and controversial areas. The essential point here is that there is no isolated system in the real Universe: the "isolated system" beloved of classical physics does not exist in reality. This is important for example in quantum theory, where it underlies the issue of decoherence induced by interaction with the environment [6], and in the claimed link between M-Theory and existence of event horizons in the Universe [38]. The underlying philosophical issue is that of reductionism versus holism — which in the end is likely to turn out to be one of the deepest issues in cosmology, even though it is barely touched by many present studies.

3: Relation of observations to theory

A major issue, which has underlain much of the previous discussion, is the relation between observations and testability on the one hand and explanatory physical theories on the other. It has two major aspects.

The first philosophical question is which has priority in one's cosmological theory: physical explanation or observation and testability? Depending on one's attitude leads to different priorities. Particularly because of the existence on the one hand of visual and causal horizons and on the other of the physics horizon as discussed above, there are many limits on what can be observationally proven in cosmology, as, for example, in the case of chaotic inflation where neither the underlying physics nor the large-scale geometric predictions are testable. Consequently in the approach adopted by many supporters of chaotic inflation, theory takes precedence over testability. A further crucial area where this tension between theory and observation matters is in regard to the issue of ensembles of universes, discussed below.

The second major question, mentioned above, is how to handle the relation between unique existing physical situations and generic physical laws, in particular those of a statistical nature. There is again a tension between the weight one gives to each of these aspects. Present day cosmology emphasizes the generic universal theory underlying our understanding of the Universe, and de-emphasizes the unique specific features of the actual Universe. A key example is the idea of "cosmic variance": the difference between what actually exists in the real Universe, and the statistical predictions made by theory based on assuming a hypothetical ensemble of universes. Cosmic variance is in effect a complaint that the real Universe is not a good representative of this ensemble we have dreamt up and take as the foundation of our analysis. But surely theory should focus on what actually exists, and query the suggested ensemble?

An interesting issue here is the increasing use of Bayesian statistics, and how this allows theory to appear more compatible with observations than frequentist interpretations of probability. There is a tension here both in determining what actually happened in the evolution of the Universe according to different sets of data, and also in relating the outcome to what the theory says is probable. An example of the first is that recently the CBR data has, taken at its face value, weakly given a Hubble constant $H_0 = 40\,\mathrm{km/(sec\,Mpc)}$ [82], whereas almost all other data suggests $H_0 \geq 60\,\mathrm{km/(sec\,Mpc)}$. Using a Bayesian prior to incorporate this latter fact in the CBR analysis makes the different data sets look more compatible than if taken at their face value. In this case the procedure is acceptable because other data constrain the Hubble constant much more strongly than the CBR data. An example of the second is that one of the best explanations of the low value of the cosmological constant is given essentially by using the anthropic principle: the existence of life is taken as a Bayesian prior and so makes compatible the extremely different values of Λ predicted by theoretical physics and actually observed [4]. This method of handling the statistics does not change the fact that the basic theory gives a result quite different from what is observed. What this does do is reduce the demand for consistency between that theory and actual observations. One is modifying the acceptability of various outcomes of a theory that makes statistical predictions by using data on what actually happened. The underlying tension here is between determining what the actual history of the Universe was, on the one hand, and trying to determine if what actually happened is probable in some well defined sense, on the other. The Bayesian view seems very appropriate in the former case but seems to allow a great variability in the latter regard. However, it does have the advantage of making quite clear what assumptions are being made in the analysis.

4: Ensembles and testability

The particular key point where this tension between theory and testability presently needs discussion is as regards the issue of the claimed physical existence of ensembles of universes, or 'multiverses'. This is being increasingly defended, apparently because it is the only purely scientific approach to solving the puzzles raised by the anthropic issue (why the Universe has the peculiar properties that allow intelligent life to exist) [3,61]. For example, as mentioned above, this argument has been used to explain the value of the cosmological constant: one assumes an ensemble of universes with a wide variety of properties, and then only some of these universes (with low values of Λ) will admit the existence of life [4]. A setting in terms of ensembles challenges the idea of the uniqueness of the Universe, and allows statistical methods to be used, at least in principle, in studying its properties.

But is this physics or metaphysics? The major problem is that any such proposal can have no link to any direct observational test. Even showing that other universes in the postulated 'multiverse' physically exist, let alone determining their properties, will forever be hypothetical; consequently one can explain anything whatever by postulating the existence of an ensemble with suitable properties. That means this concept actually has no solid predictive power (any concept that explains everything actually explains nothing). The trick often used is to say that "all that can happen, happens" in an infinite ensemble of universes, thus eliding both numerous problems to do with the handling of infinities, and also problems to do with what is possible and what is not in view of Gödel's theorem and its elaborations, for example, by Chaitin [14]. This is in fact a very ill defined idea: who defines what is 'possible' in such a context? Does logic have to be maintained, for example, and if so what logic? One can attain any result one wants by postulating various ensembles with differing properties, for example, ensembles with universes where Harry Potter would be at home, or where there is no underlying logic at all. And suppose one could properly define an ensemble that leads to the desired anthropic result, the ultimate problem has simply been deferred; the question then is why does this ensemble exist, that allows anthropic universes, rather than another ensemble, that does not? Neither observational tests nor experiment have anything to say here. One has to argue from philosophical grounds; and the major tension then comes as to the degree to which these arguments should be based only on what physics can teach us, or whether they should also take into account wider aspects of human life and understanding that some claim relate to deeper aspects of existence.

7 Conclusion

Given the problems outlined above, the essential need is to be critical but not over-critical: to adopt View 3, in the context of an integrative explanatory framework — a larger interpretative scheme taking seriously the uniqueness of the Universe and the limits on observations. What we look for here is overall explanatory power and coherence, compatible with observations. Inflation is one such theory, but has a series of problems, in particular, the ongoing lack of identification of any specific preferred physical model for the inflaton, the problems in understanding the quantum vacuum, the reliance on an ill understood pre-inflationary era for setting up initial conditions at the start of inflation, and the inability to verify some of its main claims (such as the eternal existence of many distant FL domains with quite different properties than those in our region of the Universe). It has been argued above that the Synge trick is a dominant form of argument in present day cosmology — inflation, 'dark matter', and Λ all being deduced this way — which means we don't really understand any of the dynamic epochs of the Universe except the HBB era. It has also been argued that we need to remedy the present imbalance of a dominance of argument based on highly sophisticated physics but rather unsophisticated geometry, to carry out all observational and experimental consistency tests of the standard model that are possible, and to look more closely at the tensions involved in using universal physical theory to model specific unique events, and, in particular, the unique physical Universe.

The observational situation is exciting, with for example MAP, PLANCK and LISA coming on line; theory must develop adequately to deal with the new high-precision observations that will be available. We need to be daring in speculation, but to be conservative in what we claim — particularly in relation to the public at large, honestly representing the real uncertainty in our conclusions and not mistaking our models for reality. In this context, we can recognise the major strengths of what has been achieved: whatever their faults, it is only the family of inflationary almost-FL models that is able to encompass all the observations and explain the origin of structure in a unified and reasonably convincing way. That gives it a major claim to being the best model we have for the Universe at present, and so the base from which all other models will be judged. We need to continue searching for alternatives, but for the time being it is pre-eminent.

I thank Roy Maartens and Henk van Elst for helpful comments on previous drafts.

References

1. Abroe M E *et al* 2001 *Preprint* astro-ph/0111010; Bond J R *et al* 2000 in *Proc. IAU Symposium* **201** (PASP); Douspis M *et al* 2001 *Astron. Astrophys.* **379** 1; Efstathiou G *et al* 2001 *Preprint* astro-ph/0109152; Lahav O 2001 Review talk at "X Encontro Nacional de Astronomia e Astrofisica" Lisbon 2000 eds Lemos J *et al Preprint* astro-ph/0105353.
2. Bardeen J M 1980 *Phys. Rev.* D **22** 1882; Bardeen J M, Steinhardt P J and Turner M S 1983 *Phys. Rev.* D **28** 679.
3. Barrow J and Tipler F J 1986 *The Anthropic Cosmological Principle* (Oxford: Oxford University Press).
4. Barrow J D, Sandvik H B and Magueijo J 2001 *Preprint* astro-ph/0110497; Weinberg S W 2000 *Phys. Rev.* D **61** 103505.
5. de Bernadis P *et al* 2000 *Nature* **404** 955; de Bernadis P *et al* 2001 *Preprint* astro-ph/0105296; Hanany S *et al* 2000 *Astrophys. J.* **545** L5; Stompor R *et al* 2001 *Astrophys. J.* **561** L7.
6. Blanchard Ph *et al* (eds) 1998 *Decoherence: Theoretical, Experimental and Conceptual problems* (New York: Springer-Verlag).
7. Blake C and Wall J 2002 *Nature* **416** 150.
8. Boersma J P 1998 *Phys. Rev.* D **57** 798; Buchert T and Ehlers J 1997 *Astron. Astrophys.* **320** 1; Seriu M 2001 *Class. Quantum Grav.* **18** 5329.
9. Bonnor W B and Ellis G F R 1986 *Mon. Not. R. Astron. Soc.* **218** 605.
10. Bothun G 1998 *Modern Cosmological Observations and Problems* (London: Taylor and Francis).
11. Bruni M, Dunsby P K S and Ellis G F R 1992 *Astrophys. J.* **395** 34.

12. Brustein R *et al* 1995 *Phys. Rev.* D **51** 6744; Gasperini M `http://www.to.infin.it/~gasperini`; Veneziano G 2000 *Preprint* hep-th/0002094.
13. Challinor A and Lasenby A 1999 *Astrophys. J.* **513** 1; Lewis A, Challinor A and Lasenby A 2000 *Astrophys. J.* **538** 473.
14. Chaitin G 1999 *The Unknowable* (New York: Springer-Verlag).
15. Coles P and Ellis G F R 1997 *Is the Universe Open or Closed?* (Cambridge: Cambridge University Press).
16. Copeland E J, Kolb E W, Liddle A R and Lidsey J E 1993 *Phys. Rev.* D **48** 2529; Liddle A R and Taylor A N 2002 *Phys. Rev.* D **65** 041301; Lidsey J E *et al* 1997 *Rev. Mod. Phys.* **69** 373.
17. Disney M J 2002 *Gen. Rel. Grav.* **32** 1125.
18. Cornish N J and Spergel D N 2000 *Phys. Rev.* D **62** 087304; Lachièze-Rey M and Luminet J-P 1995 *Phys. Rep.* **254** 135.
19. Dyer C C and Roeder R C 1973 *Astrophys. J.* **180** L31; Dyer C C and Roeder R C 1981 *Gen. Rel. Grav.* **13** 1157.
20. Easther R, Greene B R, Kinney W H and Shiu G 2001 *Phys. Rev.* D **64** 103502; Martin J and Brandenberger R H 2001 *Phys. Rev.* D **63** 123501; Martin J and Brandenberger R H 2001 *Mod. Phys. Lett.* **A16** 999; Martin J and Brandenberger R H 2000 astro-ph/0012031; Starobinskiĭ A A 2001 *JETP Lett.* **73** 371.
21. Ehlers J 1961 *Akad. Wiss. Lit. Mainz, Abhandl. Math.-Nat. Kl.* **11** 793; Engl. transl.: 1993 *Gen. Rel. Grav.* **25** 1225.
22. Ellis G F R 1971 in *General Relativity and Cosmology* (*Proc. 47th Enrico Fermi Summer School*) ed Sachs R K (New York: Academic Press) p 104.
23. Ellis G F R 1971 *Gen. Rel. Grav.* **2** 7; Ellis G F R and Schreiber G 1986 *Phys. Lett.* A **115** 97.
24. Ellis G F R 1984 in *General Relativity and Gravitation* (*Invited Papers and Discussion Reports of the 10th International Conference*) ed Bertotti B, de Felice F and Pascolini A (Dordrecht: Reidel) p 215.
25. Ellis G F R 1991 *Mem. Italian Astron. Soc.* **62** 553.
26. Ellis G F R 1999 *Class. Quantum Grav.* **16** A37.
27. Ellis G F R 1999 *Astrophys. Space Sci.* **269-270** 6931.
28. Ellis G F R 1999 *Astron. Geophys.* **40** 4.20.
29. Ellis G F R and Baldwin J 1984 *Mon. Not. R. Astron. Soc.* **206** 377.
30. Ellis G F R and Bruni M 1989 *Phys. Rev.* D **40** 1804.
31. Ellis G F R and van Elst H 1999 in *Theoretical and Observational Cosmology* ed Lachièze-Rey M (Dordrecht: Kluwer) p 1 *Preprint* gr-qc/9812046.
32. Ellis G F R and Madsen M 1991 *Class. Quantum Grav.* **8** 667.
33. Ellis G F R, Matravers D R and Treciokas R 1983 *Ann. Phys. (N.Y.)* **150** 455; Gebbie T, Dunsby P and Ellis G F R 2000 *Ann. Phys. (N.Y.)* **282** 321; Gebbie T and Ellis G F R 2000 *Ann. Phys. (N.Y.)* **282** 285; Maartens R, Gebbie T and Ellis G F R 1999 *Phys. Rev.* D **59** 083506.

34. Ellis G F R , Perry J J and Sievers A 1984 *Astronom. J.* **89** 1124.
35. Ellis G F R and Sciama D W 1972 in *General Relativity* (Synge Festschrift) ed O'Raifeartaigh L (Oxford: Oxford University Press) p 35; Ellis G F R 2001 *Preprint* gr-qc/0102017.
36. Ellis G F R and Stoeger W J 1988 *Class. Quantum Grav.* **5** 207.
37. van Elst H, Uggla C and Wainwright J 2002 *Class. Quantum Grav.* **19** 51.
38. Fischler W *et al* 2001 *JHEP* **0107** 003.
39. Gibbons G W, Hawking S W and Siklos S T C 1983 *The Very Early Universe* (Cambridge: Cambridge University Press).
40. Goodman J 1995 *Phys. Rev.* D **52** 1821; Maartens R, Ellis G F R and Stoeger W 1996 *Astron. Astrophys.* **309** 17.
41. Guth A H 1981 *Phys. Rev.* D **23** 347.
42. Guth A H 2001 *Preprint* astro-ph/0101507; Vilenkin A 1983 *Phys. Rev.* D **27** 2848.
43. Hawking S W 1993 *Hawking on the Big Bang and Black Holes* (Singapore: World Scientific).
44. Hawking S W and Ellis G F R 1973 *The Large Scale Structure of Space-Time* (Cambridge: Cambridge University Press).
45. Hawking S W and Penrose R 1970 *Proc. R. Soc. Lond. A* **314** 529.
46. Hubble E P 1935 *The Realm of the Nebulae* (New Haven: Yale University Press).
47. de Jong R S and Lacey C 2000 *Astrophys. J.* **545** 781.
48. Kamionkowski M and Loeb A 1997 *Phys. Rev.* D **56** 4511.
49. Kantowski R 1998 *Preprint* astro-ph/9804249; Kantowski R and Thomas R C 2000 *Preprint* astro-ph/0011176.
50. Khoury J, Ovrut B A, Steinhardt P J, and Turok N 2001 *Phys. Rev.* D **64** 123522; and *Preprint* hep-th/0109050.
51. Klypin A, Hoffman Y, Kravtsov A V and Gottlöber S 2001 *Preprint* astro-ph/0107104.
52. Kolb E W and Turner M S 1990 *The Early Universe* (New York: Wiley).
53. Kristian J and Sachs R K 1966 *Astrophys. J.* **143** 379.
54. Lena P 1988 *Observational Astrophysics* transl.: King A R (New York: Springer-Verlag).
55. Levshakov S A, Molaro P, Dessauges-Zavadsky M and D'Odorico S 2002 *Preprint* astro-ph/0201043; Srianand R *et al* 2000 *Nature* **408** 931.
56. Liddle A R and Lyth D H 2000 *Cosmological Inflation and Large-Scale Structure* (Cambridge: Cambridge University Press).
57. Lidsey J E, Wands D and Copeland E J 2000 *Phys. Rep.* **337** 343; Maartens R 2000 *Phys. Rev.* D **62** 084023; Maartens R 2001 *Preprint* gr-qc/0101059; Randall L and Sundrum R 1999 *Phys. Rev. Lett.* **83** 4690; Weinberg S *Preprint* astro-ph/0104482.
58. Lifshitz E M 1946 *J. Phys. USSR* **10** 116.
59. Linde A D 1990 *Particle Physics and Inflationary Cosmology* (Chur: Harwood

Academic).
60. Maclay G J, Fearn H and Milloni P W 2001 *Preprint* quant-ph/0105002.
61. Müller B 2001 *Preprint* astro-ph/0108259; Rees M J 2001 To appear in *New Cosmological Data and the Values of the Fundamental Parameters*, ed Lasenby A and Wilkinson A (ASP Publications); Tegmark M 1998 *Ann. Phys. (N.Y.)* **270** 1; Weinberg S 2001 *Preprint* astro-ph/0104482.
62. Mukhanov V F, Feldman H A and Brandenberger R H 1992 *Phys. Rep.* **215** 203.
63. Mustapha N, Hellaby C W and Ellis G F R 1998 *Mon. Not. R. Astron. Soc.* **292** 817.
64. Nilsson U S, Uggla C, Wainwright J and Lim W C 1999 *Astrophys. J.* **522** L1.
65. D'Odorico S, Dessauges–Zavadsky M and Molaro P 2001 *Preprint* astro-ph/0102162.
66. Peacock J A 1999 *Cosmological Physics* (Cambridge: Cambridge University Press).
67. Peebles P J E 1971 *Principles of Physical Cosmology* (Princeton, NJ: Princeton University Press).
68. Peebles P J E 1980 *The Large Scale Structure of the Universe* (Princeton, NJ: Princeton University Press).
69. Peebles P J E 1999 *Publ. Astron. Soc. Pac.* **111** 274; Rees M J 1995 *Perspectives in Astrophysical Cosmology* (Cambridge: Cambridge University Press).
70. Penrose R 1989 in *Proc. 14th Texas Symp. on Relativistic Astrophysics* ed Fergus E J (New York: New York Academy of Sciences) p 249.
71. Perlmutter S *et al* 1999 *Astrophys. J.* **517** 565; Schmidt B P *et al* 1998 *Astrophys. J.* **507** 46.
72. Rindler W 1956 *Mon. Not. R. Astron. Soc.* **116** 662.
73. Robertson H P 1933 *Rev. Mod. Phys.* **5** 62.
74. Sachs R K and Wolfe A M 1967 *Astrophys. J.* **147** 73.
75. Sandage A 1961 *Astrophys. J.* **133** 355.
76. Schneider P, Ehlers J and Falco E E 1992 *Gravitational Lenses* (Berlin: Springer-Verlag).
77. Stoeger W, Maartens R and Ellis G F R 1995 *Astrophys. J.* **443** 1.
78. Tavakol R K and Ellis G F R 1988 *Phys. Lett.* A **130** 217.
79. Tipler F J, Clarke C J S and Ellis G F R 1980 in *General Relativity and Gravitation* ed Held A (New York: Plenum) p 97.
80. Wainwright J and Ellis G F R (eds) 1997 *Dynamical Systems in Cosmology* (Cambridge: Cambridge University Press).
81. Wainright J, Coley A A, Ellis G F R and Hancock M 1998 *Class. Quantum Grav.* **15** 331.
82. Wang X, Tegmark M and Zaldarriaga M 2001 *Preprint* astro-ph/0105091.
83. Weinberg S W 1972 *Gravitation and Cosmology* (New York: Wiley).
84. Wheeler J A 1968 *Einsteins Vision* (Berlin: Springer-Verlag).
85. Zel'dovich Ya B 1970 *Comment. Astrophys. Space Phys.* **2** 12.

GRAVITATING LUMPS

D. V. GAL'TSOV
Department of Theoretical Physics,
Moscow State University, 119899, Moscow, Russia
Email: galtsov@grg.phys.msu.su

Recent progress in the study of solitons and black holes in non-Abelian field theories coupled to gravity is reviewed. New topics include gravitational binding of monopoles, black holes with non-trivial topology, Lue-Weinberg bifurcation, asymptotically AdS lumps, solutions to the Freedman-Schwarz model with applications to holography, non-Abelian Born-Infeld solutions.

1 Introduction

During the past decade many surprising phenomena were discovered in the study of gravitating solitons and black holes involving Yang-Mills (YM) fields. A key role in this development has been played by gravitational sphalerons, known as Bartnik-McKinnon (BK) particle-like solutions to the Einstein-Yang-Mills (EYM) equations. The associated black holes, violating no-hair and uniqueness theorems, attracted much attention and helped to clarify a number of beliefs persistent in the Einstein theory. Other conceptually important results were obtained in the study of self-gravitating magnetic monopoles in gauge theories with spontaneous symmetry breaking. Meanwhile, for many years this research basically served only as a theoretical laboratory to study gravity in unusual conditions. Direct physical applications were unknown until the recent progress in superstring theory culminated in the discovery of D-branes and the related gauge theory/supergravity correspondence. It turned out that some gravitating lumps have non-trivial ten-dimensional interpretation and are extremely useful in the holographic treatment of the $\mathcal{N} = 1$ supersymmetric gauge theory. The goal of this talk is to give a brief overview of the subject with an emphasis on the results obtained during the last three years. A review of the work done before 1998 was given in [1] where we address the reader for references prior to November 1998.

2 Gravitational sphalerons

Neither pure gravity, nor flat space Yang-Mills (YM) theory admit static particle-like solutions. Gravity is purely attractive, and no static equilibrium is possible, unless some repulsive matter is introduced. On the contrary, the

YM field field is repulsive: due to scale invariance, the trace of the energy momentum tensor is zero, so the positivity of the energy density implies the positivity of the sum of principal pressures. Gravity breaks scale invariance, therefore the coupled Einstein-Yang-Mills (EYM) theory is expected to contain particle-like solutions. There is also a topological argument similar to Manton-Taub explanation of the existence of the sphaleron in the Weinberg-Salam theory. In that theory the Higgs broken phase manifold is S^3. Non–triviality of the third homotopy group $\pi_3(S^3)$ is an indication of the presence of non–contractible loops in the configuration space for asymptotically flat solutions, and, hence, saddle points on the energy surface. The corresponding unstable solution is *the* sphaleron; it sits on the top of the potential barrier separating topologically distinct Yang–Mills vacua. It was argued (Gal'tsov and Volkov (1991), Sudarsky and Wald (1992)) that the same argument can be applied to the gauge group itself, in fact $SU(2) \sim S^3$, once Higgs is replaced by another attractive field of non-topological nature, like gravity or dilaton.

Historically, particle-like SU(2) EYM solutions were discovered numerically by Bartnik and McKinnon (1988) before this sphaleron interpretation was suggested. The metric is spherically symmetric and is parameterized by two functions N, σ of a radial variable:

$$ds^2 = N\,\sigma^2\, dt^2 - \frac{dr^2}{N} - r^2 d\Omega, \tag{1}$$

while the YM field is purely magnetic: $A = (1-w)\,(T_\theta \sin\theta d\varphi - T_\varphi d\theta)$, where w is a real function of r and T_θ, T_φ are spherical projections of the SU(2) generators. This is a configuration with the unit winding number. Higher winding numbers lead to non-spherical solutions. An embedded Abelian magnetic monopole with unit magnetic charge is $w \equiv 0$, while $w = \pm 1$ correspond to the neighboring topologically distinct YM vacua. Particle-like solutions are characterized by regularity at the origin

$$w = 1 - br^2 + O(r^4), \quad N = 1 - 4b^2r^2 + O(r^4), \tag{2}$$

and asymptotic flatness: $N \to 1 - 2M/r$ with finite mass M, together with $\sigma \to 1$. This implies that the YM field asymptotically stays in one of the vacua

$$w = \pm(1 - a/r) + O(1/r^2). \tag{3}$$

Parameters b, M, a can be determined via numerical matching. Solutions exist for a discrete increasing sequence b_n on the semi-open interval $[b_1 = 0.4537, b_\infty = 0.706)$, with masses M_n converging to unity as $n \to \infty$ in the units m_{Pl}/g, where g is the gauge coupling constant. The function w_n has n zeroes and tends to $(-1)^n$, solutions exist for any finite n. The limiting solution, corresponding to b_∞, is asymptotically non-flat.

For the lowest $n = 1$ BK solution, w is a decreasing function varying between $w = 1$ and $w = -1$ in exactly the same way, as in the electroweak sphaleron. Odd-n solutions all have sphaleron nature, while even-n ones are topologically trivial. A vacuum to vacuum path was constructed explicitly (Gal'tsov and Volkov (1991)), showing that the Chern-Simons number is 1/2 for odd-n solutions and zero for even n. Sphaleron interpretation is supported by the fact that the n-th solution has exactly n *odd*-parity negative modes, corresponding to excitations in the electric sector. In addition, there are also n *even*-parity negative modes exhibiting gravitational instability of BK solutions. Spherically symmetric magnetic configurations of the SU(N) YM field are described by $N-1$ real functions, the corresponding BK-type solutions have a nodal structure determined by the set of $N-1$ integers (Kleihaus, Kunz and Sood (1995-1996)).

Magnetic configurations with the winding number ν could be obtained by replacing $\varphi \to \nu\varphi$ in the YM ansatz, but from the EYM equations it follows that there are no $\nu \neq 1$ spherically symmetric solutions other than embedded Abelian. Essentially non-Abelian configurations with $\nu \geq 2$ can be axially symmetric, in which case purely magnetic field is parameterized by four real functions subject to one gauge condition. Static axially symmetric regular solutions of the BK type were found numerically for lower $\nu = 2, 3, 4$ (Kleihaus and Kunz (1997)), they are likely to exist for all ν. The SU(2) solutions are characterized by a node number n, their masses increase both with growing ν and n, approaching $M = \nu$ for $n \to \infty$. The metric is parameterized by three functions entering the Papapetrou ansatz; they exhibit toroidal or spheroidal shape of the equipotential surfaces. These solutions can be interpreted as superposition of BK particles aligned along z-axis, their Chern-Simons number is $\nu/2$. It remains unknown, whether static asymptotically flat EYM solutions exist *without* axial symmetry, like multimonopoles with $\nu \geq 3$.

3 Gravitating monopoles

3.1 Gravitational excitations

Magnetic monopole in the Georgy-Glashow model possesses a mass $M_{mon} \sim v/g$, where v is the vacuum expectation value of Higgs, and a radius $R_{mon} \sim 1/(gv)$. When the monopole gravitational radius $R_G \sim GM_{mon}$ approaches R_{mon}, one can expect new phenomena to occur. With this motivation, Lee,

Nair and Weinberg (1992), Ortiz (1992), Breitenlohner, Forgacs and Maison (1992, 1995), see also [2], have investigated the coupled EYM-Higgs (EYMH) static spherically symmetric system and found that the existence of the maximal value of the dimensionless parameter $\alpha = \sqrt{G}v$, which is proportional to the ratio of the mass of W-boson in this theory to the Planck's mass, such that self-gravitating monopoles cease to exist for $\alpha > \alpha_{cr}$. This is not surprising, since for $\alpha \sim 1$ the monopole radius and its gravitational radius become comparable, so for large α the monopole would be inside the event horizon.

An important new feature is the possibility of gravitational excitation of monopoles. Note that the YM function w in the monopole solution starts with $w = 1$ at the origin and monotonically tends to zero at infinity. There are other solutions in which w oscillates n times around zero before to enter the asymptotic regime, these correspond to higher values of the parameter b in (2), increasing with n, for each α. On the phase diagram (b, α) one observes the lower branch extending up to α_{cr} (ground state monopole) where it bifurcates giving rise to the upper branch (excited monopole with one w-node). This upper branch bifurcates in its turn giving rise to the next $n = 2$ branch and so on. All upper branches in the limit $\alpha \to 0$ tend to BK_n EYM solutions after a suitable rescaling. Solutions belonging to the main branch are stable, like flat-space monopoles, while BK-excited monopoles are unstable, like BK particles. Excited monopoles can be regarded as bound states of monopoles with BK particles inside. Solutions near the main bifurcation point were recently investigated in detail by Lue and Weinberg [3] revealing new surprising features. Other recent work on gravitating monopoles includes the study of dyons [4], SU(3) monopoles [5], EYMH solutions with non-minimal gravitational coupling [6], YMH coupled to Brans-Dicke gravity [7].

3.2 Gravitational binding of monopoles

Like monopoles at rest in the flat space-time do not interact in the BPS limit where the Higgs field is massless, and a scalar attraction is precisely compensated by a magnetic repulsion. Away from the BPS bound, the Higgs field is massive, and the scalar force falls off exponentially, so magnetic repulsion dominates, and the energy of multimonopole solutions per winding number is always larger than the energy of the spherical $\nu = 1$ monopole. For $\nu \geq 3$ the BPS multimonopoles with no axial symmetry also exist [8].

It was found that gravity can bind monopoles provided the Higgs self-interaction constant λ is small enough. It was verified [9,10] that the energy per unit winding number for $\nu = 2, 3$ aligned gravitating monopoles is smaller than the energy of a single $\nu = 1$ monopole. In the case of monopole-antimonopole pairs, gravity enlarges the space of solutions: beside the ground state pair solution, there are BK-excitations which bifurcate at critical α. The upper branch in the limit of vanishing α correspond to the (rescaled) BK_1 solution, which is spherically symmetric. An excited monopole-antimonopole pair linking to BK_2 was also found [11], this is conjectured to be valid for all n.

Other flat-space particle-like solutions include sphalerons in the Weinberg-Salam model and Skyrmions. Similarly to monopoles, gravitating sphalerons exist up to finite α and link to the BK-excitations (the bifurcation pattern is somewhat different from that in monopoles, for details see [1]. When Higgs self-interaction constant increases, new sphaleron solutions, currently called *bi-sphalerons*, bifurcate from the main branch (Kunz and Brihaye (1989),Yaffe (1990)). Their response to gravity was studied in [12]. For Skyrmions the self-gravitating problem was studied by Luckock (1987) well before the corresponding monopole problem was investigated. More detailed study was undertaken by Heusler, Droz and Straumann (1991) and Bizon and Chmaj (1992). The moduli space again exhibits bifurcations with BK-excited branches.

4 Hairy black holes

4.1 Violation of black hole rules

Soon after discovery of BK solutions, it was shown that there also exist black hole solutions with similar structure of fields outside the event horizon. For them, instead of regularity conditions (2) at the origin, the regularity of the event horizon is required. The (non-degenerate) horizon is a linear zero of $N: N(r_h) = 0,\ N'(r_h) > 0$, on which w starts with a finite value $w_h = w(r_h)$. This quantity serves as a parameter labeling solutions, another parameter is the horizon radius r_h. One finds black hole solutions for any r_h and a discrete sequence of w_h within the interval $0 < w_h < 1$ (Volkov and Gal'tsov (1989), Kunzle and Masood-ul-Alam (1990), Bizon (1990)).

The EYM black holes violate a naive no-hair conjecture, since the magnetic YM field outside the horizon is not associated with conserved charges. Also, a uniqueness property is violated: an infinite number of different solutions exists with equal ADM mass and zero Coulomb charges. Black hole counterparts to axially symmetric higher-ν solutions were found as well (Kleihaus and Kunz, 1997), these obviously violate the Israel theorem, valid for vacuum and electrovacuum, stating that static black holes should be spherically symmetric. Further, the Birkhoff theorem is no longer valid for the YM matter, since 'spin from isospin' phenomenon gives rise to field modes, which are spherically symmetric, but time-dependent. Finally, the (electro)vacuum staticity theorem does not hold (though admits a non-Abelian generalization), and the Einstein-Maxwell circularity theorem is violated. All this stimulated a widespread interest to black holes with YM hair as a laboratory to test various long-standing beliefs of the black hole physics.

An internal structure of EYM black holes turned out to be particularly interesting (Donets, Gal'tsov and Zotov (1996). There are discrete sequences of black holes which have either Schwarzschild or Reissner-Nordström (RN) type internal structure, but a generic black hole exhibits a very unusual interior. The metric function N oscillates with an exponentially growing amplitude remaining always negative, but approaching zero closer and closer in subsequent cycles. Each time when such an 'almost' Cauchy horizon is reached, the mass function starts to inflate, making N to turn back, so true Cauchy horizons never form. These oscillations never end, and the approach to the singularity is non-analytic. All solutions, except for a discrete RN-type set (of zero measure in the moduli space) satisfy the strong cosmic censorship hypothesis (no internal Cauchy horizons).

4.2 Monopole-black hole transition

Gravitating Georgy-Glashow model solutions with $\alpha > \alpha_{cr}$ contain an event horizon inside the monopole core. This phenomenon is rather generic (Kastor and Trashen (1992)), it occurs each time when the lagrangian leads to the relation $\varepsilon + p_r \sim N$ at a 'would be' event horizon (the sum of the energy density and the radial pressure). It is so for the EYMH systems with triplet or complex doublet Higgs and the Skyrme model, all these theories admit hairy black holes. In fact, the first such black hole was discovered in the Skyrme model (Luckock and Moss (1986)), it is particularly interesting as an example of a *stable* hairy black hole. Note, that purely EYM black holes are unstable, n-node solution inherits from BK $2n$ negative modes. Skyrme black holes violate even a weak form of the no-hair conjecture, ascribing the no-hair property only to stable black holes. For the recent work on gravitating Skyrme model see [13]. Dyonic EYMH black holes were discussed in [14].

A monopole-black hole transition near the bifurcation threshold was recently studied in detail by Lue and Weinberg [3], see also a review [15]. Globally regular and black hole space-times are qualitatively different, nevertheless, transition between them passes through a continuous family of configurations. Near the threshold $\alpha \sim \alpha_{cr}$ the metric function N develops a minimum at some point, attaining a value very close to zero. This regular space-time has properties close to those of black holes. One can associate with near-critical solutions an *entropy* [16], since information about the quasi-interior region is accessible with larger and larger time delay as one approaches the critical point. This entropy is likely to be compatible with the Bekenstein entropy. Lue-Weinberg bifurcation was also discussed in other models [17,4].

4.3 *Static non-spherical black holes*

Black hole counterparts to axially-symmetric multimonopole solutions were constructed [17,10] possessing a deformed event horizon with a constant surface gravity. These disprove [18] a recent modified uniqueness conjecture formulated in terms of the 'isolated horizons' language [19]. Axial dyonic black holes were constructed in [20]. An interesting case is the neutral black hole with a magnetic dipole hair [21]. It is obtained by inserting a horizon between two centers in the monopole-antimonopole solution. Black holes in higher ν magnetic monopoles may be non-axisymmetric, therefore violating Hawking's 'strong rigidity' (Ridgway and Weinberg (1995)). Though no explicit solutions (even numerical) were obtained so far, the argument showing their existence is simple: there is an instability of the magnetically charged (embedded Abelian) Reissner-Nordström solution, which may have any magnetic charge, with respect to massive vector field present in the spontaneously broken gauge theory, in modes which are not axially symmetric. Moreover, again due to 'spin from isospin', there are *no* spherically symmetric states for $\nu \neq 1$. Going beyond the linear approximation, one is able to show that for $\nu = 2$ the instability results in an axially symmetric configuration, for $\nu = 3$ — tetrahedral, for $\nu = 4$ — cubic, etc. In fact, such crystal-type structures are known in the flat space BPS multimonopole systems [8]. Note that the internal structure of hairy black holes with scalar fields is very different then described above for the EYM case: near the singularity the scalar field dominates and the mass-function exhibit a 'power-low inflation' (Gal'tsov, Donets and Zotov (1977), Breitenlohner, Lavrelashvili and Maison (1998)). For the Skyrme case see [22].

4.4 *YM hair and quantum coherence*

String infinite symmetries may be responsible for the maintenance of the quantum coherence in the black hole evaporation. These symmetries could encode all information about an enormous number of massive string excitations accumulated in the black hole state. Initially this idea was implemented using essentially two-dimensional black holes (quantum W-hair). An attempt to construct a four-dimensional model was made in [23] invoking hairy black holes with the $SU(\infty)$ YM field. The $su(N)$ algebra in the limit $N \to \infty$ can be mapped to Poisson brackets defined on a unit 'internal' sphere. The static spherically symmetric ansatz, elaborated by Kunzle (1991) for finite N, then gives rise to the YM function depending on a radial variable and an internal angle θ. In presence of the large negative cosmological constant (with respect to the energy scale of the YM field) an approximate solution was obtained describing a black hole with hair, depending on infinite number of parameters.

5 Asymptotically AdS lumps

5.1 Nodeless solutions for EYM system

Recent interest to AdS/CFT correspondence stimulated an investigation of the EYM systems in presence of the negative cosmological constant Λ. Hairy black holes in the SU(2) theory were first studied in [24], and then (together with regular solutions) in [25]. Asymptotic conditions now follow form a finiteness of the mass M, which is a limiting value of the mass function $m(r)$ defined via $N = 1 - 2m(r)/r - \Lambda r^2/3$ (also $\sigma \to 1$). This no more implies $w \to \pm 1$, instead any constant value w_∞ is allowed. This means that solutions will have non-zero magnetic charge $Q_m = (1 - w_\infty^2)$. Since the asymptotic condition on w is relaxed, one finds solutions for continuously varying parameter b in (2) for each number of nodes n including zero. Topologically, the nodeless solutions have no sphaleronic nature, so it can be expected that they do not have odd parity negative modes. This was proved to be the case both in the spherically symmetric sector [24,25,?] and for non-spherical perturbations [26]. The gravitational (even parity) perturbations were studied in the spherical sector in [25] and for non-spherical excitations in [26]. Spherical modes are stable in the $n = 0$ sector for $w_\infty < 1/\sqrt{3}$. Another news is that the 'non-Abelian baldness' (forbidding AF SU(2) dyons, Gal'tsov and Ershov (1989,1990)) is no longer valid: there are essentially non-Abelian solutions with an electric sector in the SU(2) theory. One finds dyon solutions with finite M for continuously varying b (and the corresponding parameter in the electric sector) and any number of nodes n including $n = 0$.

Although these AdS monopoles and dyons are very different from those in the AF EYMH theory, one finds again the bifurcation patterns similar to described above: branches with neighboring n get connected via bifurcation points in the parameter space. In particular, the mass of nodeless monopole solutions as a function of the magnetic charge exists till some critical charge value, then it turns back forming an upper unstable $n = 1$ branch. This feature, first observed in [25], was further studied in [27]. Axially symmetric EYM solutions with higher winding numbers were investigated for $\Lambda < 0$ in a recent paper [28]. Transition from negative to positive Λ apparently reveals a fractal structure of the moduli space [25].

5.2 Asymptotically AdS monopoles and sphalerons

Solutions of the EYMH system with triplet Higgs were studied in [29,30]. Contrary to the EYM case, no essentially new features appear as compared with the $\Lambda = 0$ theory. Like in the AF case, they exist in a limited region of the parameter Gv^2, the critical value being somewhat smaller. The theory with the doublet Higgs in presence of the negative cosmological constant was investigated in [31]. Again, the situation is rather similar to the $\Lambda = 0$ case: the electric component of the YM field is forbidden, no arbitrary w_∞ is possible, a critical value of the Newton constant is observed where solutions cease to exist. However, non-zero Λ implies a power law fall-off of the fields at infinity. An extensive study of perturbations of asymptotically AdS solitons and black holes was performed recently [26,32] confirming earlier results on stability of nodeless solutions.

5.3 Black holes with non-spherical topology

Once asymptotic flatness is abandoned, there is no more 'topological censorship', forbidding black holes with non-spherical topology of the event horizon. Namely, for the negative cosmological constant, two-dimensional section of the horizon may be an arbitrary genus Riemann surface [33,34,35]. Generalization of topological censorship for higher genus black holes was discussed in [36]. Asymptotically AdS solutions provide non-trivial examples of black holes with the winding number one which are topologically different from the BK type solutions. One has merely to replace the spherical element $d\Omega$ in (1) by flat or hyperbolic $d\Omega_k = d\theta^2 + f_k(\theta)d\varphi^2$, with $k = 1, 0, -1$ and $f_1 = \sin\theta$, $f_0 = \theta$, $f_{-1} = \sinh\theta$. Purely magnetic nodeless $n = 0$ black holes with $k = -1, 0$ were constructed numerically in [37]. For $k = 0$ the mass function remains positive definite elsewhere, so the solutions have positive mass. For the negative curvature $k = -1$, there are black holes not only with positive, but also with zero or negative mass. Solutions with $k = 0$ are stable both in odd and even parity sectors, while the hyperbolic solutions $k = -1$ are stable provided $w_\infty > 1$. Black hole solutions with non-spherical topology do not have globally regular counterparts.

6 Rotation

6.1 Failure of circularity

Mechanical equilibrium inside the flat space static compact field configurations generically will be destroyed once they are forced to rotate, so a priori it is unlikely that one could endow solitons with an arbitrary angular momentum. On the other hand, in quantum theory, magnetic monopoles generically acquire discrete angular momentum (spin) due to interaction with fermions (fermion zero modes). This is closely related to supersymmetric embeddings, so the possibility of *proper* rotational zero modes (involving bosonic fields) is an important question. It is known that there are no such modes satisfying Bogomol'nyi equation. More recently the absence of non-Bogomol'nyi rotational zero modes for EYMH particle-like solutions was proved by Brodbeck and Heusler (1997) and Brodbeck, Heusler, Straumann and Volkov (1997). One serious complication to study rotation non-perturbatively is a failure of the circularity theorem for non-Abelian matter (Heusler and Straumann, 1993). The problem is that the Ricci-circularity condition for the stationary axially-symmetric space-time is not ensured by just imposing the same symmetries on the YM field. Therefore a Papapetrou ansatz does not necessarily hold. The most general parameterizations was derived in [38]. It is a 2D dilaton gravity, with the matter sector including the 2D Yang-Mills field , two 2D Higgs fields , as well as two scalar moduli , the dilaton and two 2D Kaluza-Klein two-forms.

6.2 Rotational zero modes

Meanwhile, in the *linearized* theory, the number of odd-parity zero modes involved in rotation is considerably less. An analysis of perturbations of the EYMH system, extending previous work on this subject (see [1] for a review and references), was performed in [39,40], showing, in particular, stability of the Schwarzschild and Reissner-Nordström black holes with respect to odd-parity EYM perturbations. It was confirmed also that only $l = 1$ zero modes, consisting of three real-valued functions, are responsible for (infinitesimal) rotation of both EYM and EYMH solitons and black holes. At this level circularity *does* hold. The set includes $h_{t\varphi}$ metric perturbation and two isotopic variations of A_t. They satisfy a coupled system of linear ordinary differential equations of the second order, so the overall order of the system is *six*. A necessary condition for globally defined continuously varying zero modes to exist is a non-trivial intersection of the moduli spaces of local solutions near the origin (the event horizon black holes) and at spatial infinity. A local analysis reveals that the moduli space at infinity is *four*-dimensional for EYM (without Higgs) and *three*-dimensional for EYMH. In both cases the dimension of the moduli space near the origin (solitons) is *three*, while that at the horizon is *four*. Thus, for solitons involving scalar fields, the intersection of two three-dimensional moduli spaces inside the full six-dimensional space is trivial. For black holes inside magnetic monopoles, there is a one-parameter intersection space, so event horizons may be rotating. For EYM (BK) particle-like solutions the intersection space is also one-dimensional, and more careful analysis shows that the electric charge and angular momentum are excited simultaneously.

6.3 Rotation and charge

An essential difference between EYMH solitons and BK particles is that the binding is non-gravitational in the first case and gravitational in the second. So it is perhaps not surprising that BK solutions possess rotational zero modes, while magnetic monopoles do not. For black holes gravitational attraction is dominating, so zero modes do exist in both cases. These may be regarded as another admissible hair for static solutions. Another interesting feature is that the charge and rotational degrees of freedom are connected in a different ways for different solutions. Recall that the magnetic YM configuration in BK particles is of the dipole type. An intuition based on the properties of the Kerr-Newmann solution suggest to view the magnetic moment of charged rotating solutions as due to Faraday effect. For rotating BK the situation is inverse: magnetic dipole configuration is primary, while rotation leads to charging the particle. For EYM black holes the angular momentum and the electric charge are independent parameters, but the relationship between the angular velocity of the event horizon and the angular momentum defined at infinity is non-trivial. This is again related to the 'spin from isospin' phenomenon: there is an extra contribution to the angular momentum from the black hole hair. In particular, there exist black holes with zero angular velocity of the horizon (static), but non-zero angular momentum (and non-zero charge). Alternatively, there are black holes with rotating horizons, but zero angular momentum, as seen from infinity. A non-perturbative numerical analysis of rotating EYM black holes with an assumption of circularity was reported recently [41], confirming consistency of this assumption at non-linear level.

7 Links to superstrings

7.1 Dilaton

Since the YM field is scale-invariant, the only dimensional parameter of the EYM theory is the Planck mass. It is natural, therefore, to look for physical applications in the superstring theory. The first step to incorporate BK solutions into this framework consisted in including the dilaton. Particle-like solutions similar to BK were found in the flat YMD theory by Lavrelashvili and Maison (1992). When gravity is also taken into account, one finds a surprising equality of the ADM mass and the dilaton charge of BK-type solutions (Donets and Gal'tsov, 1993). This was shown to be is a consequence of the scaling symmetry of the system of equation leading to the relation $g_{tt} = \exp(2\phi)$ (Donets, Gal'tsov and Volkov, 1993). Another enigmatic feature was observed for the $n = 1$ EYMD solution: it turned out that the parameter b in (2) is rational, $b_1 = 1/6$, while $a_1 = 2M$ (Lavrelashvili and Maison, 1993). Efforts to explain these 'magic' numbers have led to a fascinating development; before going to this, let us mention some recent work on gravitating lumps with dilaton: axially symmetric EYMD configurations [42], Georgy-Glashow-dilaton monopoles [43] and black holes [44], dilaton binding in multimonopoles [45].

7.2 Freedman–Schwarz model

In a search of hidden supersymmetry tentatively underlying these features, Chamseddine and Volkov (1997) have investigated the $\mathcal{N} = 4\, SU(2) \times SU(2)$ gauged supergravity also known as Friedman-Schwarz (FS) model. Its bosonic sector includes two YM fields, dilaton and axion (which can be set zero for static purely magnetic configurations) with the dilaton potential, proportional to $-(g_1^2 + g_2^2)\exp(-2\phi)$. Formally one can derive the SU(2) EYMD equations for $g_2 = ig_1$. This trick does not destroy supersymmetry: the supergravity Bogomol'nyi equations survive. Later it was shown by Volkov [47] that such 'complexified' theory corresponds to the Euclidean FS model, and the associated Bogomol'nyi equations admit indeed one-node $w(r)$ as a solution [48,49] (though still not found analytically). If one passes to the Lorentzian metric in this solution (this does not return us to the initial Lorentzian FS model), one obtains the EYMD analog of the BK_1, which is, of course, not supersymmetric (and unstable), but still bears an imprint of supersymmetry of its Euclidean counterpart.

Leaving only one SU(2) field in N=4 supergravity ('half-gauged' FS model), one gets the lagrangian $L = -R + 2(\nabla\phi)^2 - e^{2\phi}F^2/2 + e^{-2\phi}$ which differs from the simple EYMD one by the dilaton potential. This theory admits an analytic solution [51,52]

$$ds^2 = dt^2 - dr^2 - R^2 d\Omega, \quad R^2 = 2r\coth(r+c) - w^2 - 1,$$

$$w = r\sinh^{-1}(r+c), \quad e^{2\phi} = R^{-1}\sinh(r+c), \tag{4}$$

where the metric is presented in the string frame, and c is a real parameter. The solution preserves 1/4 of the initial supersymmetry of the FS model for all c. Asymptotically $w \to 0$, so this is a charge one monopole. However, the metric asymptotically is not flat and not AdS. The $c = 0$ solution is globally regular, an expansion of w near the origin is of the type (2) with 'magic' $b = 1/6$. For $c \neq 0$ (c is assumed to be non-negative), the radius R shrinks at some finite r, which is a curvature singularity. In the limit $c \to \infty$ one gets an Abelian dilatonic monopole.

7.3 Holographic interpretation

This solution has a ten-dimensional counterpart in Type I (pure) supergravity; in fact, as was shown by Volkov (1997), the $D = 4$ FS model corresponds to compactification of the latter on $S^3 \times S^3$. Somewhat unexpectedly, it was found [50] that the same solution describes wrapping of a NS5 brane of the IIB theory on a shrinking S^2 . This opens a way to construct a holographic description for the $\mathcal{N} = 1$ super Yang-Mills. A particularly interesting aspect of the latter theory, which can be investigated using this duality, is the chiral symmetry breaking transition. The non-Abelian component of the gauge field serves as the order parameter: unbroken chiral symmetry corresponds to having only Abelian YM component, $w = 0$. Thus, the BK-type solutions correspond to broken chiral symmetry, while a formation of an Abelian black hole signals the symmetry restoration. So one needs a detailed information about solutions including black holes; this was the subject of the recent paper by Gubser, Tseytlin and Volkov [52]. Other solutions than (4) are non-BPS, they include globally regular solutions which start at the origin with $b \neq 1/6$. Like for asymptotically AdS BK solutions, the parameter b here is continuous, for $b < 1/6$ w-solutions are still nodeless, while for $1/6 < b < 1/2$, solutions with more and more nodes appear, with the limiting solution $n = \infty$ starting with $b = 1/2$. These solutions may be regarded as BK-excited monopoles. This part of the solution space is somewhat similar to solutions of the pure EYM theory with a negative cosmological constant. However, now an asymptotic value of w is zero, indicating the charge one monopole topology, while in the AdS case the magnetic (and electric) charges are continuously varying. An interesting new feature is that there is a discrete subspace of solutions with $b = b_n$ possessing a finite energy. Apart from these, there are non-trivial solutions with $w \equiv \pm 1, 0$, which have different ten-dimensional counterparts.

Black hole solutions are determined by the horizon value w_h. For $w_h = 0$ solutions are Abelian, while non-zero values w_h give rise to non-Abelian black holes. These contain a discrete subclass of finite-energy solutions for which w falls off exponentially at infinity. Black holes with toroidal event horizon in the $U(1) \times U(1)$ sector of the FS model were constructed in [53]

7.4 Gauss-Bonnet

One earlier attempt to incorporate 'stringy' effects was to include higher-curvature terms in the lagrangian. First α' correction to the heterotic string effective action includes the Gauss-Bonnet term, $\beta \exp(2\phi)R_{GB}$, $R_{GB} = R_{\mu\nu\lambda\tau}R^{\mu\nu\lambda\tau} - 4R_{\mu\nu}R^{\mu\nu} + R^2$, which produces a non-trivial effect in four dimensions due to the dilaton factor. Note by passing that this term does not generate derivatives higher than the second order. Remarkably, in the static spherical space-time this term does not destroy the scale invariance of the equations of motion, so a non-trivial first integral still exists (Donets and Gal'tsov, 1993) implying the equality $M = D$. However, solutions exist only for restricted values of the coefficient $\beta < \beta_n$, e.g. $n = 1$ solution disappears for $\beta_1 = 0.37$, the critical value decreases for higher n. The question was raised whether quadratic curvature terms (Gauss-Bonnet) can stabilize negative modes [46]. It is known that Gauss-Bonnet acts as a 'hair tonic' opening a channel of neutral black holes with a dilaton charge (Kanti, Mavromatos, Rizos, Tamvakis and Winstanley, 1996), and these are linearly stable in absence of the vector fields (the same authors, 1998). Actually, in the EYM(D) static spherically symmetric theory with the Gauss-Bonnet term there exists four types of black holes: i) neutral dilatonic (YM sector non-excited), ii) usual BK type, iii) holes with the unit magnetic charge (embedded Abelian, electric sector non-excited) and vi) holes with an electric charge, also embedded Abelian. It was found in [46] that those solutions whose perturbations involve the YM magnetic function (ii, iii) remain unstable (in particular, sphaleronic instabilities of ii) are untouched), while neutral and electric solutions are stable, reflecting stability of Schwarzschild and Reissner-Nordström black holes.

8 Born-Infeld

Another aspect related to superstrings consist in investigation of solitons and black holes in the non-Abelian SU(N) Dirac-Born-Infeld theory, arising as an effective theory on N D3-branes [54]. If branes are coincident, one deals with the purely massless theory, while for separated branes there is a Higgs potential with the vacuum expectation value equal to the brane separation. The simplest case is the pure YM theory with the Born-Infeld (BI) type action

$$S = \frac{\beta^2}{4\pi}\mathrm{tr}\int \left\{1 - \sqrt{-\det(g_{\mu\nu} + \beta^{-1}F_{\mu\nu})}\right\} d^4x \tag{5}$$

In the non-Abelian case, $F_{\mu\nu}$ is matrix-valued, and one can a priori take the trace in the action in different ways (see [55] for details). Two definitions are the most reliable: the symmetrized trace suggested by Tseytlin, which well covers the lower orders of the perturbative string effective action, and the 'ordinary trace-square root' form, which is a direct non-Abelian generalization of the four-dimensional U(1) BI action obtained by evaluating the determinant under the square root. For the static spherically symmetric SU(2) magnetic ansatz these definitions lead to somewhat different one-dimensional lagrangians:

$$L_{tr} = \beta^2 r^2 (1 - R) \quad \mathcal{R} = (1 + V^2 + 2K^2)^{1/2}, \tag{6}$$

where $V^2 = (1-w^2)^2/(2\beta^2 r^4)$, $K^2 = w'^2/(2\beta^2 r^2)$, and

$$L_{str} = \beta^2 r^2 \left[1 - (1+V^2)^{1/2} + K^2 \mathcal{A}(1+V^2)^{-1/2}\right] \tag{7}$$

with $\mathcal{A} = W^{-1} \arctan W$, $W = (1+V^2)^{1/2}(V^2 - K^2)^{-1/2}$.

8.1 NBI particle-like solutions (sphalerons)

The BI lagrangian breaks scale invariance introducing a new new scale parameter — BI critical field β, which should be regarded as a manifestation of non-locality of the underlying string theory. Therefore, the no-go theorem for classical glueballs in the flat space YM theory is overruled, and one is led to look for such solutions in the non-Abelian Born-Infeld (NBI) theory. The proof of existence and numerical results for flat space NBI particle-like solutions were presented in [56] for the ordinary trace (6) and in [55] for the symmetrized trace (7) lagrangians, they have qualitatively similar properties. Solutions form a sequence of the BK type, but the difference is that now the parameter b_n in the expansion (2) is rather large and rapidly growing with the node number n, for example, $b_1 = 12.7$, $b_2 = 887$ in the case (6) (these values are even larger for the model (7) [55]). Solutions have sphaleron features and are expected to be unstable. When gravity is taken into account [57,58], one finds that b_n, as a function of the dimensionless parameter $G\beta$, continuously interpolates [58] between these values for small $G\beta$, and the BK values for large $G\beta$.

8.2 Damping of gigantic oscillations inside YM black holes

Black hole solutions in the Einstein-NBI theory were also constructed [57,58], outside the horizon they are qualitatively similar to those in the EYM case. However, an *internal* structure of the ENBI black holes is drastically different. In fact, one could expect that violent oscillations inside the EYM black holes should be modified in quantum theory: already in the first inflationary cycle the mass function attains values exceeding the Planck mass. Although the

ENBI theory is still classical, it incorporates string α' quantum corrections in a non-perturbative way, so one could expect that the problem of the over-Planckian masses in the EYM black holes will be resolved. This is indeed the case [58]: the internal behavior of the mass function now is perfectly smooth, though the singularity still bears non-analytic features. Namely, the function w attains a final value w_0, and there exist a two-parameter family of solutions for which the mass function is analytic in the Schwarzschild-type singularity. This family, however, is not generic, like in the EYM case. But now a generic three-parametric solution can be obtained in the vicinity of the singularity as the series expansion in terms of $r^{1/2}$. Mass function again attains a finite value, but one finds that the singularity is much weaker than in the Schwarzschild case. The scalar curvature diverges only as $R \sim r^{-3/2}$. Note that non-analytic behavior of the mass function $m(r)$ in the singularity is also observed in the Abelian Einstein-Born-Infeld theory coupled to dilaton [59].

8.3 NBI monopoles

Adding triplet Higgs to the NBI lagrangian, one finds flat space monopoles [60,61] which exhibit features, similar to those of gravitating monopoles. In view of the existence of the flat-space NBI sphalerons, this is not surprising. Monopole (and dyon) solutions exist for β varying from infinity to a finite value β_{cr}, which was shown [62] to be a bifurcation point giving rise to the branch of excited monopoles with one-node w (higher node excitations exist as well). In [62] a non-commutative U(1) monopole (with Higgs) was also constructed following the D-brane in B-field / non-commutative gauge theory correspondence. Note that it is the symmetrized trace version of the theory which leads to the BPS equations coincident with the YMH ones. Effect of gravity on the NBI monopoles was discussed in [63,64]. Recently the NBI model with doublet Higgs was considered [65], it was found that the sphalerons in this theory interpolate between the usual electroweak sphaleron and the NBI sphalerons of [56].

9 Conclusion

Basic properties of gravitating non-Abelian solitons and hairy black holes are fairly well understood now and exhibit certain universality: character of gravitational excitations, bifurcation patterns of moduli spaces, dependence of spectra on asymptotic conditions, phenomena near 'almost' horizons. Our list of references is almost complete for the period 1999 – November 2001 within the covered topics. However, many related directions remained outside the scope of the talk: global monopoles, boson stars, solutions to the gravity coupled Abelian Higgs model, cosmological solutions with non-Abelian fields, topological inflation, critical collapse of the YM field, solutions to higher-dimensional EYM systems. More complete update to the review [1] will be given elsewhere.

Acknowledgments

The author is grateful to LOC for hospitality during the conference and to Gravity Research Foundation for support. The work was also supported in part by the Russian Foundation for Basic Research under grant 00-02-16306.

References

1. M. S. Volkov and D. V. Gal'tsov. *Phys. Rept.*, 319:1–83, 1999.
2. D. Maison. *Lect. Notes Phys.*, 539:263–278, 2000.
3. A. Lue and E. J. Weinberg. *Phys. Rev.*, D60:084025, 1999; D61:124003, 2000.
4. Y. Brihaye, F. Grard, and S. Hoorelbeke. *Phys. Rev.*, D62:044013, 2000.
5. Y. Brihaye and B. M. A. G Piette. *Phys. Rev.*, D64:084010, 2001.
6. J. J. van der Bij and E. Radu. *Nucl. Phys.*, B585:637–665, 2000.
7. T. Tamaki, Kei-ichi Maeda, and T. Torii. *Phys. Rev.*, D60:104049, 1999.
8. C. J. Houghton and P. M. Sutcliffe. *Commun. Math. Phys.*, 180:343–362, 1996 .
9. B. Hartmann, B. Kleihaus, and J. Kunz. *Phys. Rev. Lett.*, 86:1422–1425, 2001.
10. B. Hartmann, B. Kleihaus, and J. Kunz. 2001, hep-th/0108129.
11. B. Kleihaus, and J. Kunz. *Phys. Rev. Lett.*, 85:2430–2433, 2000.
12. Y. Brihaye and M. Desoil. *Mod. Phys. Lett.*, A15:889–900, 2000.
13. I. G Moss, N Shiiki, and E Winstanley. *Class. Quant. Grav.*, 17:4161–4174, 2000.
14. Y. Brihaye, B. Hartmann, and J. Kunz, and N. Tell. *Phys. Rev.*, D60:104016, 1999.
15. E. J. Weinberg. Black holes with hair. 2001, gr-qc/0106030.
16. A. Lue and E. J. Weinberg. *Gen. Rel. Grav.*, 32:2113–2118, 2000.
17. Y. Brihaye, B. Hartmann, and J. Kunz. *Phys. Rev.*, D62:044008, 2000.
18. B. Kleihaus, J. Kunz, A. Sood, and M. Wirschins. Horizon properties of Einstein-Yang-Mills black hole. 2001, gr-qc/0110084.

19. A. Corichi, U. Nucamendi, and D. Sudarsky. *Phys. Rev.*, D62:044046, 2000.
20. B. Hartmann, B. Kleihaus and J. Kunz. *Mod. Phys. Lett.*, A15:1003–1012, 2000.
21. B. Kleihaus and J. Kunz. *Phys. Lett.*, B494:130–134, 2000.
22. T.Tamaki, Kei-ichi Maeda, and T. Torii. Internal structure of Skyrme black hole. 2001, gr-qc/0106008.
23. N. E. Mavromatos and E. Winstanley. *Class. Quant. Grav.*, 17:1595–1611, 2000.
24. E. Winstanley. *Class. Quant. Grav.*, 16:1963–1978, 1999.
25. J. Bjoraker and Y. Hosotani. *Phys. Rev. Lett.*, 84:1853, 2000; *Phys. Rev.*, D62:043513, 2000.
26. O. Sarbach and E. Winstanley. *Class. Quant. Grav.*, 18:2125–2146, 2001.
27. Yutaka Hosotani. Scaling behavior in the Einstein-Yang-Mills monopoles and dyons. 2001, gr-qc/0103069.
28. E. Radu. Static axially symmetric solutions of Einstein-Yang-Mills equations with a negative cosmological constant: The regular case. 2001, gr-qc/0109015.
29. A. R. Lugo and F. A. Schaposnik. *Phys. Lett.*, B467:43–53, 1999.
30. A. R. Lugo, E. F. Moreno, and F. A. Schaposnik. *Phys. Lett.*, B473:35–42, 2000.
31. J. J. Van der Bij and E. Radu. *Phys. Rev.*, D64:064020, 2001.
32. E. Winstanley and O. Sarbach. On the linear stability of solitons and hairy black holes with a negative cosmological constant: the even-parity sector. 2001, gr-qc/0111039.
33. J. P. S. Lemos. *Class. Quant. Grav.*, 12: 1081, 1995; *Physics Letters*, B353: 46 1995.
34. J. P. S. Lemos and V. T. Zanchin, *Phys. Rev*, D54: 3840, 1996.
35. L. Vanzo. *Phys. Rev.*, D56:6475–6483, 1997.
36. G. J. Galloway, K. Schleich, D.M. Witt, and E. Woolgar. *Phys. Rev.*, D60:104039, 1999.
37. J. J. Van der Bij and Eugen Radu. New hairy black holes with negative cosmological constant. 2001, gr-qc/0107065.
38. D. V. Gal'tsov. Proc. Int. Seminar *Mathematical Cosmology*, Potsdam, April 1998. Eds: M. Rainer, H.-J. Schmidt, WS Publ., Singapore, 1998, 92-103, gr-qc/9808002.
39. O. Brodbeck, M. Heusler, and O. Sarbach. *Phys. Rev. Lett.*, 84:3033, 2000.
40. O. Sarbach, M. Heusler, and O. Brodbeck. *Phys. Rev.*, D62:084001,

2000; D63:104015, 2001.
41. B. Kleihaus and J. Kunz. *Phys. Rev. Lett.*, 86:3704–3707, 2001.
42. B. Kleihaus and J. Kunz. *Phys. Rev.*, D61:107503, 2000.
43. P. Forgacs and J. Gyurusi. *Phys. Lett.*, B441:275–284, 1998.
44. Y. Brihaye, B. Hartmann, and J. Kunz. Dilatonic monopoles and 'hairy' black holes. 2001, hep-th/0106227.
45. Y. Brihaye and B. Hartmann. Bound monopoles in the presence of a dilaton. 2001, hep-th/0110080.
46. P. Kanti and E. Winstanley. *Phys. Rev.*, D61:084032, 2000.
47. M. S. Volkov. *Nucl. Phys.*, B566:173–199, 2000.
48. M. S. Volkov and D. Maison. *Nucl. Phys.*, B559:591–602, 1999.
49. M. S. Volkov. The tale of gravitational sphaleron. 1999, hep-th/0002006.
50. M. Maldacena and C. Nunez. *Phys. Rev. Lett.*, 86:558-591 (2001).
51. Ali H. Chamseddine and Mikhail S. Volkov. *Phys. Rev. Lett.*, 79:3343-3346, 1997; *Phys. Rev.*, D57:6242-6254, 1998.
52. S. S. Gubser, A. A. Tseytlin, and M. S. Volkov. Non-Abelian 4-d black holes, wrapped 5-branes, and their dual descriptions. *JHEP*, 09:017, 2001.
53. D. Klemm. *Nucl. Phys.*, B545:461–478, 1999.
54. A. Giveon and D. Kutasov. *Rev. Mod. Phys.*, 71:983, 1999.
55. V. V. Dyadichev and D. V. Gal'tsov. Sphaleron glueballs in nbi theory with symmetrized trace. *Nucl. Phys.*, B590:504–518, 2000.
56. D. Gal'tsov and R. Kerner. Classical glueballs in non-Abelian Born-Infeld theory. *Phys. Rev. Lett.*, 84:5955–5958, 2000.
57. M. Wirschins, A. Sood, and J. Kunz. Non-Abelian Einstein-Born-Infeld black holes. *Phys. Rev.*, D63:084002, 2001.
58. V. V. Dyadichev and D. V. Gal'tsov. Solitons and black holes in non-Abelian Einstein-Born- Infeld theory. *Phys. Lett.*, B486:431–442, 2000.
59. G. Clement and D. Gal'tsov. *Phys. Rev.*, D62:124013, 2000.
60. N. Grandi, R. L. Pakman, F. A. Schaposnik, and G. A. Silva. *Phys. Rev.*, D60:125002, 1999.
61. N. Grandi, E. F. Moreno, and F. A. Schaposnik. *Phys. Rev.*, D59:125014, 1999.
62. D. Galtsov and V. Dyadichev. D-branes and vacuum periodicity. Proc. NATO ARW *Non-commutative structures in mathematics and physics*, Kiev, Ukraine, Sept. 24-28, 2000; S. Duplij and J. Wess (Eds), Kluwer, 2001, pp. 61-78 hep-th/0012059.
63. P. K. Tripathy. Gravitating monopoles and black holes in Einstein-Born-Infeld-Higgs model. *Phys. Lett.*, B458:252–256, 1999; B463:1–8, 1999.
64. P. K. Tripathy and F. A. Schaposnik. Monopoles in non-Abelian

Einstein-Born-Infeld theory. *Phys. Lett.*, B472:89–93, 2000.
65. Y. Brihaye and B. Hartmann. The Born-Infeld sphaleron. 2001, hep-th/0110053.

ENERGY INEQUALITIES FOR ISOLATED SYSTEMS AND HYPERSURFACES MOVING BY THEIR CURVATURE

GERHARD HUISKEN

Max-Planck-Institut für Gravitationsphysik, Albert-Einstein-Institut, Am Mühlenberg 1, D-14476 Golm-Potsdam, Germany

E-mail: huisken@aei.mpg.de

TOM ILMANEN

Departement Mathematik, ETH Zentrum, Rämistrasse 101, CH-8092 Zürich, Switzerland

Home-Page: http://www.math.ethz.ch/~ilmanen

The total energy of an isolated gravitating system in General Relativity is described by a geometric invariant of asymptotically flat Riemannian 3–manifolds. One–parameter families of two-dimensional hypersurfaces foliating such a manifold and obeying natural curvature conditions can be used to encode and study geometrical and physical properties of the 3–manifold such as mass, quasi-local mass, the center of mass and energy inequalities. The article describes recent results on Penrose inequalities, inverse mean curvature flow, constant mean curvature surfaces and their interconnections.

1 The mass of asymptotically flat 3–manifolds

The classical description of an isolated gravitating system such as a star, black hole or galaxy is given by a Lorentzian 4–manifold $(L^4, \bar{g})$ of signature $(- + ++)$ satisfying Einsteins equations

$$R^4_{\alpha\beta} - \frac{1}{2} R^4 \bar{g}_{\alpha\beta} = 8\pi T_{\alpha\beta} \tag{1}$$

and approaching flat Minkowski space near infinity in an appropriate sense. We denote by $\{R^4_{\alpha\beta}\}$, $0 \leq \alpha, \beta \leq 3$, and R^4 the Ricci tensor and the scalar curvature of the Lorentzian metric $\bar{g}$ respectively. The energy momentum tensor $\{T_{\alpha\beta}\}$ is often assumed to satisfy a physically motivated positivity condition such as the weak energy condition

$$T_{\alpha\beta} X^\alpha X^\beta \geq 0 \qquad \text{for timelike } X. \tag{2}$$

The standard approach to the study of $(L^4, \bar{g})$ is a decomposition of L into a family of spacelike 3–dimensional hypersurfaces, see York[31] and Christodoulou-Klainerman[9]. The 3-dimensional spacelike slices have the structure of asymptotically flat complete Riemannian 3–manifolds (M^3, g, K)

carrying an induced metric $g = \{g_{AB}\}$, $1 \leq A, B \leq 3$, and a second fundamental form $K = \{K_{AB}\}$. Rewriting the Einstein equations as a Hamiltonian system for g and K it is well known that the energy condition (2) implies a lower bound for the scalar curvature of the induced metric g of the hypersurface in view of the constraint equations.

It was shown by Bartnik[2] that under very general assumptions on L the hypersurfaces (M^3, g, K) can be chosen to be maximal, ie with vanishing trace of the second fundamental form K. In this gauge the energy condition (2) implies that the scalar curvature R^3 of the induced metric g is nonnegative, $R^3 \geq 0$. If the second fundamental K vanishes completely, the hypersurface (M^3, g) is called time-symmetric and all geometric information is encoded in the metric g. In this special case the horizon of a black hole is modelled by a two dimensional minimal surface in (M^3, g).

Having this motivation in mind we will now concentrate on one space-like slice and try to understand how much geometric information is already encoded in complete Riemannian 3–manifolds (M^3, g) of nonnegative scalar curvature approaching the geometry of Euclidean $\mathbb{R}^3$ near infinity. We use the following notions of asymptotically flat manifolds and exterior region.

Definition 1.1 *i) A 3-dimensional asymptotically flat end is a 3–manifold (M^3, g) with Riemannian metric g, which is diffeomorphic to $\mathbb{R}^3 \setminus K$ for some compact set K, such that in Euclidean coordinates $\{x^A\}$ the metric approaches the flat metric $\delta = \{\delta_{AB}\}$ for $r = |x| \to \infty$:*

$$|g_{AB} - \delta_{AB}| \leq \frac{C_0}{r}, \quad |\partial_A g_{BC}| \leq \frac{C_1}{r^2}, \quad Ric \geq -\frac{C_2}{r^2}. \tag{3}$$

ii) A complete 3–manifold (M^3, g) is called asymptotically flat if it can be decomposed as a disjoint union of a compact set K and finitely many asymptotically flat ends (M_i^3, g), $1 \leq i \leq q$:

$$M^3 = K \cup M_1^3 \cup \cdots \cup M_q^3. \tag{4}$$

iii) An exterior region is a complete, connected asymptotically flat 3-manifold (M^3, g) with one end having non-negative scalar curvature $R(g) \geq 0$, such that the boundary of (M^3, g) is compact and smooth, consisting of finitely many minimal surfaces, with no other minimal surfaces contained in (M^3, g).

Apart from empty space $\mathbb{R}^3$ the most important example of an asymptotically flat 3–manifold is the spatial Schwarzschild manifold given by $({}^S M^3, g^S) = (\mathbb{R}^3 - \{0\}, g^S)$ with

$$g^S_{AB} = \delta_{AB}(1 + \frac{m}{2r})^4, \quad 0 < m = \text{const.} \tag{5}$$

It arises as the $\{t = 0\}$–slice of the Schwarzschild space–time modelling a spherically symmetric, static vacuum black hole of mass m. It has two asymptotically flat ends for $r \to \infty$ and $r \to 0$, which are isometric under reflection in the totally geodesic 2–sphere $\{r = m/2\}$. The two regions $\{r \geq m/2\}$ and $\{r \leq m/2\}$ are both exterior regions in the sense of the definition above.

For a general asymptotically flat end (M^3, g) as in definition 1.1 the total energy or ADM–mass in each end is defined according to Arnowitt, Deser and Misner[1] by a flux integral at infinity,

$$m_{ADM}(M^3, g) := \lim_{R\to\infty} \frac{1}{16\pi} \int_{\partial B_R(0)} (g_{ii,j} - g_{ij,i}) n^j \, d\mu, \tag{6}$$

in agreement with the spatial Schwarzschild manifold above: $m_{ADM}({}^S M^3, g^S) = m$. Due to work of Bartnik[3] and Chrusciel[10] it is known that m_{ADM} is a geometric invariant of the manifold (M^3, g). It is finite precisely when the scalar curvature R^3 is bounded in L^1. Physically the mass defined in this way is a total energy and measures both the matter content of the system and the gravitational energy.

The "Positive Mass Theorem" by Schoen and Yau[29] is a crucial result showing that the local energy condition (2) is consistent with this concept of global energy, a basic version states:

Theorem 1.2 (Schoen-Yau 1979) *If (M^3, g) is an asymptotically flat 3–manifold with non-negative scalar curvature such that the mean curvature of its boundary is nonnegative with respect to the outward normal, then the mass of each end is nonnegative. If the mass is zero in one end, then (M^3, g) is isometric to flat space $(\mathbb{R}^3, \delta)$.*

The proof of Schoen and Yau is based on the fundamental insight that the stability of two-dimesional minimal surfaces in a 3-manifold can be controlled by the scalar curvature of the ambient space. The results discussed in the following sections also exploit this deep relation between the second variation of area for a two dimensional hypersurface and the scalar curvature of the surrounding 3-manifold. Asymptotically flat 3–manifolds and the question of positivity for the mass were also studied by many other authors including Geroch, Jang, Kijowski, Jezierski, Chrusciel, Witten, Choquet–Bruhat, Parker–Taubes, Reula, Tod, Bray, Herzlich and Lohkamp. See[18] and the survey article of Lee–Parker[26] for additional references to asymptotically flat 3–manifolds and the concept of total energy.

2 Hawking mass and hypersurfaces of constant mean curvature

Consider a smooth embedding of a sphere $F : S^2 \mapsto (M^3, g)$ into an asymptotically flat 3-manifold (M^3, g). From the physical information present in (M^3, g) we will try to capture as much as possible in terms of the geometry of suitable hypersurface embeddings F.

If ν is a choice of unit normal for the hypersurface $N^2 = F(S^2)$, let $A = \{h_{ij}\}, 1 \leq i,j \leq 2$, be the second fundamental form and $H = tr_g A$ be the mean curvature of N^2 such that $-H\nu$ is the mean curvature vector. In an exterior region we choose ν to be the exterior normal, such that the mean curvature of a sphere in Euclidean space is positive.

The definition of the ADM-mass (6) shows how 2-dimensional hypersurfaces can be used to define the total energy in terms of an asymptotic flux integral. Hawking[14] introduced a geometric quantity for 2-dimensional surfaces N^2 now called the Hawking-mass with the aim of capturing the energy content of the region bounded by N^2:

$$m_H(N^2) := \frac{|N^2|^{1/2}}{(4\pi)^{3/2}} \left(16\pi - \int_{N^2} H^2 \, d\mu \right). \tag{7}$$

Let us collect some of the properties of this geometric quantity: A simple computation shows that

$$m_H(\partial B_r^S(0)) \equiv m, \quad r > 0, \quad B_r(0) \subset ({}^S M^3, g^S), \tag{8}$$

i.e. m_H yields the desired result for the contained energy of all centered coordinate spheres in the spatial Schwarzschild metric. In general it can be shown[18] that asymptotically m_H yields the total mass in general asymptotically flat ends when evaluated on large coordinate spheres:

$$\lim_{r \to \infty} m_H(\partial B_r(0)) = m_{ADM}. \tag{9}$$

On the other hand m_H is rarely positive, for hypersurfaces in Euclidean space we have

$$N^2 \subset \mathbb{R}^3 : \qquad \int_{N^2} H^2 \, d\mu \geq 16\pi, \tag{10}$$

with equality only on round spheres. In fact, by introducing "wrinkles" in a given surface, one can make the Hawking mass as much negative as one likes. The geometric quantity m_H therefore can only be related to "energy" if the surface N^2 is chosen well. As was first observed by Christodoulou and Yau[8], stable surfaces of constant mean curvature are a good class of surfaces: Here a surface of constant mean curvature is called stable, if the second variation of area with respect to all volume preserving variations is non-neagtive, i.e. if it is a stable critical point of the isoperimetric problem in (M^3, g). It can be seen that a constant mean curvature surface is stable for the area functional with respect to volume preserving variations exactly when the inequality

$$\int_{N^2} (|A|^2 + Ric(\nu,\nu)) f^2 \, d\mu \leq \int_{N^2} |\nabla f|^2 \, d\mu \tag{11}$$

holds for all functions f satisfying $\int f \, d\mu = 0$. Strict stability means that the first eigenvalue of the Jacobi operator

$$Lu = -\Delta u - u\big(|A|^2 + Ric(\nu,\nu)\big) \tag{12}$$

when restricted to functions f with $\int f \, d\mu = 0$ is strictly positive.

Theorem 2.1 (Christodoulou-Yau 1986) *If $F : S^2 \mapsto (M^3, g)$ is a smooth, stable immersion of constant mean curvature in a Riemannian 3-manifold (M^3, g) of non-negative scalar curvature, then $m_H(N^2) \geq 0$.*

This result fits nicely with the reverse inequality in Euclidean space (10), since the only stable constant mean curvature surfaces in Euclidean space are round spheres. Notice that the centered coordinate spheres of the spatial Schwarzschild manifold $({}^S M^3, g^S)$ are strictly stable constant mean curvature surfaces, with smallest eigenvalue of the stability operator (12) on $\partial B_r(0)$ of order $6m/r^3$ as $r \to \infty$[21].

The existence of constant mean curvature surfaces in an asymptotically flat end of a Riemannian 3-manifold has first been adressed by Huisken and Yau [21] assuming that the end is *strongly asymptotically flat*, i.e. assuming that for large r the metric g has the form

$$g_{AB} = \delta_{AB}\big(1 + \frac{m}{2r}\big)^4 + P_{AB}, \tag{13}$$

$$|P_{AB}| \leq \frac{C_0}{r^2}, \; |\partial^j P_{AB}| \leq \frac{C_j}{r^{2+j}}, \; j = 1, \cdots, 4. \tag{14}$$

It was shown that there is a smooth foliation by unique stable constant mean curvature spheres outside some compact set, provided the mass m is strictly positive:

Theorem 2.2 (Huisken-Yau 1996) *Let (M^3, g) be a strongly asymptotically flat end of a Riemannian 3-manifold with strictly positive mass $m > 0$. Then there are constants $R_0 > 0$, $\tau_0 > 0$ depending only on the mass m and the constants in (14) together with a family of 2-spheres $\Sigma^2_\tau, 0 < \tau < \tau_0$, of constant mean curvature τ providing a regular foliation of the exterior region $M^3 \setminus B_{R_0}(0)$. Furthermore, for each $1/2 < q \leq 1$ there is a constant $\tau_1 > 0$ depending only on q, m and the constants in (14) such that for each $0 < \tau < \tau_1$*

the surface Σ^2_τ is the only stable constant mean curvature surface contained in $M^3 \setminus B_{\tau-q}(0)$.

The proof of the existence result uses the L^2-gradient flow of the area functional subject to a volume constraint to deform an initial coordinate sphere to the desired constant mean curvature surface. The uniqueness result exploits that the stability inequality (11) yields a priori estimates for the difference of the two principal curvatures on a constant mean curvature surface. Notice that Ye[30] has an alternative approach to the existence part of the theorem.

The existence and uniqueness result above extends the notion of "center of mass" from Newtonian mechanics to the relativistic setting: The family of 2-spheres Σ_τ provides an asymptotic center of mass for the infinitely far observer. Such a family of unique 2-spheres seems to be the most natural concept of a "center", since no distinguished Euclidean structure can be refered to on (M^3, g) in a natural way and since already the spatial Schwarzschild manifold illustrates that a "center" cannot be provided by a point of the manifold. The family Σ_τ thus breaks the translation invariance in the asymptotically flat region and provides a natural geometric radial coordinate there. It can easily be employed to define a canonical coordinate system near spatial infinity using a harmonic gauge in angular direction.

The relation of constant mean curvature surfaces to the Hawking-mass has been further clarified by Bray [6], who studies the isoperimetric problem in (M^3, g) from the variational point of view and observes in particular that the Hawking-mass is monotone along a foliation of stable constant mean curvature surfaces.

Theorem 2.3 (Bray 1997) *Let (M^3, g) be a Riemannian 3-manifold of nonnegative scalar curvature. Then the Hawking mass is monotonically increasing along a foliation of expanding stable constant mean curvature spheres in (M^3, g).*

In the special case where (M^3, g) is an exterior region admitting a foliation by stable constant mean curvature spheres between a single boundary component and spatial infinity, the monotonicity above and (9) yield an inequality between the ADM-mass and the area of the boundary known as the Penrose inequality. In general such a foliation of constant mean curvature spheres will not exist in (M^3, g). The next section shows how a different family of 2-spheres with monotone Hawking-mass can be constructed for general exterior regions.

3 Inverse mean curvature flow and the Penrose inequality

Let (M^3, g) be an exterior region as in definition 1.1. In theorem 1.2 Schoen and Yau show that the total energy of an isolated system given by the ADM-mass is non-negative and that in some sense Minkowski space is the absolute ground state of minimum energy. It was already conjectured earlier by Penrose [28] in 1973 that a stronger result holds, namely there should be a positive lower bound for the total energy in terms of the size of the outermost black hole present in the system, with the Schwarzschild metric as the sharp limiting ground state for such an inequality.

In recent work by the authors[17,18] such a lower bound for the ADM-mass is proved in general exterior regions:

Theorem 3.1 (Huisken-Ilmanen 1997) *Let (M^3, g) be an asymptotically flat exterior region. Then the total mass m of (M^3, g) is non-negative and*

$$16\pi m^2 \geq |\Sigma^2|,$$

where $|\Sigma^2|$ is the area of any connected component of ∂M^3. Equality holds if and only if M^3 is one-half of the spatial Schwarzschild manifold $({}^S M^3, g^S)$.

Notice that the exclusion of other compact minimal surfaces in the definition of an exterior region is necessary to account for the possibility of large black holes being hidden behind small ones. Also, the theorem provides an alternative proof of the positive mass theorem 1.2 as a special case. An approach to this result based on 2–dimensional surfaces flowing along the inverse of the mean curvature was originally suggested by Geroch[11], see also[22,23]. Other approaches are due to Gibbons[12,13], Herzlich[16], Bartnik[4], Jezierski[24,25] and Bray[6] as mentioned in the previous section. See[17,18] for additional references and a more detailed account. Recently Bray[7] has used a method of conformal deformations and area-minimizing boundaries in conjunction with the positive mass theorem to estimate the ADM-mass from below by the sum of the areas of the boundary components, improving the estimate above.

The approach suggested by Geroch employs smooth solutions $F : S^2 \times [0,T) \mapsto (M^3, g)$ of the inverse mean curvature flow

$$\frac{d}{dt}F(p,t) = \frac{1}{H}\nu(p,t), \quad p \in S^2,\, t \in [0,T), \tag{15}$$

which is a parabolic system for the evolving surfaces $N_t^2 = F(\cdot,t)(S^2)$ if the mean curvature H is strictly positive. Geroch observes that the flow acts monotone on the Hawking-mass (7) if the scalar curvature of the ambient manifold is non-negative:

$$\frac{d}{dt}m_H(N_t^2) \geq 0. \tag{16}$$

Near an initial minimal surface the Hawking-mass yields the appropriate area term and for large coordinate spheres it approaches the ADM-mass as in (9), such that the desired inequality is proven provided the existence of a solution to (15) is established.

Without extra assumptions on the ambient manifold and the initial data it is clear that singularities will occur in general. For example, the solution evolving from a thin symmetric torus cannot exist forever and similar examples can be constructed in the class of 2–spheres, showing that the flow usually has no smooth solutions in general exterior regions.

In[18] the problem of singularities is overcome by using a level–set formulation of the flow where the surfaces N_t^2 solving (15) are replaced by level–sets

$$N_t^2 = \partial\{x \in M^3 | u(x) < t\}$$

of a scalar function $u : M^3 \mapsto \mathbb{R}$. Intuitively $u(x)$ is the time where the expanding surface passes through $x \in M^3$. In the smooth case the function u satisfies a degenerate elliptic boundary value problem

$$\operatorname{div}\Big(\frac{Du}{|Du|}\Big) = |Du|, \quad u\Big|_{N_0^2} = 0 \tag{17}$$

replacing the parabolic equation (15), note that the LHS is the mean curvature of a level-set and the RHS is the inverse speed. To obtain a concept of weak solution suitable for our purposes, we introduce a global variational principle for u requiring that u minimizes an energy depending on u itself,

$$J_u(v) = J_u^K(v) := \int_K |\nabla v| + v|\nabla u|\, dx, \tag{18}$$

amongst all competing functions $v \in C^{0,1}$ with the same initial data N_0^2 and agreeing with u outside some compact set K. This concept of weak solution implies (17) in an integral form as Euler-Lagrange equation, it allows spatial jumps of the surfaces N_t^2 where the mean curvature $H = |Du|$ becomes zero and the level-sets of u "fatten", and it ensures that the surfaces N_t^2 have continuously varying area while allways enclosing the maximal possible volume for the area available at time t. Using elliptic regularisation

$$\operatorname{div}\Big(\frac{Du}{\sqrt{\epsilon^2 + |Du|^2}}\Big) = \sqrt{\epsilon^2 + |Du|^2} \tag{19}$$

it is possible to construct for $\epsilon \to 0$ a unique weak solution of (17) with these properties that still possesses the crucial monotonicity property for the Hawking-mass (16), leading to the proof of theorem 3.1. The proof of the monotonicity of the mass employs geometric measure theory and is one of the main analytic difficulties in this approach, see[18] for the theory of inverse mean curvature flow. The regularised equation (19) has been used by Pasch[27] to compute weak solutions of the flow numerically.

In [18] we show that the unique weak solution u of inverse mean curvature flow is of class $C^{0,1}$ with level-sets N_t^2 in $C^{1,\alpha}, 0 < \alpha < 1$, of non-negative bounded mean curvature. Heidusch[15] has recently shown that the level sets N_t^2 of the weak solution are allways at least of class $C^{1,1}$, which is the best possible regularity result for the flow. In new work[20] we show that solutions of inverse mean curvature flow in all dimensions are smooth as long as the mean curvature is strictly positive and then go on to show that such a strictly positive lower bound for the mean curvature is allways true on "starshaped" surfaces in Euclidean space. In a forthcoming paper we show that solutions of inverse mean curvature flow in asymptotically flat ends are smooth outside some compact set. If the end is strongly asymptotically flat, the flow approaches the center of mass given by the constant mean curvature foliation in theorem 2.2 at an exponential rate.

4 Quasi-local mass and the Bartnik capacity

Given an open 3-manifold (Ω, g) with compact boundary and non-negative scalar curvature, it is important to have a rigoros mathematical notion of quasi-local mass motivated by the energy content attributable to a subset $(\Omega, g) \subset (M^3, g)$ of an asymptotically flat 3-manifold modelling a spacelike slice of an isolated gravitating system.

One of the first suggestions for a quasi-local mass was the Hawking-mass of the boundary, $m_H(\partial\Omega)$. But as we have seen in the previous sections the Hawking-mass, despite having very desirable properties on certain special two-spheres such as stable constant mean curvature surfaces or solutions of the inverse mean curvature flow, is negative on generic boundaries $\partial\Omega$ and is therefore not suitable as a general measure of energy content. A good concept of quasi-local mass Q should have the following properties[5]:

(i) (Monotonicity) If $\tilde{\Omega} \subset \Omega$, then $Q(\tilde{\Omega}) \leq Q(\Omega)$.
(ii) (Positivity) $Q(\Omega) \geq 0$, and $Q(\Omega) > 0$ unless Ω is locally isometric to $\mathbb{R}^3$.
(iii) (Exhaustion) If Ω_i is a sequence of bounded subsets of an exterior region M^3 such that $\chi_{\Omega_i} \to \chi_M$ locally uniformly, then $\lim Q(\Omega_i) = m_{ADM}(M^3)$.

The following definition is inspired by Bartnik[5] and also used in this form in[18]:

Definition 4.1 *Let (Ω, g) be an open Riemannian 3-manifold with compact boundary and non-negative scalar curvature. We say that an exterior region*

(M^3, g) is an admissible extension of (Ω, g) if there is an isometric embedding $(\Omega, g) \to (M^3, g)$. If Ω has an admissible extension the Bartnik gravitational capacity of Ω is defined as

$$c_B(\Omega) := \inf\{m_{ADM}(M^3, g) \mid (M^3, g) \text{ is an admissible extension of } \Omega\}.$$

Notice that with this definition the extension (M^3, g) (and hence (Ω, g)) can have horizons on an inner boundary, which is less restrictive than Bartniks original definition, making our definition less than or equal to Bartniks. The monotonicity property (i) above is satisfied trivially and the weak positivity in (ii) is a consequence of the positive mass theorem 1.2. Using the inverse mean curvature flow and the precise form of the monotonicity for the Hawking-mass (16) we can show[18] that all the properties (i)-(iii) listed above are satisfied for the Bartnik capacity:

Theorem 4.2 (Huisken-Ilmanen 1997) *The Bartnik capacity c_B satisfies the positivity property (ii) and the exhaustion property (iii).*

From the proof it can be seen that in specific models, eg for stars having certain density profiles, the presence of the scalar curvature in the monotonicity formula for the Hawking-mass allows the computation of concrete lower bounds for the Bartnik capacity. It is an open problem whether an admissible domain (Ω, g) with $c_B(\Omega) = 0$ is actually isometric to a subset of $\mathbb{R}^3$. Other interesting questions concern the limiting behaviour of sequences of extensions realizing the infimum in the definition of c_B.

References

1. R. Arnowitt, S. Deser and C. Misner, *Coordinate Invariance and Energy Expressions in General Relativity*, Phys. Rev. **122** (1961) 997–1006.
2. R. Bartnik, *Existence of Maximal Surfaces in Asymptotically Flat Spacetimes*, Comm. Math. Phys. **94** (1984) 155–175.
3. R. Bartnik, *The Mass of an Asymptotically Flat Manifold*, Comm. Pure Appl. Math. **39** (1986) 661–693.
4. R. Bartnik, *Quasi-Spherical Metrics and Prescribed Scalar Curvature*, J. Diff. Geom. **37** (1995) 31–71.
5. R. Bartnik, *New Definition of Quasi-Local Mass*, PhysRevLett **62** (1989) 2346.
6. H. Bray, *The Penrose Inequality in General Relativity and Volume Comparison Theorems Involving Scalar Curvature*, thesis, Stanford University, (1997).

7. H. BRAY, *Proof of the Riemannian Penrose Inequality using the Positive Mass Theorem*, J. Diff. Geom., to appear.
8. D. CHRISTODOULOU AND S.-T. YAU, *Some Remarks on the Quasi-Local Mass*, Mathematics and General Relativity (Santa Cruz CA 1986), edJ. Isenberg, Contemporary Mathematics #**71**, AMS, (1988) 9–14.
9. D. CHRISTODOULOU AND S. KLAINERMAN, *The Global Nonlinear Stability of Minkowski Space*, Princeton Math. Series #41, Princ. Univ. Press, Princeton, 1993.
10. P. T. CHRUŚCIEL, *Boundary Conditions at Spatial Infinity from a Hamiltonian Point of View*, Topological Properties and Global Structure of Space-Time (Erice, 1985), NATO Adv. Sci. Inst. Ser. B: Phys. #138, Plenum, New York, 1986, 49–59.
11. R. GEROCH, *Energy Extraction*, Ann. New York Acad. Sci. **224** (1973) 108–17.
12. G. W. GIBBONS, thesis, Cambridge University, 1973.
13. G. W. GIBBONS, *Collapsing Shells and the Isoperimetric Inequality for Black Holes*, Class. Quant. Grav. **14** (1997) 2905-2915, also hep-th/9701049.
14. S. W. HAWKING, *Gravitational Radiation in an Expanding Universe*, J. Math. Phys. **9** (1968) 598–604.
15. M. HEIDUSCH, *Zur Regularität des Inversen Mittleren Krümmungsflusses*, Dissertation Tübingen University (2001).
16. M. HERZLICH, *A Penrose-like Inequality for the Mass of Riemannian Asymptotically Flat Manifolds*, Comm. Math. Phys. **188** (1997) 121-133.
17. G. HUISKEN AND T. ILMANEN, *The Riemannian Penrose Inequality*, International Math. Res. Not. **20** (1997) 1045–1058, also from http://www.math.nwu.edu/~ilmanen.
18. G. HUISKEN AND T. ILMANEN, *The inverse mean curvature flow and the Riemannian Penrose inequality*, to appear J.Diff.Geom.
19. G. HUISKEN, T. ILMANEN, *A Note on Inverse Mean Curvature Flow*, Proc. Workshop on Nonl. Part. Diff. Equ. (Saitama University, Sept. 1997), available from Saitama University, also from http://www.math.nwu.edu/~ilmanen.
20. G. HUISKEN AND T. ILMANEN, *Higher regularity of the inverse mean curvature flow*, preprint.
21. G. HUISKEN AND S.-T. YAU, *Definition of Center of Mass for Isolated Physical Systems and Unique Foliations by Stable Spheres with Constant Mean Curvature*, Invent. Math. **124** (1996) 281–311.

22. P. S. JANG, *On the Positive Energy Conjecture*, J. Math. Phys. **17** (1976) 141–145.
23. P. S. JANG, *On the Positivity of Energy in General Relativity*, J. Math. Phys. **19** (1978) 1152–1155.
24. J. JEZIERSKI, *Positivity of Mass for Certain Spacetimes with Horizons*, Class. Quant. Grav. **6** (1989) 1535–1539.
25. J. JEZIERSKI, *Perturbation of Initial Data for Spherically Symmetric Charged Black Hole and Penrose Conjecture*, Acta Phys. Pol. B **25** (1994) 1413–17.
26. J. M. LEE AND T. H. PARKER, *The Yamabe Problem*, Bull. Amer. Math. Soc. **17** (1987) 37–91.
27. E. PASCH, *Numerische Verfahren zur Berechnung von Krümmungsflüssen*, Ph.D. thesis, Tübingen Univ. (1998).
28. R. PENROSE, *Naked singularities*, Ann.New York Acad.Sci. **224** (1973) 125–134.
29. R. SCHOEN AND S.-T. YAU, *On the Proof of the Positive Mass Conjecture in General Relativity*, Comm. Math. Phys. **65** (1979) 45–76.
30. R. YE, *Foliation by constant mean curvature spheres on asymptotically flat manifolds*, in: Geometric Analysis and the Calculus of Variations, Ed.J.Jost, International Press, 1996.
31. J. W. YORK JR., *Kinematics and Dynamics of General Relativity*, Sources of Gravitational Radiation, ed. L. Smarr, Camb. Univ. Press, 1979, p.83.

CONSTRUCTING SOLUTIONS OF THE EINSTEIN CONSTRAINT EQUATIONS

JAMES ISENBERG
Department of Mathematics
and
Institute of Theoretical Science
University of Oregon
Eugene, OR 97403 USA
E-mail: jim@newton.uoregon.edu

The first step in the building of a spacetime solution of Einstein's gravitational field equations via the initial value formulation is finding a solution of the Einstein constraint equations. We recall the conformal method for constructing solutions of the constraints and we recall what it tells us about the parameterization of the space of such solutions. One would like to know how to construct solutions which model particular physical phenomena. One useful step towards this goal is learning how to glue together known solutions of the constraint equations. We discuss recent results concerning such gluing.

1 Introduction

The constraint equations of Einstein's theory of the gravitational field are familiar to most of those who work in general relativity: If we choose a set of appropriate local spacetime coordinates (x^a, t) and if we write Einstein's field equations for the gravitational metric field g together with a coupled matter source field[a] Ψ in the tensorial form[b]

$$G_{\mu\nu}(g) = \kappa T_{\mu\nu}(g, \Psi), \tag{1}$$

then the four equations

$$G_{t\nu}(g) = \kappa T_{t\nu}(g, \Psi), \tag{2}$$

involve no time derivatives of higher than first order. These are the Einstein constraint equations.

What makes the constraint equations important in the study of the physics of the gravitational field is their role in the initial value formulation of

[a] We presume that the matter source field is not derivative-coupled, and that the coupled field equations constitute a well-posed PDE system.
[b] We use MTW^1 conventions on signs of the curvature and on indices: Greek letters run from 0 to 3 (spacetime) while Latin letters run from 1 to 3 (space). Here κ is a coupling constant.

Einstein's equations. We recall how the initial value formulation works: Say we want to construct a spacetime solution $\{g, \Psi\}$ of the Einstein equations (1) on a manifold $M^4 = \Sigma^3 \times R$, where Σ^3 is a smooth three-dimensional manifold. The first step in doing this is to find a set of *initial data* consisting of

γ_{ab} - a Riemannian metric on Σ^3
K^{cd} - a symmetric tensor field on Σ^3
ψ - space covariant pieces of the matter field Ψ on Σ^3

with $\{\gamma, K, \psi\}$ satisfying the constraint equations (2). In terms of the data $\{\gamma, K, \psi\}$ the constraint equations take the form

$$R + (trK)^2 - K^{cd}K_{cd} = 2\kappa\rho(\gamma, \psi) \tag{3}$$

$$\nabla_a K_b^a - \nabla_b(trK) = \kappa J_b(\gamma, \psi) \tag{4}$$

where ∇_a is the Levi-Civita[c] covariant derivative corresponding to γ, R is its scalar curvature, ρ is a scalar function of γ and ψ, and J_b is a vector field function of γ and ψ.[d]

Once we have obtained constraint-satisfying initial data $\{\gamma, K, \psi\}$, we may evolve this data in time by choosing everywhere on Σ^3 a scalar field N (the "lapse function") and a vector field M^a (the "shift vector") and then solving the Einstein evolution equations in the form

$$\frac{d}{dt}\gamma_{ab} = -2NK_{ab} + \pounds_M \gamma_{ab} \tag{5}$$

and

$$\frac{d}{dt}K_d^c = N(R_d^c + trK K_d^c) - \nabla^c\nabla_d N + \kappa T_d^c + \pounds_M K_d^c \tag{6}$$

where T_d^c are the spatial components of the stress-energy tensor corresponding to the matter field Ψ. We note that the time evolution of the lapse and shift are not determined by the Einstein equations; they may be chosen freely, and they must be chosen for all t if one wishes to evolve γ and K in time. The role of N and M is to determine the coordinates relative to the evolving spacetime.

---

[c] i.e., metric compatible and torsion-free

[d] If, for example, the "matter source field" is Maxwell's electromagnetic field, then the spacetime covariant field Ψ is the spacetime one-form field A_u, the space covariant fields ψ may be chosen to be the magnetic spatial one-form field B_a and the electric spatial vector field E^a, and one calculates $\rho = \frac{1}{2}(E^a E^b \gamma_{ab} + B^a B^b \gamma^{ab})$ and $J_c = \eta_{cmn} E^m B^n$, where η_{cmn} is the alternating symbol $\eta_{123} = 1$, etc. In addition, the data must satisfy the Maxwell constraint equations $\nabla^a B_a = 0$ and $\nabla_a E^a = 0$.

After $N(x,t)$ and $M^c(x,t)$ have been chosen and the evolved data $\gamma_{ab}(x,t)$ and $K^{cd}(x,t)$ and $\psi(x,t)$ have been determined[e], we may reconstruct the spacetime metric (on the manifold $M^4 = \Sigma^3 \times I$, where I is the interval for which the evolution can be carried out) via the formula

$$g(x,t) = -N^2 dt^2 + \gamma_{ab}(dx^a + M^a dt)(dx^b + M^b dt). \tag{7}$$

We similarly reconstruct Ψ. It then follows (from the Gauss-Codazzi-Mainardi 3+1 decomposition equations for the spacetime curvature) that g and Ψ satisfy the Einstein equation (1) on M^4.

Does the initial value formulation work? This general question is best considered by breaking it up into a number of more specific questions:

1. Does every choice of initial data $\{\gamma, K, \psi\}$ which satisfies the constraint equations evolve into a spacetime solution $\{g, \Psi\}$?
2. How do we find data $\{\gamma, K, \psi\}$ which satisfy the constraint equations (3)-(4)?
3. What is the space of solutions of the constraint equations?
4. If we wish to use Einstein's equations to model the physical behavior of a particular physical situation (e.g., a black hole collision), how do we build that physical situation into the choice of initial data?
5. What criteria should we use in making a choice of the lapse and shift?
6. What do we know about the long time behavior of solutions?
7. Is the initial value formulation a practical way to construct and study solutions of Einstein's equations?

In this essay, we shall focus our attention on questions 2, 3, and 4, which concern finding and understanding solutions to the constraint equations. We start in Section 2 by discussing the conformal method for constructing solutions to the constraints. After recalling the basic steps of the conformal method, we state the results which tell us the extent to which the method is known to work. Based on these results, we discuss (also in Section 2) what is known about the function spaces which parameterize the set of solutions.

Knowing how to construct constraint-satisfying initial data sets from conformal data sets, and knowing how to parameterize the space of all such initial data sets, does not tell us much about how to use solutions constructed via the

[e] The coupled Einstein-matter source field equations determine evolution equations for ψ.

initial value formulation for studying physical questions. With observational data from LIGO and other gravitational radiation detectors expected to start arriving within the next few years, it is crucial that we learn how to do this. Although progress to date on this front has been fairly limited, recent results on the gluing of initial data sets could be useful. We discuss some of these gluing results and the ideas involved in Section 3.

Only if initial data can be evolved is it worthwhile to construct it. We comment briefly on evolution in Section 4, touching on questions 1, 5, 6, and 7. We make some concluding remarks in Section 5.

2 The Conformal Method and the Space of Solutions of the Constraint Equations

For the past thirty years, the conformal method has been the most successful procedure for producing and studying solutions of the Einstein constraint equations, both numerically and analytically. The key idea of this method, which was developed primarily by Lichnerowicz, Choquet-Bruhat and York[2] is to split the initial data $\{\gamma, K, \psi\}$ into two parts: The first part is to be chosen freely, while the second part is to be determined by the constraints. As a consequence of the split, the constraint equations take the form of a determined PDE system of four equations for four unknowns (equations (12)-(13) below), rather than the form of an underdetermined system of four equations for twelve unknowns (equations (3)-(4)).

To illustrate how this method works explicitly with matter source fields as well as gravitational fields, let us consider the Einstein-Maxwell equations, for which the initial data consists of $\{\gamma_{ab}, K^{cd}, B_a, E^a\}$ and for which the full set of constraint equations takes the form (presuming no charges present)

$$R + (trK)^2 - K^{cd}K_{cd} = \kappa(B^aB_a + E^aE_a) \tag{8}$$

$$\nabla_a K^a_b - \nabla_b(trK) = \kappa\eta_{bmn}B^mE^n \tag{9}$$

$$\nabla^a B_a = 0 \tag{10}$$

$$\nabla_a E^a = 0 \tag{11}$$

For this system, the split of the data is as follows:

Free ("Conformal") Data:

λ_{ab} - a Riemannian metric (specified up to conformal factor)

σ^{cd} - a divergence-free[f] and trace-free symmetric tensor field
($\nabla_c \sigma^{cd} = 0$) ($\lambda_{cd} c \sigma^{cd} = 0$)

τ - a scalar field

β^a - a divergence-free vector field

ε^a - a divergence-free vector field

Determined Data

ϕ - a positive definite scalar field

W^a - a vector field

Determining Equations

$$\nabla_a (LW)^a_b = \frac{2}{3}\phi^6 \nabla_b \tau + \kappa \eta_{bmn} \beta^m \varepsilon^n \tag{12}$$

$$\begin{aligned}\nabla^2 \phi = \frac{1}{8} R\phi - \frac{1}{8}(\sigma^{ab} + LW^{ab})(\sigma_{ab} + LW_{ab})\phi^{-7} \\ - \frac{\kappa}{8}(\beta^a \beta_a + \varepsilon^a \varepsilon_a)\phi^{-3} + \frac{1}{12}\tau^2 \phi^5\end{aligned} \tag{13}$$

Here the covariant derivative, the Laplacian, the scalar curvature, and the index manipulations all correspond to the metric λ_{ab}. Note that L is the conformal Killing operator, defined by

$$LW_{ab} := \nabla_a W_b + \nabla_b W_a - \frac{2}{3}\lambda_{ab} \nabla_c W^c. \tag{14}$$

The determining equations (12)-(13) are to be solved for ϕ and W^a. If, for a given set of conformal data $\{\lambda_{ab}, \sigma^{cd}, \tau, \beta^a, \varepsilon^a\}$ one can indeed solve (12)-(13), then the reconstructed data

$$\gamma_{ab} = \phi^4 \lambda_{ab} \tag{15}$$

$$K^{cd} = \phi^{-10}(\sigma^{cd} + LW^{cd}) + \frac{1}{3}\phi^{-4}\lambda^{cd}\tau \tag{16}$$

$$B_a = \phi^{-2}\lambda_{ab}\beta^b \tag{17}$$

$$E^a = \phi^{-6}\varepsilon^a \tag{18}$$

---

[f] In the conformal data, the divergence-free condition is defined using the Levi-Civita covariant derivative compatible with the conformal metric λ_{cd}.

constitute a solution of the constraint equations (8)-(11).

Does the conformal method always work in the sense that, for any choice of conformal data $\{\lambda_{ab}, \sigma^{cd}, \tau, \beta^a, \epsilon^a\}$ one can always solve (12)-(13) for $\{\phi, W^a\}$ and thereby obtain a solution $\{\lambda_{ab}, K^{cd}, B_a, E^a\}$ for the constraints (8)-(11)? It is easy to see that this is not the case, as illustrated by the following simple example:

Let us choose the manifold Σ^3 to be the three-dimensional sphere S^3, and let us choose the conformal data to consist of the round sphere metric $\lambda = g_{(round)}$ with $R = 1$, together with $\sigma^{cd} = 0$, $\varepsilon^a = 0$, $\beta^a = 0$, and $\tau = 1$. With τ constant and with $\eta_{bmn}\varepsilon^m\beta^n$ equal to zero, the right hand side of equation (12) vanishes. Since the operator $\nabla \cdot L$ is self-adjoint and injective (up to conformal Killing vector fields), it readily follows that the solutions W^a to (12) for this choice of conformal data all satisfy the condition $LW_{ab} = 0$ (which is the conformal Killing equation). Using this result, we find that the remaining equation (13) takes the form

$$\nabla^2\phi = \frac{1}{8}\phi + \frac{1}{12}\phi^5. \tag{19}$$

Since we seek solutions ϕ which are positive definite, for any such solution the right hand side of equation (19) is positive define. But the maximum principle tells us that there are no functions ϕ on a compact manifold for which $\nabla^2\phi > 0$ (or < 0). Hence, for this choice of conformal data, there is no solution $\{\phi, W^a\}$ to equations (12)-(13), and therefore no corresponding solution to the Einstein-Maxwell constraint equations.

This example shows that there are choices of conformal data which are not mapped to solutions of the constraint equations by the conformal method. What happens generally? This question is difficult to address in any complete sense because there is a very wide variety of cases of physical and mathematical interest, and each must be handled separately. The various cases may be classified using the following criteria:

Manifold and Asymptotic Geometry

- Closed Σ^3
- Asymptotically Euclidean
- Asymptotically hyperbolic
- Compact Σ^3 with boundary conditions

Mean Curvature

- Maximal

- Constant non maximal (CMC)
- Non Constant (non CMC)

Fields

- Einstein vacuum
- Einstein - standard matter source field[g]

Function Spaces

- C^α (analytic)
- C^∞ (smooth)
- $C^{k+\beta}$ (Hölder)
- H^p_k (Sobolev)
- $H^p_{k,\delta}$ (weighted Sobolev)

For a number of cases (labeled using these criteria), we have a fairly complete understanding of which sets of conformal data map to solutions and and which do not. This is true for the following:

1. Closed Σ^3; CMC; Einstein vacuum or Einstein-standard matter source field; $\lambda \in C^3$ and σ, ε, $\beta \in H^p_2 (p > 3)$.
2. Asymptotically Euclidean; Maximal; Einstein vacuum or Einstein-standard matter source field; Data in weighted Sobolev spaces.
3. Asymptotically hyperbolic: CMC; Einstein vacuum or Einstein-standard matter source field; Data in weighted Sobolev spaces.

For these three cases, one finds that all sets of conformal data which are not readily disqualified by simple sign criteria or by the maximal principle are mapped to solutions by the conformal method. To explain the sense in which this holds, let us focus for the moment on the first of these cases; that with Σ^3 closed:

The Yamabe classification of metric conformal classes is a key tool for determining which sets of conformal data on a closed manifold Σ^3 are mapped

[g] We label as "standard matter source field" any field theory which couples to Einstein's theory without derivative coupling. Included are Einstein-Maxwell, Einstein-Yang-Mills, Einstein-Dirac, Einstein-Klein-Gordon, Einstein-Cartan, and Einstein fluids. For all these theories, there are conformal splittings such that, if $\nabla\tau = 0$, the constraint equations semi-decouple. See Isenberg and Nester[4].

to solutions of the constraints. This classification is based on the Yamabe Theorem[3] which says that every metric λ_{ab} on a closed manifold Σ^3 is conformally related to another which has constant scalar curvature either $+1$, 0, or -1; and moreover, the sign of R for constant scalar curvature metrics conformally related to a given λ_{ab} is unique. Thus the set of all metrics (and metric conformal classes) on a closed Σ^3 is partitioned into the three Yamabe classes $\mathcal{Y}^+(\Sigma^3)$, $\mathcal{Y}^0(\Sigma^3)$, and $\mathcal{Y}^-(\Sigma^3)$ depending on this sign.

The Yamabe theorem and Yamabe classification are important for CMC conformal data on a closed Σ^3 because solubility of (12)-(13) is invariant under the conformal transformation $\{\lambda_{ab}, \sigma^{cd}, \tau, \beta^a, \varepsilon^a\} \to \{\theta^4 \lambda_{ab}, \theta^{-10}\sigma^{cd}, \tau, \theta^{-6}\beta^a, \theta^{-6}\varepsilon^a\}$ for any positive θ. (This is *not* true if $\nabla\tau \neq 0$). Thus, to determine for which sets of conformal data (12)-(13) admit a solution, one may without loss of generality restrict attention to conformal metrics with $R(\lambda) = -1, 0$, or $+1$.

As a consequence of this conformal invariance and the resulting simplification, one immediately sees from (13) and the maximum principle that the following categories of conformal data do not lead to solutions[h] $\{\mathcal{Y}^+; (\sigma, \rho) \equiv 0; \tau = 0\}$, $\{\mathcal{Y}^-; (\sigma, \rho) \equiv 0; \tau = 0\}$, $\{\mathcal{Y}^+; (\sigma, \rho) \equiv 0; \tau \neq 0\}$, $\{\mathcal{Y}^0; (\sigma, \rho) \equiv 0; \tau \neq 0\}$, $\{\mathcal{Y}^0; (\sigma, \rho) \not\equiv 0; \tau = 0\}$, $\{\mathcal{Y}^-; (\sigma, \rho) \not\equiv 0; \tau = 0\}$. Much less evident is the fact that for *all other sets of conformal data on closed* Σ^3 (12)-(13) can be solved, and we obtain a solution of the constraints. The proof of this result (with one case left out) is discussed in the review paper of Choquet-Bruhat and York[2]; the complete proof appears in the work of the author[5].

Analogous results, with a few important modifications, hold for the other two cases mentioned above. Specifically, for the asymptotically Euclidean case with $\tau = 0$, we first note that there is a Yamabe-type partition of asymptotically Euclidean metrics into two classes $\mathcal{Z}^+$ and $\mathcal{Z}^-$, with the $\mathcal{Z}^+$ conformal geometries admitting a conformal transformation to metrics with $R = 0$, and the $\mathcal{Z}^-$ conformal geometries not admitting such a transformation. It follows from work of Brill and Cantor[6] that asymptotically Euclidean metrics with non-negative scalar curvature, as well as those satisfying a certain integral condition, are contained in $\mathcal{Z}^+$. The work of Cantor[7] (See also[8]) shows that a set of asymptotically Euclidean conformal data $\{\lambda_{ab}, \sigma^{cd}, \tau, \beta^a, \varepsilon^a\}$ (in appropriate function spaces) is mapped to a solution by the conformal method if and only if $\lambda \in \mathcal{Z}^+$ (The behavior of the other fields is irrelevant.).

For the asymptotically hyperbolic case (with $\tau \neq 0$), there are important

---

[h] Here $(\sigma, \rho) \equiv 0$ means that σ^{cd} and ρ are identically zero everywhere on Σ^3, while $(\sigma, \rho) \not\equiv 0$ means that there exists at least one point on Σ^3 at which σ^{cd} or ρ is not zero.

questions regarding the asymptotic behavior of solutions; but ignoring those, the results are very simple: For every set of asymptotically hyperbolic conformal data, equations (12)-(13) can be solved, and we obtain a solution to the constraints[9 10].

These results together tell us that if we are interested in constant mean curvature (or maximal) initial data, the conformal method works, apart from a few readily identified special cases. This is useful for two reasons. The first is that since equations of the form of the determining equations (12)-(13) can be handled by modern computer algorithms (at least when they are semi-decoupled, which holds so long as we restrict to conformal data with constant τ), the conformal method is an effective way to produce CMC initial data sets for numerical relativity. The second is that, as a consequence of uniqueness theorems relating conformal data and solutions of the constraints[5], together with the results noted above, certain function spaces of conformal data provide an effective parameterization of the set of solutions of the constraints[11]. In light of the uniqueness of CMC foliations of spacetime solutions of Einstein's equations[12], one may use these function spaces of conformal data to parameterize the set of spacetime solutions as well (ignoring those solutions which do not admit CMC or maximal foliations[13 14]).

These results are very useful, but they are limited in a number of ways, and so there is a lot more that needs to be done. First, one would like to know which *non CMC* conformal data sets lead to solutions. While there are some results concerning this question, they all presume that $|\nabla\tau|$ is small. Indeed, for small $|\nabla\tau|$(with certain restrictions on the zeros of τ) our results show that solutions to the coupled system (12)-(13) always exist[15 16 17 8 18]. Unfortunately, the iteration methods we have used to prove the results just cited do not appear to be able to handle general $|\nabla\tau|$, so new ideas are likely needed.

Next, we would like to be able to generate solutions to the constraint equations on manifolds with boundaries. The initial value-boundary value problem has been studied for Einstein's theory in Friedrich's formulation[19], but there has not been much mathematical work on it in the standard $\{\gamma, K\}$ formulation discussed here. In view of the recent numerical work on spacetimes with excised black holes, a mathematical treatment, in terms of the $\{\gamma, K\}$ formulation, of solutions with boundaries would be useful.

A third issue regarding the construction of solutions of the constraint equations which should be addressed further is the loosening of differentiability requirements for solutions, including weak solutions. The motivation for doing this comes from the progress that has recently been made by Klainerman and Rodnianski[20] and by Smith and Tataru[21] in proving local well-posedness for the Einstein equations for data of lower differentiability. The goal (based on energy considerations) is $H^2_{3/2}$; so far $H^2_{2+\epsilon}$ has been obtained. This leads one to try to construct initial data sets in $H^2_{2+\epsilon}$ or even $H^2_{3/2}$. The results to date[5 10 8] produce initial data sets which are considerably more restricted (i.e., smoother) than this, but work is proceeding toward obtaining solutions of the constraint equations with this low differentiability.

The question that remains is how to construct initial data sets which can be used to model physical phenomena of interest. We discuss this in the next section.

3 Physical Modeling and the Gluing of Solutions of the Constraint Equations

Say we want to study the gravitational radiation produced by the inspiral collision of a neutron star and a black hole. The first step of a numerical modeling of this phenomenon is to find initial data $\{\gamma, K, \psi\}$ which constitutes a snapshot of the two objects and their ambient spacetime at some moment prior to the collision. It is important that the initial data accurately represent the two objects and their surroundings in a relatively quiet pre-collision state, without any extraneous unrelated radiation, or the modeling will be of little use.

The electromagnetic analogue of this modeling problem is familiar and easily handled, since the background space is fixed and independent of the electromagnetic fields in Newtonian-Maxwell theory, and since the representation of electromagnetic radiation is well understood in this theory as well. So, too, if we work with a post-Newtonian approximation to Einstein's theory, the choice of initial data for phenomena like the inspiral collision is understood to some extent[22 23 24].

However, attempts to find physically accurate initial data sets for phenomenon such as this in terms of the full Einstein theory have been stymied, despite many years of significant effort, both numerical and analytical. The conformal method can of course be used to generate candidate sets of data, and such sets are used in most numerical modeling studies. The lack of a priori control of the conformal factor and therefore of the ambient space, together with the difficulty in representing and recognizing gravitational radiation in the conformal data $\{\lambda, \sigma, \tau, \psi\}$ for massive relativistic objects renders these sets somewhat suspect, however.

One approach that has been discussed to try to control the choice of initial data is to work with sets of post-Newtonian initial data in the very early pre-radiation stages of the collision and then try to evolve these into the later, more relativistic stage. This approach has some promise, but can be very expensive in numerical evolution time.

Another possible approach that could be tried relies on "gluing." The idea here is to start with well-understood initial data sets for separate portions of the spacetime–say, one for the black hole and one for the neutron star–and develop a procedure for joining these together into one set. Mathematically, the idea is to join two or more solutions of the constraint equations into a single solution in such a way that, away from the joining region, the data is largely unchanged. Physically, the challenge is to understand and control the physical effects of the joining region.

There has been substantial progress during the past couple of years on the mathematical aspects of gluing solutions of the Einstein constraint equations. Mazzeo, Pollack, and the author (IMP) have proven theorems which prescribe a procedure for gluing connected sums of sets of initial data[25], while Corvino and Schoen[26 27] have shown that rather general asymptotically flat initial data sets can be glued to exact Schwarzschild or exact Kerr exteriors outside a transition region. We first discuss the IMP work, and then briefly describe the results of Corvino and Schoen.

3.1 Connected Sum Gluing

We start with a pair of constant mean curvature solutions $\{\Sigma_1^3, \gamma_1, K_1\}$ and $\{\Sigma_2^3, \gamma_2, K_2\}$ of the vacuum constraint equations (3)-(4). (We presume for now that there are no matter source fields ψ, and so $\rho = 0$ and $J = 0$). Each of these solutions may be asymptotically Euclidean, asymptotically hyperbolic, or one may have either or both of Σ_1^3 and Σ_2^3 closed. The two solutions need not have matching asymptotic properties, but they do need to have the same constant mean curvature τ.

If we now pick a pair of points $p_1 \in \Sigma_1^3$, $p_2 \in \Sigma_2^3$ on each manifold, then the gluing procedure produces a one-parameter family of solutions $\{\tilde{\Sigma}^3, \tilde{\gamma}_s, \tilde{K}_s\}$ with the following properties: (i) $\tilde{\Sigma}^3 = \Sigma_1^3 \# \Sigma_2^3$, with Σ_1^3 joined to Σ_2^3 by a "neck" ($S^2 \times R$) connecting a neighborhood of p_1 to a neighborhood of p_2 (both now excised); (ii) the data $\{\tilde{\gamma}_s, \tilde{K}_s\}$ can be made to be arbitrarily close to the original data $\{\gamma_1, K_1\}$ and $\{\gamma_2, K_2\}$ on appropriate regions of $\tilde{\Sigma}^3$ away from the neck by choosing s sufficiently large; (iii) $tr\tilde{K} = \tau$; and (iv) the geometry on the neck (i.e., lengths, curvature) can degenerate for $s \to \infty$, but its behavior is controlled by exponential bounds in s.

This gluing procedure works for quite general classes of data. The only restrictions on $\{\Sigma_1^3, \gamma_1, K_1\}$ and $\{\Sigma_2^3, \gamma_2, K_2\}$ besides the CMC matching condition $trK_1 = trK_2$ and certain regularity restrictions (stated in terms of weighted Hölder spaces [25]) are that if either of the sets of data has Σ^3 closed, that set of data must have K^{cd} not vanishing identically, and must be free of nontrivial conformal Killing vector fields[i]. Neither of these latter conditions are needed for asymptotically Euclidean or asymptotically hyperbolic data, and we note that if a set of data $\{\Sigma^3, \gamma, K\}$ fails to satisfy either condition, it will generally satisfy them both after a small perturbation is made.

The procedure allows a number of interesting sets of initial data to be produced. One can, for example, add a sequence of small black holes to any asymptotically flat spacetime $\{M^4, g\}$ by gluing a sequence of copies of Minkowski data to asymptotically Euclidean data for $\{M^4, g\}$ and then evolving the glued data. These spacetimes (somewhat reminiscent of the Misner multi-black hole solutions[28], but far more general) may be constructed with the interiors either independent or connected. Another set of spacetimes one obtains via this gluing procedure allows one to study the question of black holes in cosmological spacetimes: One can glue asymptotically Euclidean or asymptotically hyperbolic data to cosmological data on closed Σ^3 and then see if black hole-like physics develops[j]. This gluing procedure also permits us to add one or more wormholes to most solutions of Einstein's equations: If one chooses the pair of points p_1 and p_2 on a single connected initial data set $\{\Sigma^3, \gamma, K\}$ and carries out the gluing procedure, one obtains initial data much like $\{\Sigma^3, \gamma, K\}$ except with an $S^2 \times R$ wormhole glued on.

To what extent can one carry out these gluings in practice, say, numerically? To answer this question, we shall now describe the basic steps of the connected sum gluing procedure:

Step 1: One starts by performing a conformal transformation of the data $\gamma_{ab} \to \psi^{-4}\gamma_{ab}, \sigma^{cd} \to \psi^{10}\sigma^{cd}$, (Here σ^{cd} is the trace-free part of K^{cd}.) which is trivial ($\psi = 1$) away from a neighborhood of each of the points p_1 and p_2, and which is singular at each of the points (with a specified type of singular behavior). The effect of this conformal blowup is to replace a ball surrounding p_1 with an $S^2 \times R$ infinite length half-cylinder which is asymptotically a standard round half-cylinder, and to do the same in a neighborhood of p_2. Note that there are standard explicit formulas one can choose for ψ to carry out this first step.

Step 2: Each of the newly added half cylinders extends out infinitely far. If one cuts each at a distance $s/2$ (measured using the local conformally

[i] Actually, a Killing vector field can be present, so long as it does not vanish at the chosen points p_1 or p_2.

[j] Note that there is no accepted definition of a black hole in a spacetime without some sort of "asymptotic infinity" structure.

transformed metric) out from the neighborhoods of p_1 and p_2, and identifies the two cut half cylinders on the joining sphere, then one has the manifold $\tilde{\Sigma}^3 = \Sigma_1^3 \# \Sigma_2^3$, with its connecting s-parametrized "neck". Then, using a pair of cut-off functions χ_1 and χ_2, one may patch together the metric $\hat{\gamma}_s = \gamma_1\chi_1 + \gamma_2\chi_2$ and the traceless apart of the second fundamental form $\hat{\sigma}_s = \sigma_1\chi_1 + \sigma_2\chi_2$. Both $\hat{\gamma}_s$ and $\hat{\sigma}_s$ are well-defined on $\tilde{\Sigma}^3$, and both depend on the gluing distance parameter s. One also glues together the conformal function $\hat{\psi}_s = \psi_1\xi_1 + \psi_2\xi_2$, using a different set of cut-off functions. (For details, see IMP[25]). This step, like the first, can be done using standard explicit formulas for χ_1, χ_2, ξ_1, and ξ_2.

Step 3: A careful choice of the cut-off functions results in $\hat{\sigma}_s$ being everywhere traceless with respect to $\hat{\gamma}_s$. The conformal transformation $\sigma^{cd} \to \psi^{10}\sigma^{cd}$ guarantees that away from the neck, $\hat{\sigma}_s$ is divergence-free with respect to $\hat{\gamma}_s$. However, in the neck, $\hat{\nabla}_c \hat{\sigma}_s^{cd} \neq 0$. So one replaces $\hat{\sigma}_s^{cd}$ by $\hat{\mu}_s^{cd}$, which is obtained by solving

$$\hat{\nabla}_c(\hat{L}Y_s)^{cd} = \hat{\nabla}_c \hat{\sigma}_s^{cd} \tag{20}$$

for the vector field Y_s^c, and setting

$$\hat{\mu}_s^{cd} = \hat{\sigma}_s^{cd} - (\hat{L}Y_s)^{cd}. \tag{21}$$

The tensor field $\hat{\mu}_s^{cd}$ is divergence-free everywhere on $\tilde{\Sigma}^3$ with respect to $\hat{\gamma}_s$. This step is straightforward to carry out in practice. However, to be able to proceed further with the gluing procedure, one needs to know not only that the divergence-free field $\hat{\mu}_s^{cd}$ exists, but also that $\| \hat{\mu}_s - \hat{\sigma}_s \|_{C^{k+\beta}}$ is very small for sufficiently large s. It is shown in IMP that this estimate always holds.

Step 4: $\{\hat{\gamma}_s, \hat{\mu}_s, \tau\}$ is a standard set of conformal data on $\tilde{\Sigma}^3$. So one may set up and attempt to solve

$$\hat{\nabla}^2\phi_s = \frac{1}{8}\hat{R}_s\phi_s - \frac{1}{8}\hat{\mu}_{scd}\hat{\mu}_s^{cd}\phi_s^{-7} + \frac{1}{12}\tau^2\phi_s^5 \tag{22}$$

for ϕ_s. Although the Yamabe class of the metric $\hat{\gamma}_s$ constructed in Step 2 is not evident, the estimates obtained in Step 3 allow one to show that (22) has a unique solution. Further, with a considerable amount of work, one shows that $\| \phi_s - \psi_s \|_{C^{k+\beta}}$ is very small for sufficiently large s. As noted above, once one knows that a solution to (22) exists, one can readily obtain that solution numerically.

Step 5: Using formulas analogous to (15)-(16), one constructs $\tilde{\gamma}_s$ and $\tilde{K}_s$ from ϕ_s and $\{\hat{\gamma}_s, \hat{\mu}_s, \tau\}$. One immediately verifies that for each value of the parameter s, $\{\tilde{\Sigma}^3, \tilde{\gamma}_s, \tilde{K}_s\}$ solves the vacuum constraint equations. As a consequence of the estimates of Steps 3 and 4, one verifies that away from the neck, $\hat{\gamma}_s$ and $\hat{K}_s$ approach the original data for sufficently large s.

We see from the description of these five steps that in practice, this connected sum gluing procedure involves making a few choices of standard cut off functions, plus solving familiar elliptic equations. It should be a useful procedure for studying solutions of Einstein's equations numerically.

We note that, to date, the IMP connected sum gluing procedure has been proven to work for the vacuum Einstein case only. It is expected that it can be implemented for Einstein-Maxwell, Einstein-Yang-Mills, and the other Einstein-standard matter source fields as well; work is under way to show this. It would be nice to also extend the procedure to non CMC data, but this is likely to be difficult.

3.2 Exterior Schwarzschild and Kerr Gluing

The starting point for the exterior Schwarzschild gluing procedure is any asymptotically Euclidean, time symmetric ($K^{cd} = 0$) initial data set $\{\Sigma^3, \gamma\}$. One chooses a compact region Λ^3 in Σ^3. One then shows [26] that there is an asymptotically Euclidean time symmetric solution $\{\Sigma^3, \tilde{\gamma}\}$ such that (i) inside a ball B_r which contains Λ^3 as a subset, $\tilde{\gamma}|_{B_r} = \gamma$; and (ii) outside the ball B_{2r}, we have $\tilde{\gamma}|_{\Sigma^3 \setminus B_{2r}} = \gamma_{Sch(m)}$, where $\gamma_{Sch(m)}$ is the spatial Schwarzschild metric for some positive mass m.

We note three important features of this type of gluing: First, when the gluing is completed, the region inside B_r as well as the exterior region outside B_{2r} are *unchanged*. This is not the case for the connected sum gluing, in which the original solutions are slightly changed (away from the neck) by the gluing procedure. Second, the analysis used to prove this result does not rely in any way on the conformal method. That is, one does not construct the solution $\{\Sigma^3, \tilde{\gamma}\}$ by first specifying a conformal metric everywhere and then solving for the conformal factor. Indeed, one can carry out exterior Schwarzschild gluing only if one works with the constraints as an underdetermined system to be solved for $\tilde{\gamma}$, rather than as a determined system to be solved for ϕ. Third, the proof that this gluing works does not readily translate into a step by step procedure that one can carry out numerically. The proof guarantees existence, but does not prescribe a procedure for constructing $\{\Sigma^3, \tilde{\gamma}\}$

The result is quite surprising and remarkable. It says that for any compact portion of a time symmetric asymptotically Euclidean solution, one can arrange the gravitational fields in an annular region around it so that outside this region, the gravitational field is exactly Schwarzschild. Note that the analogous result does not hold true for Newtonian gravity. Note also that, combining this result with the work of Friedrich[29], one may be able to produce a very large family of solutions of Einstein's equations with a complete

$\mathcal{I}$ (null asymptotic infinity).

One large but unsurprising restriction on the type of data which can be glued to an exterior Schwarzschild spatial slice is that the data have $K^{cd} = 0$. More recent results of Corvino and Schoen [27] allow one to remove this restriction provided that one replaces the Schwarzschild exterior by a slice of the more general Kerr (rotating black hole). They do impose restrictions on the data $\{\Sigma^3, \gamma, K\}$ which can be patched to an exterior Kerr slice: They require certain decay behavior in γ and K which are essentially just enough so that the momentum and angular momentum integrals at infinity are well-defined. These restrictions appear to be minor, in view of the result. Again, one consequence of these results might be an even larger family of solutions with complete $\mathcal{I}$.

4 Some Comments on Evolution

Of the seven questions raised in the introduction, we have focussed in this essay on those three – 2,3, and 4 – which deal with the constraint equations. We shall now briefly comment on questions 1,5, and 6, which mostly concern evolving sets of initial data into spacetime solutions.

The first question addresses the issue of local existence and well-posedness of the initial value formulation of the Einstein equations. To a large extent, this issue was settled fifty years ago by Choquet-Bruhat[30] who proved (using harmonic coordinates) that for smooth initial data (C^l, with large l) the Einstein system is well-posed. While this result is sufficient for most purposes, there have been strong attempts in recent years to prove well-posedness for more general classes of data. The goal is to prove it for data in the Sobolev space $H^2_{\frac{3}{2}}$. The recent work of Klainerman and Rodnianski[20], and of Smith and Tataru[21] achieves it for $H^2_{2+\epsilon} (\epsilon > 0)$ data.

Well-posedness theorems show (among other things) that data in the specified function space can always be evolved to a spacetime solution for a sufficiently small time duration into the future and into the past. They say little about how long a spacetime solution will last, or what its behavior will be far into the future and far into the past. While these issues are very difficult and far from resolved, there has been substantial progress on them in recent years. We note in particular the work of Christodoulou and Klainerman[31], which verifies the stability of Minkowski spacetime (and thereby proves that there is a nontrivial family of asymptotically flat spacetimes which extend for infinite proper time into the future and into the past); that of Andersson and Moncrief[32] which does the same for the stability of the expanding

$k = -1$ vacuum Friedman-Robertson-Walker cosmological spacetimes; and the work of Christodoulou[33] which verifies weak cosmic censorship[k] for spherical collapsing Einstein-scalar field spacetimes. There has also been a collection of works by Berger, Chrusciel, Garfinkle, Kichenassamy, Moncrief, Rendall, Ringstrom, Wainwright, Weaver, and the author which provide increasing evidence for the presence of asymptotically velocity dominated behavior and mixmaster behavior in cosmological spacetimes, and further tends to support the validity of the strong cosmic censorship conjecture[l] in these spacetimes. (See, for example, the review papers by Berger[34] and Rendall[35].)

The construction or study of spacetime solutions of Einstein's equations via the initial value formulation requires one to make a choice of spacetime foliation (i.e., a specification of the t = constant hypersurfaces which fill the spacetime) and of spacetime threading (i.e., a specification of the x^a = constant observer world lines which are everywhere transverse to the foliation, and fill the spacetime). The choice of foliation is controlled infinitesmally by the lapse function N, and the choice of threading is similarly controlled by the shift vector M^a.

In constructing a spacetime from specified initial data, one may always make the simple choice $N = 1$ and $M^a = 0$ ("Gaussian normal coordinates"). However, this choice generally leads to a non physical and premature breakdown in the evolution (a coordinate singularity). Consequently, the maximal development[37] of the initial data may not be obtained. So, one criterion for a good choice of the lapse and shift is that the choice effectively avoids such coordinate singularities. Also important is that the choice be relatively easy to implement in practice, and that it not obscure the gravitational physics of the spacetime (e.g., by simulating gravitational radiation which is not physically present).

Constant mean curvature or maximal slicing, coupled with some sort of coordinate shear minimizer, is often cited as a choice of foliation and threading which avoids coordinate singularities and clarifies the physics. However, implementing this choice requires that one solve a set of elliptic equations on each time slice. This is a large expense in computer time for numerical constructions, and it precludes explicit forms for the lapse and shift in analytical studies. For special families of solutions such as the Gowdy space-

---

[k] The weak cosmic censorship conjecture says that in asymptotically Euclidean spacetime solutions, the singularities which develop during gravitational collapse are generically shielded from the view of observers at infinity by the development of an event horizon.

[l] The strong cosmic censorship conjecture[36] says that in generic solutions developed from Cauchy data on a compact Cauchy surface, a Cauchy horizon (with its attendant causality difficulties) does not form.

times or the $U(1)$ Symmetric solutions , there are certain choices of lapse and shift picked out by the geometry (areal for Gowdy[38], harmonic time for $U(1)$ Symmetric[39]); but these are special cases. More generally, the choice of lapse and shift remains a difficult issue.

5 Concluding Remarks

We have discussed a number of the challenges that one encounters in constructing and studying spacetimes via the initial value formulation. These occur both in finding sets of initial data which satisfy the Einstein constraint equation, and in evolving these sets of data. Some are fairly difficult. However, paraphrasing Winston Churchill's description of democracy, one finds that the initial value formulation is the most impractical way to work with solutions of Einstein's equations, except all of those other ways which have been tried from time to time[40]. The fact that most numerical studies of solutions are carried out using the initial value formulation attests to the relative practicality of this approach.

6 Acknowledgments

I thank the Scientific Committee of GR16 for inviting me to speak, and I thank the local Organizing Committee and the local South African hosts for running a very fine conference. Portions of the work discussed have been supported by NSF grant PHY 0099373.

References

1. C. Misner, K. Thorne and J. Wheeler, *Gravitation* (Freeman, San Franciso, 1973).
2. Y. Choquet-Bruhat and J. York in *General Relativity and Gravitation*, ed. A. Held (Plenum, New York, 1980).
3. R. Schoen, *J. Diff. Geom.* **20**, 479 (1984).
4. J. Isenberg and J. Nester, *Ann. Phys.* **107**, 368 (1977).
5. J. Isenberg, *Class. Qtm. Grav.* **12**, 2249 (1995).
6. D. Brill and M. Cantor, *Comput. Math.* **43**, 317 (1981).
7. M. Cantor, *Comm. Math. Phys.* **57**, 83 (1977).
8. Y. Choquet-Bruhat, J. Isenberg and J. York, *Phys. Rev.* D **61**, 084034 (2000).
9. L. Andersson, P. Chrusciel and H. Friedrich, *Comm. Math. Phys.* **149**, 587 (1992).

10. L. Andersson and P. Chrusciel, *Dissert. Math.* **355**, 1 (1996).
11. J. Isenberg, *Phys. Rev. Lett.* **59**, 2389 (1987).
12. D. Brill and F. Flaherty, *Ann. Inst. H. Poincare* **28**, 335 (1978).
13. D. Brill in *Proc Third Marcel Grossman Meeting*, ed. H. Ning (North-Holland, Amsterdam, 1982).
14. R. Bartnik, *Comm. Math. Phys.* **117**, 615 (1998).
15. Y. Choquet-Bruhat, J. Isenberg and V. Moncrief, *C. R. Acad. Sci. Paris* **315**, 349 (1992).
16. J. Isenberg and V. Moncrief, *Class. Qtm. Grav.* **13**, 1819 (1996).
17. J. Isenberg, *Fields Inst. Comm.* **15**, 59 (1997).
18. J. Isenberg and J. Park, *Class. Qtm. Grav.* **14**, A189 (1997).
19. H. Friedrich and G. Nagy, *Comm. Math. Phys.* **210**, 619 (1999).
20. S. Klainerman and I. Rodnianski, preprint: math.AP/0109173.
21. H. Smith and D. Tataru, preprint: Sharp local well-posedness results for nonlinear wave equation.
22. T. Damour in *300 Years of Gravitation*, eds. S. Hawking and W. Israel (Cambridge Univ Press, New York, 1987).
23. A. Buonanno and T. Damour, *Phys. Rev.* D **62**, 06401 (2000).
24. K. Alvi, *Phys. Rev.* D **64**, 104020 (2001).
25. J. Isenberg, R. Mazzeo, and D. Pollack, preprint: gr-qc/0109045.
26. J. Corvino, *Comm. Math. Phys.* **214**, 137 (2000).
27. J. Corvino and R. Schoen, unpublished.
28. C. Misner, *Ann. Phys.* **24**, 102 (1963).
29. H. Friedrich in *Gravitation and Relativity at the Turn of the Millennium*, eds. N. Dadhich and J. Narlikar (Inter-University Centre, Pune, 1998).
30. Y. Bruhat, *Acta Math* **88**, 141 (1952).
31. D. Christodoulou and S. Klainerman, *The Global Nonlinear Stability of the Minkowski Space* (Princeton Univ Press, Princeton, 1993).
32. L. Andersson and V.Moncrief, unpublished.
33. D. Christodoulou, *Ann. Math.* **149**, 183 (1999).
34. B. Berger, *Living Reviews* **5**, 2002-1 (2002).
35. A.Rendall, *Living Reviews* **3**, 2000-1 (2000).
36. R. Penrose in *General Relativity-An Einstein Centenary Survey*, eds. S. Hawking and W. Israel (Cambridge Univ Press, New York, 1979).
37. Y. Choquet-Bruhat and R. Geroch, *Comm. Math. Phys.* **14**, 329 (1969).
38. R. Gowdy, *Ann. Phys.* **83**, 203 (1974).
39. V. Moncrief , *Ann. Phys.* **167**, 118 (1986)
40. W. Churchill, House of Commons, 11 Nov., 1947.

THE CURRENT STATUS OF THE GROUND-BASED LASER INTERFEROMETRIC GRAVITATIONAL WAVE ANTENNAS

S. KAWAMURA

National Astronomical Observatory of Japan, 2-21-1 Osawa, Mitaka, Tokyo, 181-8588, Japan

E-mail: seiji.kawamura@nao.ac.jp

Several medium- to large-scale ground-based interferometric gravitational wave antennas have been constructed around the world; some of them are already partially or even fully operational. Within a few years all of the detectors are expected to be completed and the international network of gravitational wave antennas will be realized. Although these antennas of the first generation could detect gravitational waves, it is necessary to improve the sensitivity of the detector significantly with much more advanced techniques to ensure more frequent detection of gravitational waves. Some of the advanced techniques are already implemented in some detectors, and others are currently being investigated for the second-generation detectors. We expect that such advanced detectors will eventually open a new window to the universe and establish a new field "gravitational wave astronomy".

1 Introduction

The detection of gravitational waves was pioneered by Weber in the early 1960s. He developed the first resonant bar detectors and announced that he had detected gravitational waves [1]. Although no convincing evidence for the detection was provided, his pioneering work inspired others to search for gravitational waves. In addition to the subsequent efforts on cooling bar detectors, Weiss and several other researchers had come to the idea of using laser interferometers to detect gravitational waves [2]. Since then several groups over the world have developed laser interferometric detectors and most experimentalists have come to believe that interferometers are the most promising means for detecting gravitational waves.

There are at present several on-going projects in the world for building medium- to large-scale (300m to 4km) laser interferometric gravitational wave detectors: LIGO [3] by the United States, Virgo [4] by France and Italy, GEO600 [5] by Germany and the United Kingdom, and TAMA300 [6] by Japan. In addition the Australian group (ACIGA) is preparing for AIGO [7].

Currently some of the projects are in the process of improving the sensitivity down to the level with which the detection of gravitational waves could be possible. Others are still implementing basic elements of the interferometer to obtain the first sensitivity curve. It can be expected that soon there will be an international network of gravitational wave detectors, which will eventually lead to the detection of gravitational waves. However, it is necessary to improve the sensitivity of the detectors significantly with advanced techniques in order to ensure frequent detection of gravitational waves. Some of the advanced techniques are already implemented in some of the existing detectors. Others are being investigated intensively. These techniques are intended for planned second-generation detectors that have the goal of establishing the field of gravitational wave astronomy.

In this article the most exciting recent progress of each project is briefly mentioned in section 2. The basics of a laser interferometric gravitational wave detector are reviewed in section 3. The standard techniques used for most exsisting detectors are described in section 4. Some of the advanced techniques are introduced in section 5. Finally prospects for gravitational wave detection are summarized in chapter 6.

2 Recent Progress of Each Project

2.1 TAMA300

The sensitivity of TAMA300 has been further improved, especially around 100 Hz, by a factor of 10 compared with the sensitivity obtained in summer of 2000. The current sensitivity of the detector is about $5\times10^{-21}\,\mathrm{Hz}^{-1/2}$ around 1 kHz and $5\times10^{-19}\,\mathrm{Hz}^{-1/2}$ around 100 Hz in terms of strain. Figure 1 shows the evolution of the sensitivity of TAMA300.

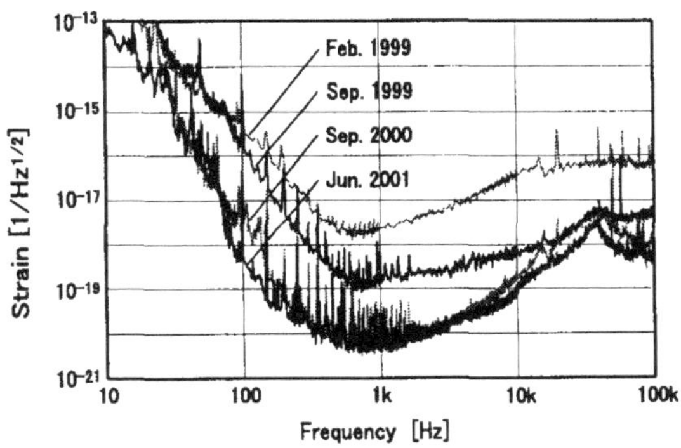

Figure 1. Evolution of the sensitivity of TAMA300.

This improvement resulted in a significant improvement of the sensitivity to chirps from heavier-mass binary coalescence. The current signal-to-noise ratio of the detector for gravitational waves emitted from a binary neutron star coalescence in the center of our galaxy, using the matched filtering analysis, is about 50 and that for ten solar-mass black holes is about 30, as shown in Fig. 2.

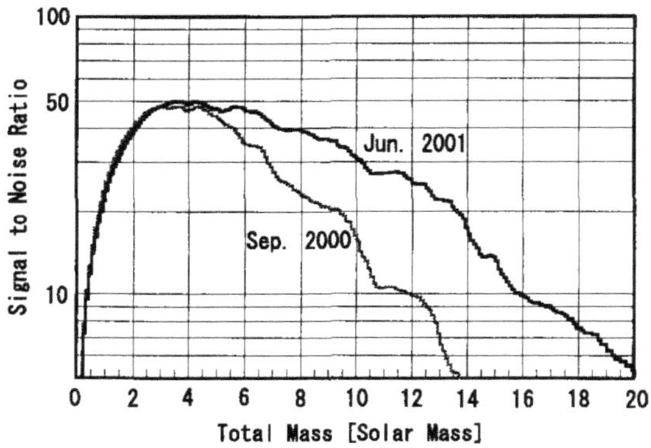

Figure 2. Signal to noise ratio of TAMA300 for gravitational waves emitted from a binary coalescence in the center of our galaxy.

The stability of TAMA300 has been improved significantly by the installation of the active isolation system and modification of the suspension system. The longest continuous lock time of TAMA300 is more than 24 hours.

2.2 LIGO

The 2km interferometer at the Hanford site was successfully operated in its full optical configuration, that is, a power-recycled Fabry-Perot Michelson interferometer. The recycling gain of 15 was achieved and the lock was quite stable. LIGO is now trying to improve the sensitivity by implementing a fine frequency stabilization system.

2.3 Virgo

A short-arm Michelson interferometer with their advanced seismic isolation system was successfully locked. The seismic motion of the mirrors was measured to be reduced from $1\times10^{-6}\,\mathrm{m}/\sqrt{\mathrm{Hz}}$ of the ground motion level to $1\times10^{-9}\,\mathrm{m}/\sqrt{\mathrm{Hz}}$, by a factor of 1000 around 1 Hz, by this isolation system.

Figure 3 shows the displacement noise of the Virgo short-arm Michelson interferometer at low frequencies.

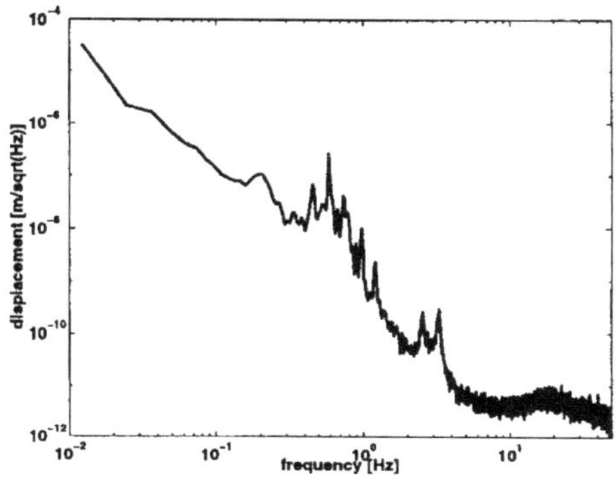

Figure 3. Displacement noise of the Virgo short-arm Michelson interferometer at low frequencies. (Provided by Virgo)

2.4 GEO600

The two sequential mode cleaners of GEO600 were both locked and operated very stably with all the degrees of freedom under automatic alignment control or drift control. The longest continuous lock-holding time is about 40 hours. GEO is now trying to lock their power-recycled long-arm Michelson interferometer.

2.5 ACIGA

Although the ACIGA group has not yet obtained a budget for a large interferometer, their Gingin facility is now being used as a high power optical test facility to diagnose the cavity performance at 0.3 MW circulating power. This work is for future advanced detectors in collaboration with LIGO.

3 Laser Interferometric Gravitational Wave Detector

3.1 Detection Principle

A Michelson laser interferometer with suspended mirrors is used to detect gravitational waves. As shown in Fig. 4, when gravitational waves with an appropriate polarization pass the interferometer vertically, the lengths of the two arms are modulated differentially. It results in changes in the intensity of the interfering light, which is detected by a photodetector. In the actual interferometer the intensity of the light at the detector port will be kept dark by controlling the arm length. Therefore the feedback signals applied to the mirrors to control the arm length should contain gravitational wave signals.

The beamsplitter and mirrors should be suspended so that they can free-fall horizontally at frequencies well above their pendulum frequencies. This is necessary for the interferometer to respond properly to gravitational waves.

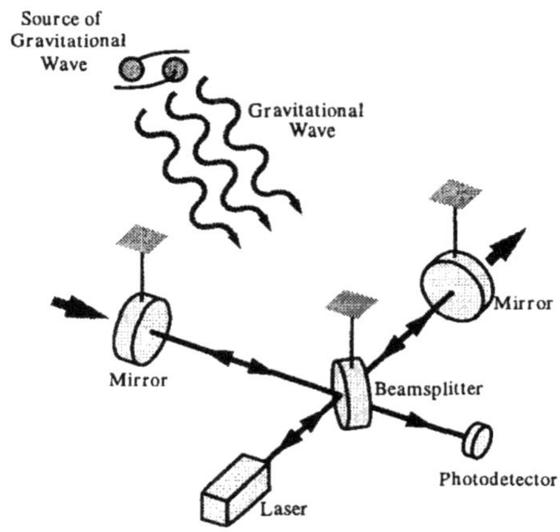

Figure 4. Detection of gravitational waves using a Michelson laser interferometer with suspended mirrors.

Laser interferometers are ideally suited for detecting gravitational waves, because the length change caused by gravitational waves can be increased by simply scaling up the arm length. Therefore the larger interferometer will provide better sensitivity to gravitational waves. However, the signal would be cancelled if the arm length is too long. For example if the arm length is equal to a half the wavelength of the gravitational wave to be detected, the light experiences modulation of the apparent speed of light positively during half the time of its stay in the arm and negatively during the rest of the time, resulting in a complete cancellation of the total phase change of the modulated light. It can be derived that the optimum arm length is a quarter of the wavelength of the gravitational wave. For example an arm length of 750km is optimum for a 100Hz gravitational wave.

3.2 Noise Sources

Among various noise sources existing in an interferometer, the shot noise, seismic noise and thermal noise are regarded as three largest fundamental noise sources, while others are considered to be practical noise sources.

Shot noise exists because a photon has finite energy. Statistical fluctuations in the photon number in the detection process result in shot noise. In order to reduce the shot noise level, the interferometer should be operated in such a way that the intensity of the light at the detection port is dark, which gives the lowest counting fluctuations. In addition to this fringe control, it is desirable to have higher incident laser power to improve the signal-to-noise ratio for the shot noise. This is because the intensity change at the photodetector caused by gravitational waves is proportional to the incident laser power, while the intensity fluctuations caused by shot noise is proportional to the square root of the incident laser power; thus the signal to shot noise ratio is improved by the square root of the incident laser power.

Seismic noise is the mirror motion caused by ground motion. It can be reduced by isolating the mirrors from the ground motion. A pendulum or a spring can work as an isolator. The isolation factor increases quadratically in frequency above the pendulum or spring resonant frequency.

Any mass cannot be exempted from Brownian motion. According to thermodynamics, any mode of a system immersed in another system with a finite temperature is given energy determined by the temperature of the outer system as a result of the energy exchange between the two systems. There are two important modes relevant to the suspended mirrors: the pendulum mode and internal mode. In both modes, thermal noise at the relevant frequencies can be reduced by making a quality factor of the corresponding mode higher. Since a mode with a higher quality factor has more energy concentrated around its resonance, lower thermal motion is engaged off the resonant frequency. This is because the root-mean-square thermal motion integrated over the frequency should be constant.

The sensitivity of an ideal interferometer is limited by these three fundamental noise sources. The seismic noise is dominant at lower frequencies, the shot noise at higher frequencies, and the thermal noise at intermediate frequencies. The noise spectrum of such detectors is schematically shown in Fig. 5.

It is important to emphasize that the above-mentioned three fundamental

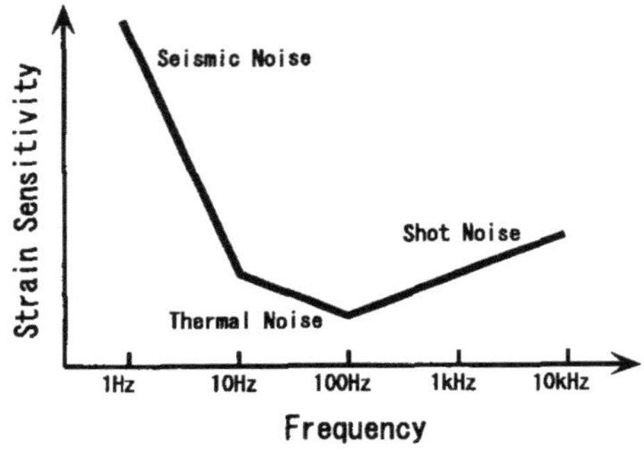

Figure 5. Noise spectrum of an ideal interferometer. The seismic noise is dominant at lower frequencies, the shot noise at higher frequencies, and the thermal noise at intermediate frequencies.

noise sources are not the only noise existing in the interferometer. There are hundreds of other noise sources, most of which are practical ones. Actually we spend most of our time trying to reduce such practical noise. Some of them are rather fundamental, some are extremely complicated in their coupling mechanisms, and others are totally trivial but have sources that can be difficult to identify . Good sensitivity can be attained only after all the noise sources are suppressed.

3.3 Standard Optical Configuration

The actual interferometric gravitational wave detector is far more complicated than a simple Michelson interferometer. Fig. 6 shows an interferometer which is currently considered to have a standard optical configuration. It has three crucial optical elements: Fabry-Perot arm cavities, power recycling, and a mode cleaner.

Since it is unrealistic to build a 750km interferometer on earth, the idea of using Fabry-Perot arm cavities has been brought up. In the Fabry-Perot arm cavity the light is bounced back and forth many times when it is on resonance in the cavity. Thus the phase change of the light caused by gravitational waves is amplified by the effective number of bounces. Since most sensor noise does not explicitly depend on the number of bounces, the signal-to-noise ratio of the interferometer can be improved by use of arm cavities. However, this advantage does not hold for the mirror motion noise, since the mirror motion noise is also amplified by the arm cavity exactly in the same manner as the

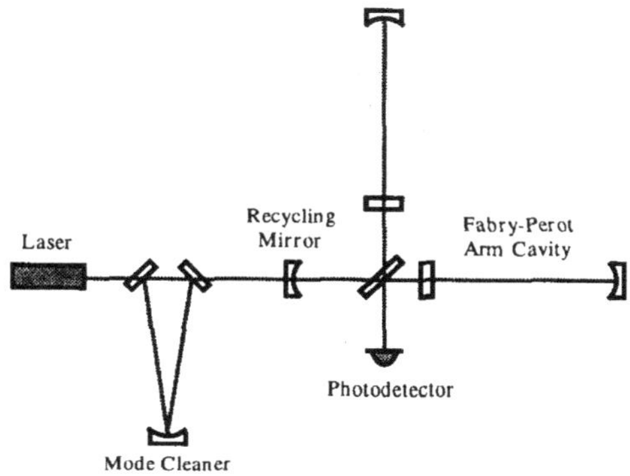

Figure 6. Standard optical configuration of an interferometric gravitational wave antenna. It is a power-recycled Fabry-Perot Michelson interferometer with a mode cleaner.

gravitational wave signal.

Since the interferometer is controlled in such a way that the intensity of the interfering light at the detection port is dark, all the light is reflected back to the laser and unused. The idea of power recycling is to reflect this unused light coherently back to the interferometer so that the laser power is effectively increased, which leads a better signal-to-noise ratio for the shot noise.

The laser geometrical fluctuations, such as beam jitter and pulsation, could make significant noise in the interferometer by coupling with the asymmetry of the interferometer. A mode cleaner is used to suppress the laser geometrical fluctuations by cleaning up the transverse mode of the laser light before it enters the interferometer. The mode cleaner also serves as a reference cavity for the coarse frequency stabilization as well as a passive low pass filter, whose corner frequency is determined by the finesse of the mode cleaner, for frequency and intensity of the light .

4 Standard Interferometer Techniques

In this section some of the standard techniques of the laser interferometric gravitational wave detectors that are actually used for some of the exsisting large antennas are described. It should be noted that the standard techniques described below are not necessarily the currently-best or the mostly-used techniques. They should be regarded just as some examples of currently-used basic techniques.

4.1 Signal Extraction

In order to operate the recycled Fabry-Perot Michelson interferometer, the length of all the cavities and the Michelson interferometer should be properly controlled. There are four important lengths in the recycled Fabry-Perot Michelson interferometer: the common-mode arm cavity length, the differential arm cavity length, the recycling cavity length, and the differential Michelson length. The method to extract these four signals independently using only one phase modulation was developed by LIGO [8]. As shown in Fig. 7, in order to extract the signals, the light should be picked off before the recycling mirror and between the recycling mirror and the beamsplitter in addition to the ordinary detection port. The outputs from those photodetectors are demodulated with the in-phase and quadrature-phase local signals to produce the four independent length signals.

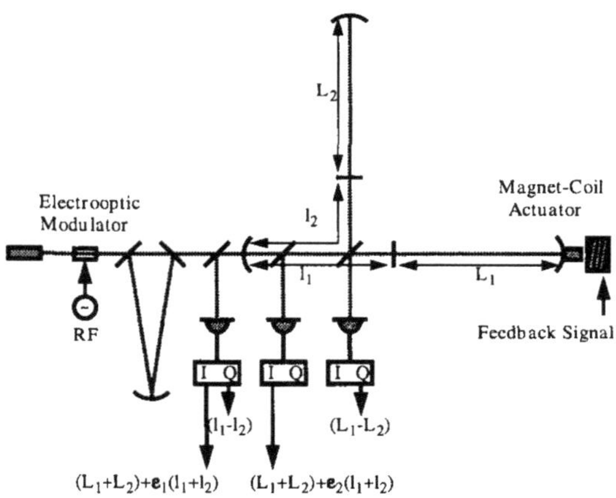

Figure 7. Length signal extraction method for the recycled Fabry-Perot Michelson interferometer using a single phase modulation.

4.2 10W-class Laser

An injection-locked laser-diode pumped Nd:YAG laser is used for GEO600 to provide higher laser power and low frequency noise [9]. An 800mW non-planar ring oscillator master laser is injected into a slave laser (Fig. 8). The slave laser consists of two Nd:YAG crystals, each pumped by a fiber-coupled laser diode, and four-mirror ring cavity, mounted on rigid Invar resonator spacers. The piezo-electric transducer attached to one of the cavity mirrors is used to control the length of the slave cavity to keep the slave laser frequency within the injection locking range. The system is locked very stably for days. The output power of 12W is obtained with very little higher-order mode power. The frequency noise of the injection-locked laser was measured and shown to be dominated by that of the master laser.

Figure 8. Outer picture of the GEO slave laser used for the injection-locked laser system. (Provided by GEO)

4.3 Frequency Stabilization System

Frequency noise of the laser couples with the imbalance of the storage time between two arms in the interferometer, appearing as noise at the output of the interferometer. The frequency noise of a standard high-power laser is not good enough to provide the required sensitivity of the interferometer for a typical imbalance of the interferometer. LIGO employs a multi-stage frequency stabilization system to satisfy the requirement [10]. The laser frequency is stabilized by a small pre-mode cleaner in the first stage, a fixed reference cavity in the second stage, a 12m mode cleaner in the third stage, and a common-mode arm cavity length in the final stage.

4.4 *Suspension/Isolation System*

The LIGO suspension system is a single pendulum suspension with one loop of wire (Fig. 9). The pitch and yaw resonant frequencies of the suspension are made to be lower than 1Hz to prevent noise imposed by the alignment control [11]. All the free motions of the suspension mirrors are damped by the optical sensors and the coil-magnet actuators.

Figure 9. Suspension system used for LIGO. (Provided by LIGO)

The LIGO seismic isolation system is a stack consisting of several layers of metal plates and springs (Fig. 10). The springs are made of metal with damping materials sealed inside. This isolation system is effective above 10Hz.

Figure 10. Seismic isolation system used for LIGO. (Provided by LIGO)

4.5 Practical Noise Sources

There are many kinds of practical noise sources, which could prevent the interferometer from operating with a sensitivity limited by only the three fundamental noise sources. The first step to reduce practical noise is to estimate the contribution of the noise to the interferometer. The standard technique for estimation is to measure the origin of the noise, to measure or analyze the coupling transfer function from the noise origin to the interferometer output, and then mutiply the noise origin by the transfer function to obtain the noise estimate. If a particular kind of noise is found to be limiting the exsisting sensitivity of the interferometer, either the noise origin or the coupling transfer function must be reduced, so that the detector sensitivity can be improved to a new sensitivity limited by another kind of noise. TAMA has been repeating this process to gradually improve its sensitivity. The current noise estimate of TAMA300 is shown in Fig. 11.

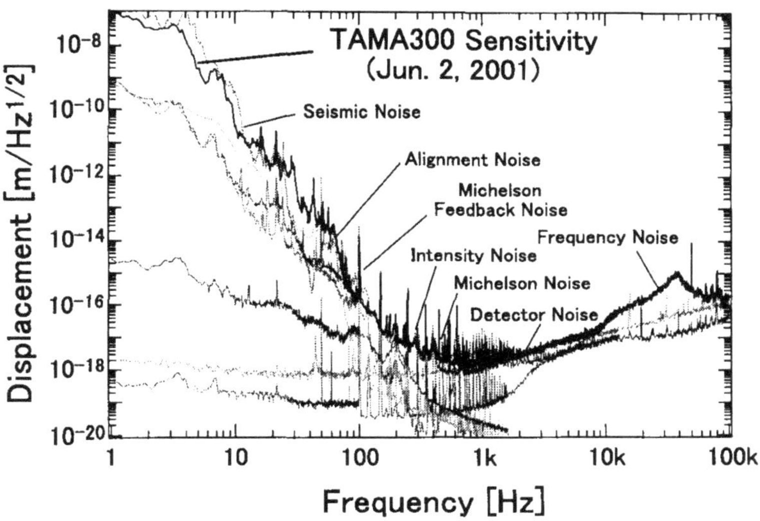

Figure 11. Estimate of various noise sources in the TAMA300 sensitivity.

4.6 Matched Filtering Analysis

Inspiraling compact binaries consisting of neutron stars or black holes are considered to be the most promising sources of gravitational waves. This is not only because they are expected to occur in the universe relatively frequently but also because gravitational waves from the inspiraling binaries can be theoretically calculated in their waveforms to a very high accuracy. The waveform of such gravitational waves is a chirp-like signal, sweeping in frequency and increasing in amplitude. Thus it is possible to improve the signal-to-noise ratio of the detector to gravitational waves from inspiraling binaries by applying a template with a few parameters of the binary system to the actual interferometer output. This is called a matched filtering technique. The TAMA data of 6 hours obtained in 1999 was analyzed using the matched filtering technique [12].

5 Advanced Interferometer Techniques

In order to improve the sensitivity of the detectors, it is necessary to further reduce the three kinds of the fundamental noise. In order to reduce the shot noise, the laser power should be increased. An alternative way of reducing the shot noise is to implement a variable-bandwidth interferometer. Seismic noise should be reduced by a stronger isolation system. As for the thermal noise, the quality factor of the mirror and the suspension system should be increased or the cryogenic mirror/suspension system should be implemented as the final means.

In this section some examples of the advanced techniques of laser interferometric gravitational wave detectors that are either already used for some of the existing large antennas or are being investigated or developed for future use are described.

5.1 100W-class Laser

The ACIGA group is developing a 100W-class laser. They use a side-pumped, side-cooled Nd:YAG slab with zigzag mode in the plane of the pumping (Fig. 12). This zigzag configuration minimizes thermal and stress induced focusing while allowing extraction of energy from the edge of the slab where the pump density is highest. They have already attained 75 W and found that the power was limited by nonuniform pumping of the slab, causing loss of transverse mode control, and possibly by the non-optimal orientation of the crystallographic axes in the slab. These limitations are currently being investigated and corrected.

Figure 12. Prototype of the 100W-class laser developed by the ACIGA Group. (Provided by ACIGA)

5.2 Dual Recycling

The optical configuration of the interferometer for GEO600 is very different from that of the other large antennas. It has a signal recycling mirror placed at the anti-symmetric port and has no arm cavities [13]. In a Michelson interferometer, gravitational wave signals leak into an anti-symmetric port when the fringe there is kept dark. Therefore the mirror placed at the anti-symmetric port can recycle the gravitational wave signals and enhance the interferometer response to the gravitational wave signals. With a slight displacement of the signal recycling mirror, the interferometer response can be made narrow-band with a center frequency deviated from DC. As a result the shot noise can be improved within a limited frequency range.

5.3 Resonant Sideband Extraction

Resonant sideband extraction is another advanced optical configuration. It has a signal extraction mirror placed at the anti-symmetric port and high-finesse arm cavities. The signal extraction mirror can extract the gravitational wave signals from the arm cavities before the gravitational wave signals would be averaged out by the high-finesse arm cavity. This configuration makes it possible to reduce the circulating power inside the power recycling cavity while maintaining the optimized shot noise. This is useful to prevent the undesirable heating effect at the beamsplitter and the bulk of the input mirrors of the arm cavities. Like dual recycling, this configuration can also detune the interferometer response to improve the shot noise within a limited frequency range. The ACIGA group performed a table-top experiment and demonstrated the satisfactory performance of this interferometer (Fig. 13).

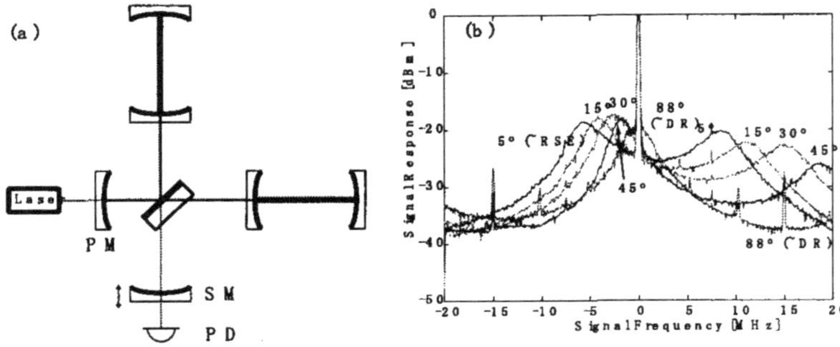

Figure 13. Schematic view of the resonant sideband extraction interferometer and the results obtained by the ACIGA group. (Provided by ACIGA)

5.4 Seismic Isolation

The Virgo group employs an advanced seismic isolation system for their 3km detector, which is called the "Super Attenuator" [14]. It is cascaded five 1.2-m pendulums attached at the top of an inverted pendulum (Fig. 14). The vertical spring of each pendulum gives a very low frequency resonance because of the pre-loaded blades made of Maragin steel and the anti-spring effect of the magnets. The last stage of the pendulum is a mirror and a reaction mass, both suspended from an upper mass called a marionette. The marionette is used to align the mirror.

5.5 Monolithic Suspension

Metal wires are lossy, thus a mass suspended by a metal wire shows large pendulum thermal noise. Fused silica fiber was found to have a much higher quality factor. The GEO group is using a monolithic suspension system for GEO600 [15] (Fig. 15). The fused silica fibers are welded onto specially designed ears with flat surfaces, which are attached to flats on the mass by the hydroxy-catalysis bonding. This bonding technique has been shown to produce negligible additional mechanical loss in the test mass.

Figure 14. VIRGO advanced seismic isolation system. (Provided by VIRGO)

Figure 15. GEO monolithic suspension system. (Provided by GEO)

5.6 Cryogenic Mirrors

The Japanese group is developing cryogenic mirrors and suspensions for their future large antenna project LCGT [16]. The final means to reduce the thermal noise is to cool the mirror and the suspension system because the thermal kinetic energy of the system itself is smaller at lower temperature. In addition they demonstrated that a quality factor of a sapphire mirror, which is a promising candidate for the future antennas, is higher than at room temperature [17] (Fig. 16). Therefore the thermal noise of Sapphire can be even lower at lower temperature.

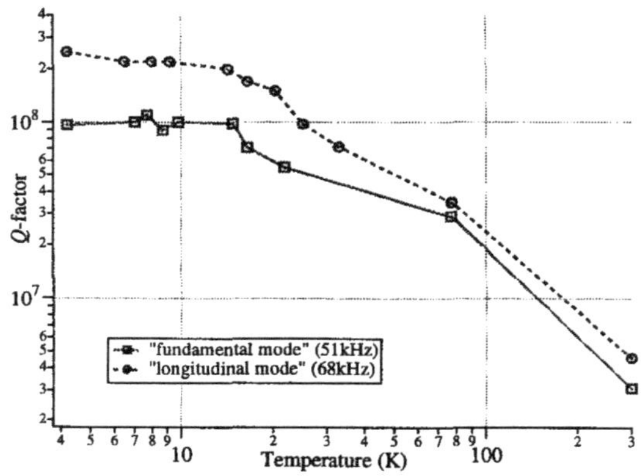

Figure 16. Temperature dependence of the quality factor of Sapphire.

6 Prospect of the Laser Interferometric Gravitational Wave Antenna

Several medium- to large-scale ground-based interferometric gravitational wave antennas have been constructed around the world, and some of them are partially or even fully operational. In a few years most of them will be operational, and it is expected that an international network of the antennas will be formed. There is even a chance that such an international network will be able to detect gravitational waves, although according to the estimate of frequency of the known sources, the chances are not extremely high. Detection or not, it is important to further improve the sensitivity of the detectors significantly in order to ensure frequent detection of gravitational waves. Actually various kinds of advanced techniques necessary for such advanced detectors are being intensively investigated over the world. We expect that those advanced detectors will eventually open a new window to the universe and establish a new field - gravitational wave astronomy.

References

1. J. Weber, *Phys. Rev. Lett.* **22**, 1320 (1969).
2. R. Weiss, *MIT Quarterly Progress Report (Research Laboratory of Electronics)* **105**, 54 (1972).
3. M. Coles, in *Proc. of 3rd Amaldi Conference*, ed. S. Meshkov (American Institute of Physics, New York, 2000) 101.
4. F. Marion for the VIRGO collaboration, in *Proc. of 3rd Amaldi Conference*, ed. S. Meshkov (American Institute of Physics, New York, 2000) 110.
5. H. Lück, *et al.*, in *Proc. of 3rd Amaldi Conference*, ed. S. Meshkov (American Institute of Physics, New York, 2000) 119.
6. M. Ando, *et al.*, *Phys. Rev. Lett.* **86**, 3950 (2001).
7. D. E. McClelland, *et al.*, in *Proc. of 3rd Amaldi Conference*, ed. S. Meshkov (American Institute of Physics, New York, 2000) 140.
8. M. W. Regehr, *et al.*, *Optics Lett.* **20**, 1507 (1995).
9. O. S. Brozek, *et al.*, in *Proc. of 2nd TAMA Workshop*, ed. S. Kawamura and N. Mio (Universal Academy Press, Tokyo, 2000) 295.
10. P. Fritschel, in *Proc. of 2nd TAMA Workshop*, ed. S. Kawamura and N. Mio (Universal Academy Press, Tokyo, 2000) 217.
11. S. Kawamura and M. Zucker, *Appl. Opt.* **33**, 3912 (1994).
12. H. Tagoshi, *et al.*, *Phys. Rev. D* **63**, 062001 (2001).
13. B. Willke, *et al.*, in *Proc. of 2nd TAMA Workshop*, ed. S. Kawamura and N. Mio (Universal Academy Press, Tokyo, 2000) 25.
14. E. Majorana for the VIRGO Collaboration, in *Proc. of 2nd TAMA Workshop*, ed. S. Kawamura and N. Mio (Universal Academy Press, Tokyo, 2000) 15.
15. J. Hough, *et al.*, in *Proc. of 2nd TAMA Workshop*, ed. S. Kawamura and N. Mio (Universal Academy Press, Tokyo, 2000) 311.
16. K. Kuroda, *et al.*, *Int. J. Mod. Phys. D* **5**, 557 (1999).
17. T. Uchiyama, *et al.*, *Phys. Lett. A* **261**, 5 (1999).

NUMERICAL RELATIVITY: STATUS AND PROSPECTS

LUIS LEHNER*

Department of Physics and Astronomy, UBC & Pacific Institute for the Mathematical Sciences

E-mail: luisl@physics.ubc.ca

We are entering an era where the numerical construction of generic spacetimes is becoming a reality. The use of computer simulations, in principle, allows us to solve Einstein equations in their full generality and unravel important messages so far hidden in the theory. Despite the problems still found in present applications, significant progress is being achieved in a number of areas where simulations are on the verge of providing useful physical information. This article reviews recent progress made in the field, paying particular attention to black hole spacetimes and discuss open problems and prospects for the future.

1 Introduction

¿From its beginning a few decades ago, Numerical Relativity has enticed researchers with the promise to finally unravel what General Relativity has kept hidden for almost a century. However, as any other discipline in its infancy, a costly price has been paid to understand how to construct spacetimes numerically and decipher its physical content. Despite significant results obtained in a number of areas, like the discovery of *critical phenomena* in G.R.[1], some criticisms to the field exist due to progress not being as rapid as some expected.

Although it is certainly true that the pace has not been as fast as hoped, progress has been steady and the current status of the discipline gives confidence for even the near future. Just as much has been learned from the theory from analytical ('pencil and paper') techniques in its first century, the second century promises to considerably advance our knowledge due to our ability to numerically probe interesting scenarios. In this article, I describe the current status of the field in a number of fronts and highlight the progress achieved during the past three years (from the previous G.R. meeting[2,3]). It will be impossible to cover all areas in this contribution, thus I will concentrate on black hole simulations since it is precisely in strongly gravitating/highly dynamical scenarios where this field has its *'niche'*. Also, for the most part, I will avoid non-vacuum scenarios as these will be discussed by Shibata in the present volume.

*CITA NATIONAL FELLOW

2 The Arena

Achieving a numerical implementation of a PDE system is a complicated task as one tries to model a 'continuum' with a discrete representation. Of all systems to be modeled, arguably, the most difficult one is given by Einstein equations. Lack of a preferred frame; gauge issues; a system of ten quasi-linear tightly coupled equations; singularities; variables "being" the spacetime rather than living 'on top of it'; etc. complicate the implementation task. In fact, even the equations are not a priori expressed in a convenient form for a numerical evolution; i.e. $E_{ab} \equiv G_{ab} - 8\pi T_{ab} = 0$ does not have an explicit initial value problem in general (IVP). That is, given the initial state of the system at a given "time", the solution to the future is obtained by integrating the equations. The seminal work of Choquet-Bruhat[4] 50 years ago showed the path to follow to achieve a convenient treatment and since then, many works have extended this and presented different variations on how to define an IVP for Einstein equations.

At present, there are roughly three main 'venues' or approaches pursued in numerical relativity to cast the problem in a convenient way: the *Cauchy*, *characteristic* and *conformal approaches*. Several important milestones have been achieved in each of these approaches and I would present them in separate sections, stressing where each one stood just few years ago and where they are now which illustrate they have come a considerable way. Yet, there are still many hurdles left to be crossed for a robust implementation and where appropriate I will include some points that would need to be addressed for this purpose.

2.1 Singularity handling

Of all the aforementioned problems, a crucial one is handling the singularity which can be already present on the initial hypersurface or might be formed at a later time by gravitational collapse. At the analytical level, a singularity marks the demise of the theory; but at the numerical one, even points away from it but "close" enough will have fields whose gradients will likely be too large for the implementation. Thus, if at all possible, one would like to stay away from the singularity and neighboring points. For years, the prevalent strategy was to use the freedom in choosing the foliation of the spacetime, so as to not evolve the equations in that region[5]. By 'piling-up' the slices so that the proper-time along the normal direction is zero, one 'avoids' the singularity. Although this strategy works for short time evolutions, it induces large gradients in long evolutions (due to the slices being considerably 'bent').

Recently, the prevalent strategy has become the one known as *singularity excision*[6]; here the idea is to simply cut-short the evolution inside the event horizon. For this strategy to be physically valid, two conditions are required: First, the validity of cosmic censorship and second a method to estimate the location of the event horizon with only partial information of the spacetime. The second one is addressed by locating apparent horizons (with the caveat that they might not exist even where there is an event horizon[7]) while the first one is still an assumption believed to hold in 'reasonable' spacetimes.

2.2 The Formulations

As mentioned, there are many ways to cast Einstein equations in convenient form for numerical purposes. These can be roughly distinguished by the level set of hypersurfaces adopted to foliate the spacetime and whether this is the physical one or a 'larger' one where the physical description is obtained a *posteriori* by a rather straightforward restriction. A notion of "time" is defined by a chosen foliation of the spacetime (physical or not) which is parametrized by a parameter λ. The character of the normal to the $\lambda = const$ surfaces ($n_a = \nabla_a \lambda$) characterizes the approach. A Cauchy/characteristic approach corresponds to n_a being timelike/null[a].

In all approaches, the evolution equations are obtained by projecting Einstein equations onto the hypersurfaces. The projections with respect to the normal (either one or both components of E_{ab}) correspond to constraint equations which are preserved by the evolution equations if they are satisfied at the initial hypersurface by virtue of the Bianchi identities. Note however, that this allows one to arbitrarily modify the evolution equations by adding the constraints to them. Although this defines a totally equivalent system, the behavior of the numerical implementation can be drastically different as we will discuss later.

Just as is customary in analytical investigations of the theory, there is far from a preferred approach; we have long learned that in General Relativity *"one size does not fit all"*. Indeed, different approaches are naturally suited for different problems and I will describe the main focus of these in current efforts.

Cauchy approach: As mentioned, in this case the spacetime is foliated with spacelike hypersurfaces, as a result, all $\lambda = const$ surfaces 'end' at i_o. This brings in a problem as only in a limited number of cases, the numerical implementation can cover the whole hypersurface due to limited

[a] Of course one can deal with more generic conditions but this has not been pursued agressibly yet.

computational power[b]. Hence the hypersurfaces are terminated at an outer boundary defining a finite domain. The most widely known formulation of Einstein equations within this approach is the so-called ADM system[8]; however, difficulties observed in the implementation of this system have spurred substantial efforts to consider different reformulations that might aid in the implementation (and in several cases there is evidence for this being the case[9,10]). Aside from having spacelike leaves, the choice of foliation is largely arbitrary; as a result this approach is quite flexible and is certainly the most popular one in numerical relativity.

Characteristic approach: In this formulation, the hypersurfaces are characteristics of the spacetime. In the particular case of an isolated system, these hypersurfaces can be chosen to reach future null infinity. Since gravitational waves propagate along these hypersurfaces the phase of the waves approaches a constant in the far regions. The spacetime can be compactified (as suggested by Penrose[11] to study asymptotic properties) without numerical problems and enabling the rigorous evaluation of fields at future null infinity. The applicability of this approach is basically restricted to cases where null hypersurfaces do not cross[c] (i.e. free of caustics or crossovers), otherwise the coordinate system becomes singular. In caustic-free regions, this approach is quite useful and simplifies the description when compared to the Cauchy approach as much of the gauge freedom is fixed and the variables are intimately related to physical quantities[14,15,16].

Conformal approach: This formulation is in a way the most general one as it does not assume any particular condition on the hypersurfaces, it is thus amenable for a Cauchy, characteristic or more generic implementations. The peculiarity of this approach is that it first obtains the description of an unphysical spacetime and the physical one is obtained a posteriori[17]. The computational boundaries are therefore outside the physical spacetime, and hence causally disconnected from it when all leaves of the foliation cross future null infinity.

---

[b]One can compactify the spacetime and do include infinity in the numerical grid, however, special care must be taken as the effective discretization length grows to infinity which can easily render the numerical implementation useless.

[c]There exist proposals to go around this problem in a variety of cases[12,13] though they might be difficult to implement numerically.

3 Status

3.1 Cauchy approach

Recent Past: A few years ago numerical simulations within this approach presented a series of remarkable results. Among these, the discovery of critical phenomena in GR in spherical symmetry (1D)[1,18], axisymmetric (2D) head-on collisions of black holes[19] and collapse of Vlasov systems[20]. 1D simulations were implemented with impressive accuracy thanks to the use of adaptive mesh refinement (AMR), [see for instance[21]]. 2D simulations on the other hand, barely had enough resolution to resolve the essential features and since singularity excision was used they could not proceed too far into the future. The need to produce generic simulations capable of producing waveforms templates for gravitational wave data analysis shifted the attention onto three-dimensional (3D) implementations. In a way, this was a bit regrettable as many interesting physical systems which are truly (or basically) two dimensional in nature were left aside. Currently, activities in both 2D and 3D are aggressively carried out, a few examples are:

2D: Recently, a few groups have undertaken the effort to produce 2D simulations with the necessary numerical techniques to obtain the desired accuracy to address problems like critical collapse in axisymmetry; head-on collisions of black holes and accretion onto a black hole. Among these AMR and singularity excision are the most notable ingredients. The implementation in [22] is concentrating on studying critical phenomena in Einstein-Klein-Gordon systems to investigate critical behavior in more generic settings. The ADM formulation is being used with a partially constrained implementation (i.e. some of the constraints are used to update the field variables) which apparently helps to remove exponential growing modes typical of free evolution schemes (where the constraints are just used to monitor the quality of the solution) and encouraging results are being obtained. For instance, simulations describing two separated 'pulses' of scalar field in a slightly super-critical configuration are followed. As time progresses, each pulse collapses forming two separated black holes which later merge into one which then settles into an almost stationary black hole. At late times the accuracy deteriorates but this will likely be solved once AMR gets fully incorporated.

3D: Three dimensional simulations still have a number of unresolved issues. Poor resolution; lack of understanding of 'ideal formulations'; boundary treatment and initial data definitions are among the outstanding problems being studied at present. Despite these difficulties, significant advances have been presented recently. Next I comment on some of these,

Initial Data: On the initial data problem, work is advancing towards obtaining realistic initial data. Within this approach, the problem of posing consistent initial data reduces to providing the intrinsic metric of the initial hypersurface and its extrinsic curvature satisfying the constraints (Hamiltonian and momentum constraints). A procedure to handle this problem was developed by Lichnerowicz[23] and further refined by York[24] which involves specifying a conformally related 'seed' metric (see [25] for a detailed description).

For decades the prevalent approach was to solve this Lichnerowicz-York[24] problem assuming a conformally flat seed metric (supplemented by an appropriate seed for a quantity related to the extrinsic curvature). However, since the Kerr-Newman spacetime can not be obtained this way[26], spinning black holes which can be obtained under this assumption must contain spurious gravitational radiation. A series of proposals have been presented to give away with conformal flatness[27,28,29,30] and implementations of these are already in place or being developed[31,32,33]. Reasonable arguments indicate that these initial should data contain less spurious radiation, but it is yet unclear how much there is. This issue will be fully resolved by numerical evolutions. A key ingredient that needs to be worked out for treating binary systems is to choose the initial data achieving a smooth transition from the Post-Newtonian phase description of the inspiraling phase. This will guide the choice of 'seed variables' in a way consistent with the description of the problem and aid in minimizing spurious radiation. Work in this area is just beginning.

Single Back Hole evolutions: Singularity excision has also allowed for extending simulations of single black hole spacetimes by several of orders of magnitude when compared to similar spacetimes modeled with singularity avoiding techniques (see figure 1). At least for a number of cases, there are now simulations that can evolve single perturbed black holes for considerable lengths of time, certainly longer than needed for studying a variety of scenarios; probing regimes beyond what we know from perturbation theory and study the produced waveforms. The work presented in[34], illustrates how singularity excision, coupled with a carefully designed coordinate condition that minimizes changes in particularly sensitive variables in the

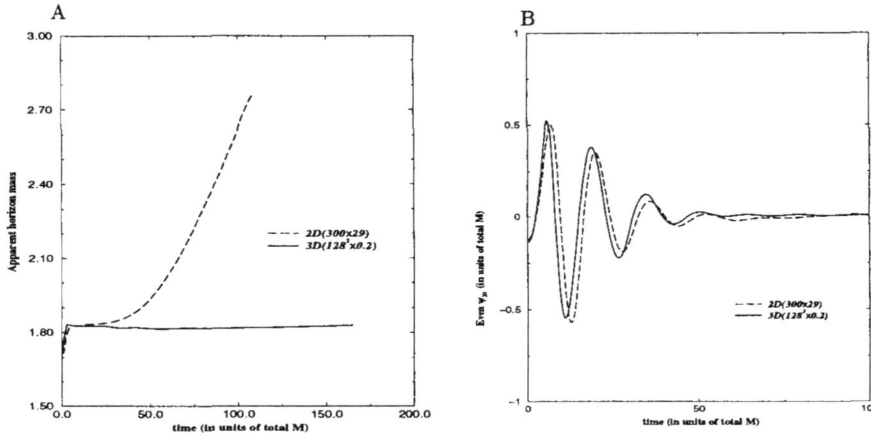

Figure 1. Plots of: (A) the apparent horizon area obtained with the 2D (dashed line) and 3D codes (solid line) vs. time, the problems induced by the use of singularity avoidance techniques are apparent in the growth of the apparent horizon which is not present in the 3D code's result. (B) Gravitational wave extraction with both codes, for the length of simulations considered with the 2D codes, the results are in good agreement.

BSSN[d] formulation[35,9], enables practically unlimited evolutions of spacetimes describing black holes perturbed by 'Brill waves'. Not only do these results outperform the very best ones obtained with 'old' 2D codes (with singularity avoidance), but it is used to accurately extract the expected waveforms.

Binary Back Hole evolutions: Binary black hole systems are among the 'hottest' problems for numerical relativity. This system is a prime candidate for producing gravitational radiation to be measured by the new generation of gravitational wave detectors. An accurate description of the system requires numerical simulations of the full Einstein equations especially for the merger phase. In dealing with such a problem numerically, one must handle two initially separated black holes which will merge and settle into a single perturbed black hole (which can then be treated by black hole perturbations[36,37]). The total length of simulations required is at least in the

[d]BSSN is a variant of the ADM formulation which has become quite popular due to its better performance in several situations when compared to the ADM one.

order of hundreds of M's (from about the inner most stable circular orbit to the point where perturbations around a single black hole can be safely used). There is still no implementation capable of addressing these requirements but some hopeful preliminary investigations have been presented. In particular, a proof-of-principle implementation was presented demonstrating the ability to simulate the system containing two initially separated singularities to past the merger phase with the use of singularity excision. The results in[38] illustrated an evolution of a system whose initial data described two-separated (by about $10M$) equal mass (M) black holes (judging from the locations of the apparent horizons) and followed the dynamics of the system to well past the merger. Figure 2 illustrates the location of the apparent horizons at different times. Although the simulation crashes after a few tens of Ms, this work showed that singularity excision techniques can handle the binary black hole problem.

Towards the attainment of the 'ultimate' reformulation: Considerable efforts are being spent to understand why different formulations of Einstein equations yield simulations behaving quite different. The difficulties observed in numerical implementations spurred a large number of works that presented re-formulations of Einstein equations in explicit symmetric hyperbolic form (see [39] for a detailed list). These formulations write the system in first order form (and hence require introducing additional variables to express spatial derivatives) and manipulate the equations through the addition of constraints to guarantee the existence of a complete set of eigenvectors (of the matrices describing the the principal part of the system). Several benefits are gained after all this work[e]: the standard mathematical machinery for PDE's can be used to determine the well posedness of the problem under study; whether the characteristic speeds of the system are physical (lie inside the null cones) and furthermore, determine which combination of variables are ingoing and outgoing with respect to a given boundary (which plays an important role when imposing boundary conditions). In a well posed problem, the solution at a later time, $S(t)$, satisfies $|S(t)| \leq \alpha e^{\kappa t}|S(t=0)|$ (with $\alpha, \kappa \in \mathbf{R}$). This property guarantees continuous dependence on the initial data which is important as the initial data provided in a numerical implementation is, at best, the desired one only up to round-off errors. The fact that the solution is bounded by an exponential function though, is not sufficient information for the numerical implementation, as one would like to rule out exponential modes in some given problems. This is the case, for instance, when modeling a black hole spacetime in an appropriately chosen gauge which, even when

[e] See Rendall's contribution to this volume.

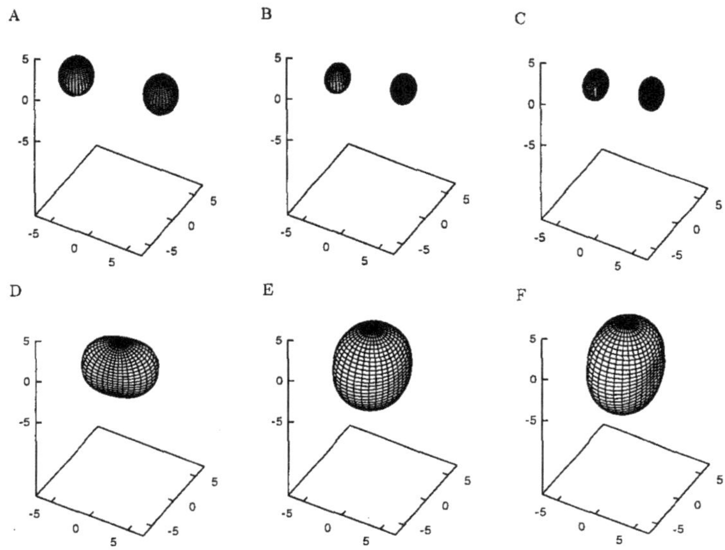

Figure 2. Snaphots of apparent horizon locations during the evolution (A-F corresponding to $t = 0, 2M, 4M, 6M, 10M, 12M$). At $t = 0$ (A) two distinct horizons, separated by $10M$, at t $5M$ a single one is already formed which acquires a more spherical shape as time proceeds (and the evolution is still well behaved).

perturbed, should relax to a stationary black hole.

Clearly, given any formulation one can write infinitely many others by simply adding constraints to the right hand sides with free multiplicative parameters. Although at the analytical level all these formulations are exactly equivalent; at the numerical one they might display considerable differences. Examples of such behavior have been illustrated in 3D [40] and in 1D [41]. For instance, in [40] a two-parameter hyperbolic family of reformulations of Einstein equations has been presented which have *exactly* the same principal part. Therefore, well posedness of systems treated by all members of the family follows in a totally equivalent manner. This work presented results obtained by simulating Schwarzschild spacetimes with different members of this family under the *same* numerical implementation. Total evolution times ranged from as short as $6M$ to $1500M$. Figure 3 illustrates the norm of a

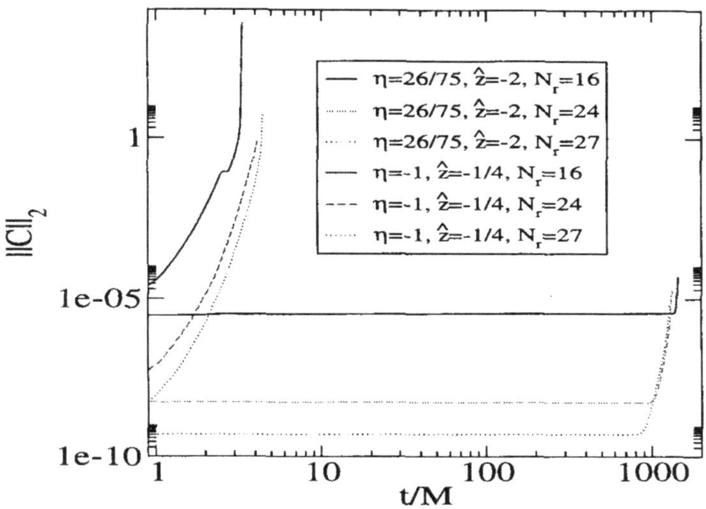

Figure 3. Solution behavior for two members of the family of formulations (parametrized by η, $\hat{z}$). For the same resolutions (given by N_r) the results of both differ considerably.

representative constraint vs. time for two different members of the family of formulations (under the same discretization and resolution). While for one, the violations become very severe at earlier times, for the other only at quite late times do these become apparent. The message from these and other works is that for numerical simulations, *well posedness is not enough!* There is an urgent need for a method to discern from the "zoo" of formulations those that might simplify the attainment of a stable implementation. Formulating such a method is certainly non-trivial, as one would need to be able to separate "gauge modes" from physical ones; understand the behavior of constraints off the constraint surface and the influence of non-principal part terms in the evolution as well as that of boundaries. Nevertheless, its impact on the field would likely be profound.

Boundary conditions: Outer boundary conditions are difficult to define in any theory containing propagating waves. It is particularly difficult in relativity due to backscattering, gauge and constraint issues. Approximate boundary conditions have been used in the past with different degrees of effectiveness (see [42] for a list). However, not only is it not clear how much spurious reflections or incoming waves these introduced but most importantly whether are these even consistent. At the core of this question lies the fact that 'standard' boundary conditions provide values to all variables and whether these conditions satisfy the constraints. Recently, insight into this problem and its physical interpretation has been presented in[43] where a careful analysis was presented that studied the question of well posedness of the problem considering the presence of boundaries. The conclusion is that only *two* pieces of information can be given (related to the polarizations of incoming radiation) while the rest are determined by the Einstein equations themselves.

The need for having correct boundary conditions respond to several reasons: (i) Placing boundaries as 'close as possible' to minimize computational costs; (ii) Eliminate constraint violating mode sources, (iii) Eliminate spurious reflections . Recently a number of efforts (in linearized gravity in 3D[44] or full gravity in spherical symmetry[45]) have shown that carefully treated boundaries can satisfy these requirements benefiting the numerical evolution.

3.2 Characteristic approach

Recent Past: Numerical implementations within this formulation have come a long way despite it being 'newer' than the Cauchy ones. In the recent past the characteristic formulation has been demonstrated to be a useful tool by producing the first unlimited evolution of single black hole spacetimes[46]. Work within this formulation is concentrating on scenarios or spacetime regions which are expected to be free of caustics. I will here mention a few examples both in the vacuum and non-vacuum cases (I include the latter since it is not being covered by Shibata's contribution to this volume).

Unfortunately, because of caustic formation, this formulation can not address all problems of interest, in particular binary black holes or binary neutron stars. However for certain systems 'back of the envelope' calculations indicate they should be free of this problem .

Motivated by these problems a series of works were devoted to understand the incorporation of matter within the formulation. Most notably the implementation of high resolution schemes[47] and the simulation of weak pressure fluids in full 3D black hole spacetimes[48].

Present focus

2D: Much activity is being carried out with *2D* characteristic codes[49,47]. In particular, studies of neutron star perturbations in full general relativity are being carried out where the dynamics of the fluid is described by high resolution shock capturing schemes[47]. In these works, the star is perturbed

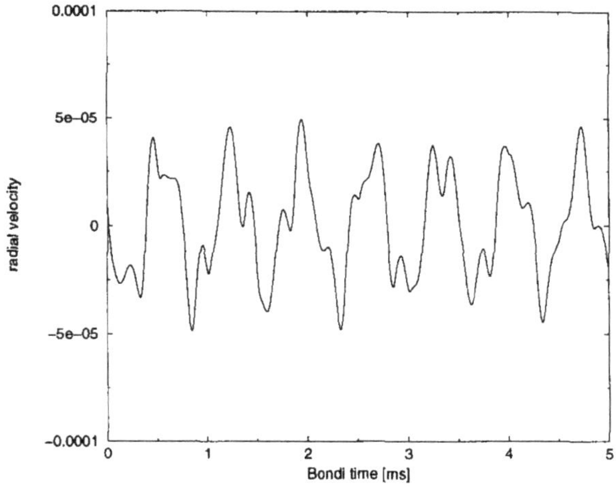

Figure 4. Central density oscillations of a polytrope perturbed by an incoming gravitational wave.

by incoming gravitational waves. The oscillation of the star are accurately modeled and compared with perturbative results and the gravitational radiation is read-off at future null infinity. Figure 4 illustrates the oscillations in the central density of a polytrope star which has been perturbed by an incoming wave. Mass conservation is verified to an impressive degree of accuracy which coupled to a series of other successful tests indicate the code is ready to tackle interesting scenarios.

3D: Non-vacuum 3D simulations within this approach are targeting spacetimes containing a black hole and a compact star. Proof of principle tests to examine the feasibility of extending the vacuum code described in[50,51] to the non-vacuum case proved successful in very weak pressure cases[48]. Efforts are now concentrated in providing correct initial data describing a star "close" to the black hole. The main reason for this is to avoid caustics. A 'back of the envelope' estimate using the lens equation implies that the spacetime will be free of caustics provided $D < R_s^2/(4m_s)$ (where R_s, m_s are the are radius and mass of the star which is at a distance D of a unit mass black hole). This defines a window of applicability which does contain interesting sources for gravitational wave detectors. Since neutron star-black hole systems are also likely sources for gravitational wave detection, simulations of these systems

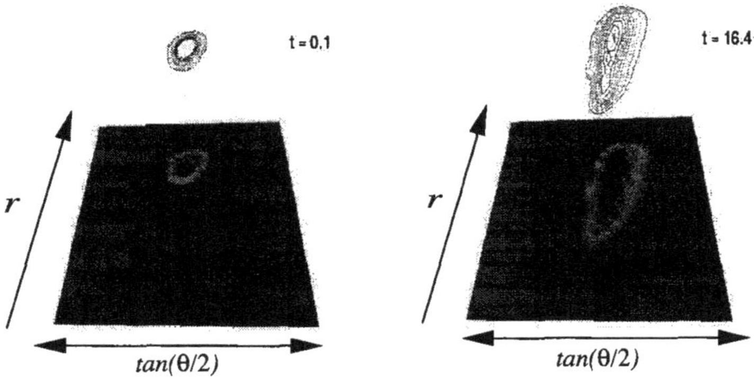

Figure 5. Two density snapshots in the at $\phi = const$, as time progresses the tidal distortion of the polytrope becomes evident.

will have immediate practical uses.

The implementation to treat this problem follows closely the $2D$ one as far as treatment of the fluid equations is concerned but, being still a serial code, it can not achieve as high resolutions in full 3D and some numerical problems are still being addressed. Once the level of resolution needed is achieved, the simulation should perform as well as the one in the 2D case. Figure 5 shows a $\Gamma = 2$ polytrope with radius $5M$ which is placed at a distance of $9M$ from a mass M black hole and moves in the θ direction. Since a portion of the star lies inside ISCO, as time progresses it becomes quite distorted. Results obtained with this simulation are qualitatively correct but they still have inadequate accuracy for producing useful gravitational wave templates.

On the vacuum front, a project aimed towards obtaining gravitational radiation of a binary black hole spacetime is also under development[52]. Here, the spacetime is envisioned in a time-reversed point of view. This is motivated by the possibility of posing a double null problem whose inner boundary corresponds to a fissioning white hole[53,54] (which in a time reversed point of view corresponds to merging black holes) and the other corresponds to $\mathcal{I}^-$. An inverse scattering process can be formulated to obtain the radiation produced by a binary black hole collision[55]. Preliminary investigations of this approach have targeted a "close limit approximation" yielding excellent results[56]. Of course, since the theory is non-linear this procedure is far from direct. However, note the data needed for this problem is intimately related to the radiation fields at both scri's which are expected to be quite small (in fact should be 'zero' at $\mathcal{I}^-$ and about 5% of the total mass at $\mathcal{I}^+$); thus, one can formulate (partially hand-waving) arguments to give support to this approach in terms of an iterative procedure.

3.3 Conformal approach

As mentioned, within this approach the description of an unphysical spacetime is obtained and a restriction of the solution yields the physical spacetime (and the metric of the latter is conformally related to that of the former). All implementations within this approach have used a Cauchy initial value problem formulation thus have many similarities to what was included section 3.1. There is an extra gauge issue here since the relation between the conformal metrics is essentially arbitrary.

Recent Past: The conformal implementation of Einstein equations has proceeded at a steady pace since it was first used to investigate critical phenomena in 1D[57] and subsequently extended to higher dimensional spacetimes[58,59].

Present: Preliminary proof-of-principle simulations of Schwarzschild black holes in this formulation have been presented in [59]. These targeted the Kruskal extension and simulations proceeded well for moderate amounts of times.

More recently, linear waves on Minkowski spacetimes have been successfully evolved reconstructing the spacetime all the way to i^+ [60]. Since i^+ is safely inside the computational domain, this formulation allows for studying the regularity and structure of the whole spacetime. Figure 6 illustrates three different timelike geodesics in flat spacetime perturbed linearly. As expected, all end at the same point in the future which meets $\mathcal{I}^+$, it is confirmed that i^+ is regular by studying the spacetime structure in its neighborhood. Although more experience still needs to be gained to exploit this formulation (particularly on gauge related issues), it is becoming clear that this formulation is indeed a useful tool in numerical relativity and the ideal one to study global properties of spacetimes.

4 Future Problems

It is certainly true that most current efforts in the field are directed towards systems which will likely produce gravitational waves detectable by the new generation of detectors. However, this is just the tip of the iceberg. The

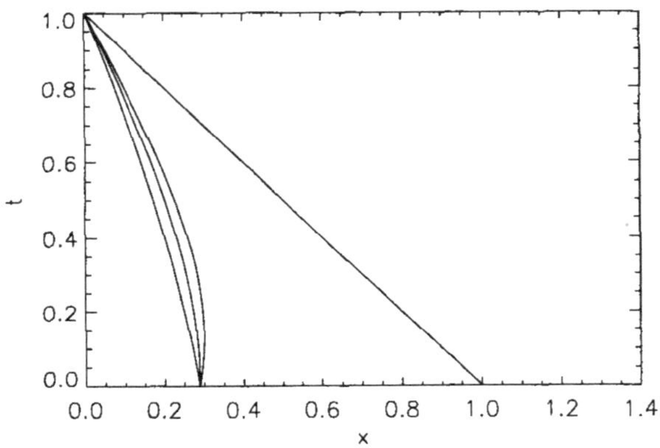

Figure 6. Timelike geodesics in a linearly perturbed flat spacetime and $\mathcal{I}^+$ meet at i^+.

ability to solve Einstein equations in generic scenarios will open the door for studying a variety of problems, a few examples being

- *Astrophysically relevant systems:* A series of spectacular and yet poorly understood phenomena "illuminate" the sky with powerful energetic emissions. Active galactic nuclei, gamma ray bursts, quasars, micro-quasars, give away huge amounts of energy in relatively short periods of time (but not so short that "explosive" mechanisms can be called into place; see Piran's contribution to this volume) that it is believed that a compact central object, a fairly massive surrounding plasma and magnetic fields are key ingredients of the central engine powering these emissions. Modeling such systems accurately require solving Einstein equations in the highly dynamical/strong field scenarios. Future research in this field will certainly have strong emphasis towards astrophysically motivated problems.

- *Spacetimes on the large:* Numerical simulations are also opening the door to understand the structure of spacetimes in the large. Under certain assumptions the asymptotic behavior of spacetimes is well understood but only in a limited number of cases does one have an explicit relation between sources and fields at infinity. Numerical implementations can be used to search for this relation and to examine asymptotic fields behavior without past adopted assumptions to render the problem analytically tractable.

- *Quantum Gravity:* Another area where simulations might provide valuable insights on prospective quantum gravity theories. First attempts in this direction are studying consequences of the AdS/CFT duality[61], and the final fate of black strings[62] which will provide insights in the classical or low energy limit of possible theories. By analyzing this limit, one can foresee that unacceptable results will help in ruling out some candidates while others might help us understand acceptable ones.

5 Final Comments

As has been evident from the contributions to the numerical relativity sessions in this meeting and recently published results notable advances are taking place in this field. It is certainly impossible to do justice to all efforts in numerical relativity in this contribution[f]. I hope this article gives a current glimpse of the activity in the field and, especially, highlights that considerable progress has been achieved in the past few years. Undoubtedly, we still face many problems to realize the full potential of this activity and researchers are working hard towards their resolutions. Numerical techniques, complementing the analytical ones are paving the way for understanding the full implications of the theory; certainly, the second century of General Relativity promises to be an exciting one.

6 Acknowledgements

I would like to thank N. Bishop and F. Pretorius for helpful comments on early versions of the manuscript. I thank M. Alcubierre, S. Husa, M. Scheel and F. Siebel for providing some of the figures included in this text. Computations of some of the results presented here were performed on the vn.physics.ubc.ca Beowulf cluster which was funded by the Canadian Foundation for Innovation (CFI). I want to express my gratitude to the California Institute of Technology for its hospitality where the last parts of this work were completed. Finally it is a great pleasure to thank the organizers of GR16 for such an exciting conference.

References

1. M.W. Choptuik. *Phys. Rev. Lett.*, 70:9–12, 1993.

[f] For more details please refer to [42,18,63,65,66].

2. E. Seidel. In N. Dahdich and J. Narlikar, editors, *Proceedings of the 15th International Conference on General Relativity and Gravitation*, Pune, India, December 1998. IUCAA Press.
3. M.W. Choptuik. In N. Dahdich and J. Narlikar, editors, *Proceedings of the 15th International Conference on General Relativity and Gravitation*, Pune, India, December 1998. IUCAA Press.
4. Y. Choquet-Bruhat. *Acta Math*, 88:141, 1952.
5. T. Piran. In N. Deruelle and T. Piran, editors, *Gravitational Radiation*, Amsterdam, 1982. North Holland.
6. J. Thornburg. *Class. Quantum Grav.*, 4:1119, 1987.
7. R.M. Wald and V. Iyer. *Phys. Rev. D.*, 44:3719, 1991.
8. C.W. Misner, K. S. Thorne, and J.A. Wheeler. *Gravitation.* W. H. Freeman and Co., San Francisco, 1973.
9. T. Baumgarte and S. Shapiro. *Phys. Rev.*, D59:024007, 1999.
10. M. Alcubierre et al. *Phys. Rev.*, D62:044034, 2000.
11. R. Penrose. *Phys. Rev. Letters*, 10:66, 1963.
12. H. Friedrich and J. M. Stewart. *Proc. R. Soc. A*, 385:345, 1983.
13. L. Lehner. *Int. J. Mod. Phys.*, D9:459, 2000.
14. H. Bondi, M.G.J. van der Burg, and A.W.K. Metzner. *Proc. R. Soc. London Ser. A*, (270):103, 1962.
15. R.K. Sachs. *Proc. R. Soc. A*, 270:103, 1962.
16. L. Tamburino and J. Winicour. *Phys. Rev.*, 150:1039, 1966.
17. H. Friedrich. *Proc. R. Soc. London A*, 375:169, 1981.
18. C. Gundlach. *Living Rev. Rel.*, 2:4, 1999.
19. P. Anninos, et. al. *Phys. Rev.*, D52:2044–2058, 1995.
20. S. L. Shapiro and S. A. Teukolsky. *Phys. Rev.*, D45:2739–2750, 1992.
21. M.J. Berger and J. Oliger. *J. Comp. Phys.*, 53:484, 1984.
22. M.W. Choptuik, et. al. *Private communication*, 2002.
23. A. Lichnerowicz. *J. Math. Pures et Appl.*, 23:37–63S, 1944.
24. J. York. *J. Math. Phys.*, 14:456–4641, 1973.
25. G. Cook, et. al. *Phys. Rev. D.*, 50:5025, 1994.
26. A. Garat and R. Price. *Phys. Rev.*, D61:124011, 2000.
27. N. Bishop, et. al. *Phys. Rev.*, D57:6113–6118, 1998.
28. R. Matzner, M. Huq, and D. Shoemaker. *Phys. Rev.*, D59:024015, 1999.
29. P. Marronetti et al. *Phys. Rev.*, D62:024017, 2000.
30. S. Dain. *Preprint gr-qc/0012023*, 2000.
31. P. Marronetti and R. Matzner. *Phys. Rev. Lett.*, 85:5500–5503, 2000.
32. P. Diener, et. al. *Class. Quant. Grav.*, 17:435–451, 2000.
33. H. Pfeiffer, S. Teukolsky, and G. Cook. *Phys. Rev.*, D62:104018, 2000.
34. M. Alcubierre, et. a.. *Phys. Rev.*, D64:061501, 2001.

35. M. Shibata and T. Nakamura. *Phys. Rev.*, D52:5428–5444, 1995.
36. R. Price and J. Pullin. *Phys. Rev. Lett.*, 72:3297–3300, 1994.
37. J. Baker, et. al. *Class. Quant. Grav.*, 17:L149–L156, 2000.
38. S. Brandt, et al. *Phys. Rev. Lett.*, 85:5496–5499, 2000.
39. O. Reula. *Living Rev. Rel.*, 1:3, 1998.
40. L. Kidder, M. Scheel, and S. Teukolsky. *Phys. Rev.*, D64:064017, 2001.
41. B. Kelly et al. *Phys. Rev.*, D64:084013, 2001.
42. L. Lehner. *Class. Quant. Grav.*, 18:R25–R86, 2001.
43. H. Friedrich and G. Nagy. *Comm. Math. Phys.*, 204:691–707, 1999.
44. B. Szilagyi, et. al. *Phys. Rev.*, D62:104006, 2000.
45. G. Calabrese, L. Lehner and M. Tiglio. *Phys. Rev. D., to appear*, 2002.
46. R. Gomez, L. Lehner, R. L. Marsa, and J. Winicour. *Phys. Rev.*, D57:4778–4788, 1998.
47. P. Papadopoulos and J. Font. *Phys. Rev.*, D61:024015, 2000.
48. N. Bishop, et. al. *Phys. Rev.*, D60:024005, 1999.
49. R. A. d'Inverno, M. R. Dubal, and E. A. Sarkies. *Class. Quant. Grav.*, 17:3157, 2000.
50. N. Bishop, et. al. *Phys. Rev.*, D56:6298–6309, 1997.
51. N.T. Bishop, R. Gómez, L. Lehner, and J. Winicour. *Phys. Rev. D.*, 54:6153, 1996.
52. J. Winicour. *Prog. Theor. Phys. Suppl.*, 136:57, 1999.
53. L. Lehner,et. al. *Phys. Rev.*, D60:044005, 1999.
54. S. Husa and J. Winicour. *Phys. Rev.*, D60:084019, 1999.
55. R. Gomez, S. Husa, and J. Winicour. Complete null data for a black hole collision. *Prerint gr-qc/0009092*, 2000.
56. M. Campanelli, et. al. *Preprint gr-qc/0012107*, 2000.
57. P. Hubner. *Phys. Rev.*, D53:701–721, 1996.
58. J. Frauendiener. *Phys. Rev.*, D58:064003, 1998.
59. P. Huebner. Cauchy formulation boundary conditions. In D. Eardley, ed., *Colliding Black Holes: Mathematical Issues in Numerical Relativity*, Santa Barbara, January 2000. http://online.itp.ucsb.edu/online/numrel00/.
60. S. Husa. Semiglobal numerical calculations of asymptotically minkowski spacetimes. gr-qc/0112005 (2001).
61. V. Husain and M. Olivier. *Class. Quant. Grav.*, 12:L1 2001.
62. M Choptuik, et.al. The final fate of the black string. *In preparation.*
63. Berger B, 1998, *Living Rev. Rel.* **7**, 1.
64. Font J A, 2000, *Living Rev. Rel.* **3**, 2.
65. Cook G, 2000, *Living Rev. Rel.* **5**, 1.
66. Anninos P, 1998, *Living Rev. Rel.* **9**, 4.

STRINGS, GRAVITY AND PARTICLE PHYSICS

JUAN MALDACENA
Institute for Advanced Study, Einstein Dr. Princeton, NJ 08540, USA
E-mail: malda@ias.edu

We describe briefly some central ideas about string theory. We sketch how string theory gives rise to the black hole entropy and we describe the relation between string theory in anti-de-Sitter spaces and conformal field theories on the boundary.

1 Introduction

Our present world picture is based on two theories: the standard model of particle physics and general relativity, the theory of gravity. These two theories have scored astonishing successes. It is therefore quite striking when one learns that this picture of the laws of physics is inconsistent. The inconsistency comes from taking a part of theory, the standard model, as a quantum theory while the other, gravity, as a classical theory. At first sight it seems that all we need to do is to quantize general relativity. If one performs an expansion in terms of Feynman diagrams, the technique used in the standard model, then one finds infinities that cannot be absorbed in a renormalization of the Newton constant (and the cosmological constant). In fact as one goes to higher and higher orders in perturbation theory one would have to include more and more counterterms. In other words, the theory is not renormalizable. So the principle that is so crucial for constructing the Standard model fails. It is now more surprising that an inconsistent theory could agree so well with experiment!. What happens is that quantum gravity effects are usually very small due to the weakness of gravity relative to other forces. Since the effects of gravity are proportional to the mass, or the energy of the particle, they grow at high energies. At energies of the order of $E \sim 10^{19} Gev$ gravity would have a strength comparable with that of the other Standard Model interactions[a]. We should also remember that physics as we understand it now cannot explain the most important "experiment" that ever happened: the Big Bang. It is quite interesting that the Big-Bang theory links high energy physics and cosmology and in order to understand what happened in the beginning it seems that we need to understand quantum gravity. There are also esthetic reasons for wanting a theory beyond the Standard Model. We

[a] This is an energy scale where we should definitely see new physics, it is nevertheless possible that quantum gravity becomes relevant at much lower energies, $E \sim$ 1-10 Tev [9].

would like to explain the origin of the gauge group, the relations between the three couplings, the rest of the parameters of the standard model, why we have three generations, etc. It is very suggestive that if one extrapolates the running of the couplings they seem to meet at energies that are close to the energy where quantum gravity becomes important, suggesting that a grand unified field theory, based on a bigger gauge group would in any case lie close to the Planck scale. The present situation is analogous to the situation particle physics was in when we only had Fermi's theory of weak interactions. It was a theory that agreed well with experiments performed at low energies but it was not a consistent theory. Renomalizability, or mathematical consistency, was a crucial clue for the discovery of the Standard Model.

The challenges we face can be separated, in degree of difficulty, in the following three:

I) Formulate an internally consistent theory of quantum gravity. By this we mean a theory which reduces at low energies, $E \ll 10^{19} Gev$, to general relativity but in which we can perform quantum calculations to any order we wish. This theory should solve some fundamental problems with quantum gravity such as explaining the origin of black hole entropy, etc. These are questions about gravity which do not involve directly the fact that we also have the particle physics that we see in nature.

II) Be capable of incorporating the Standard Model. So the theory should be such that at low energies it can contain chiral gauge fields, fermions, etc.

III) Explain the Big-Bang and the parameters of the Standard Model. We should understand the resolution of the initial singularity in cosmology and we should understand why we have the Standard Model. We should understand how the standard model parameters arise, which parameters are related, and which parameters (if any) arise as a "historical" accident.

We know have a theory, called string theory (or M-theory), which has been able already to provide a solution to the first two challenges. Unfortunately we do not know yet how to solve the third challenge. Maybe string theory is the solution and we just have to understand it better or maybe we have to modify it in some way. String theory is a theory under construction. We know several limits and aspects of the theory but we still do not know the fundamental axioms of the theory that would enable us to approach the third challenge.

String theory is based on the idea that fundamental objects are not point particles, as in particle theories, but one dimensional objects called strings. Let us first review how we construct a theory of interacting particles. We start with a set of free particles, for example electrons, photons, quarks, gluons. These particles can have different states; e.g. they could have the spin pointing up or down. Then we consider interactions. These are introduced by allowing particles to split into two other particles with some probability amplitude g. g will be the interaction strength or coupling constant. For example and electron could emit a photon, etc. So in order to compute a scattering amplitude we have to sum over all paths of the particles and all ways in which they could emit other particles, etc. These sums are performed via Feynman diagrams. In figure 1(a,b) we see some examples of Feynman diagrams. String theory is constructed in a completely analogous way. We first start with free strings. Strings can be open or closed. Let us just consider a theory with only closed strings. The strings are "relativistic", meaning that their tension is equal to their mass per unit length. So that if we had a stretched string an oscillation would propagate at the speed of light along it. The tension is a dimensionful quantity which we can parameterize in terms of a distance scale $T = 1/l_s^2$. Strings can oscillate. These oscillations can be decomposed in normal modes. Since strings are quantum mechanical objects each normal mode will have a certain occupation number. The total energy of the oscillating string will be quantized. And the total mass of an oscillating string will be equal to the total energy contained in the oscillations. When we view this oscillating string from far away it looks like a pointlike object. These different oscillatory states of the string are analogous to the different polarization states of the particles, now that the mass of the string state itself depends on the "polarization" state. Some of these oscillatory states of the string will have zero energy and will thus be massless particles. There is one state with spin two which can be viewed as the graviton. The masses of the massive string states are of the order $m \geq 1/l_s$. String interactions are introduced by allowing strings that touch to recombine into one string. These are splitting and joining interactions as shown in figure 1 (c,d). The amplitude for this process defines the string coupling g. In order to compute any process in string theory we have to sum over all possible splitting and joining interactions. The simplest string theories are those that live in ten dimensions and are supersymmetric. The sum over string theory Feynman diagrams can be performed and yields finite results. At low energies, energies lower than the mass of the massive string states $E \ll 1/l_s$ the only excitation we will have are gravitons and other massless particles. The interactions of these particles are those of Einstein gravity plus some other massless fields. In this way string theory manages to quantize gravity. What we have described here amounts to a perturbative quantization of the theory, in the same way that the Feynman diagram expansion in particle physics is a perturbative quantization of the field theory. But there are non-perturbative aspects of the theory that are not captured by the perturbative theory. One example is soliton solutions, like magnetic monopoles of grand unified theories.

These are collective excitations that are stable, typically for some topological reason. Their masses go as $m \sim 1/g^2$ and in the weak coupling approximation we can study them as solutions of the classical field theory action. In field theories we can also have extended solitons, like cosmic strings or domain walls. In string theory we also have solitons. These solitons are called D-branes. D-branes are solitons with different dimensionalities. They can be pointlike (D-0-brane), one dimensional (D-1-brane), two-dimensional (D-2-brane), etc. These solitons have a very precise description in string theory [10]. Their excitations are described by open strings that end on them. At low energies some of the open string modes are massless, have spin one and give rise to gauge fields. When we put many branes together the open strings have two indices i, j labeling the brane where they start and the brane where they end see figure 2. These two indices become the indices of non-abelian $U(N)$ gauge fields.

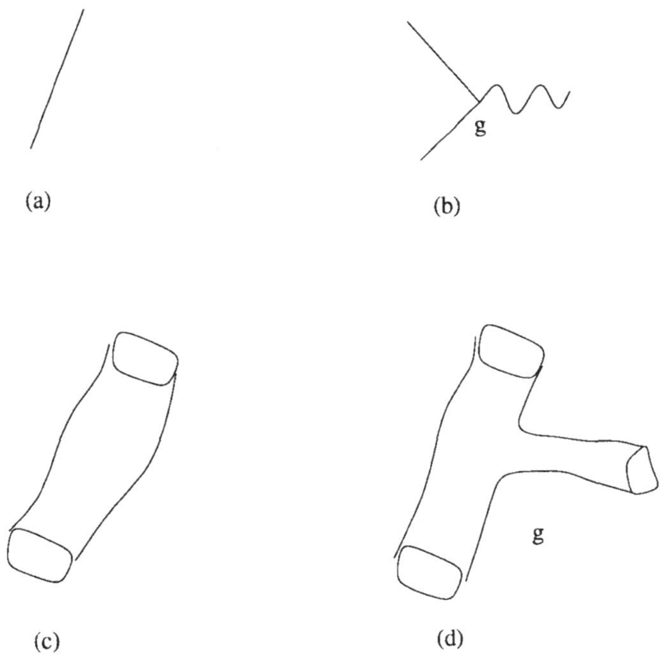

Figure 1. Feynman diagrams for strings and particles. (a) Free particle propagation. (b) Interaction vertex between particles. (c) Free string propagating. (d) String interacting with other strings.

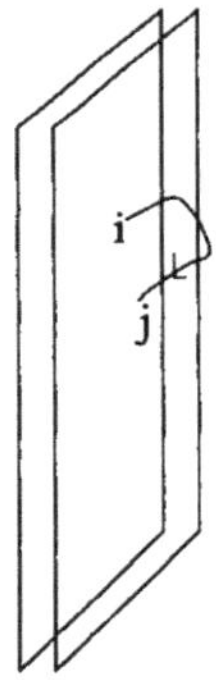

Figure 2. D-branes are extended solitons whose excitations are described by open strings which end on the brane. A string going from brane i to brane j is labeled by two indices. These give rise to non abelian gauge fields living on the D-brane.

It might be surprising that we were discussing a ten dimensional theory while our world is "obviously" four dimensional. What we really see is that the world is four dimensional at long distances, we do not really know the dimensionality of the world at short enough distances. In string theory we assume that we are living in a world that has four large dimensions (the ones we see) and six very small dimensions, see figure 3. It is a familiar phenomenon in condensed matter that if an electron is confined to move on a very narrow layer then the electron will behave as if it was moving in only two dimensions. Similarly particles that move in a ten dimensional space with six small dimensions will behave at low energies as if they were moving only in four dimensions. What is the size of these extra dimensions? The traditional view is that they are all small, of the order of $10^{-33}cm$. But recently it was realized that some dimensions could be as big as $1mm$ [9]. In that case all the standard model fields would have to be confined to live on a D-brane that is extended along the four extended dimensions that we see but transverse to the large extra dimensions. Picking different manifolds or brane configurations we can have different particles at low energies. In both cases the Standard Model parameters would depend on the details of the internal manifold or brane configuration. Compactifications which preserve 8,4 or 2 supersymmetries at low energies are fairly well understood. The case where we preserve only one supersymmetry is not so well understood and we do not understand how supersymmetry can be broken, as it is in the real world, without generating a huge cosmological constant, of the order of the supersymmetry breaking scale. This seems to be the most important obstacle in understanding precisely how the Standard Model is embedded in string theory.

Figure 3. Picture of a ten dimensional geometry of the form $R^4 \times M^6$ where M^6 is a small six dimensional space and R^4 represents the four large dimensions that we see, 3 spatial and time.

Recent progress in string theory was based on the idea of dualities. It is familiar that classical electromagnetism is invariant under the interchange of electric and magnetic fields $\vec{E} \to \vec{B}$, $\vec{B} \to -\vec{E}$. This exchanges electric charges with magnetic charges. In field theories electric charges are carried by fundamental particles and magnetic charges by solitons. So this duality exchanges elementary particles with solitons. This can be achieved by changing the coupling constant as $g \to 1/g$ so that solitons that were heavy become light, like elementary particles. In many string theories we have dualities of this type. When the coupling becomes strong in terms of some variables the theory has an equivalent description in terms of some dual variables that can be weakly coupled. In this fashion many string theories are connected. These dualities are hard to check since one has to solve the strongly coupled theory in order to show that it is the same as some dual weakly coupled theory. In supersymmetric cases there are several quantities that one can calculate which do not depend on the coupling. They can be calculated at weak coupling, extrapolated to strong coupling, and then compared with the corresponding result in the dual theory. For this reason dualities have been checked mostly in supersymmetric theories. Examples of quantities that are protected by supersymmetry and can be calculated are: 1) Low energy effective action, 2) The number and masses of various "protected" special states, these are typically some charged particles. These states could be elementary on one theory and solitons on the dual theory. Books on string theory include [1,2]

2 Black hole entropy

Black holes are one of the most intriguing objects that general relativity predicts. In classical general relativity black holes have a horizon, which is a surface in spacetime such that if somebody crosses it he/she cannot come back out. This surface, however, does not look special at all for the observer who is falling in. Black holes became even more intriguing when Hawking showed that they emit thermal radiation due to quantum effects. So a black hole is more like a black body, which will emit thermal radiation. For a black hole in four dimensions the temperature is inversely proportional to its radius, $T \sim 1/r \sim 1/(G_N M) \sim 10^{-4} K(M_{sun}/M)$. We see that for black holes that are produced for normal astrophysical processes, whose mass is always bigger than the mass of the sun, this temperature is too small to be detected, since it much lower that the temperature of the cosmic microwave background radiation. The fact that they are thermal objects raises very interesting and very important theoretical puzzles, solving these puzzles is one of the challenges of a theory of quantum gravity. We are used to the fact that when we encounter a thermal object we can explain its temperature as arising from the motion of the internal constituents. So the question becomes, what are the internal constituents of the black hole that explain its temperature? This question is often phased in terms of explaining the microscopic origin of the entropy. The entropy can be defined through the first law of thermodynamics as $dM = TdS$. The entropy comes of to be $S = A_H/(4G_N)$. In other words, the entropy is proportional to the area of the horizon in Planck units. Any theory of quantum gravity, such as string theory, should explain this entropy. In string theory it is hard to calculate this entropy directly since strings describe small fluctuations around flat space while a black hole represents a large deviation from Minkowski space. Recently, when the dynamics of D-branes was understood, it became possible to calculate this entropy for some special cases [3]. Consider a compactification of string theory down to four dimensions that preserves two supersymmetries. In such a theory we could consider charged black holes. In general charged black holes should satisfy a constraint on the mass that looks like $M \geq Q$ in order to avoid naked singularities, i.e. singularities which are not covered by a horizon. In these theories with two supersymmetries this constraint coincides with the so called BPS bound. This is a bound coming from the supersymmetry algebra. The charge Q appears in the right hand side of the supersymmetry algebra and the BPS bound comes from demanding unitarity of the representations. Furthermore, the states with $M = Q$ lie in a smaller representation of the supersymmetry algebra and the number of states in such representations does not depend on the coupling or other continuous parameters in the theory[b]. Black holes with $M = Q$ are also special from the point of view of the gravity theory, they are

[b]Most precisely the number of states that cannot be combined into larger representations, such as the ones with $M > Q$, remains invariant.

called extremal black holes and for them the Hawking temperature vanishes. In these supersymmetric theories it is possible to change parameters so that the black hole configuration becomes a weakly coupled system of D-branes and strings whose entropy one can calculate fairly easily, see figure 4. The answer, of course, comes out to be the same as the area of the corresponding black hole solution. Since the number of BPS states does not change when we do this transformation this provides a derivation of black hole entropy for these special black holes in these supergravity theories. The entropy of more general black holes, including near extremal black holes ($M > Q$ but $M - Q \ll Q$), can be computed using the AdS/CFT correspondence as it will be explained below. The entropy of general black holes in completely general string backgrounds cannot be calculated with the present techniques.

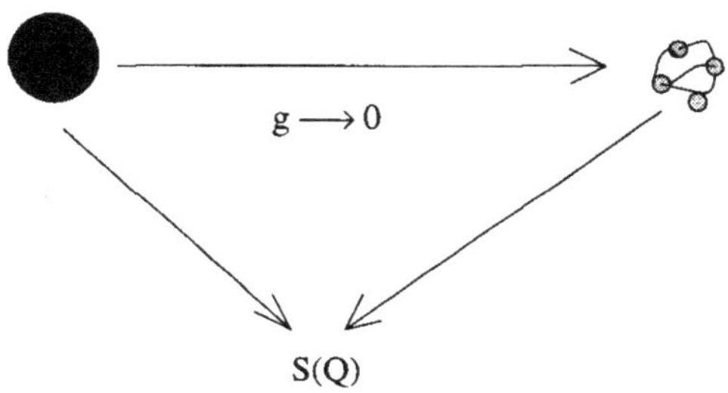

Figure 4. Starting from an extremal charged black hole we change the value of the coupling and the black hole becomes a weakly coupled gas of D-branes and strings. In supersymmetric theories these are BPS states and their number does not depend on the coupling, so we can calculate the black hole entropy by counting the number of states in the gas of D-branes and strings.

3 Conformal field theories and Anti-de-Sitter spacetimes

Although string theory was described above as a theory of quantum gravity, it originated as an attempt to describe hadrons. The string description explained some features of the hadron spectrum such as Regge trajectories, etc.. We now know that hadrons are described by QCD, but it is still quite hard to do computations at low energies due to strong coupling problems. In fact we expect confinement. Confinement is thought to arise from the fact that the color electric field lines form narrow bundles in going from a quark to an anti-quark, see figure 5. These fluxes look at low energies like strings and one might expect that at low energies a description in terms of string might be valid. It was shown by 't Hooft [8] that the proper way to understand these strings is to take the limit of a large number of colors $N \to \infty$, keeping $g_{YM}^2 N$ fixed. In this limit only planar Feynman diagrams survive. These planar Feynman diagrams seem to be giving a discretization of a string worldsheet. However it was not clear what kind of string theory that would be. For the

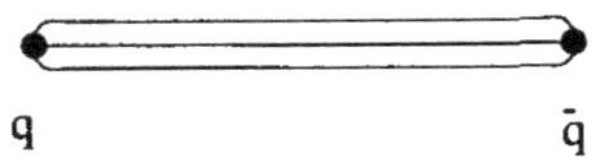

Figure 5. Fluxes of the color electic field forms a narrow bundle leading to a linear potential and confinement.

case of Yang Mills theory with $\mathcal{N} = 4$ supersymmetries and a large number of colors N it has been conjectured that these gauge strings are the same as the fundamental strings described above but moving in a particular curved spacetime: the product of five-dimensional Anti-de-Sitter space and a five-sphere [4]. Five dimensional AdS has a boundary which is four dimensional. The field theory is defined on this four dimensional boundary. In figure 6 we can see the Penrose diagram of AdS.

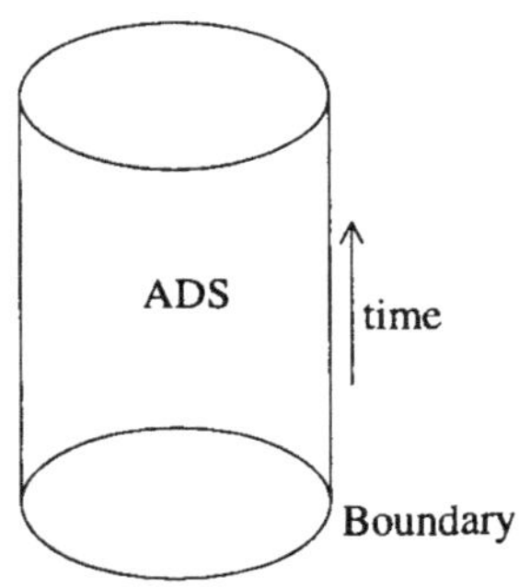

Figure 6. Penrose diagram of Anti-de-sitter which shows its causal structure. It is a solid cylinder. Time runs in the vertical direction. The boundary of AdS is the boundary of the cylinder. In the case of AdS_5 the boundary is $R \times S^3$.

The radius of curvature of this spacetime is proportional to $(g_{YM}^2 N)^{1/4}$ and the string coupling is $g_s = g_{YM}^2$ which goes like $1/N$ for fixed $g_{YM}^2 N$ as expected from 't Hooft's general argument. There have been a large number of checks for this correspondence. Many checks are possible due to the large number of supersymmetries. The simplest check is the observation that both theories have the same symmetries. $\mathcal{N} = 4$ supersymmetric Yang Mills theory is scale invariant, the coupling does not run, it is independent of the energy. Non supersymmetric Yang Mills is classically conformal invariant but quantum corrections introduce a length scale through the running of the coupling. In conformal theories the symmetry group includes translations, rotations, scale transformations and the so called special conformal transformations. All these form the group mathematically known as $SO(2,4)$. This group is the group of isometries of AdS. Similarly the Yang Mills theory has an $SO(6)$ global symmetry group, which is the same as the group of rotations of S^5. In fact when we consider string theory on $AdS - 5 \times S^5$ we also have the same supersymmetries as the gauge theory. Other checks include the comparison of the spectrum of BPS particles, renormalization group flows that partially break supersymmetry, etc. Many extensions of this correspondence were suggested, for other theories with less or no supersymmetry and for theories that are not conformal invariant. For a detailed account of the checks and a more extensive list of references on this subject see [7]. A puzzling aspect is that the bulk theory contains gravity while the field theory does not contain gravity. Gravity in the bulk is related to the stress tensor of the boundary theory in the following way. Correlation functions for the stress tensor in the field theory are equated with the amplitude for propagation of gravitons between prescribed points at the boundary. Correlation functions of operators in the field theory can be calculated, using the correspondence, as the amplitudes that particles propagate through the bulk between prescribed points at the boundary [5,11], see figure 7. Each operator corresponds to a particle, or more precisely a string mode, propagating in AdS. The mass of the particle in AdS is related to the conformal dimension Δ of the field through

$$\Delta = 2 + \sqrt{4 + (mR_{AdS})^2}$$

Similarly we can calculate the the quark antiquark potential by considering a string that goes between two points on the boundary see figure 8. In this picture we can see that the strings that move in the ten dimensional space are precisely the strings representing fluxes of color fields. We can thus say that strings made of gluons look very much like ordinary fundamental strings when $g_{YM}^2 N$ is large. It is a general feature of dualities that some excitations that look fundamental in one picture are collective excitations in the dual theory. If we compute the quark-antiquark potential by computing the energy of the string in this curved background we find that it goes like $V(L) \sim -1/L$ where L is the separation in the four dimensional theory. This potential is not confining and it should not be, since the field theory is conformal. It is possible to deform the field theory in such a way that one destroys conformal invariance

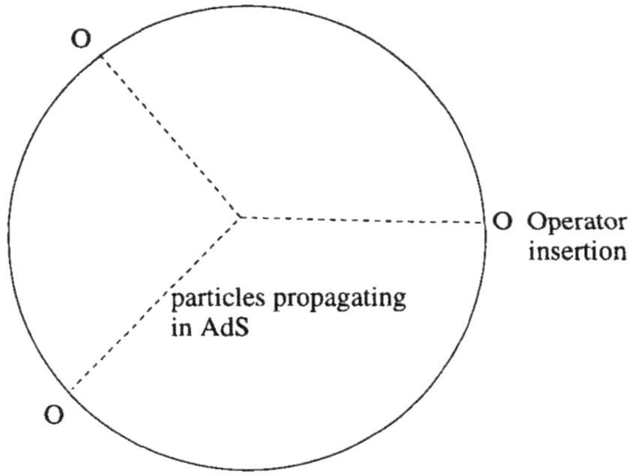

Figure 7. The correlation function of operators in the boundary theory can be calculated as the probability amplitude that the particles associated to the operators travel between the points at the boundary where the operators are inserted.

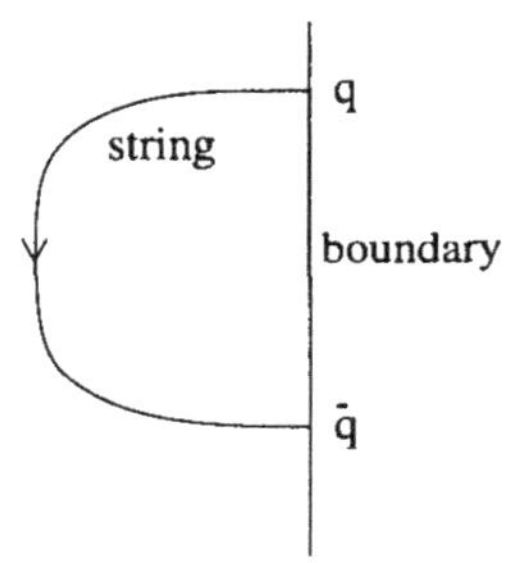

Figure 8. A quark anti-quark configuration. The flux of color field joins the quark and anti-quark and it is described by a string that moves in the higher dimensional geometry which is dual to the field theory. The potential can be calculated as the proper energy of this string in the curved geometry.

and supersymmetry at low energies. The resulting theory is expected to be confining. Indeed one can find the corresponding supergravity solution and find that the geometry is deformed and now the quark-antiquark potential is indeed confining and the theory has a mass gap [6]. Even though the theory is confining it is not pure Yang-Mills, it is a strongly coupled version of it. In order to find the large N limit of pure Yang-Mills one needs to consider strings propagating in a curved spacetime whose curvature is of the order of the string scale. In this situation the gravity approximation would not be good enough. Treating strings is these small spaces is a challenging problem, which is being explored. What we have seen so far are applications of the correspondence to the study of large N, strongly coupled field theories. The correspondence can also be used to learn about gravity. It is harder to use it in this direction since the field theory is strongly coupled and therefore hard to solve. There are, however, some general statements that one could make. One of the most mysterious objects in a gravity theory is a black hole. One can consider a black hole in AdS. This black hole is, in principle, described by some thermal state in the boundary theory. If we have a big black hole, of the order of the size of the AdS radius, its entropy will be inversely proportional to the Newton constant which is of order $1/N^2$. So the entropy will be of order N^2, which agrees with the number that we would naively expect in the field theory since we have N^2 gluons. In the case of AdS_3 it is possible to precisely calculate the entropy in the field theory and the result agrees with gravity precisely, see the review [7]. An aspect of gravity that is manifest in via this correspondence is holography [13,12]. Holography says that in a quantum theory of gravity we should be able to describe physics in some region of space by a theory with at most one degree of freedom per unit Planck area. Notice that the number of degrees of freedom would then increase with the area and not with the volume as we are normally used to. Of course, for all physical systems that we normally encounter the number of degrees of freedom is much smaller than the area, since the Planck length is so small. It is called "holography" because it would be analogous to a hologram which can store a three dimensional image in a two dimensional surface. In this case we represent the physics of the five dimensional Anti-de-Sitter spacetime with a theory that lives on its boundary. It is a concrete example of holography. Understanding it better might lead to more insights about quantum gravity.

4 Conclusions

We have learned that the laws of physics as we know them are not consistent, since we do not treat gravity using quantum mechanics. Formulating the correct theory of quantum gravity would enable us to understand the big bang and most probably the parameters of the standard model. String theory is enables us to study a whole class of phenomena that we expect in a theory of quantum gravity. The theory is not yet developed to the point of making definite experimental predictions, but is understood well enough to explain some quantum gravity phenomena like black hole entropy and evaporation, topology change, etc. Supersymmetric phases of string theory are fairly well explored. One of the main challenges is to find a supersymmetry breaking mechanism that does not generate a large cosmological constant. Many connections were found between different string theories and between string theory and field theory. It has been shown that string theory reduces, in certain circumstances, to ordinary four dimensional field theories and vice-versa. We hope that in the near future we will understand the theory better so as to make contact with experiment.

Acknowledgments

I am grateful to the organizers and participants for an interesting conference. This work was supported in part by DOE grant DE-FG02-91ER40654.

References

1. J. Polchinski, *String theory*, Vol 1 and 2, Cambridge University press 1998.
2. M. Green, J. Schwarz and E. Witten, *Superstring theory*, Vol 1 and 2, Cambridge University Press 1987.
3. A. Strominger and C. Vafa, Phys. Lett. **B379**, 99 (1996) [hep-th/9601029].
4. J. Maldacena, Adv. Theor. Math. Phys. **2**, 231 (1998) [hep-th/9711200].
5. S. S. Gubser, I. R. Klebanov and A. M. Polyakov, Phys. Lett. **B428**, 105 (1998) [hep-th/9802109].
6. E. Witten, Adv. Theor. Math. Phys. **2**, 505 (1998) [hep-th/9803131].
7. O. Aharony, S. S. Gubser, J. Maldacena, H. Ooguri and Y. Oz, hep-th/9905111.
8. G. 't Hooft, Nucl. Phys. **B72**, 461 (1974).
9. N. Arkani-Hamed, S. Dimopoulos and G. Dvali, Phys. Lett. **B429**, 263 (1998) [hep-ph/9803315].
10. J. Polchinski, Phys. Rev. Lett. **75**, 4724 (1995) [hep-th/9510017].
11. E. Witten, Adv. Theor. Math. Phys. **2**, 253 (1998) [hep-th/9802150].
12. L. Susskind, J. Math. Phys. **36** (1995) 6377 [hep-th/9409089].
13. G. 't Hooft, gr-qc/9310026.

THE CMB - CONTEMPORARY MEASUREMENTS AND COSMOLOGY

A. D. MILLER*

Department of Astronomy and Astrophysics, The University of Chicago, 5640 S. Ellis Ave., Chicago, IL 60637, USA

E-mail: amber@oddjob.uchicago.edu

Since the discovery of the Cosmic Microwave Background (CMB) in 1965, characterization of the CMB anisotropy angular power spectrum has become somewhat of a holy grail for experimental cosmology. Because CMB anisotropy measurements are difficult, the full potential of the CMB is only now being realized. Improvements in experimental techniques and detector technology have yielded an explosion of progress in the past couple of years resulting in the ability to use measurements of the CMB to place meaningful constraints on cosmological parameters. In this review, I discuss the theory behind the CMB but focus primarily on the experiments, reviewing briefly the history of CMB anisotropy measurements and focusing on the recent experiments that have revolutionized this field. Results from these modern experiments are reviewed and the cosmological implications discussed. I conclude with brief comments about the future of CMB physics.

1 Introduction

The Cosmic Microwave Background (CMB) consists of photons emitted in the early hot, dense phase of the Universe before the formation of structure by gravitational collapse. CMB photons are the oldest remaining photons in the universe; in effect a snapshot of the initial conditions for structure formation. The CMB is therefore a valuable tool for connecting the structure we see in the universe today with theories predicting its origin.

At times earlier than $z \approx 1000$, the universe consisted of a hot dense plasma in which electrons were electromagnetically coupled to baryons, photons were coupled to electrons via Compton scattering, and the universe was opaque to radiation. When the universe cooled enough to form the first neutral hydrogen atoms, the number density of free-electron scatterers dropped precipitously and the photons effectively decoupled from the baryons. Small inhomogeneities in the initial density field, which existed before decoupling, seeded the gravitational collapse of baryonic mater, resulting in the formation of all observable structures in the current universe. The photons were left to free stream and cool, interacting very little for the remaining history of the universe. These CMB photons therefore provide the cleanest known

---

*Hubble Fellow

probe of the early universe; anisotropies in the observed CMB temperature field directly correspond to the tiny fluctuations in the initial density field.

2 Anisotropy in the CMB - Theoretical Overview

The CMB is remarkably uniform, but small inhomogeneities in the density field of the early universe are revealed by temperature fluctuations of order one part in 10^5 in the currently observable CMB. Information from measurements of these fluctuations is quantified by constructing the angular power spectrum of the fluctuations discussed below. Since different models of structure formation predict different power spectrum characteristics, observations can be used to test cosmological predictions.

2.1 The Angular Power Spectrum

CMB temperature fluctuations on the sky can be described by an expansion in spherical harmonics,

$$T(\theta,\phi) = \sum_{l,m} a_{lm} Y_{lm}(\theta,\phi). \tag{1}$$

For Gaussian fluctuations, the multipole moments of the temperature field are fully described by their angular power spectrum,

$$\langle a^*_{lm} a_{l'm'} \rangle = \delta_{ll'} \delta_{mm'} C_l \tag{2}$$

The angular power spectrum can be related to the autocorrelation function, $C(\theta)$, as

$$C(\theta) = \frac{1}{4\pi} \sum_l (2l+1) C_l P_l(\cos\theta) \tag{3}$$

where the P_l's are the Legendre Polynomials. Since the Legendre Polynomials are orthogonal, the individual values of C_l can be recovered by multiplying $C(\theta)$ by a Legendre Polynomial and integrating over $\cos\theta$.

It is customary to plot the angular power spectrum, C_l, in terms of $(\delta T_l)^2$, the power per logarithmic interval in l,

$$(\delta T_l)^2 = \frac{1}{2\pi} l(l+1) C_l \tag{4}$$

or as the square root of this quantity,

$$\delta T_l = \sqrt{\frac{l(l+1)C_l}{2\pi}}. \tag{5}$$

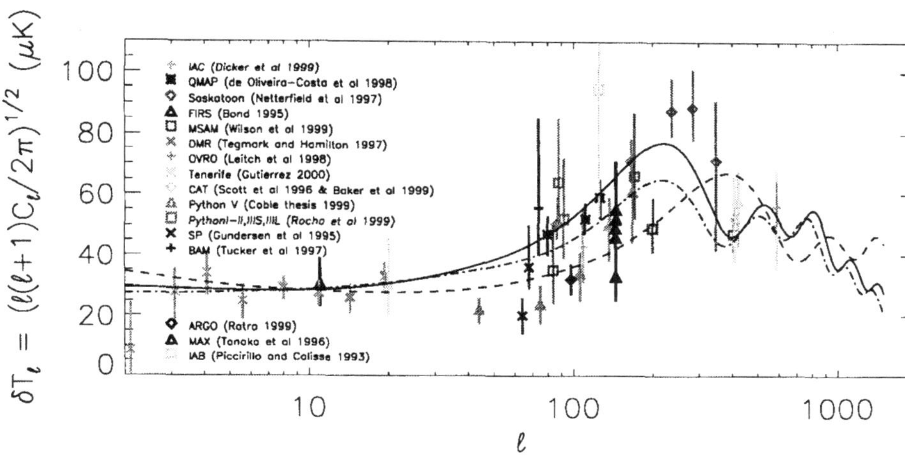

Figure 1. Models for the CMB angular spectrum computed with CMBfast[73] plotted with CMB data published as of 1999. The solid line is a ΛCDM model with $\Omega_m = 0.33$, $\Omega_b = 0.041$, $\Omega_\Lambda = 0.67$, and $h = 0.65$. The dotted line is a standard CDM model with $\Omega_m = 1$, $\Omega_b = 0.05$, and $h = 0.65$. The dashed line is an open model with $\Omega_m = 0.3$, $\Omega_b = 0.05$, and $h = 0.65$. Plotted CMB bandpowers do not include calibration errors, which are typically at least $10 - 20\%$.

2.2 Primary Anisotropies

Three representative models for the CMB angular spectrum are shown in Figure 1. A typical model is characterized by a flat plateau at large angular scales rising with l to a first prominent peak, and followed by peaks of decreasing amplitude at smaller angular scales. The series of peaks and valleys is referred to as "Doppler" or "acoustic" peaks. The primary anisotropies discussed in this section are those occurring on the surface of last scattering. We distinguish them from secondary anisotropies discussed in the next section, which are due to photons scattering along the line of sight between the surface of last scattering and the observer. The three fundamental effects that produce primary anisotropies can be summarized by the equation [52,68],

$$\frac{\delta T}{T}(\mathbf{r}) = \phi(\mathbf{r}) + \frac{1}{3}\delta(\mathbf{r}) - \delta\mathbf{v} \cdot \hat{\mathbf{r}}. \tag{6}$$

The first term represents a redshift due to the gravitational potential, ϕ. Photons in overdense regions must climb out of potential wells in order to reach an observer. Dense regions, therefore appear cooler by an amount, $\delta T/T = \phi(\mathbf{r})$.

The second term is the adiabatic contribution. The physical mechanism behind this term is presented by Hu, Sugiyama, and Silk [31] and is explained in detail there. The basic idea is the following. Before decoupling, the photons and baryons are tightly coupled (due to Compton scattering between electrons and photons and Coulomb interactions between electrons and baryons) and can be treated as a single fluid. In the context of a cold dark matter (CDM) model, Gaussian random fluctuations in the dominant CDM component create potential wells into which the photon-baryon fluid falls. The potential wells compress the fluid, the photon temperature increases, and the radiation pressure rises enough to overcome the compression and to cause a rarefaction. These oscillations proceed until recombination when the photon-baryon coupling breaks down. The denser regions on the last scattering surface will therefore be hotter while the less dense regions are cooler. Following inflation but before recombination, the horizon size increases and oscillation modes of increasing size enter the horizon. These modes are frozen in at last scattering, leaving peaks in the angular spectrum corresponding to modes caught at extrema at that epoch. The smallest wavelength modes have had the most time to evolve before being frozen. Odd peaks represent the compression maxima while even ones result from rarefaction maxima. The first "Doppler Peak" represents the mode that has just reached its first compression maximum at the time of last scattering. The density perturbations can be parameterized in terms of the density contrast,

$$\delta = \frac{\delta\rho}{\rho}, \tag{7}$$

where ρ is the average density and $\delta\rho$ is the local density perturbation. Assuming that the universe was matter-dominated at the epoch of decoupling, the density $\rho \propto T^3$ so

$$\frac{\delta T}{T}_{\text{adiabatic}} = \frac{1}{3}\delta. \tag{8}$$

The longest wavelength modes (> a few degrees) will not have had time to evolve significantly by the time of decoupling and will yield information about the primordial potential fluctuations. Assuming adiabatic initial conditions, $\delta \approx -2\phi$ [30], or

$$\frac{\delta T}{T}_{\text{adiabatic}} = -\frac{2}{3}\phi \tag{9}$$

In this case, the first and second terms partially cancel yielding

$$\frac{\delta T}{T} = \frac{\phi}{3}. \tag{10}$$

This is called the Sachs-Wolfe effect and is responsible for the plateau in the angular spectrum seen in Figure 1 at large angular scales (low l).

The third term represents the Doppler contribution, which is due to the combination of the Earth's motion with respect to the CMB and the motion of the matter on the surface of last scattering. In a universe in which the baryon contribution is large enough to be consistent with nucleosynthesis, the velocity contribution is subdominant [30].

2.3 Secondary Anisotropies

In addition to the primary anisotropies that imprint temperature fluctuations on the CMB before decoupling, several processes also exist after last scattering that imprint detectable secondary anisotropies. These can be broadly divided into two categories; those due to gravitational effects (early and late Integrated Sachs-Wolfe (ISW) effects and the Rees-Sciama effect), and those due to reionization at later times in the history of the universe. The mechanisms governing secondary CMB anisotropies are discussed elsewhere [30,68] but some observational consequences of these effects are presented in the next section.

2.4 Observational Signatures

Measurements of the CMB angular spectrum are a powerful probe of the underlying cosmology because the detailed shape of the power spectrum depends on the specifics of the cosmological model. In particular, parameters such as the energy density of the universe relative to the critical value, Ω, the fraction of that density in baryons, Ω_b, matter, Ω_m, curvature, Ω_k and due to a cosmological constant, Ω_Λ, as well as the Hubble constant, H_0, the tilt, n, and the reionization fraction, τ each leave an observable signature on the CMB angular power spectrum. Discussed below are some of the cosmological parameters which, according to the most popular class of adiabatic models, can be estimated by careful measurement of the angular power spectrum.

• Ω_k: The location of the acoustic peaks in $l-$space depends most sensitively on the curvature of the universe through a simple projection effect [33]. The scale of the sound horizon in the photon-baryon fluid before decoupling sets the maximum size of a causally connected region, corresponding to the angular size of the first acoustic peak. The radiation density is fixed by the

FIRAS measurements of T_{CMB} today [21] so the physical size of the sound horizon depends primarily on $\Omega_m h^2$ and can be used as a standard ruler. The measured angular scale at which this peak is detected therefore sheds light on the background geometry. The measured angular scale for an object of fixed size will be larger for a universe with positive curvature than for a flat universe and will be smaller for a universe with negative curvature. When detailed calculations are performed, the first acoustic peak is shown to be located at $l \approx 220$ for a large class of flat models and $l \approx 350 - 400$ for a large class of open models.

• Ω_m: The primary signature on the CMB of the matter density is a shift in the position of the peak due to the curvature effect discussed above. A lower value of Ω_m in the absence of a compensating higher value of Ω_Λ will shift the peak to higher l values. In the context of a flat universe, a lower value of Ω_m implies a larger Ω_Λ or Ω_{DE}[a].

• **Ω_Λ or Ω_{DE}:** If the universe is dominated by a negative pressure component such as a cosmological constant or some other form of dark energy, the lowest multipoles will be enhanced by the late ISW effect. The presence of a negative pressure component provides the necessary energy density to accommodate a flat universe in which $\Omega_m \approx 0.3$, consistent with other measurements [22]. In a flat universe, the early ISW effect leads to an increase in amplitude of the first peak with increasing cosmological constant. This dependence is due to the fact that the baryon-to-photon ratio before decoupling is lower in models that are dominated at the present epoch by some form of dark energy than it is in flat models in which the energy density is due entirely to matter. This is the same effect as described above for a lower value of Ω_m but peak amplitude is boosted without the accompanying shift to smaller angular scales characteristic of open models.

• Ω_b: Under the assumption that the universe is dominated by dark matter at early times, the photon-baryon fluid oscillates under the driving force of gravity and the restoring force of photon pressure. The effective mass term of the photon-baryon fluid depends on the baryon-to-photon ratio. As the baryon contribution is enhanced, the additional mass provided by the baryons increases the compression of the fluid in the well and enlarges the compression peaks (odd peaks). In addition, the photons diffuse in the photon-baryon fluid by an amount determined by the ionization history and the baryon content [31]. The effect of this diffusion is a damping of structure on small scales as photons from over dense regions mix with those from under

[a] Ω_{DE} refers to a component of dark energy with negative pressure but in which the equation of state, $w > -1$ in contrast for that of Λ for which $w = -1$ [8,32]

dense regions. The extent to which the angular spectrum is suppressed at higher multipoles will, therefore depend on the baryon content.

• H_0: During radiation domination, radiation pressure induces a decay of the gravitational potential, boosting the height of the acoustic peaks. If H_0 is relatively lower, matter-radiation equality happens later and the peaks are enhanced. In other words, one interpretation of a high first acoustic peak is a relatively lower value of the Hubble constant which implies a delay in matter-radiation equality. The location of the peaks also depends weakly on the Hubble constant. The value of the Hubble constant affects the distance to the last scattering surface. A larger distance to the last scattering surface will result in features of known physical extent appearing at smaller angular scales, which translates to a shift in the peaks to higher l.

The currently favored class of models are inflation-inspired adiabatic models in which potential wells influence all particle species in the same way; the fractional fluctuation in the number density of each species is independent of species. Most of the above discussion assumes models of this type. Isocurvature models, on the other hand, arise from initial conditions where there are no potential wells. Fluctuations in the photon density balance fluctuations in the other particle species in order to maintain zero curvature in the initial state. As in the adiabatic case, oscillations are produced in the photon-baryon fluid but they are 90° out of phase with respect to those expected from adiabatic initial conditions. The first acoustic peak is therefore expected to be found at $l \approx 350\ \Omega^{-1/2}$ rather than $l \approx 200\ \Omega^{-1/2}$ as is expected for adiabatic models [31]. Isocurvature conditions can arise from cosmological defects (strings, monopoles, and textures) [41,11]. An essential signature of most defect models is that temporal phase coherence for Fourier modes of a given wavenumber, which is responsible for driving the acoustic peaks, is destroyed by the non-linear evolution of the defect network. The result is that the fluctuation power is smeared out and the series of acoustic peaks is converted into a single broad peak, wider than the observed first Doppler peak and shifted to higher l. Simple isocurvature models are not consistent with contemporary CMB measurements.

3 CMB Anisotropy Measurements - Experimental Challenges

Many experiments have been designed to measure the anisotropy in the CMB. These measurements are difficult due to the small relative amplitude of the effect. Fluctuations exist at the level of a few parts in 10^5 of a ≈ 3 K signal that is already difficult to distinguish from atmospheric noise, emission from Galactic and extragalactic sources, and systematic errors inherent to the experiments. A successful CMB experiment must have high sensitivity and low and well understood systematic errors.

There are three general approaches that have been taken in order to measure anisotropy in the CMB. The first is to make a map of the fluctuations and to extract the power spectrum from the map. Examples of map-making experiments include satellite experiments such as COBE [65], NASA's recently launched MAP, and ESA's upcoming Planck. Mapping has also been a popular technique for balloon-borne experiments such as QMAP [15], BOOMERanG [13], and MAXIMA [38]. CMB maps are characterized by complex scan patterns, facilitating the comparison of each pixel (containing a temperature measurement and an associated uncertainty) with each neighboring pixel. Mapping instruments are generally made up of a single collecting telescope dish as a primary mirror with secondary and sometimes tertiary mirrors used to focus the radiation into the receiver. Receivers are single or multiple element radiometers generally constructed using HEMT [58] detectors at frequencies below 90 GHz and bolometers at higher frequencies.

The second class of CMB experiments are those making use of a scanning or beam switching technique to compare signals at difference locations on the sky. These experiments are constructed in much the same way as are map-making instruments (also generally single-dish telescopes using HEMTs and bolometers) but the scan strategy is optimized for probing the power spectrum directly without the intermediate step of making a map. This is an approach often taken by ground-based experiments, simplifying the scan to a constant elevation in order to keep the atmospheric temperature as uniform as possible over the scan. Examples of this type of experiment are Saskatoon [76],MSAM [20], MAX [67] Tenerife [63], PYTHON [10], VIPER [55], and MAT [44,71].

The third general class of CMB experiments are those using interferometry. Since the natural data products of an interferometer are components of the Fourier transform of the sky brightness distribution, interferometry is a natural way to measure the angular power spectrum of the CMB. Interferometers also have the advantage of low systematic errors, easy removal of point sources, and good atmospheric rejection. Interferometers require a phase-preserving detector so generally HEMT amplifiers or SIS mixers are used rather than the more sensitive bolometers. Examples of interferometric CMB experiments are CAT [3], DASI [40], IAC[29], and CBI [48].

4 Early CMB Experiments

The Hot Big Bang model was introduced by Gamow in 1946 [23] and the existence of a cosmic radiation background was first predicted by Alpher and

Herman in 1948 [1]. These ideas were largely ignored, however until the sixties when Dicke and Peebles worked out a new hot big bang model and predicted the existence of a resulting thermal background relic. Roll and Wilkinson were preparing an experiment to search for this radiation when it was detected by Penzias and Wilson in 1964 as excess noise in their communications antenna. The discovery [53] and the explanation [16] were published in companion papers in 1965 and the field of CMB physics was officially on the map. The CMB spectrum was definitively measured over the frequency range between 70 and 630 GHz by the Cosmic Background Explorer (COBE) Satellite's Far Infrared Absolute Spectrophotometer (FIRAS) instrument and found to be that of a thermal blackbody with a temperature of 2.728 ± 0.004 K at the 95% confidence level [21].

While the measurement of the CMB and the identification of the radiation as such a strikingly good black body was a triumph for the big bang model, information about the formation history of the large scale structure in the universe is contained not in the absolute measurement of the CMB but in measurements of the anisotropy. Had no anisotropy been detected, the entire model of structure formation growth by the collapse of density inhomogeneities would have to have been reformulated. The COBE DMR detection of these anisotropies on angular scales of several degrees [65] therefore marked another large milestone in the field and encouraged other experimentalists to look further.

COBE measured fluctuations on scales that reflect the primordial fluctuation field; those fluctuation modes that had not yet had time to complete an oscillation in the potential well before crossing the horizon. These measurements therefore normalize the CMB spectrum and can be used to connect with measurements of large scale structure.

Since the location in l-space of the first peak contains the most direct known measurement of the geometry of the universe and the location and height of this peak contain a wealth of information about other cosmological parameters, the next challenge was to determine the shape of the angular power spectrum on scales near the location of the predicted first peak. Since the COBE DMR detection, measurements of the CMB have largely focused on this goal. While descriptions of individual experiments will be confined to recently published results, it is important to note the immense amount of progress that was made by early experiments. Figure 1 shows CMB data published before 1999. It is clear that there is evidence for a rise in the spectrum up to $l \approx 200$ and a fall above $l \approx 400$ but no single instrument had detected and localized the peak. Since each instrument has a calibration error on the order of 10% or more and the error bars are large, measurements at different angular scales made with different instruments are difficult to compare in a quantitative way. It was therefore important that single instruments (with the same systematic errors and the same calibration uncertainties) measured the angular spectrum over the entire range of multipoles spanning the predicted location of the peak.

5 Modern CMB Experiments

The dream of the CMB being used to place meaningful constraints on cosmological parameters has come true over the past couple of years. The location of the first Doppler peak has been conclusively measured, the second and arguably third peaks have been detected, and constraints on parameters have been inferred from the location and relative heights of these peaks. It is therefore useful to discuss in more detail these measurements and the instruments that have made them possible. Data from modern CMB experiments is plotted in Figures 2a-2c and Table 1 shows experimental parameters and the best fit cosmological models for each of the experiments.

The first detection of the first Doppler peak with a single instrument came from the Mobile Anisotropy Telescope on Cerro Toco (MAT/Toco) [44], a single dish scanning experiment using HEMT [58] and SIS-based [35] receivers. MAT operated at the base of Cerro Toco mountain at an altitude of 17,000 ft. (5200 m) in the Atacama desert in Northern Chile. Data published in 1999 from this experiment (see Figure 2a) localized the peak in the spectrum to $l \approx 200$, giving support to adiabatic, inflationary models and ruling out the most common class of isocurvature and texture models. The amplitude of the first peak as measured by Toco was also suggestively high. As seen in figure 2 the data does not fit the standard CDM model very well and is a better fit to a ΛCDM model.

The next experiment to detect a peak in the CMB power spectrum was the BOOMERanG experiment [13]. BOOMERanG is also a single dish experiment with multiple detectors operating at higher frequency and higher angular resolution than the Toco experiment. It uses bolometers rather than HEMTs or SISs and operates from a balloon platform. BOOMERanG was the first CMB experiment to fly on a long duration balloon from Antarctica. Before flying in Antarctica, the instrument underwent a test flight in North America. Data from the test flight (BOOM/NA) [42] was published early in 2000 and showed striking agreement with the Toco data in the location of the peak (see Figures 2a and b). The amplitude of the peak is lower than that measured by Toco but agrees within the uncertainties. Also published in 2000, data from the BOOMERanG long duration flight (BOOM/LDB) [12] measured, for the first time, entire rang of l-space covering the predicted locations of the first and second peaks in the angular power spectrum. This data caused some speculation at the time of publication because while power was clearly detected in the region of the second peak, the amplitude was lower than predicted by the cosmological models that were at the time becoming standard.

Also published in 2000, data from the MAXIMA experiment again confirmed the location of the first peak and measured the region of l-space covering the second peak [27]. MAXIMA [38] is a balloon-borne mapping experiment that flew from North America. While MAXIMA saw slightly more power in the region of the second peak than BOOMERanG did, joint fits to the MAXIMA and BOOMERanG data yielded a value of $\Omega_b h^2$ higher than predicted from Big Bang Nucleosynthesis [70,6]. Analyses suggested that within the standard cosmology and standard BBN, this value could not be accommodated [7], and alternative theories were proposed such as non-standard reionization histories [28], and the possibility of evolution in the electromagnetic entropy since the time of BBN [34].

Data published in 2001 resolved this discrepancy. Three data sets were released in the same weekend, covering the range of l-space spanning the first three Doppler peaks, which agreed remarkably well. Figure 2c shows data from the first year of DASI observations along with re-anlyses of the BOOMERanG and MAXIMA 2000 data. Differences between the 2000 and 2001 analyses for these experiments are outlined in the caption for Figure 2.

DASI (Degree Angular Scale Interferometer) is a very different instrument from those discussed above. Rather than being a single dish mapping or scanning telescope, it is a compact interferometer designed specifically for the purpose of measuring anisotropy in the CMB. Like Toco, it uses HEMT amplifiers and operates from the ground (in this case the South Pole). DASI was designed to operate as a sister experiment to another interferometer, CBI (Cosmic Background Interferometer) with larger dishes and larger dish separation than DASI, therefore designed to probe smaller angular scales (see Figure 2c). CBI operates from the a desert plateau in the Northern Atacama desert very close to the Toco site.

Focusing on figure 2a-c, there are a couple of subtleties to note. First, the data plotted in figures 2a-2c do not include calibration uncertainties. These uncertainties are listed in parentheses in the upper right of each figure. One is free to shift each entire data set up or down by the calibration uncertainty. In addition, the BOOMERanG and MAXIMA data sets have uncertainties in the effective beam size that are not included in the plotted error bars. These uncertainties depend on l (grow with increasing l) so the effect is to tilt the spectrum. See Netterfield *et al.* for an illustration of the amplitude of the tilt corresponding to the 1 sigma uncertainty in the BOOMERanG beam of ±13%. Beam and pointing errors for the MAXIMA experiment are listed as a function of l in Lee *et al.* and range from 0% at $l = 77$ to $^{+25}_{-18}$% and ±10.2% at $l = 1147$ for beam and pointing respectively. Caveats stated, agreement between the three independent measurements covering a large range of angular scales and the CBI result with the smallest error bars at the smallest angular scale is impressive. In addition, it is reassuring to note that while BOOMERanG and MAXIMA are similar instruments employing similar scan strategies, DASI is a completely different type of instrument producing indistinguishable results.

Figure 2d shows CMB data as combined by Wang, Tegmark, and Zaldarriaga [74]. Data have been compressed from an original 105 bandpowers to the 24 bandpowers shown in the Figure 2d. Clearly evident are three peaks in the CMB power spectrum, the first of which appears at $l \approx 200$, followed by a decrease in power at smaller scales. Cosmological implications are discussed below.

5.1 Cosmological Parameters and Implications

As discussed above, much of the focus of CMB experiments following COBE has gone into characterizing the first Doppler peak in order to constrain the geometry of the universe. Using the Toco data and the BOOMERanG/NA data alone, Knox and Page performed a model-independent fit to the data solving for the location and height of the first Doppler peak [36], localizing it to be at $175 < l < 243$ (Toco) and $151 < l < 259$ (BOOM/NA) both at the 95% confidence level, as predicted by adiabatic inflationary models. Models predicting the peak location to be at $l \approx 350 - 400$ (isocurvature and texture models) are therefore ruled out at the many sigma level by each data set independently. Combining this data with recent low measurements of the matter density [22] we can infer that there must be a component of dark energy in the universe. This agrees well with supernova measurements [54,61], which suggest that the cosmological constant (or other form of dark energy) plays a significant role in the dynamics of the universe today.

Beyond the location of the first peak, the detection of the other peaks along with the characterization of their locations and heights relative to each other and to the first peak shed light on cosmological parameters. Table 1 shows the values of several cosmological parameters as measured by each of the CMB experiments both independently and when combined into a single set of bandpowers [74]. It is clear that individual determinations are consistent with each other and with the general picture of a flat, low Ω_m universe that

Table 1. **Top: Experimental parameters** References at top refer to papers describing the instruments. **Bottom: Best fit cosmological parameters** Column two shows the best fit values for TOCO+DMR assuming a prior of $\Omega_b h^2 = 0.019 \pm 0.003$ and $h = 0.65 \pm 0.1$[18]. No formal errors were calculated. Column three shows cosmological parameters extracted from BOOM/LDB 2001 data. Error bars are one sigma and assumed priors are $0.45 < h < 0.9$, and age > 10Gyrs[47]. Similar fits were done to the BOOM/LDB 2000 data assuming the same priors resulting in $\Omega_{TOT} = 1.15^{+0.1}_{-0.09}$, $\Omega_b h^2 = 0.036^{+0.006}_{-0.005}$, $\Omega_{CDM} h^2 = 0.24^{+0.08}_{-0.09}$, $\Omega_m = 0.84 \pm 0.29$, $\Omega_\Lambda < 0.83$, and $n_s = 1.04^{+0.1}_{-0.09}$[37]. Fits to BOOM/NA data result in $0.85 < \Omega_{TOT} < 1.25$ at the 68% confidence level and when combined with COBE results in a best fit model of $\Omega_{CDM} = 0.46$, $\Omega_b = 0.05$, $\Omega_\Lambda = 0.5$, $n_s = 1$, and $h = 70$[43]. Column four shows the best fit parameters for the MAXIMA data released in 2000 combined with COBE[2]. Error bars are at the 95% confidence level with $0.4 < h < 0.9$, age > 10Gyr, $\Omega_m > 0.1$. The values of Ω_Λ and Ω_m were calculated combining MAXIMA data with Supernova results. Under the same assumptions, fits to the MAXIMA 2001 data combined with COBE result in (at the 95% confidence level) $\Omega_b h^2 = 0.0325 \pm 0.0125$, $\Omega_{CDM} h^2 = 0.17^{+0.16}_{-0.07}$, $\Omega_{TOT} = 0.9^{0.18}_{-0.16}$, and $n_s = 0.46\tau_c = (0.99 \pm 0.14)$ assuming that $\tau_c \leq 0.5$[66]. Column five contains the best fit cosmological parameters to the DASI data[59]. Errors are 68% confidence and priors are $h > 0.45$ and $0.0 < \tau_c \leq 0.4$. Constraints on Ω_m, and Ω_Λ also assume $h = 0.72 \pm 0.08$. Column six tabulates the best fit parameters from combined bandpowers[74]. Fits are to the CMB data alone with 95% confidence limits.

	Toco[45]	BOOM[13]	MAXIMA[38]	DASI[40]	Comb.
Platform	ground	balloon	balloon	ground	–
Type	beam-switch	mapping	mapping	int	–
Detector	HEMT,SIS	Bolo	Bolo	HEMT	–
Freq (GHz)	30-144	90-400	150-410	26-36	–
resolution	$0.2^\circ - 1^\circ$	$0.17^\circ - 0.23^\circ$	$10'$	$20'$	–
Sky $(\mathrm{deg})^2$	500	1800	124	400	–
Ω_{TOT}	–	1.03 ± 0.06	0.90 ± 0.15	1.04 ± 0.06	–
Ω_K	0	–	–	–	$-0.06^{+.13}_{-0.59}$
Ω_m	–	0.50 ± 0.20	$0.25 - 0.5$	0.4 ± 0.15	–
Ω_Λ	0.57	$0.52^{+0.24}_{-0.19}$	$0.4 - 0.76$	0.6 ± 0.15	$0.60^{+0.32}_{-0.55}$
$\Omega_{CDM} h^2$	0.16	0.12 ± 0.05	0.13 ± 0.10	0.14 ± 0.04	$0.10^{+0.07}$
$\Omega_b h^2$	0.019	$0.021^{+0.004}_{-0.003}$	0.025 ± 0.01	$0.022^{+0.004}_{-0.003}$	$0.02^{+0.06}_{-0.01}$
n_s	1.18	$0.93^{+0.1}_{-0.08}$	0.99 ± 0.09	$1.01^{+0.08}_{-0.06}$	$0.93^{+0.75}_{-0.16}$

is currently dominated by some form of dark energy.

6 The Future of the CMB

While tremendous progress has been made using the CMB to measure cosmological parameters, much remains to be done. In terms of primary anisotropies the error bars and calibration uncertainties over the range of measured angular scales need to be reduced, the detailed shape of the damping tail at small angular scales has yet to be characterized, and the range of angular scales between the COBE measurement and the more recent measurements have yet to be connected by a single instrument. In addition, more work needs to be done in order to fully understand the extent to which foregrounds such as point sources, dust, and synchrotron radiation are capable of contaminating CMB results. Other foregrounds such as the Sunyaev-Zel'dovich effect (SZE) are capable of providing powerful probes of the early universe in their own right. Furthermore, the same processes responsible for creating the temperature anisotropies discussed throughout this paper should also produce polarization anisotropies. The spectrum of these anisotropies will provide a complementary look at the early universe and will allow the removal of degeneracies inherent to measurement of the temperature anisotropy spectrum alone.

Experiments designed to measure primary anisotropies, polarization anisotropies, and the Sunyaev-Zel'dovich effect are underway. NASA's MAP satellite, launched on June 30th, 2001 will provide detailed information about the spectrum of primary anisotropies out to $l \approx 1000$. It is also expected to detect the polarization of the CMB over a wide range of multipoles. ESA's Planck satellite, scheduled for launch in 2007 will measure the spectrum of both primary and polarization anisotropies up to $l \approx 2000$. Planck also expects to survey the Sunyaev-Zel'dovich effect in clusters of galaxies. In the mean time, several groups are analyzing data already obtained and many balloon-based and ground-based experiments are being designed to measure CMB polarization and to conduct non-targeted SZE surveys. Detector technology continues to improve and new techniques are rapidly being developed. The near future in this field promises to be as exciting as the recent past.

Acknowledgments

The author would like to thank Erik Leitch for a thorough reading of the text, Mark Devlin and Andreas Berlind for helpful comments, and is supported by Hubble Grant ASTR/HST-HF-0113.

References

1. Alpher, R. A., and Herman, R. C., *Nature,* **162**, 774 (1948).
2. Balbi, A., *et al., ApJ,* **545**, L1-L4; Erratum-ibid. **558**, L145-L146 (2000).
3. Baker, J. C., *et al., MNRAS,* 308, 1173 (1999).
4. Barreiro, R. B., *New Astron.Rev.* **44** 179-204 (2000).
5. Bond, J. R., Phys. Rev. Lett., **74**, 4369 (1995).

6. Bond, J. R., *et al.*, *Proc. IAU Symposium*, **201** (PASP), CITA-2000-65 (2000).
7. Burles, S., Nollett, K., and Turner, M., *Phys. Rev.* D, **63**, 063512 (2001).
8. Caldwell, R. R., Dave, R., Steinhardt, P. J., 1998, *Phys. Rev Lett.*, **80**, 1582 (1998).
9. Coble, K, *PhD Thesis*, The University of Chicago (1999).
10. Coble, K., *et al.*, *ApJ*, **519**, L5 (1999).
11. Crittenden, R. G., & Turok, N. G., *Phys. Rev Lett.*, **75**, 2642 (1995).
12. de Bernardis, P., *et al.*, *Nature*, **404**, 955-959 (2000).
13. de Bernardis, P., et al., astro-ph/9911461 (1999).
14. de Oliveira-Costa, A., Devlin, M. J., Herbig, T., Miller, A. D., Netterfield, C. B., Page, L. A., and Tegmark, M., *ApJ*, **509**, L77 (1998).
15. Devlin, M. J., de Oliveira-Costa, A., Herbig, T., Miller, A. D., Netterfield, C. B., Page, L., & Tegmark, M., *ApJ*, **509**, L73 (1998).
16. Dicke, R. H.; Peebles, P. J. E.; Roll, P. G.; Wilkinson, D. T., *ApJ* **142**, 414-419 (1965).
17. Dicker, S., *et al.*, *MRAS*, **309**, 750 (1999).
18. Dodelson, S., *private communication*, (2000).
19. Dodelson, S., and Knox, L., *Phys. Rev. Lett.*, **84**, 3523 (2000).
20. Fixsen, D. J., *et al.*, *ApJ*, **470**, 63 (1996).
21. Fixsen, D. J., *et al.*, *ApJ*, **473**, 576 (1996).
22. Freedman, W. L., *Phys. Scripta* **T85**, 37-46 (2000).
23. Gamow, G., *Phys Rev.*, **71**, 273 (1946).
24. Gundersen, J., *et al.*, *ApJ*, **443**, L57-L60 (1995).
25. Gutiérrez, C. M., Rebolo, R., Watson, R. A., Davies, R. D., Jones, A. W., and Lasenby, A. N., *ApJ*, **529**, 47 (2000).
26. Halverson, N., *et al*, astro-ph/0104489 (2001).
27. S. Hanany, *et al.*, *ApJ*, **545**, L5 (2000).
28. Hannestad, S., and Scherrer, R., astro-ph/0011188 (2000).
29. Harrison, D. L., *et al.*, astro-ph/0004357 (2000).
30. Hu, W., *PhD Thesis*, The University of California at Berkeley (1995).
31. Hu, W., Sugiyama, N., & Silk, J., *Nature*, **386**, 37 (1997).
32. Huey, G., Wang, L., Dave, R., Caldwell, R. R., & Steinhardt, P. J., *Phys. Rev* D, 59 (1998).
33. Jungman, G., Kamionkowski, M., Kosowsky, A., and Spergel, D., Phys.Rev.Lett., **76**, 1007-1010 (1996).
34. Kaplinghat, M., and Turner, M., *Phys. Rev. Lett.*, **86**, 385 (2001).
35. Kerr, A. R., Pan, S. K., Lichtenberger, A. W., LLoyd, F. L, & Horner, N., in *4th International Symposium on Space Terahertz Technology* (1993).
36. Knox, L., & Page, L., *Phys. Rev. Lett.*, **85**, 1366-1369 (2000).

37. Lange, A., et al., *Phys. Rev.* D, **63**, 042001 (2001).
38. Lee, A., *et al.*, in *EC-TMR Conference Proceedings 476, 3K Cosmolgy*, ed. L. Maiani, F. Melchiorro, and N. Vittorio (Woodburym New York: AIP, 224, 1999).
39. Lee, A., *et al, ApJ*, **561**, L1-L5 (2001).
40. Leitch, E. M., Readhead, A. C. S., Pearson, T. J., Myers, S. T., Gulkis, S., and Lawrence, C. R., *ApJ*, **532**, 37 (2000).
41. Magueijo, J., & Brandenberger, R., astro-ph/0002030 (2000).
42. Mauskopf, P., *et al., ApJ*, **536**, L59-L62 (2000).
43. Melchiorri, A., *et al., ApJ*, **536**, L63-L66 (2000).
44. Miller, A. D., *et al., ApJ*, **524**, L1-4 (1999).
45. Miller, A. D., in preparation, astro-ph/0108030 (2001)
46. Netterfield, C. B., Devlin, M. J., Jarosik, N., Page, L., and Wollack, E. J., *ApJ*, **474**, 47 (1997).
47. Netterfield, C. B., *et al.*, astro-ph/0104460 (2001).
48. Padin, S., *et al.*, astro-ph/0110124 (2001).
49. Padin, S., et al., *ApJ* **549** L1-L5 (2001).
50. Page, L. A., Conference Proceedings, astro-ph/9911199 (1999).
51. Partridge, B., *3K: The Cosmic Microwave Background*, (Cambridge University Press, Cambridge, 1995).
52. Peacock, J. A., *Cosmological Physics*, (Cambridge University Press, Cambridge, 1999).
53. Penzias, A. A., and Wilson, R. W., *ApJ*, **142**, 419-421 (1965).
54. Perlmutter, S., *et al., Bull Am. Astron. Soc.*, **29**, 1351 (1997).
55. Peterson, J. B., *et al.*, astro-ph/9910503 (1999).
56. Piccirillo, L., and Calisse, P., *ApJ*, **411**, 529 (1993).
57. Podariu, S., Souradeep, T., Gott, R., Ratra, B., amd Vogeley, M., *Ap.J.* **559**, 9 (2001).
58. Pospieszalski , M. W., Weinreb, S., Norrod, R. D., & Harris, R., *IEEE Trans. Microwave Theory Tech., MTT-36*, **3**, 552 (1998).
59. Pryke, C., *et al.*, asro-ph/0104490 (2001).
60. Ratra, B., Ganga, K., Stompor, R., Sugiyama, N., de Bernardis, P, and Górski, K. M., *ApJ*, **510**, 11 (1999a).
61. Riess, A., *et al., Astron. J.*, **116**, 1009-1038 (1998).
62. Rocha, G., Stompor, R., Ganga, K., Ratra, B., Platt, S. R., Sugiyama, N., and Górski, K. M., *ApJ*, **517**, 549 (1999b).
63. Romeo, G., Ali, S., Femenia, B., Limon, M., Piccirillo, L., Rebolo, R., and Schaefer, R., *ApJ*, **548**, L1-L4 (2000).
64. Scott, P. F., *et al., ApJ*, **461**, L1 (1996).
65. Smoot, G., *et al., ApJ*, **396**, L1 (1992).

66. Stompor, R., *et al.*, astro-ph/0105062 (2001).
67. Tanaka, S. T., *et al., ApJ*, **468**, L81 (1996).
68. Tegmark, M, "Doppler Peaks and all that: CMB Anisotropies and what they can tell us". *Proc. Enrico Fermi, Course CXXXII*, (Varenna, 1995).
69. Tegmark, M., & Hamilton, astro-ph/9702019 (1997).
70. Tegmark, M., and Zaldarriaga, M., *Phys. Rev. Lett.*, **85**, 2240 (2000).
71. Torbet, E. T., *et al., ApJ*, **521**, L79 (1999).
72. Tucker, G. S., Gush, H. P., Halpern, M., and Shinkoda, I., *ApJ*, **475**, L73 (1997).
73. Seljak, U., and Zaldarriaga, M., *Ap. J.*, **469**, 437 (1996).
74. Wang, X., Tegmark, M., and Zaldarriaga, M., astro-ph/0105091 (2001).
75. Wilson, G. W., *et al., ApJ*, **532**, 57 (2000).
76. Wollack, E. J., Jarosik, N., Netterfield, C. B., Page, L., & Wilkinson, D., *Ap. J.*, **419**, L49 (1993).

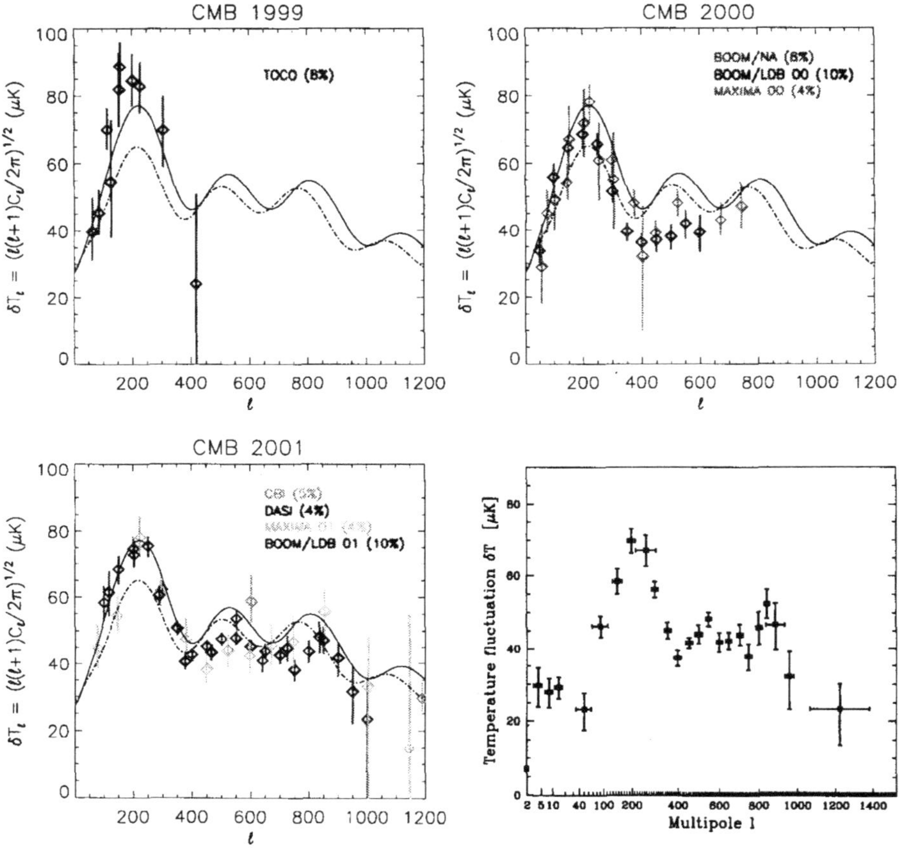

Figure 2. **a** (upper left): The MAT/TOCO data[44,71] **b** (upper right):Data from BOOM/NA[42], BOOM/LDB[12], and MAXIMA[27] published in 2000. **c** (lower left): DASI[26], CBI[49], the BOOM/LDB 2001 analysis[47], and the MAXIMA 2001 analysis[39]. The BOOM/LDB 2001 analysis differs from the 2000 analysis in that the number of channels analyzed were increased from one to four, the data from the second half of the flight was added, the effective beam size was adjusted from $10.0 \pm 0.1'$ to $12.9 \pm 1.4'$, and the overall calibration was raised by 10% in C_l[47]. The MAXIMA 2001 analysis differs from the 2000 analysis in that, the number of channels used was dropped from four to three (the 240 GHz channel was removed leaving three channels at 150 GHz). Also, the 2000 analysis made use of the entire map binned into pixels of size $5' \times 5'$ while in 2001 analysis was restricted to the central portion of the map and the data was re-binned into pixels of size $3' \times 3'$[66]. **d** (lower right) CMB combined bandpowers[74] obtained by combining much of the data from Figure 1 and all data from Figure 2 with the exception of BOOM/LDB 2000 and MAXIMA 2000. Combined bandpowers include calibration and beam uncertainties from individual experiments. Note, models in each panel are the two $\Omega_{TOT} = 1$ models from Figure 1 and numbers in parentheses are calibration uncertainties.

GAMMA-RAY BURSTS - A PRIMER FOR RELATIVISTS

TSVI PIRAN
Racah Institute for Physics, The Hebrew University, Jerusalem, 91904, Israel
E-mail: tsvi@phys.huji.ac.il

Gamma-Ray Bursts (GRBs) - short bursts of 100-1MeV photons arriving from random directions in the sky are probably the most relativistic objects discovered so far. Still, somehow they did not attract the attention of the relativistic community. In this short review I discuss briefly GRB observations and show that they lead us to the fireball model - GRBs involve macroscopic relativistic motion with Lorentz factors of a few hundred or more. I show that GRB sources involve, most likely, new born black holes, and their progenitors are Supernovae or neutron star mergers. I show that both GRB progenitors and the process of GRB itself produce gravitational radiation and I consider the possibility of detecting this emission. Finally I show that GRBs could serve as cosmological indicators that could teach us about the high redshift ($z \approx 5 - 15$) dark ages of the universe.

1 Introduction

Gamma-Ray bursts - GRBs, short and intense bursts of γ-rays arriving from random directions in the sky were discovered accidentally more than thirty years ago. During the last decade two detectors, BATSE on CGRO and BeppoSAX have revolutionized our understanding of GRBs. BATSE has demonstrated [1] that GRBs originate at cosmological distances in the most energetic explosions in the Universe. BeppoSAX discovered X-ray afterglow[2]. This enabled us to pinpoint the positions of some bursts, locate optical [3] and radio [4] afterglows, identify host galaxies and measure redshifts to some bursts [5].

The high energy release and the rapid time scales involved suggested very early that GRBs might be associated with relativistic compact objects. The discoveries of BATSE and BeppoSAX confirmed these expectations. These observations have established the Fireball model demonstrating that GRBs are the most relativistic objects known so far:

- GRBs involve macroscopic ultrarelativistic flows with Lorentz factors $\Gamma \geq 100$.
- GRBs involve accretion onto a newborn compact object, most likely a black hole. GRBs are the birth cries of these black hole.
- GRBs could emit significant amounts of gravitational radiation; both in the process of the formation of the black hole and during the acceleration of the matter to relativistic velocities.

- GRB can serve as beacon lights that would help us to explore the high redshift universe.

I review, here, our understanding of GRBs, emphasizing, as appropriate for this conference, their relativistic nature. I begin in with a very brief exposition of the properties of GRBs. I continue in section 3 with a short exposition of the Fireball model (see [6,7] for details), focusing on realization that GRBs must involve ultrarelativistic motion and on the observational proofs of such motion. In 4 I summarize the implication of the Fireball model to the "inner engines" and showing their association with black holes. I also discuss the possible progenitors of GRBs. In section 5 I discuss GRBs as sources of Gravitational radiation. I will not discuss here how GRBs could be used to explore the early high z universe and I refer the reader to recent reviews [8,9] on this issue. Concluding remarks, predictions and open questions are discussed in section 6.

2 Gamma Ray Bursts Observations

Gamma-Ray Bursts are short and intense bursts of soft gamma-rays, arriving from random directions in the sky. Most observational features of GRBs are nicely summarized in various reviews see e.g. [10,11]. Some of the basic features of the prompt burst are:

- The bursts originate from cosmological distances and arrive from random directions in the sky.

- The overall observed fluences range from 10^{-4}ergs/cm^2 to 10^{-7}ergs/cm^2 (the lower limit depends of course on the characteristic of the detectors and not on the bursts themselves). This corresponds to isotropic emission of 10^{53}ergs. However, we know today that most GRBs are narrowly beamed and the corresponding energies are around 10^{51}ergs [12,13,14].

- The spectrum is non thermal. The energy flux peaks at few hundred keV and there is a long high energy tail extending in cases up to GeV.

- The duration of the bursts ranges from less than 0.01sec to more than 100sec. GRBs are divided to short ($T < 2$sec) and long ($T > 2$sec) according to their duration.

- The light curves show rapid variability, at times on scales less than 10msec.

- The present local rate of long observed GRBs is $\approx 2\text{Gpc}^{-3}\text{yr}^{-1}$ [15,16]. The rate of GRBs seems to follow the star formation rate, namely this rate was higher by a factor of 10 at $z > 2$. The rate will of course be higher by a factor $2/\theta^2$, if GRBs are beamed with an opening angle θ.

- The rate of short bursts is uncertain. There are indications that short bursts are weaker than long ones and hence the observed short bursts are nearer to us that the long ones [17,18]. The best estimate so far suggests that all short bursts are at $z < 0.5$. So far afterglow was not detected from any short burst and there is no redshift measurement to any short burst hence there is no independent confirmation of this estimate. Given about 250 short bursts per year and assuming that they all come from within $z < 0.5$ we find that the current rate of observed short bursts, $R_{iso-short} = 2 \times 10^{-8}\text{Mpc}^{-3}\text{yr}^{-1}$, is about ten times larger than the rate of long GRBs.

Until 1997 there where no known counterparts to GRBs in other wavelengths. On Feb 28 1997 the Italian-Dutch satellite BeppoSAX detected x-ray afterglow from GRB 970228. The exact position given by BeppoSAX led to the discovery of optical afterglow [3]. Radio afterglow was detected in GRB 970508 [4]. By now more than thirty x-ray afgterglows have been observed. About half of these have optical and radio afterglow and in many of those the host galaxy has been discovered and in a dozen or so cases their redshift has been measured. The observed redshifts range from 0.46 to a record of 4.5.

3 Ultra-Relativistic Motion and the Fireball Model

3.1 The Compactness Problem and Ultra-Relativistic Motion

The need for ultrarelativistic motion in GRBs arise from the conflict between the large energy released, the short time scale and the non thermal spectrum. With a fluence of $\sim 10^{-6}\text{ergs/cm}^2$ and a cosmological distance $\sim 10^{28}$cm the energy of GRBs is $\sim 10^{51}$ergs. This energy is released within a few seconds. Using the usual causality limit on the size of an object, R, given a variability time scale, ΔT, $R \leq c\Delta T$ to estimate the density of photons one finds that the optical depth for pair production $\gamma\gamma \rightarrow e^+e^-$ would be $\sim 10^{15}$ (Piran, 1999). Such a source cannot emit the observed nonthermal spectrum.

Already in 1975 Ruderman [19] pointed out that relativistic effects could eliminate this problem. First if the source is moving relativistically with a Lorentz factor Γ towards us than the usual causality limit is replaced by: $R \leq c\Delta T/\Gamma^2$. Additionally if the source is moving towards us the observed photons have been blue shifted. At the source they have lower energy which would be insufficient for pair production. Together this leads to a decrease in the estimated optical depth by a factor of $\Gamma^{2+2\alpha}$, where $\alpha \sim 2$ is the spectral index, namely the exponent of the photon number distribution. Various estimates [20,21,22,23], of the Lorentz factor Γ based on the compactness problem lead to comparable values, $\Gamma \geq 100$, (see [23] for a critical review). Today we have independent direct observational evidence for such ultra-relativistic motion.

3.2 The Fireball Model

While we don't expect celestial objects to roam around at ultra-relativistic velocities one can imagine spherical explosions or jets in which matter is ejected from a central source ultra-relativistically. This leads us to the Fireball model.

The Fireball model asserts that GRBs are produced when the kinetic energy (or Poynting flux) of a relativistic flow is dissipated by shocks. These shocks accelerate electrons and generate strong magnetic fields. The relativistic electrons emit the observed γ-rays via synchrotron or Synchrotron-self Compton. There are two variants of this model: The External Shocks model [24] assumes that the shocks are between the relativistic flow and the surrounding circumstellar matter. The Internal Shocks model [25,26] assumes that the flow is irregular and the shocks take place between faster and slower shells within the flow. The rapid time variability seen in most GRBs cannot be produced by external shocks [27,28]. This leaves internal shocks as the only viable model for the production of the GRB. As internal shocks do not dissipate all the kinetic energy, the remaining energy will be dissipated later by interaction with the surrounding matter and produce an afterglow. This leads us to the Internal-External shocks model [27]. According to this model Internal shocks are responsible for the GRB while external shocks produce the longer lasting afterglow (see Fig. 1). The shocks occur at relatively large distances ($10^{13} - 10^{14}$cm for internal shocks and $10^{14} - 10^{18}$cm for external shocks) from the source that generates the relativistic flow. The observed radiation from the GRB or from the afterglow reflects only the conditions within these shocks. We have only indirect information on the nature of the "inner engines".

Fig. 1 depicts a schematic picture of the Internal-External shocks model. An inner engine produces an irregular wind. The wind varies on a scale δt and its overall duration is T. The variability scale δt corresponds to the variability scale observed in the GRB light curve [29] thus, $\delta t \leq 1$ sec. Internal shocks take place at $R \approx \delta t \Gamma^2 \approx 3 \cdot 10^{14}$cm $(\delta t/1\ sec)(\Gamma/100)^2$. External shocks become significant at $\sim 10^{16}$cm (see [30,6] for details). Initially, there is also a

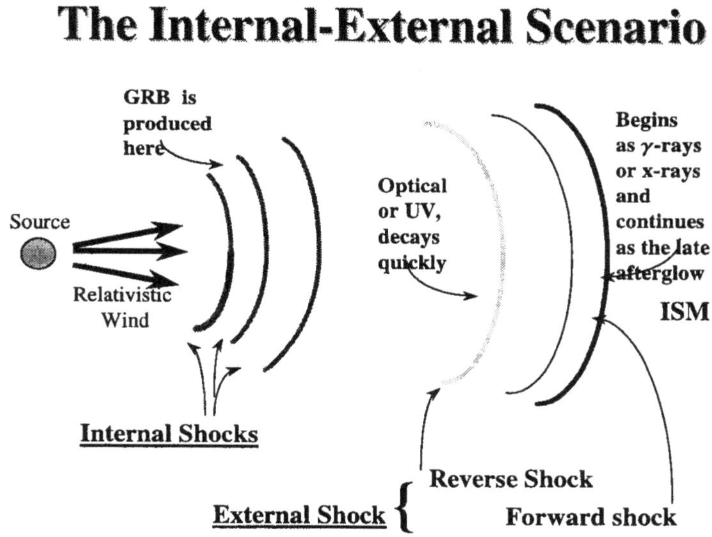

Figure 1. **The Internal-External Shocks Model.**

short lived reverse shock that propagates into the ejecta. This reverse shock is responsible to the prompt optical emission observed in GRB990123 that we discuss later

The forward shock propagates into the surrounding matter producing the afterglow. It turns out to a blast wave that is well described by the Blandford-McKee self similar solution. The theory of the afterglow is well understood [31]. Blandford and McKee have worked out (already in the seventies!) the theory of an adiabatic relativistic blast wave. This is the relativistic analog of the well known Sedov-Taylor solution. Electrons are accelerated to relativistic velocities by the shocks and their interaction with the magnetic field leads to synchrotron radiation. This provides an excellent model for the observed emission [32]. Overall we have a simple theory characterized by five parameters: the total energy, E_0, the ambient density, n_0, the ratio of the electrons and magnetic fields energy density to the total energy density, ϵ_e, ϵ_B and the exponent of the electrons' energy distribution function p. An additional sixth parameters, the exponent of the circumstellar density distribution, n, arises in cases when the external matter density is inhomogeneous ($\rho \propto r^{-n}$). Most notable is $n = 2$ corresponding to a pre-GRB wind expected in some models [33]. This rather simple theory predicts a robust relations between α and β the exponents describing the flux as a function of frequency, $F_\nu \propto t^{-\alpha}\nu^{-\beta}$. At the high frequencies, above the cooling frequency, we have (for $n = 0$), $\alpha = (3p - 2)/4$ and $\beta = p/2$.

3.3 Observations of Relativistic Motion

The radio afterglow observations of GRB 970508 provided the first verification of relativistic motion. The radio light curve (in 4.86Ghz) varied strongly during the first month. These variations died out later. Even before this transition Goodman [34] interpreted these variations as scintillations. The observation of a transition after one month enabled Frail et al., [4] to estimate the size of the afterglow at this stage as $\sim 10^{17}$cm. It immediately follows that the afterglow has expanded relativistically. Additionally, the source is expected to be optically thick in radio [35] leading to a ν^2 rising spectrum at these frequencies. The observed flux from the source enables us (using the black body law) to estimate the size of the source. As predicted the radio spectrum increases like ν^2. The size estimated with this method agrees [4] with the one derived by the scintillations estimate implying as well a relativistic motion.

The radio emission from GRB970508 showed relativistic motion in its afterglow. However, these observations took place one month after the burst and at that time the motion was only "mildly" relativistic with a Lorentz factor of order a few. Are there "direct" observations for the ultrarelativistic motion that exists earlier?

In Jan 23 1999 ROTSE recorded six snapshots of optical emission from GRB990123[36] . Three of those were taken while the burst was still emitting γ-rays. The other three snapshots spanned a couple of minutes after the burst. The second snapshot, taken 70sec after the onset of the burst corresponds to a 9th magnitude signal. A comparison of these optical observations with the γ-rays and X-rays light curves (see e.g. [7]) shows that the optical emission does not correlate with the γ-rays pulses. The optical photons and the γ rays are not emitted by the same photons [37,36].

How can one explain this flash? The collision between the ejecta and the surrounding medium produces two shocks. The outer forward shock propagates into the ISM. This shock develops later into the self similar Blandford-McKee [31] blast wave that drives the afterglow. A second shock, the reverse shock, propagates into the flow. This reverse shock is short lived. It dies out when it runs out of matter as it reaches the inner edge of the flow. While it is active, it is a powerful source of energy. Comparable amounts of energy are dissipated by the forward and by the reverse shocks [30]. Sari and Piran [38] predicted (just a few month before GRB990123) that this reverse shock will produce an intense (brighter than 11th magnitude) prompt optical flash.

The observations of an optical flash demonstrated a relativistic motion with $\gamma \sim 200$ during the burst. There are three independent estimates of the Lorentz factor at the time that the ejecta hits first the ISM [37]. First the time delay between the GRB and the optical flash suggests $\Gamma \sim 200$. The ratio between the emission of the forwards shock (x-rays) and the reverse shock (optical) gives another estimate of $\Gamma \sim 70$. Finally the fact that the maximal synchroton frequency of the reverse shock was below the optical band led to $\Gamma \sim 200$. The agreement between these three crude and independent estimates is reassuring.

3.4 Jets, Beaming and Energetics

With redshift measurements it became possible to obtain exact estimates of the total energies involved. While the first burst GRB970508 required a modest value of $\sim 10^{51}$ergs, the energies required by other bursts were alarming, 3×10^{53}ergs for GRB981226 and 4×10^{54}ergs for GRB990123, and unreasonable for any simple compact object model. These values suggested that the assumed isotropic emission was wrong and GRBs are beamed. Significant beaming would of course reduce, correspondingly the energy budget.

Beaming was suggested even earlier as it arose naturally in some specific models. For example the binary neutron star merger has a natural funnel along its rotation axis and one could expect that any flow would be emitted preferably along this axis. The Collapsar model, in which the GRB is produced during the collapse of a massive star, also requires beaming. Only a concentrated beamed energy could drill a whole through the stellar envelope that exists in this model.

Consider a relativistic flow with an opening angle θ. As long as $\theta > \Gamma^{-1}$ the forwards moving matter doesn't "notice" the angular structure and the hydrodynamics is "locally" spherical [39]. The radiation from each point is beamed into a cone with an opening angle Γ^{-1}. It is impossible to distinguish at this stage a jet from a spherical expanding shell. When $\theta \sim \Gamma^{-1}$ the radiation starts to be beamed sideways. At the same time the hydrodynamic behaviour changes and the material starts expanding sideways. Both effects lead to a faster decrease in the observed flux, changing α, the exponent of the decay rate of the flux to: $\alpha = p/2$. Thus we expect a break in the light curve and a new relation between α and β after the break [40,41,42]. The break is expected to take place at $t_{jet} \approx 6.2(E_{52}/n_0)^{1/3}(\theta/0.1)^{8/3}$hr [41]. The magnitude of the break and the time of the transition are different if the jet is expanding into a wind with r^{-2} density profile [43]. Recently numerical simulation [44] have shown that the break appears in a more realistic calculations, even though the numerical results suggest that the analytical model developed so far are probably too simple.

GRB980519 had unusual values for $\alpha = 2.05$ and $\beta = 1.15$. These values do not fit the "standard" spherical afterglow model[a]. However, these values are in excellent agreement with a sideway expanding jet [41]. The simplest interpretation of this data is that we observe a jet during its sideway expansion phase (with $p = 2.5$). The jet break transition from the spherical like phase to this phase took place shortly after the GRB and it was not caught in time. The light curves of GRB990123 shows, however, a break at $t \approx 2$days [45]. This break is interpreted as a jet break, corresponding to an opening angle $\theta \sim 5^o$. Another clear break was seen in GRB990510 [46,47].

The brightest bursts, GRB990123 and GRB980519 gave the first indications for jet like behaviour [41]. This suggested that their apparent high energy was due to the narrow beaming angles. A compilation of more bursts with jet breaks suggests that all bursts have a comparable energy $\sim 10^{51}$ergs and the variation in the observed energy is mostly due to the variation in the opening angles θ [12,13,14].

3.5 *Variability, Internal Shocks and a "NO GO Theorem"*

According to the internal shocks model the observed light curve corresponds to the temporal activity of the "inner engine" [29]. Further indication for this understanding arise from the observations that the distribution of pulse width and pulse separations are similar and that pulse widths are correlated with the intervals preceding of following them [48].

These results imply that there must be two different time scales operating within the "inner engine". The short time scale is the variability time scale. The duration of the observed burst corresponds to the time that the "inner engine" is active. This time scale is up to 5 orders of magnitude longer then the short, variability scale. This leads us to a NO GO Theorem. One cannot produce a variable GRB by a single explosion in which all the energy is released at once. This NO GO theorem rules out dozens of GRB models (such as evaporating black holes, vaccum instability, transition from a neutron star to a strange star etc...) which involve sudden energy release.

[a] A possible alternative fit is to a wind (n=2) model but with a unsual high value of $p = 3.5$

4 Black Holes, the Inner Engines and GRB Progenitors

The Fireball model tells us how GRBs operate. However, it does not answer the most interesting astrophysical questions: what produces them and how? I turn to this issues now.

GRBs must involve compact objects. There is no other way to extract so much energy, $\sim 10^{51}$ergs, so quickly. We have seen earlier that the temporal behaviour puts some stronger limits on the source. The short time scale, which is as short as a few ms, also suggest that we are dealing with a compact object. The long duration of the bursts shows that the source must be active much longer than its dynamical time scale. This suggests that GRBs arise due to accretion and this time scale is the duration of the accretion process. The energy involved requires a massive ($\sim 0.1 - 0.2 m_{\odot}$) disk. Such a massive disk can form only as debris during the formation of the compact object itself. With such a massive disk the most likely compact object is therefore a newborn black hole. A black hole is also the natural consequence of the two most common progenitors: Collapsars and neutron star mergers.

Accretion is needed to produce the two different time scales, and in particular the prolonged activity. A massive ($\sim 0.1 m_{\odot}$) disk is required because of the energetics. We expect that such a massive disk can form only simultaneously with the formation of the compact object. This leads to the conclusions that GRBs accompany the formation of black holes. This model is supported by the observations of relativistic (but not as relativistic as in GRBs) jets in AGNs, which are powered by accretion onto black holes. This system is capable of generating collimated relativistic flows even though we don't understand how.

An important alternative to accretion is Usov's model [49] in which the relativistic flow is mostly Poynting flux and it is driven by the magnetic and rotational energies of a newborn rapidly rotating neutron star. However this model seems to fall short by an order of magnitude of the energy required. Additionally, there is no indication of the slowing down pattern (that would be expected in such a case) in the light curves of GRBs.

4.1 GRB progenitors

Several scenarios could lead to a black hole - massive accretion disk system. This could include mergers (NS-NS binaries [50,25], NS-BH binaries [51] WD-BH binaries [52], BH-He-star binaries [53]) and models based on "failed supernovae" or "Collapsars" [54,55,56]. Narayan et al. [57] have recently shown that accretion theory suggests that from all the above scenarios only Collapsars could produce long bursts and only NS-NS (or NS-BH) mergers could produce short bursts.

Additional indications on the astrophysical nature of the sources arise from afterglow observations. One has to use these clues with care. Not all GRBs have afterglow (for example, so far afterglow was not detected from any short burst) and it is not clear whether these clues are relevant to the whole GRB populations. These clues seem to suggest a GRB-SN connection:

• **SN association:** Possible association of GRB980425 with SN98bw [58] and possible SN signatures in the afterglows of GRB970228 [59] and GRB980326 [60].

• **Iron lines:** have been observed in some x-ray afterglows [61]. Any model explaining them requires a significant amounts of iron at rest near those GRBs.

• **Association with Star formation:** GRBs seem to follow the star formation rate. GRB are located within star forming regions in star forming Galaxies [55].

• **GRB distribution:** GRBs are distributed within galaxies. There is no evidence for GRBs kicked out of their host galaxies [62] as would be expected for NS-NS mergers [25].

All these clues point out towards a SN/GRB association and towards the Collapsar model. However, the situation is not clear cut. The association of GRB980425 with SN98bw is uncertain. There are alternative explanations to the bumps in the afterglows of GRB970228 and GRB980326 [63]. Iron is produces in Supernovae. But there is no simple explanation what is iron at rest doing around the GRB (A model that explains the formation of iron lines requires that the supernova took place several month before the burst [64]. This would be incompatible with the reported SN bumps which coincide with the GRB). One should bear in mind that association with star formation and the distribution of GRBs within galaxies indicates that GRB stellar progenitors are short lived. This is compatible with massive stars. However, one cannot rule out a short lived binary NS population [65] which would mimic this behaviour.

We stress that there are some indication that seem incompatible with the SN association:

• **No Windy Afterglow:** No evidence for a wind (n=2) in any of the afterglow light curves. Such winds are expected from massive progenitors. Furthermore, most fits for the afterglow parameters show low ambient density [13].

• **No Jets:** Some GRBs dont show evidence for a jet or have very wide opening angles [13], this would be incompatible with the Collapsar model.

5 Gravitational Radiation from GRBs

The appearance of relativistic nonspherical (jets) motion and the association with black holes suggests that GRBs are be potential sources of gravitational radiation. There are two phases in which gravitataional radiation can arise in GRBs. First from the process that lead to the formation of the black hole and second from the fireball process itself.

I consider first gravitational radiation that arises before the GRB, as part of the formation of the "inner engine". Here I consider the two main progenitors candidates: Collapsar and neutron star mergers.

5.1 Mergers

I consider here both binary neutron star mergers and black hole-neutron star mergers under the single category of mergers. These sources are the "canonical" sources of gravitational radiation emission. Both LIGO and VIRGO aim in detecting these sources. Specifically the goal of these detectors is to detect the characteristic"chirping" signals arising from the in-spiraling phase of these events. The possibility of detection of such signals has been extensively discussed (see e.g. [66]) and we won't repeat this here. Such events could be detected up to a distance of $\sim$ 20Mpc with LIGO I and up to $\sim 300-600$Mpc with LIGO II.

The detection of the chirping merger signal is based on fitting the gravitational radiation signal to pre-calculated templets. Kochaneck and Piran [67] suggested that the detection of a merger gravitational radiation signal would requite a lower S/N ratio if this signal coincides with a GRB. This could increase somewhat the effective sensitivity of LIGO and VIRGO to such events.

It is expected that mergers (either binary neutron star or a black hole-neutron star mergers) produce the short GRBs (see [57]). Considering the isotropic rate of short GRBs estimated earlier we find that there should be one short burst per year within $\sim$ 450Mpc. This is just at the sensitivity level of LIGO II. As already mentioned it is not clear if short GRBs are beamed. If they are beamed, with the same beaming factor as long GRBs we should expect several hundred mergers events per a single observed burst. This would put one merger event per year at $\sim 80\theta_{0.1}^2$Mpc.

The corresponding distances to long GRBs are much longer. The nearest (long) GRB detected within a year would be a t 1Gpc. This is far beyond the sensitivity of even LIGO II. Beaming puts the nearest (long) event much nearer, at $135\theta_{0.1}^2$Mpc, well within the sensitivity of LIGO II. However, this burst would be, most likely, directed away from us and won't be observed as a GRB. Still a GRB that is beamed away from us is expected to produce an "orphan" afterglow and the gravitational radiation signal could trigger a search for this afterglow.

5.2 Collapsars

The Collapsar model [54,56] is based on the collapse of the core of a massive star to a black hole surrounded by a thick massive accretion disk. The accretion of this disk onto the black hole, is accompanied by the acceleration of ultra relativistic jets along the rotation axis and powers the GRB. The jets first have to punch a hole in the stellar envelope. The GRB forms only after the jets have emerged from the envelope. Due to the relatively long time that it takes for the jets to punch a hole in the envelope it is expected that Collapsars can produce only long bursts.

As far as gravitational radiation is concerned this system is very similar to a regular supernova. Rotating gravitational collapse has been analyzed by Stark and Piran [68]. They find that the gravitational radiation emission emitted in a rotating collapse to a black hole is dominated by the black hole's lowest normal modes, with a typical frequency of $\sim 20c^3/GM$. The total energy emitted is:

$$\Delta E_{GW} = \epsilon M c^2 = \min(1.4 \cdot 10^{-3} \mathrm{a}^4, \epsilon_{\max}) \mathrm{M c}^2 \ , \tag{1}$$

where a is the dimensionless specific angular momentum and ϵ_{max} is a maximal efficiency which is of the order a few $\times\, 10^{-4}$. The expected amplitude of the gravitational radiation signal, h, would be of the order of $\sqrt{\epsilon}GM/c^2 d$ where d is the distance to the source. Even LIGO II won't be sensitive enough to detect such a signal from a distance of 1Gpc or even from 100 Mpc. Furthermore, this signal would be rather similar to a supernova gravitational radiation signal. As regular supernovae are much more frequent it is likely that a supernova gravitational radiation signal would be discovered long before a Collapsar gravitational radiation signal.

5.3 Gravitational Radiation Emission from the GRB Itself

I turn now to examine the gravitational radiation that would arise from the GRB process itself. According to the fireball model the "inner engine" accelerates a mass of $M = E/\Gamma c^2$ to a Lorentz factor Γ. The most efficient generation of gravitational radiation could take place here during the acceleration phase, in which the mass is accelerated to a Lorentz factor Γ. To estimate this emission we follow Winberg's [69] analysis of gravitational radiation emitted from a relativistic collision between two particles.

I consider the following simple toy model. Two particles at rest with a mass M are accelerated instantly at $t = 0$ to a Lorentz factor Γ and energy E. Conservation of energy requires that some (actually most) of the rest mass was converted to kinetic energy during the acceleration and the rest mass of the accelerated particle is $m = E/\Gamma = M/\Gamma$. Using the formalism developed by Weinberg [69] to estimate the gravitational radiation generated in particle collisions, we calculate the gravitational radiation emitted by this system. Prior to the acceleration the two particles has momenta $m_0(1, 0, 0, 0)$. After the acceleration the particles' momenta is $m\Gamma(1, \pm\vec{\beta})$. The energy emitted per unit frequency per unit solid angle in the direction at an angle α relative to $\vec{\beta}$ is:

$$\frac{dE}{d\Omega d\omega} = \frac{GM^2\beta^2}{c\pi^2}\left[\frac{\Gamma^2(\beta^2 - \cos^2\alpha)}{(1 - \beta^2\cos^2\alpha)^2} + \frac{\cos^2\alpha}{\Gamma^2(1 - \beta^2\cos^2\alpha)^2}\right] . \tag{2}$$

The result is independent of the frequency, implying that the integral over all frequency diverges. This nonphysical divergence arises from the nonphysical assumption that the acceleration is instantaneous. In reality this acceleration takes place over a time δt, which is of order 0.01sec. This would produce a cutoff $\omega_{max} \sim 2\pi/\delta t$ above which Eq. 2 is not valid. The angular distribution found in Eq. 2 is dissppointing. The EM emission from the ultrarelativistic source is beamed forwrds into a small angle $1/\Gamma$, enhancing the emission in the forwards direction by a large factor (Γ^2). We find here that the gravitational radiation from this relativistic ejecta is spread rather uniformly in almost all 4π directions. Instead of beaming we have "anti-beaming" with no radiation at all emitted within the forward angle Γ^{-1} along the direction of the relativistic motion.

Integration of the energy flux over different directions yields:

$$\frac{dE}{d\omega} = \frac{GM^2}{c\pi^2}[2\Gamma^2 + 1 + \frac{(1 - 4\Gamma^2)}{\Gamma^2\beta}ArcTang(\beta)] . \tag{3}$$

As expected the total energy emitted is proportional to $m^2\Gamma^2$. Further integration over frequencies up to the cutoff $2\pi/\delta t$ yields:

$$E \approx \frac{2GM^2\Gamma^2}{c\pi\delta t} . \tag{4}$$

In reality the situation is much more complicated than the one presented here. First, the angular width of the emitted blobs is larger than Γ^{-1}. The superposition of emission from different directions washes out the no emission effect in the forward direction. Additionally according to the internal shocks model the acceleration of different blobs go on independently. Emission from

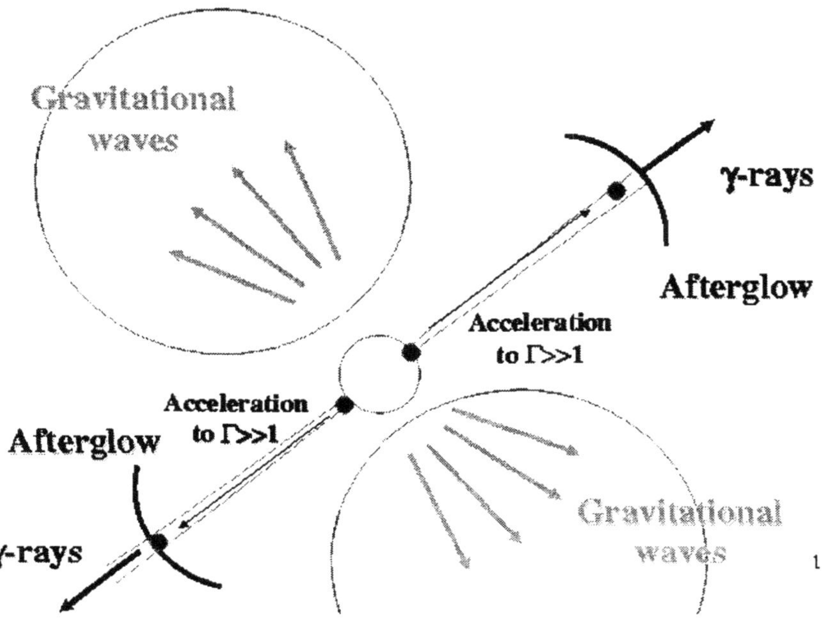

Figure 2. **Gravitational waves are mostly orthogonal to the direction of the GRB and the afterglow.**

different blobs should be combined to get the actual emission. Both effects *reduce* the effective emission of gravitational radiation and makes the above estimate an upper limit to the emission that is actually emitted.

The gravitational signal is spread in all directions (apart from a narrow beam along the direction of the relativistic motion the GRB). It ranges in frequency from 0 to $f_{max} \approx 100$Hz. The amplitude of the gravitational radiation signal at the maximal frequency, $f_{max} \approx 100$Hz, would be: $h \approx (GM\Gamma^2/c^2d)$. For typical values of $E = M\Gamma = 10^{51}$ ergs, $\delta t = 0.01$ sec and a distance of $500Mpc$, we find $h \approx .5 \times 10^{-25}$. This is far below the sensitivity of planned gravitational radiation detectors. Even if we consider a burst which is ten times nearer this "direct" gravitational radiation signal would still be undetectable.

6 Conclusions, Predictions and Open Questions

There is an ample observational support For the Fireball model. Still there are many open questions. The most interesting ones, from the point of view of this conference is how does the black hole based, inner engine, operate. What is the energy source. How does it convert the energy to ultra-relativistic flow and how does it collimate the flow to narrow jets. These interesting issues might be related to the Blandford-Znaek mechanism of extracting energy from a rotating black hole via magnetic processes [70,71,72,73]. It is a unique challenge to relativists to explore the electrodynamics of black hole and determine under what conditions this process can operate effectively.

We know how GRBs are produced. We are less certain what produces them. We can trace backwards the evolution at the source from the observations of the emitting regions to an accretion disk - black hole system. The traces from this point backwards are less clear. Theoretical considerations [57] suggest that only Collapsars can produce the disk-black hole systems needed for long bursts while only NS-NS (or possibly NS-BH) mergers can produce the systems needed for short bursts. These conclusions are supported by afterglow observations that suggest SN/GRB association for the long burst population. However, the picture is far from clear yet.

Acknowledgments

This research was supported by a US-Israel BSF grant.

References

1. C. A. Meegan *et al.*, Nature, **355**, 143 (1992).
2. E. Costa *et al.*, Nature, **387**, 783 (1997).
3. J. van Paradijs *et al.*, Nature, **386**, 686 (1997).
4. D. A. Frail *et al.*, Nature, **389**, 261 (1997).
5. M. R. Metzger *et al.*, Nature, **387**, 879 (1997).
6. T. Piran, Physics Reports, **314**, 575 (1999).
7. T. Piran, Physics Reports, **333**, 529 (2000).
8. D. Q. Lamb, American Astronomical Society Meeting, **198**, 3511 (2001).
9. D. Q. Lamb *et al.*, Ap. J., , **536**, 1 (2000).
10. , G. J. Fishman *et al.*, Ann. Rev. Astron & Astrophys., **33**, 415 (1995).
11. J. van Paradijs *et al.*, Ann. Rev. Astron & Astrophys., **38**, 379 (2000).
12. D. A. Frail *et al.*, Ap. J. Lett., , **562**, L55 (2001).
13. A. Panaitescu and P. Kumar, Ap. J. Lett., **560**, L49, (2001).

14. T. Piran, *et al.*, Ap. J. Lett., , **560**, L167, (2001).
15. M. Schmidt, Ap. J. Lett., **535**, 117 (1999).
16. M. Schmidt, talk given at the second Rome GRB meeting, Oct. 2000.
17. S. Mao *et al.*, Ap. J., , **420**, 171 (1994).
18. J. I. Katz and L. M. Canel, Ap. J., **471**, 915 (1996).
19. M. Ruderman, New York Academy Sciences Annals **262**, 164 (1975).
20. E. E. Fenimore, R. I. Epstein, and C. Ho, Astron. & Astroph. Supp., **97**, 59 (1993).
21. E. Woods and A. Loeb, Ap. J., **453**, 583 (1995).
22. T. Piran, in *Some Unsoved Problems in Astrophysics*, Eds. J. N. Bahcall and J. P. Ostriker, (Princeton University Press,1996).
23. Y. Lithwick and R. Sari, Ap. J., , 555, 540 (2001).
24. P. Meszaros and M. J. Rees, Ap. J., **405**, 278 (1993).
25. R. Narayan, B. Paczynski, and T. Piran, Ap. J. Lett., **395**, L83 (1992).
26. M. J. Rees and P. Meszaros, Ap. J. Lett., **430**, L93 (1994).
27. R. Sari and T. Piran, Ap. J., **485**, 270 (1997).
28. E. E. Fenimore, C. D. Madras, and S. Nayakshin, Ap. J., **473**, 998 (1996).
29. S. Kobayashi, T. Piran, and R. Sari, Ap. J., **490**, 92 (1997).
30. R. Sari and T. Piran, Ap. J. Lett., **455**, L143 (1995).
31. R. D. Blandford and C. F. McKee, Physics of Fluids **19**, 1130 (1976).
32. R. Sari, T. Piran, and R. Narayan, Ap. J. Lett., **497**, L17 (1998).
33. R. A. Chevalier and Z. Li, Ap. J., **536**, 195 (2000).
34. J. Goodman, New Astronomy **2**, 449 (1997).
35. J. I. Katz and T. Piran, Ap. J., **490**, 772 (1997).
36. C. W. Akerlofy *et al.*, Nature, **398**, 400 (1999).
37. R. Sari and T. Piran, Ap. J., **520**, 641 (1999).
38. R. Sari and T. Piran, Astron. & Astroph. Supp., **138**, 537 (1999).
39. T. Piran, in *AIP Conf. Proc. 307: Gamma-Ray Bursts* (1994), pp. 495.
40. J. E. Rhoads, Ap. J., **525**, 737 (1999).
41. R. Sari, T. Piran, and J. P. Halpern, Ap. J. Lett., **519**, L17 (1999).
42. A. Panaitescu and P. Mészáros, Ap. J., **526**, 707 (1999).
43. A. Panaitescu and P. Kumar, Ap. J., **543**, 66 (2000).
44. J. Granot *et al.*, astro-ph/0103038 (2001).
45. S. R. Kulkarni *et al.*, Nature, **398**, 389 (1999).
46. F. A. Harrison *et al.*, Ap. J. Lett., **523**, L121 (1999).
47. K. Z. Stanek *et al.*, Ap. J. Lett., **522**, L39 (1999).
48. E. Nakar and T. Piran, Mon. Not. RAS, in press, astro-ph/0103210 (2001).
49. V. V. Usov, Nature, **357**, 472 (1992).

50. D. Eichler, M. Livio, T. Piran, and D. N. Schramm, Nature, **340**, 126 (1989).
51. B. Paczynski, Acta Astronomica **41**, 257 (1991).
52. C. L. Fryer, S. E. Woosley, M. Herant, and M. B. Davies, Ap. J., **520**, 650 (1999).
53. C. L. Fryer and S. E. Woosley, Ap. J. Lett., **502**, L9 (1998).
54. S. E. Woosley, Ap. J., **405**, 273 (1993).
55. B. Paczynski, Ap. J. Lett., **494**, L45 (1998).
56. A. I. MacFadyen and S. E. Woosley, Ap. J., **524**, 262 (1999).
57. R. Narayan, T. Piran, and P. Kumar, Ap. J., **557**, 949 (2001).
58. T. J. Galama *et al.*, Nature, **395**, 670 (1998).
59. D. E. Reichart, Ap. J. Lett., **521**, L111 (1999).
60. J. S. Bloom *et al.*, Nature, **401**, 453 (1999).
61. L. Piro *et al.*, Science **290**, 955 (2000).
62. J. S. Bloom, S. G. Djorgovski, and S. R. Kulkarni, Ap. J., **554**, 678 (2001).
63. A. A. Esin and R. Blandford, Ap. J. Lett., **534**, L151 (2000).
64. M. Vietri *et al.*, Ap. J. Lett., **550**, L43 (2001).
65. A. V. Tutukov and L. R. Yungelson, Mon. Not. RAS, **268**, 871 (1994).
66. C. Cutler *et al.*, Phys. Rev. Lett. **70**, 2984 (1993).
67. C. Kochanek, and T. Piran, Ap. J. Lett., **417**, 17 (1993).
68. R. F. Stark and T. Piran, Phys. Rev. Lett. **55**, 891 (1985).
69. W. Weinberg, Gravitation and Cosmology, Wiley, New York (1972).
70. R. D. Blandford, and R. L. Znajek, , Mon. Not. RAS, **179**, 433 (1977).
71. H. K. Lee, C. H. Lee, and M. H. P. M. van Putten, Mon. Not. RAS, **324**, 781 (2001).
72. M. H. P. M. van Putten, Phys. Rev. Lett. **84**, 3752 (2000).
73. H. K. Lee, R. A. M. J. Wijers, and G. E. Brown, Physics Reports, **325**, 83 (2000).

APPLICATIONS OF THE THEORY OF EVOLUTION EQUATIONS TO GENERAL RELATIVITY

ALAN D. RENDALL
Max-Planck-Institut für Gravitationsphysik, Am Mühlenberg 1, 14476 Golm, Germany
E-mail: rendall@aei-potsdam.mpg.de

The theory of evolution equations has been applied in various ways in general relativity. Following some general considerations about this, some illustrative examples of the use of ordinary differential equations in general relativity are presented. After this recent applications of Fuchsian equations are described, with particular attention to work on the structure of singularities of solutions of the Einstein equations coupled to a massless scalar field. Next the relations between analytical and numerical studies of the Einstein equations are discussed. Finally an attempt is made to identify fruitful directions for future research within the analytic approach to the study of the Einstein equations.

1 Introduction

When general relativity is applied to model a physical system a connection must be made on some level between solutions of the Einstein equations, in general coupled to suitable matter fields, and the behaviour of the system concerned. As a consequence, a natural goal in general relativity is to obtain an overview of solutions of the Einstein equations which is as good as possible according to the criteria of generality, reliability and physical relevance. There are four main approaches to this goal which may be called algebraic, heuristic, numerical and analytical. In this article attention will be centred on the fourth of these but this subject will first be put in context by discussing the classification just indicated.

The approach referred to here as 'algebraic' is what is more commonly known as the field of 'exact solutions'. A reasonable strategy when confronted with unfamiliar differential equations is to look for solutions in terms of elementary functions. The basic work to be done consists of algebraic manipulations, hence the terminology chosen. For sufficiently complicated equations (e.g. the Einstein equations) the solutions obtained in this way are only a small subset of all solutions and this is why other approaches are necessary. In particular, it is not clear a priori how typical these explicit solutions are. At the same time a solution which is given explicitly in terms of an algebraic expression involving elementary functions may be less useful in understanding the nature of the dynamics than a qualitative analysis.

There are many heuristic approaches. One is to find a direct analogy between the equation being studied and the equations describing a simple mechanical system. A well-known example is the analogy between the time evolution of a homogeneous and isotropic cosmological model and the motion of a particle in a potential. Another procedure is to discard terms which are believed to be small in a particular situation and study the resulting simplified equation. One case frequently considered is that of small perturbations of an explicit solution. Here the stability of the original solution, and thus its physical relevance, is being tested. A related but more general approach is to try a particular ansatz and to see whether it is formally consistent. The strength of heuristic approaches is that they often furnish a wealth of detailed information. A problematic aspect is that of their reliability. The success of a perturbative analysis in obtaining valid results often depends sensitively on the intuition of the person carrying it out.

An attractive feature of numerical calculations of solutions of differential equations is that they are guaranteed to give results. These results may or may not be correct. Again we have a problem of reliability. Numerical investigations can provide a quick way of checking a conjecture and thus of helping to put a research project on the right track. In some situations it may turn out that there is no alternative to numerical calculations and then tests of the reliability of the results which are internal to the numerical calculations themselves are of particular importance. These may take the form of convergence tests or of the comparison of two quite different numerical calculations which are expected to give the same answer. A good illustration of the latter procedure is the work of Gundlach[34] which provides one of the most convincing pieces of evidence that certain aspects of critical phenomena in gravitational collapse are not just numerical artefacts. It should be borne in mind that certain dynamical features of solutions of partial (or even ordinary) differential equations might not be accessible to numerical calculation due to limitations in resolution.

The application of the theory of partial differential equations is what will be referred to as the 'analytical' approach in the following. Frequently in textbooks phrases are encountered such as 'no exact solutions are available and so we are forced to resort to approximate methods'. This sweeping judgement overlooks the value of qualitative information of the kind which can be obtained by the analytical approach. This article is concerned with explaining what the analytical approach has to offer and comparing it with the other approaches already mentioned. Where possible the fruitful interactions between the different ways of proceeding will be emphasized. The discussion is limited to the Einstein evolution equations and due to the limitations of what can be contained in one article the important topic of the constraints is hardly touched on here. It is the subject of the article of Isenberg in these proceedings.

In the analytical study of the Einstein equations (or other partial differential equations such as those of hydrodynamics) the concepts of existence theorems and well-posedness play important roles[32]. Here only those aspects of these ideas which are required in the sequel will be discussed. The properties which make up the definition of well-posedness of the initial value problem for an evolution equation are:

- The existence of a solution corresponding to any initial datum appropriate to the problem at hand.
- The uniqueness of the solution for a fixed initial datum.
- The continuous dependence of the solutions on the data. (A small change in the intial datum leads to a small change in the corresponding solution.)

The realization that well-posedness is important goes back to Hadamard in the first half of the twentieth century and it apparently took some time for mathematicians to be convinced. Hadamard discusses the concept in his book [35].

Ordinary differential equations with regular coefficients always have a well-posed initial value problem. This is by no means true for partial differential equations. It can be hard to prove well-posedness then. (It should be noted in passing that if the coefficients of an ODE are more or less singular then well-posedness cannot be taken for granted.)

A complication in studying well-posedness for partial differential equations is that it depends on the function spaces chosen for the data and the solutions. Related to this is the question of the sense in which the mapping from data to solutions is supposed to be continuous. This depends on the topology of the function spaces used. The type of evolution equations of most importance in the study of the Einstein equations are hyperbolic equations. The standard spaces used in this context are the Sobolev spaces of L^2 type, i.e. the spaces of functions which are square integrable together with their derivatives up to a specified order s. This property defines the Sobolev space H^s. In the limit of large s we obtain the C^∞ (smooth) functions. Well-posedness in any of these spaces with s sufficiently large implies that the solutions are controlled pointwise by the initial data. Another class of functions which often plays a role is that of analytic (C^ω) functions, i.e. those which can be expanded in a convergent power series about any point. It is relatively easy to obtain existence theorems when working in the class of analytic functions using the theorem of Cauchy-Kowaleskaya. Unfortunately this does not prove that the solution can be controlled pointwise by pointwise bounds on the data and finitely many of its derivatives. Another disadvantage of analytic functions in this context is their unique continuation property which means that relativistic causality is obscured[32]. There is a class of spaces intermediate between analytic and smooth functions which are sometimes encountered. These are the Gevrey spaces[40]. They share the disadvantage of the analytic functions that bounds on infinitely many derivatives of the data is required to estimate solutions. On the other hand they do not have the unique continuation property.

The Einstein equations have the special properties that existence is only obtained for initial data satisfying the constraint equations and that uniqueness only holds up to diffeomorphism. The notion of continuous dependence of solutions on data must be correspondingly modified[32].

The plan of the article is as follows. In section 2 various results concerning ordinary differential equations are discussed. The third section explains the results of Lars Andersson and the author on constructing general spacetimes with singularities of a particular type. It also describes other applications of Fuchsian techniques in general relativity. Section 4 is concerned with the relations between analytical results and numerical relativity. The paper concludes with some ideas about the future of the field.

2 Ordinary differential equations

Although the main concern of this paper is with applications of partial differential equations, it should not be thought that all problems in general relativity which require the analysis of ordinary differential equations have been solved. Moreover, examples concerning ordinary differential equations can serve to illustrate issues which are also relevant for partial differential equations.

An interesting example of the interaction between numerical and analytical methods is the case of the Bartnik-McKinnon solutions. These are static spherically symmetric solutions of the Einstein-Yang-Mills equations. The problem of finding solutions of this type can be formulated in terms of a system of ordinary differential equations in a radial variable. Solutions which are asymptotically flat were found numerically by Bartnik and McKinnon[4]. These came as a surprise to many people. In this way a numerical calculation led to a conceptual enrichment. Some time later the first existence proof for one of these solutions was given[53]. Unfortunately it does not appear that this existence proof led to new physical insights. It does not tell us more about why the solutions exist. A later alternative approach[14] seems to be better in this respect. Another point of view, put forward by Corlette and Wald[22], also offers an explanation, although it has not been made into an existence proof for the Bartnik-McKinnon solutions. (In related problems existence can be obtained in that way.)

For spatially homogeneous spacetimes the Einstein evolution equations reduce to ordinary differential equations. The dynamics of the solutions is very complicated, as exemplified by solutions of Bianchi type IX, also known as the Mixmaster model. There is an extensive literature where the Mixmaster model is studied by heuristic and numerical means. This work led to a consensus on what the main characteristics of the Mixmaster dynamics were but until recently there were no rigorous proofs. This changed with the work of Ringström[49,50] where the key qualitative statements, i.e. the presence of infinitely many oscillations in the approach to the singularity and the unboundedness of the curvature there in every solution except the Taub-NUT solution, were established. Thus finally, after thirty years, the analysis of the Mixmaster model came into the fold of rigorous mathematics. In the picture of singularities in general spacetimes due to Belinskii, Khalatnikov and Lifshitz (BKL)[6] the Mixmaster solutions serve as the model for the general case. For this reason the fact of having a mathematical understanding of this model acquires a special significance.

There have also been other significant developments in the area of the rigorous treatment of Bianchi models recently. Examples are the analysis of the expanding phase of models of type VII_0 with perfect fluid by Wainwright *et al.*[57,42] and that of the expanding phase of Bianchi type VIII vacuum models by Ringström[51]. The analytic investigation of spatially homogeneous spacetimes now appears to be reaching a certain maturity. At the same time there are still many interesting open problems in this area. For instance the spacetimes of Bianchi type $\mathrm{VI}_{-1/9}$, which are believed to exhibit Mixmaster-like behaviour, have not yet been analysed rigorously[56]. Also, while the work of Ringström provides a lot of knowledge about untilted Bianchi models with perfect fluids, models with tilted fluid have only been studied in special cases[36]. A similar remark applies to Bianchi models with collisionless matter as source[47,48].

Another interesting case where ordinary differential equations can be applied to general relativity is in the study of self-similar spherically symmetric spacetimes. A good example is the proof by Christodoulou[18] of the existence of spherically symmetric asymptotically flat solutions of the Einstein equations coupled to a massless scalar field which develop naked singularities. (His later proof[19] that these solutions are non-generic in the class of spherically symmetric solutions of course uses partial differential equations.) For other models less is known about self-similar solutions. The work of Carr *et al.*[15] and references therein provides a lot of information about this in the context of a convenient dynamical systems formulation but this has not yet been exploited to provide a full mathematical analysis of the problem.

3 Fuchsian systems and spacetime singularities

Lars Andersson and the author have constructed a class of solutions of the Einstein equations coupled to a massless scalar field which are very general and which have initial singularities which can be described in detail[3]. (In the following their paper will be denoted by the abbreviation AR.) They provide the strongest mathematical support so far for the heuristic analysis of spacetime singularities due to Belinskii, Khalatnikov and Lifshitz. Near the singularity in these spacetimes the evolution at different spatial points decouples, and the full evolution can be approximated by a solution of a system of ordinary differential equations, the velocity dominated system. The spatial coordinates play the role of parameters. It is worth to remember that the BKL analysis is an outgrowth of the earlier heuristic analysis of Lifshitz and Khalatnikov[39] which gave the wrong answer (generic absence of singularities) and was shown to be in error by the rigorous mathematical arguments of the singularity theorems.

The statements of BKL concerned spacetimes depending on the maximum number of free functions, and that is precisely the sense in which the solutions whose existence is proved in AR are general. In fact the free functions in AR are required to be C^ω, a restriction which does not seem natural from a physical point of view. It would be desirable to replace C^ω by C^∞ and, as will be discussed below, there are indications that this should be possible. Counting functions is a crude way of assessing the generality of a class of solutions of a partial differential equation. A better characterization of generality is given in terms of open sets of initial data. It should be shown that the class of solutions constructed include all solutions arising from a non-empty open set of initial data on a regular Cauchy surface. The possibilities of doing this will be described below.

The method used in AR is that of Fuchsian equations. Fuchsian equations are a class of differential equations whose coefficients have singularities of a certain type. They can be used to construct singular solutions of equations with regular coefficients. Suppose that a regular system of equations, such as the Einstein-scalar field system, is given. If an asymptotic form for singular solutions of this system can be guessed an ansatz can be made where the solution u of the original system is expressed in terms of an explicit singular function u_0 and a remainder u_1 supposed to be regular. In favourable cases writing the original equations in terms of u_1 leads to a Fuchsian system, for which theorems on the existence of regular solutions are available. In general, once the ansatz containing some free functions has been introduced, the remainder u_1 is determined uniquely.

In the case of the Einstein-scalar equations the ansatz for the singular part is made in terms of a solution of the velocity dominated system. Velocity dominated quantities are denoted by a left upper index zero. The starting point is a solution of the velocity-dominated constraints. This consists of tensors ${}^0g_{ab}$, ${}^0k^a{}_b$, ${}^0\phi$ and $\partial_t{}^0\phi$ in three dimensions which satisfy the equations

$$-{}^0k_{ab}{}^0k^{ab} + (\mathrm{tr}{}^0k)^2 = 8\pi(\partial_t{}^0\phi)^2 \qquad (1)$$

$$\nabla^a({}^0k_{ab}) - \nabla_a(\mathrm{tr}{}^0k) = -8\pi\partial_t{}^0\phi\nabla_a{}^0\phi \qquad (2)$$

In these equations raising and lowering of indices and covariant derivatives correspond to the metric ${}^0g_{ab}$. This system of partial differential equations can be solved in analogy to the full Einstein constraints by using the conformal method. Now the velocity dominated evolution equations should be solved for this initial data. This can be done essentially explicitly, with the result that

$${}^0\phi(t,x) = A(x)\log t + B(x) \qquad (3)$$

$${}^0k^a{}_b(t,x) = K^a{}_b(x)t^{-1} \qquad (4)$$

after which the velocity dominated metric can be obtained by solving the equation $\partial_t{}^0g_{ab} = -2{}^0g_{ac}{}^0k^c{}_b$. This can be done using matrix exponentials.

The next step is to find a unique solution of the full Einstein-scalar field equations which approaches the given velocity dominated solution as $t \to 0$. This can be achieved by applying the Fuchsian theory, provided all the eigenvalues of K are positive. The resulting solution satisfies

$${}^0g^{ac}g_{cb} = \delta^a_b + o(1) \qquad (5)$$

$$k^a{}_b = {}^0k^a{}_b + o(t^{-1}) \qquad (6)$$

$$\phi = {}^0\phi + o(1) \qquad (7)$$

Corresponding statements are obtained for space and time derivatives. With this information the properties of the singularity can be computed by straightforward algebra. Consider for example the curvature invariant $R_{\alpha\beta}R^{\alpha\beta}$. By the field equations this is equal up to a constant factor to $(\nabla_\alpha\phi\nabla^\alpha\phi)^2$ The latter quantity is equal to A^4t^{-4} up to an error of lower order in t. As a consequence of the constraints A can never vanish. Hence the square of the

Ricci tensor blows up uniformly as t tends to zero and the singularity is a curvature singularity.

Fuchsian techniques have been applied to a number of other problems in general relativity and have the potential to be applied to many more. In AR results were obtained for the Einstein equations coupled to a stiff fluid which are closely analogous to those for a massless scalar field. The results of AR were more recently extended to the case of a scalar field with mass or potential[46]. The mass and potential terms become negligible near the singularity and the velocity dominated system is identical to that obtained for a massless scalar field. While the results just mentioned are the only ones of this degree of generality obtained so far Fuchsian techniques have also been applied to various situations with symmetry.

In AR a theorem on Fuchsian systems due to Kichenassamy and the author[38] was applied. The theorem there concerned only the analytic case. It was applied in that paper to Gowdy spacetimes, which are by now a standard testbed for applying new analytic techniques in general relativity. The theorems on the construction of Gowdy spacetimes with prescribed singularities were later generalized to the C^∞ case[45]. Some of the steps in the argument did not depend on the details of the example being considered and should be much more generally applicable. With luck it will be possible to use them to handle the C^∞ case for the Einstein-scalar system without symmetry. Some obstacles which have to be overcome have been identified[45]. Going beyond this to the question of obtaining an open set of initial data without making symmetry assumptions, no case involving the Einstein equations has been handled yet. There is, however, a treatment of this kind by Kichenassamy[37] of the nonlinear wave equation $\nabla_\alpha\nabla^\alpha u = -e^u$ in Minkowski space. It was shown using the Nash-Moser theorem that there is an open set of initial data giving rise to singularities of a particular form. It is to be hoped that this technique can be extended to the Einstein equations. There are already two results covering open sets of initial data for inhomogeneous spacetimes with particular symmetries[21,44].

In the case of spacetime singularities which show behaviour of Mixmaster type Fuchsian techniques cannot be expected to apply and we have no method of constructing general spacetimes with this type of singularity. In AR the scalar field was necessary and the approach would not work for a vacuum spacetime. It is worthwhile looking at what goes wrong. If the scalar field is absent then it is like having a scalar field with $A = 0$. In that case it follows from the velocity dominated Hamiltonian constraint that one of the eigenvalues of K must be negative and the assumptions of the theorem cannot be satisfied. The significance of the sign condition on the eigenvalues of K

is that it allows the growth of the spatial curvature as t tends to zero to be estimated, thus showing that it eventually has a negligible effect. It may be noted that it might be possible to handle some negative eigenvalues and hence some vacuum solutions in higher dimensions (spacetime dimension ≥ 11) since the algebra works out differently[24].

There has recently been significant progress in understanding Mixmaster behaviour in inhomogeneous spacetimes by heuristic and numerical techniques[58,8,7]. Putting the results of this work on a rigorous mathematical footing is an fascinating challenge for the analytical approach to general relativity.

The discussion in this section up to now has been from the point of view of the BKL picture of spacetime singularities. In the past decade it has been suggested that the singularities of the kind studied by BKL, which are in some sense spacelike, are not sufficient to describe the singularities inside black holes[43]. Instead another type of singularities, called weak null singularities, are supposed to play an important role. Until recently the evidence for this was essentially heuristic or numerical in nature. This left some room for scepticism and so it is important that there is now a rigorous result by Dafermos[25] showing the development of weak null singularities from regular initial data and the occurrence of the associated effect of mass inflation.

That this kind of mathematical confirmation is not superfluous is shown by the story of the supposed stability of Cauchy horizons in the presence of a positive cosmological constant. The original suggestion[41] led to several papers coming to similar conclusions over a period of several years. However it turned out in the end[13] that these conclusions were incorrect. The original work had observed that one mechanism leading to the instability of Cauchy horizons did not seem to work in that context. It turns out, however, that a different mechanism does appear to work. This throws some light on the difference between heuristic arguments and rigorous ones. In a mathematical argument all objections to a particular statement must be effectively ruled out. This means that a mathematical approach to a given problem is often much harder than a heuristic one. Once, however, a mathematical argument has been successfully completed it rules out the presence of any other effect that was not thought of the first time around.

4 Numerical relativity

This section is concerned with the relation of the analytical approach to the Einstein equations to numerical relativity. We have already seen one kind of interaction, which is that numerical results suggest a theorem which is later proved. Now the other direction will be considered, i.e. the question whether analytical results can be of use to numerical relativists in their choice of methods.

A lot of numerical relativity has been done using what are often called the ADM equations. What is meant by this is the Einstein evolution equations written in $3+1$ form in terms of the induced metric and second fundamental form of a foliation with a fixed choice of lapse and shift. In fact these equations were in the literature long before the works of Arnowitt, Deser and Misner[26]. Despite this the term 'ADM equations' will be used in this now widespread sense in the following, whereby the lapse and shift are restricted to take the simplest form that they are identically one and identically zero respectively. There is also an ambiguity due to the fact that the evolution equations are only defined modulo the constraints. Only the vacuum equations are considered and the condition that the spatial components of the spacetime Ricci tensor vanish is taken to define the evolution equations. In fact the definition of the second fundamental form will be substituted into its evolution equation to produce a second order equation for the metric with the leading order expression $\partial_t^2 g_{ab} = -2R_{ab}$. Here only the terms containing second derivatives of the metric have been written out.

Is the Cauchy problem for the ADM equations as just defined well-posed in the space of C^∞ functions or in Sobolev spaces? Apparently nobody knows. It is known that there is a smooth solution of the ADM equations corresponding to any smooth initial datum satisfying the constraints and that the solutions depend continuously on the initial data. This is proved indirectly, for instance by the use of harmonic coordinates. This is all that is needed for the analytic theory of the Einstein equations. In numerical relativity, however, it cannot be expected that the initial data satisfy the constraints *exactly*. Thus it is of interest to know whether the ADM equations are well-posed as a system of evolution equations, independently of whether the constraints are satisfied or not. It has been shown by Choquet-Bruhat[16] that they are well-posed in Gevrey spaces, but the corresponding question for functions which are merely smooth, or belong to a Sobolev space, is open.

How can it be shown that a system of evolution equations is well-posed? This is related to the notion of hyperbolicity, which is at least intuitively equivalent to well-posedness of the initial value problem for appropriate initial data. It is perhaps useful to start by looking at the notion of ellipticity. Suppose that a nonlinear system of partial differential equations is given. Its ellipticity can be determined as follows. The system is elliptic if and only if its linearization about any given solution is elliptic. With this rule it then remains to define when a linear system is elliptic. To do this take the principal part (the part containing the derivatives of the highest order) and replace each derivative ∂_i by ξ_i. This gives a homogeneous matrix-valued polynomial in ξ, called the principal symbol. The system is called elliptic if the principal symbol is invertible in the sense that for any $\xi \neq 0$ the values of the principal symbol are invertible matrices. More generally the set of non-zero ξ for which the symbol is not invertible is called the characteristic set. Thus a system is elliptic if the characteristic set of its linearization is always empty.

Analogously to the elliptic case a non-linear system can be defined to be hyperbolic if its linearization about any solution is hyperbolic. Then it remains to define hyperbolicity in the linear case. There are criteria for hyperbolicity in terms of the principal symbol, but unfortunately, in contrast to the elliptic case, there is no simple and general condition. The characteristics are given as the roots of the characteristic polynomial, which is the determinant of the principal symbol. In general, for a system of k equations of order n the characteristic set has kn sheets in the complex domain, counting multiplicity. If a system is hyperbolic then there must be kn real sheets. In this case it is said that all characteristics are real. It was shown by Friedrich [30] that all the characteristics of the ADM equations are real. This is a positive sign but unfortunately does not suffice to show that the equations are hyperbolic.

Note that in order to check that a system is hyperbolic it is not enough to verify that the Cauchy problems for all one-dimensional reductions are well-posed as is shown by the following example[23]. The equation is

$$u_{tttt} - 2u_{ttxx} - 3u_{ttyy} + u_{xxxx} + 3u_{xxyy} + 2u_{yyyy} + u_{xxy} = 0 \qquad (8)$$

It has an ill-posed Cauchy problem. Now consider solutions of the special form $u(t,x,y) = \tilde{u}(t, x\cos\alpha + y\sin\alpha)$. For each α this reduces the original equation to an equation for $\tilde{u}$ in one space dimension. This reduced equation has a well-posed Cauchy problem for each fixed value of α. An important feature of this system is that the characteristics are of variable multiplicity. This example concerns a single higher order equation but presumably similar examples can be constructed for systems of first order equations.

To understand some of the issues involved in well-posedness it is useful to consider the trivial equation $u_{tt} = 0$ for a function $u(t,x)$ of two variables. With initial data $u(0,x) = f(x)$ and $u_t(0,x) = g(x)$ the solution is $u(t,x) = f(x) + tg(x)$. If the equation were hyperbolic, like the wave equation, it would be natural to take f in the Sobolev space H^s and g in H^{s-1}. Then the restriction of the solution to any other hypersurface of constant t would be in H^s. However the explicit solution above is only in H^{s-1} at later times. The equation is hyperbolic, but in a weaker sense than the wave equation since it loses one derivative. What is worse is that the hyperbolicity can be destroyed by adding lower order terms. The equation $u_{tt} = u_x$ (sideways heat equation) is not hyperbolic. It is not possible to bound the H^s norm at a given time $t_0 > 0$ in terms of Sobolev norms of the initial data. This can be seen by doing a Fourier transform in x.

If the ADM equations are linearized about flat space the result has some similarity to the equation $u_{tt} = 0$ and is weakly hyperbolic. If the full ADM equations are well-posed then it is reasonable to expect that their linearization about any background will have the same property. Some examples of linearization about simple non-trivial backgrounds will now be considered. The equations bear some resemblance to the equation $u_{tt} = u_x$ above. This approach has so far led to no evidence for ill-posedness of the ADM equations but it has not been carried very far. Perhaps a different choice of background would lead to a different result. Some possible directions to pursue will now be indicated.

In the examples to be considered here the metric can be written in the form

$$ds^2 = -dt^2 + e^{2\alpha(t,x)}dx^2 + e^{2\beta(t,x)}dy^2 + e^{2\gamma(t,x)}dz^2 \tag{9}$$

Ideally we would like a background metric such that the linearized Einstein equations have constant coefficients. Then taking a Fourier transform in space and time reduces the study of the growth of linearized solutions in time to an algebraic problem. It is natural to seek solutions with this property among spatially homogeneous solutions. As a first example, consider the Milne model[56]:

$$ds^2 = -dt^2 + t^2(dx^2 + e^{2x}(dy^2 + dz^2)) \tag{10}$$

This is of the form (9) with $\alpha = \log t$ and $\beta = \gamma = \log t + x$. Linearizing about this solution and denoting the linearized quantities corresponding to α, β and γ by $\hat{\alpha}$, $\hat{\beta}$ and $\hat{\gamma}$ respectively leads to a system with coefficients depending explicitly on t. However this dependence on t is sufficiently simple that introducing a new time coordinate by $\tau = \log t$ reduces the linearized Einstein equations to a system with constant coefficients. A special case consistent with the equations is obtained by setting $\hat{\beta} = \hat{\gamma}$. With $D = \partial/\partial\tau$ and $' = \partial/\partial x$ the system becomes

$$-D^2\hat{\alpha} = 3D\hat{\alpha} + 2D\hat{\beta} + 2\hat{\alpha}' + 4\hat{\alpha} - 2\hat{\beta}'' - 4\hat{\beta}' \tag{11}$$

$$-D^2\hat{\beta} = D\hat{\alpha} + 4D\hat{\beta} + \hat{\alpha}' + 4\hat{\alpha} - \hat{\beta}'' - 4\hat{\beta}' \tag{12}$$

Note that while the equation for $\hat{\beta}$ contains the second spatial derivative of $\hat{\beta}$ the equation for $\hat{\alpha}$ does not contain the second spatial derivative of $\hat{\alpha}$ but does contain its first spatial derivative. This is a reason to suspect that well-posedness might fail as a consequence of the model equation discussed above. Now the equation will be Fourier transformed in space and time. Let ω and ξ be Fourier variables corresponding to τ and x respectively. Looking for solutions of the form $e^{i(\omega\tau+\xi x)}$ leads to consideration of the roots of the cubic equation for ω with coefficients depending on ξ

$$\omega^3 - 7i\omega^2 + (-\xi^2 + 2i\xi - 14)\omega + (i\xi^2 + 2\xi + 8i) = 0 \tag{13}$$

By inspection $\omega = i$ is a root. It then only remains to solve a quadratic equation for ω to get the other two roots. These are given by $\omega = 2i + \xi$ and $\omega = 4i - \xi$. The question of interest for well-posedness is whether the imaginary parts of the roots of the cubic equation for ω are bounded functions of ξ. In this example this is the case and no evidence for ill-posedness is obtained.

Consider next the Joseph solution[56]:

$$ds^2 = \sinh 2\tau[-d\tau^2 + dx^2 + A(\tau)e^{2x}dy^2 + A^{-1}(\tau)e^{2x}dz^2] \tag{14}$$

where $A(\tau) = (\tanh t)^{\sqrt{3}}$. Introduce a new time coordinate t by means of the relation $dt = (\sinh 2\tau)^{1/2}d\tau$. Then the form (9) can be obtained by setting $\sinh 2\tau = e^{2\alpha(t)}$, $A(\tau)e^{2x}\sinh 2\tau = e^{2\beta(t,x)}$ and $A^{-1}(\tau)e^{2x}\sinh 2\tau = e^{2\gamma(t,x)}$. In this case the coefficients of the linearized equations depend on t, although the dependence is not too complicated. One way to try to get an idea about well-posedness is to freeze the coefficients at a fixed time t and do a Fourier tranform of the resulting system with constant coefficients. This has not been carried out. The conclusions to be drawn from these examples is that there are many calculations which might be done which could lead to a better understanding of the well-posedness or otherwise of the ADM equations. If the system is ill-posed, one single example could be the basis for a proof of ill-posedness of the full system. Unfortunately it not clear at this time what is the best place to look for an example of this kind.

Since a failure of well-posedness means a failure of continuous dependence of the solutions on the initial data and the initial data in a numerical problem are never exact the significance of results obtained using a numerical calculation with an ill-posed system is quite unclear. On the other hand it is known that a discretization of a partial differential equation may lead to a scheme which approximates a different equation better than the original one. (A well-known example of this is the introduction of numerical viscosity leading to a good approximation to a parabolic equation.) The ramifications of the well-posedness or otherwise of the ADM equations are still rather unclear.

In numerical relativity it is not just the well-posedness of the pure initial value problem which is important. It is necessary to impose boundary conditions and this leads to new difficulties. Friedrich and Nagy[31] have proved the well-posedness of certain initial boundary value problems for the (vacuum) Einstein equations. Much more remains to be done although there is further recent work on the subject[55].

An interesting puzzle on the interface between numerical and analytical work is provided by the BSSN system[52,5]. It is apparently much better for numerical calculations than the ADM equations in some situations. It is not known to be hyperbolic although certain deformations of it have been shown to give rise to symmetric hyperbolic systems [32,33]. Different possible explanations of its superiority have been discussed in the literature[1,54].

5 Outlook

This section collects a few comments on some research directions which may become important in the near future, without any claim to completeness.

A topic which can be expected to play a major role in the applications of partial differential equations to general relativity in the future is that of the global stability of important explicit solutions. Until recently the only results of this kind were the theorems of Friedrich[29] on the stability of de Sitter space and Christodoulou and Klainerman[20] on the stability of Minkowski space. Now there are new developments in this area. Andersson and Moncrief have proved the stability of the Milne model in the expanding direction[2] while Choquet-Bruhat and Moncrief[17] have obtained a similar result for a vacuum model of Bianchi type III under the restriction that the perturbations are assumed to have one Killing vector. The Killing vector helps, but the Milne model is easier in other ways. It should be possible to generalize these results in many ways (different backgrounds and matter models).

The results of AR discussed in section 3 point the way to the possibility of proving the stability of the singularity in Friedmann models with a scalar field. Combining this with the results on the expanding phase mentioned above could lead to theorems on the global stability of open Friedmann models with suitable matter content. The stability of the singularity alone should allow conclusions about the stability of closed Friedmann models.

A subject of capital interest is that of the stability of the Schwarzschild and Kerr black holes. No nonlinear results are available at present although there has been recent progress concerning linear stability[9,28]. The nonlinear stability of the Kerr solution is one of the outstanding open questions on the applications of partial differential equations to the Einstein equations.

The theme of critical phenomena has given rise to a considerable literature in general relativity. All of it is based on numerical calculations. The subject has been too tough to yield to analytical attack up to now. A promising approach to changing this unsatisfactory state of affairs is to study model problems. In fact it turns out that critical behaviour in gravitational collapse is not an isolated phenomenon within the field of partial differential equations. It is more likely to be typical of a kind of phenomenon which is widespread among solutions of nonlinear evolution equations. For instance Bizoń, Chmaj and Tabor[11] have studied wave maps in two space dimensions while Bizoń and Tabor[12] have studied the spherically symmetric Yang-Mills equations in space dimensions greater than three. It has not yet been possible to handle the dynamics analytically, but at least the critical solution itself has been proved to exist in one case[10].

The interaction between numerical and analytical work, some aspects of which were described in section 4, has not really taken off, in the sense that a lot of practical computations in numerical relativity are little influenced by analytical developments. A notable exception is the numerical implementation of the conformal field equations, which has recently been reviewed by Frauendiener[27]. It will be interesting to see whether this approach to numerical relativity eventually proves superior to more pedestrian methods.

Acknowledgements

I am grateful to Piotr Bizoń for discussions.

References

1. Alcubierre, M., Allen, G., Brügmann, B., Seidel, E. and Suen, W.-M. 2000 Towards an understanding of the stability properties of the 3+1 evolution equations in general relativity. Phys. Rev. D62, 124011.
2. Andersson, L. 1999 The global existence problem in general relativity. Preprint gr-qc/9911032.
3. Andersson, L. and Rendall, A. D. 2001 Quiescent cosmological singularities. Commun. Math. Phys. 218, 479-511.
4. Bartnik, R. and McKinnon, J. 1988 Particle-like solutions of the Einstein-Yang-Mills equations. Phys. Rev. Lett. 61, 141-144.
5. Baumgarte, T. and Shapiro, S. L. 1999 On the numerical integration of Einstein's field equations. Phys. Rev. D59, 024007.
6. Belinskii, V. A., Khalatnikov, I. M. and Lifshitz, E. M. 1982 A general solution of the Einstein equations with a time singularity. Adv. Phys.

31, 639-667.
7. Berger, B. K., Isenberg, J. and Weaver, M. 2001 Oscillatory approach to the singularity in vacuum spacetimes with T^2 isometry. Phys. Rev. D64, 084006.
8. Berger, B. K. and Moncrief, V. 1998 Evidence for an oscillatory singularity in generic U(1) symmetric cosmologies on $T^3 \times R$. Phys. Rev. D58, 064023.
9. Beyer, H. 2001 On the stability of the Kerr metric. Commun. Math. Phys. 221, 659-676.
10. Bizoń, P. 1999 Equivariant self-similar wave maps from Minkowski spacetime into 3-sphere. Preprint math-ph/9910026.
11. Bizoń, P., Chmaj, T. and Tabor, Z. 2000 Dispersion and collapse of wave maps. Nonlinearity 13, 1411-1423.
12. Bizoń P. and Tabor, Z. 2001 On blowup for Yang-Mills fields. Preprint math-ph/0105016.
13. Brady, P. R., Moss, I. G. and Myers, R. C. 1998 Cosmic censorship: as strong as ever. Phys. Rev. Lett. 80, 3432-3435.
14. Breitenlohner, P., Forgacs, P. and Maison, D. 1994 Static spherically symmetric solutions of the Einstein-Yang-Mills equations. Commun. Math. Phys. 163, 141-172.
15. Carr, B. J., Coley, A. A., Goliath, M., Nilsson, U. S. and Uggla, C. 2000 The state space and physical interpretation of self-similiar spherically symmetric perfect-fluid solutions. Class. Quantum Grav. 18, 303-324.
16. Choquet-Bruhat, Y. To appear.
17. Choquet-Bruhat, Y. and Moncrief, V. 2001 Future complete Einsteinian space times with $U(1)$ isometry group. C. R. Acad. Sci. 322, 137-144.
18. Christodoulou, D. 1994 Examples of naked singularity formation in the gravitational collapse of a scalar field. Ann. Math. 140, 607-653.
19. Christodoulou, D. 1999 The instability of naked singularities in the gravitational collapse of a scalar field. Ann. Math. 149, 183-217.
20. Christodoulou, D. and Klainerman, S. 1993 The global nonlinear stability of the Minkowski space. Princeton University Press, Princeton.
21. Chruściel, P. T. 1991 On uniqueness in the large of solutions of Einstein's equations (Strong Cosmic Censorship). Proceedings of the CMA Vol. 27. Australian National University.
22. Corlette, K. and Wald, R. 2001 Morse theory and infinite families of harmonic maps between spheres. Commun. Math. Phys. 215, 591-608.
23. Courant, R. and Lax, A. 1955 Remarks on Cauchy's problem for hyperbolic partial differential equations with constant coefficients in several independent variables. Commun. Pure Appl. Math. 8, 497-502.

24. Demaret, J, Henneaux, M. and Spindel, P. 1985 Non-oscillatory behaviour in vacuum Kaluza-Klein cosmologies. Phys. Lett. B 164, 27-30.
25. Dafermos, M. 2001 Stability and instability of the Cauchy horizon for the spherically symmetric Einstein-Maxwell-scalar field equations. Preprint
26. Fourès-Bruhat, Y. 1948 Sur l'intégration du problème des conditions initiales en mécanique relativiste. C. R. Acad. Sci. (Paris) 226, 1071-1073.
27. Frauendiener, J. 2000 Conformal infinity. Living Reviews in Relativity, article 2000-4.
28. Friedman, J. and Morris, M. S. 2000 Schwarzschild perturbations die in time. J. Math. Phys. 41, 7529-7534.
29. Friedrich, H. 1986 Existence and structure of past asymptotically simple solutions of Einstein's field equations with positive cosmological constant. J. Geom. Phys. 3, 101-117.
30. Friedrich, H. 1996 Hyperbolic reductions for Einstein's equations. Class. Quantum Grav. 13, 1451-1469.
31. Friedrich, H. and Nagy, G. 1999 The initial boundary value problem for Einstein's vacuum field equations. Commun. Math. Phys. 201, 619-655.
32. Friedrich, H. and Rendall, A. D. 2000 The Cauchy problem for the Einstein equations. In B. G. Schmidt (ed) Einstein's Field Equations and Their Physical Implications. LNP 540. Springer, Berlin.
33. Frittelli, S. and Reula, O. 1999 Well-posed forms of the 3+1 conformally-decomposed Einstein equations. J. Math. Phys. 40, 5143-5156.
34. Gundlach, C. 1995 The Choptuik spacetime as an eigenvalue problem. Phys. Rev. Lett. 75, 3214-3217.
35. Hadamard, J. 1952 Lectures on Cauchy's problem in linear partial differential equations. Dover, New York.
36. Hewitt, C. G., Bridson, R. and Wainwright, J. 2001 The asymptotic regimes of tilted Bianchi II cosmologies. Gen. Rel. Grav. 33, 65-94.
37. Kichenassamy, S. 1996 The blow-up problem for exponential nonlinearities. Commun. Part. Diff. Eq. 21, 125-162.
38. Kichenassamy, S. and Rendall, A. D. 1998 Analytic description of singularities in Gowdy spacetimes. Class. Quantum Grav. 15, 1339-1355.
39. Lifshitz, E. M. and Khalatnikov, I. M. 1963 Investigations in relativistic cosmology. Adv. Phys. 12, 185-249.
40. Mascarelli, M. and Rodino, L. 1997 Partial differential equations with multiple characteristics. Akademie Verlag, Berlin.
41. Mellor, F. and Moss, I. G. 1990 Stability of black holes in de Sitter space. Phys. Rev. D41, 403-409.
42. Nilsson, U. S., Hancock, M. J. and Wainwright, J. 2000 Non-tilted Bianchi VII_0 models - the radiation fluid. Class. Quantum Grav. 17, 3119-3134.

43. Poisson, E. and Israel, W. 1990 Internal structure of black holes. Phys. Rev. D41, 1796-1809.
44. Rein, G. 1996 Cosmological solutions of the Vlasov-Einstein system with spherical, plane or hyperbolic symmetry. Math. Proc. Camb. Phil. Soc. 119, 739-762.
45. Rendall, A. D. 2000 Fuchsian analysis of singularities in Gowdy spacetimes beyond analyticity. Class. Quantum Grav. 17, 3305-3316.
46. Rendall, A. D. 2000 Blow-up for solutions of hyperbolic PDE and spacetime singularities. Proceedings of Journées Equations aux Dérivées Partielles, (Eds.) N. Depauw, D. Robert, X. Saint-Raymond. Groupement de Recherche 1151 du CNRS, 2000, XIV, 1-12.
47. Rendall, A. D. and Tod, K. P. 1999 Dynamics of spatially homogeneous solutions of the Einstein-Vlasov equations which are locally rotationally symmetric. Class. Quantum Grav. 16, 1705-1726.
48. Rendall, A. D. and Uggla, C. 2000 Dynamics of spatially homogeneous locally rotationally symmetric solutions of the Einstein-Vlasov equations. Class. Quantum Grav. 17, 4697-4714.
49. Ringström, H. 2000 Curvature blow up in Bianchi VIII and IX vacuum spacetimes. Class. Quantum Grav. 17, 713-731.
50. Ringström, H. 2000 The Bianchi IX attractor. Ann. H. Poincaré 2, 405-500.
51. Ringström, H. 2001 The future asymptotics of Bianchi VIII vacuum solutions. Class. Quantum Grav. 18, 3791-3823.
52. Shibata, M. and Nakamura, T. 1995 Evolution of 3-dimensional gravitational waves - harmonic slicing case. Phys. Rev. D52, 5428-5444.
53. Smoller, J., Wasserman, A., Yau, S.-T. and McLeod, J. 1991 Smooth static solutions of the Einstein-Yang-Mills equations. Commun. Math. Phys. 143, 115-147.
54. Szilagyi, B., Gomez, R., Bishop, N. T. and Winicour, J. Cauchy boundaries in linearized gravitational theory. 1999 Phys. Rev. D62, 104006.
55. Szilagyi, B., Schmidt, B. and Winicour, J. 2001 Boundary conditions in linearized harmonic gravity. Preprint gr-qc/0106026.
56. Wainwright, J. and Ellis, G. F. R. (eds.) 1997 Dynamical systems in cosmology. Cambridge University Press, Cambridge.
57. Wainwright, J., Hancock, M. J. and Uggla, C. 1999 Asymptotic self-similarity breaking at late times in cosmology. Class. Quantum Grav. 16, 2577-2598.
58. Weaver, M., Isenberg, J. and Berger, B. K. 1998 Mixmaster behavior in inhomogeneous cosmological spacetimes Phys.Rev. Lett. 80, 2984-2987.

MERGER OF BINARY NEUTRON STARS IN FULLY GENERAL RELATIVISTIC SIMULATION

MASARU SHIBATA[1] AND KŌJI URYŪ[2]

[1] *Graduate School of Arts and Sciences, University of Tokyo, Komaba, Meguro, Tokyo 153-8902, Japan*

[2] *Department of Physics, University of Wisconsin-Milwaukee, P.O. Box413, Milwaukee, WI 53201, USA*

We review recent progress in the numerical study on the last phase of coalescing binary neutron stars in the general relativistic framework. First, current status for computation of quasiequilibrium states for binary neutron stars in close circular orbits is reviewed. Then, current status of numerical simulation for binary neutron star merger is summarized. Some numerical results recently computed are also presented.

1 Introduction

Binary neutron stars such as the Hulse-Taylor binary pulsar [1] adiabatically inspiral as a result of radiation reaction of gravitational waves, and they merge eventually. The latest statistical study suggests that such merger could happen one per year in a distance of about 30Mpc in the most optimistic scenario [2]. Even according to the most conservative scenario, it would happen one per year in a distance of about 400Mpc [2]. This implies that the coalescence of binary neutron stars is one of promising sources for kilo-meter-size laser interferometric detectors such as LIGO, TAMA, GEO600 and VIRGO [3].

The evolutionary track of binary neutron stars is roughly divided into three phases; (i) inspiraling phase in which the characteristic orbital separation a is much larger than the radius of neutron stars R and in which the orbital separation decreases adiabatically due to emission of gravitational waves, (ii) intermediate phase in which the binary still evolves in the adiabatic manner but a/R is so small ($\sim 2-4$) that the effects of tidal deformation and general relativity become important, and (iii) the final merger phase in which the two neutron stars merge on the dynamical timescale. The phase (i) has been treated by an analytic method using a post Newtonian approximation and equations of motion for point particles, resulting in recent successful developments [4]. On the other hand, phases (ii) and (iii) have to be treated numerically taking into account effects of the finite size, tidal deformation, hydrodynamic interaction and full general relativity.

The signal of gravitational waves in phases (ii) and (iii) is expected to give a wide variety of information about neutron stars and strong gravitational fields: At the innermost stable circular orbit (ISCO) at which the orbit of binaries becomes dynamically unstable, the signal is likely to change abruptly from a quasi periodic waveform to a burst waveform. In the frequency domain, such a change will be recognized as a clear cliff [5]. The characteristic frequency at the edge of the cliff may be used for extracting information about the nature of equations of state of high density neutron star matter, because the location of the ISCO could depend sensitively on the equations of state (or compactness) of the neutron stars [6].

The signal in the merger phase can also bring a variety of information about neutron star physics. During the merger, a transient massive merged object is likely formed before it settles down to a massive neutron star or a black hole. This transient object oscillates in a dynamical and nonaxisymmetric manner and emits quasi-periodic gravitational waves which are associated with its fundamental oscillation modes. Numerical simulations in the framework of Newtonian [7,5], post-Newtonian [8,9,10,11], semi-relativistic [12] and fully general relativistic (GR) gravity [13] have indicated that the frequency is approximately between ~ 2 and ~ 3kHz depending on equations of state and compactness of neutron stars. This frequency is too high for the first generation laser interferometric detectors to detect, but it is not extremely high and may be in the frequency range for the advanced laser interferometric detectors such as LIGO II [14] or the resonant mass detectors in near future. This implies that from gravitational waves induced by such fundamental oscillations, it may be able to obtain information on the dynamical merger stage and nature of the transient massive neutron star, which will be useful to understand the nature of neutron stars.

Interest has also been stimulated by a hypothesis about the central engine of γ-ray bursts (GRBs) [15]. Recently, it has revealed that GRBs are likely to be of cosmological origin. In cosmological GRBs, the central sources must supply a large amount of the energy $\gtrsim 10^{51}$ ergs in a very short timescale of order from a millisecond to a second. It has been suggested that the merger of binary neutron stars is a likely candidate for the powerful central source, in particular for short-duration events. In the typical hypothetical scenario, the final product should be a system composed of a rotating black hole surrounded by a disk of certain mass, which could supply the large amount of the energy by neutrino processes or by extracting the rotational energy of the black hole. Detailed numerical works in the framework of Newtonian gravity have been performed by Ruffert and Janka for investigation of this hypothesis [9].

Numerical hydrodynamic simulations in full general relativity provide the best approach to understanding the properties of binary neutron star merger.

Over the last few years, numerical methods for solving the coupled equations of the Einstein equations and hydrodynamic equations have been developed [16,13,17,18], and now such simulations are feasible. In this article, we review current status for such GR simulation.

The paper is organized as follows. In Sec. 2, we review current status for computation of quasiequilibrium states in general relativity. In Sec. 3, we describe implementations necessary for performing numerical relativity simulation, and their present status. In Sec. 4, we present some numerical results recently obtained using a large scale supercomputer. Sec. 5 is devoted to a summary.

Throughout this paper, we adopt the following notation; ρ: baryon rest-mass density; ε: specific internal energy; P: pressure; u^μ: four velocity; $v^i = u^i/u^t$: three velocity; $h = 1 + \varepsilon + P/\rho$; $g_{\mu\nu}$: spacetime metric; γ_{ij}: spatial metric; $\tilde{\gamma}_{ij} = \{\det(\gamma_{ij})\}^{-1/3}\gamma_{ij}$; α: lapse function; β^k: shift vector; ∇_μ: covariant derivative with respect to $g_{\mu\nu}$; and D_i: covariant derivative with respect to γ_{ij}. We adopt geometrical units $G = c = 1$ where G and c denote the gravitational constant and speed of light, respectively. Latin and Greek indices denote spatial components $(1-3)$ and space-time components $(0-3)$, respectively. $\delta_{ij}(=\delta^{ij})$ denotes the Kronecker delta. We use Cartesian coordinates $x^k = (x, y, z)$ as the spatial coordinates; t denotes coordinate time.

2 Computation of quasiequilibrium : Current status

To investigate the last inspiraling stage of binary neutron stars as well as to perform realistic numerical relativity simulation, it is necessary to compute realistic binary neutron stars in close circular orbits.

Even just before the merger, binary neutron stars are in a quasiequilibrium state because the timescale of gravitational radiation reaction in the Newtonian order $\sim 5/\{64\Omega(M\Omega)^{5/3}\}$ [19], where M and Ω denote the Newtonian total mass of the system and the orbital angular velocity of binary neutron stars, is several times longer than the orbital period. Thus for computation of intermediate phase between inspiral and merger phases, the quasiequilibrium approximation is justified.

In studying quasiequilibrium states, we may assume the existence of a helical Killing vector as

$$\ell^\mu = \left(\frac{\partial}{\partial t}\right)^\mu + \Omega\left(\frac{\partial}{\partial \varphi}\right)^\mu. \tag{1}$$

Since the emission of gravitational waves violates the helical symmetry, this assumption does not strictly hold in reality, although it is a good approximation as mentioned above.

Realistic velocity field is found to be irrotational for binary neutron stars because of negligible viscous effects inside neutron stars [20]. Assuming the irrotational velocity field in the existence of the helical Killing vector, the Euler equation can be integrated to give the first integral as [21]

$$\frac{h}{u^t} + hu_k V^k = \text{const.}, \tag{2}$$

where $V^k = v^k - \ell^k$. For the irrotational velocity field, u_i is written as

$$u_i = h^{-1} D_i \Phi, \tag{3}$$

where Φ denotes the scalar velocity potential which satisfies an elliptic-type PDE [21]

$$D_i(\rho\alpha h^{-1} D^i \Phi) - D_i[\rho\alpha h^{-1}(\ell^i + \beta^i)] = 0. \tag{4}$$

These two equations are basic fluid equations for binary neutron stars to be solved with appropriate boundary conditions.

For the spacetime geometry of quasiequilibria, most of groups currently assume conformally flat three-metric for simplicity as

$$\gamma_{ij} = \psi^4 \delta_{ij}. \tag{5}$$

The lapse, the shift vector, and the conformal factor are taken as geometric variables to be solved. Their equations are derived from the Hamiltonian constraint, the momentum constraint and a trace of Einstein equations projected onto three hypersurface, respectively. The last equation is regarded as an equation for a slicing condition. Thus, instead of solving full sets of Einstein's field equations, a reduced set is chosen and solved for determination of the gravitational field. We refer to this framework as the Isenberg-Wilson-Mathews (IWM) formalism [22]. The set of equations turns out to be the five elliptic-type equations and is solved consistently imposing an asymptotic flatness condition at spatial infinity with fluid equations described above.

Numerical methods for solving these coupled equations are independently developed by Uryū and Eriguchi [23] and French-Meudon group [24]. They have implemented numerical codes which enable to compute quasiequilibrium states with high accuracy for polytropic equations of state of any polytropic indices and for a wide range of compactness of neutron stars. A detailed study for their properties has been done by Uryū et al. recently [25]. They find that ISCO exists before two stars contact only for neutron stars of relatively stiff equations of state. For soft equations of state, two neutron stars reach the Lagrange point at which mass transfer from the inner edges of neutron stars sets in. Also, they clarified that ISCO for neutron star binaries with a realistic compactness is determined by the hydrodynamic instability not by GR effect in contrast to binary black holes or to stellar-mass compact objects orbiting massive black holes.

In determining the location of ISCO along a quasiequilibrium sequence of binaries, we rely on the theorem with regard to the turning point method. Recently, it has been proved that quasiequilibria of GR binaries satisfy a relation which is similar to the first law of the thermodynamics [26]. For a system which has an approximate asymptotic region, the relation is written as

$$dM_{\rm ADM} = \Omega dJ \tag{6}$$

for a sequence of constant baryon mass, entropy and circulation. Here, $M_{\rm ADM}$ and J are the ADM mass and the total angular momentum of the binary, respectively. When this relation is satisfied, the theorem tells us that a simultaneous minimum of $M_{\rm ADM}(d)$ and $J(d)$, where d is an appropriate parameter of the sequence, is a point at which the stability changes. We may regard the minimum along the sequence as the ISCO. Laws of binary mechanics for more generic spacetime with a helical symmetry have been also derived by Friedman et al. [26]

As mentioned above, we have adopted the IWM formalism to compute quasiequilibria. However, it gives exact solutions only for static, spherical symmetric spacetimes or in first post Newtonian approximations. Thus, in general, the solution contains a certain error for GR neutron star binaries with $(M/R)_\infty \gtrsim 0.14$. Quantitative discussion for the error in the IWM formalism and a proposal for a new formalism to compute quasiequilibria including realistic geometry are found, e.g., in Ref. 28.

In addition, methods for computation of gravitational waves from quasiequilibrium states are also proposed and a couple of computations have been already done to show that these methods are promising [27,28].

3 Implementations necessary in numerical relativity and current status

The following elements are the keys to perform successful hydrodynamic simulations in full general relativity and to draw scientific results from them; 1) a realistic initial condition. For binary neutron star merger, quasiequilibrium states in close circular orbits provide appropriate initials. 2) a formulation and an implementation of numerical scheme for stable and accurate evolution of Einstein's field equations. 3) an appropriate choice of slicing and spatial

	Problem	Status
1)	Initial condition	Approximately OK
2)	Solving Einstein's equations	OK
3)	Gauge conditions	OK
4)	Solving hydrodynamic equations	OK
4)	Microphysics	No effort in GR simulation yet
5)	Apparent horizon finder	OK
6)	Wave extraction	Approximately OK
7)	Horizon boundary	Not yet

Table 1. Current status of simulation for binary neutron star merger in numerical relativity. "OK" means that with these implementations, we are able to perform a simulation to get certain scientific results.

gauge conditions. To perform a longterm numerical simulation, it is necessary to suppress growth of gauge modes. 4) an implementation for solving GR hydrodynamic equations stably and accurately. For problems in which shock heating plays an important role, a shock-capturing scheme should be prepared to perform accurate simulations. Also, it is necessary to take into account certain microphysical process such as neutrino cooling ultimately. 5) an implementation for a black hole finder. In many simulations of numerical relativity, black holes will be formed during a simulation or exist from the beginning of a simulation. To confirm the existence of black holes and to capture their location, apparent horizon finder and/or event horizon finder are necessary. 6) an implementation for extraction of gravitational waveforms in the wavezone. One of main purposes in numerical relativity is to compute gravitational waveforms. To extract accurate waveforms in the wavezone which is located near the outer boundaries of computational domain, a certain formulation and technique to distinguish true dynamical degrees of freedom from gauge modes are necessary. Also, to compute gravitational waves in the wavezone, it is necessary to prepare computational domain extending at least to the local wavezone. To do so, we need a large scale supercomputer or a numerical technique such as a nested grid technique. Or alternatively, we have to prepare a special implementation for the wave extraction [29,32]. 7) an implementation for accurately computing a region around black holes. After a formation of black holes, the geometric variables become very steep around them whenever we use a singularity-avoidance time slicing. To handle this steep gradient, we need a numerical technique such as adaptive mesh refinement or excision technique inside the black hole horizons.

Over the last several years, we have achieved significant progress in this field. Current status for these issues are summarized in Table 1 from our point of view. The status for preparing initial conditions has been discussed in Sec. 2 [30]. The point to be improved for this issue is to develop a formulation beyond the so-called IWM formalism. Let us describe some of recent progress for other issues.

It is widely recognized that it is necessary to use a certain formulation for Einstein's field equation in order to solve it stably and accurately. Earlier, standard formalism in numerical relativity has been the ADM (Arnowitt-Deser-Misner) formalism [31]. However, with this formalism, it is difficult to perform a stable and longterm numerical simulation. As a result many alternative methods have been proposed in the last several years [32]. One of promising prescriptions was originally proposed by Nakamura in 1987 [33], and subsequently modified, e.g., by Shibata [34] and by Baumgarte and Shapiro [35]. In this formalism, they slightly modified the ADM's equations introducing auxiliary variables such as $F_i \equiv \delta^{jk}\partial_k\tilde{\gamma}_{ij}$ and $\mathrm{tr}(\gamma_{ij})$, in addition to the three metric and extrinsic curvature. The formulation has been tested for a longterm evolution of linear gravitational waves [34,35] as well as used for a wide variety of purposes such as to simulate nonlinear evolution of gravitational waves [34,36], evolution of a rotating black hole [37], grazing collision of two black holes [38], longterm simulation of rotating neutron stars [16,39,18] and merger of binary neutron stars [13,43], demonstrating its robustness.

Appropriate choice of gauge conditions is also crucial to perform a longterm and stable numerical simulation. With regard to the slicing condition, the maximal slicing condition, in which the trace of the extrinsic curvature is zero, has been recognized to be an appropriate choice for singularity avoidance as well as for stable numerical simulation. However, an appropriate spatial gauge condition in 3D numerical relativity had not been investigated until quite recently. In 1978, Smarr and York proposed the minimal distortion gauge condition by which growth rate of the coordinate distortion in the three metric is likely to be minimized [40]. Community of numerical relativity has recognized that this would be an appropriate gauge condition, but it has not been tried to impose because of complexity of its equation and because of its expensive computational cost. Recently, several attempts for imposing this condition have been done by several groups [41,17,42]. In particular, Shibata [41] demonstrated that a kind of minimal distortion gauge is very useful to perform a longterm simulation of binary systems suppressing the coordinate distortion.

In addition to the above theoretical and numerical progress, computational resources have been significantly improved in the last few years. Currently, it is possible to use a supercomputer of several 100 GBytes for memory and several 100 GFlops in computational speed. For example, in Japan, we currently use FACOM VPP5000 in which the maximum available memory is about 700 GBytes. With this resource, it is possible to perform a large-scale numerical simulation with a grid size as $(500, 500, 500)$ for (x, y, z) in about 100 CPU hours. Even using a uniform grid with covering the neutron star diameter by about 30 grid points, the outer boundaries are located at $\sim 50M$, implying that we can locate outer boundaries in the local wavezone.

4 Recent numerical results

We have performed numerical simulations for a wide variety of initial data sets adopting a Γ law equation of state in the form $P = (\Gamma - 1)\rho\varepsilon$, where Γ is an adiabatic constant. Initial conditions are prepared using the polytropic equations of state of adiabatic index Γ as $P = \kappa\rho^{\Gamma}$. Here, we present results for some models of $\Gamma = 2.25$ for example. More detailed results are presented in our latest paper [43].

In Table 2, we list several quantities which characterize quasiequilibrium states of irrotational binary neutron stars used as initial conditions. All quantities are dimensionless appropriately scaled with respect to κ in units $c = G = 1$. The scaled quantities are denoted with bar. We choose binaries at innermost orbits for which the Lagrange points appear at the inner edge of neutron stars [25]. Since the orbits of such binaries are stable in the framework of IWM formalism and our choice of parameters, we reduce the angular momentum slightly (by $\lesssim 1\%$) to accelerate merger. Note that the frequency of gravitational waves for these quasiequilibria is computed as

$$f_{\rm QE} = \frac{\Omega_0}{\pi} \approx 960{\rm Hz}\left(\frac{2.8M_{\odot}}{M_{\rm ADM0}}\right)\left(\frac{C_0}{0.12}\right)^{3/2}, \tag{7}$$

where Ω_0 denotes the angular velocity of quasiequilibrium states. Thus, the orbital period of the quasiequilibria, $P_{t=0}$, is about 2msec in assuming the total mass as $2.8M_{\odot}$. Here, C_0 is a compactness parameter of orbits defined as $C_0 \equiv (M_{\rm ADM0}\Omega_0)^{2/3} \sim M_{\rm ADM0}/a$, where a denotes initial orbital separation.

Simulations have been performed using a fixed uniform grid and assuming reflection symmetry with respect to the $z = 0$ plane (i.e., the equatorial plane is chosen as the orbital plane). The typical grid size is (505, 505, 253) for the (x, y, z) directions, respectively. The largest grid size so far is (633, 633, 317). The grid covers the region $-L \leq x, y \leq L$ and $0 \leq z \leq L$ where L is a constant. The grid spacing is determined from the condition that major diameter of each star is covered with 33 grid points initially. With (505, 505, 253) grid size, the computational memory required is about 120 GBytes, and the computational time for one model is about 100 CPU hours for about 10000

Model	$(M/R)_\infty$	$\bar{\rho}_{\max}$	$\bar{M}_*$	$\bar{M}_{\rm ADM}$	q	$\hat{P}_{t=0}$	C_0	R_τ	$R_{\rm mass}$	Prod.
(A)	0.12	0.139	0.186	0.173	1.02	233	0.0897	5.1	1.15	NS
(B)	0.14	0.169	0.216	0.198	0.97	182	0.106	3.4	1.33	BH
(C)	0.16	0.202	0.244	0.220	0.93	145	0.124	2.3	1.51	BH

Table 2. A list of several quantities for quasiequilibria of irrotational binary neutron stars with $\Gamma = 2.25$. $(M/R)_\infty$: compactness of each star in isolation; $\bar{\rho}_{\max}$: the maximum density; $\bar{M}_*$: total baryon rest-mass; $\bar{M}_{\rm ADM0}$: ADM mass at $t = 0$; $q = J_0/M_{\rm ADM0}^2$; $\hat{P}_{t=0} \equiv P_{t=0}/M_{\rm ADM0}$; $C_0 \equiv (M_{\rm ADM0}\Omega_0)^{2/3}$; $R_\tau = 5(M_{\rm ADM0}\Omega_0)^{-5/3}/128\pi$: ratio of the emission timescale of gravitational waves to the orbital period in the Newtonian order; $R_{\rm mass} \equiv M_*/M_{*\,\max}^{\rm sph}$: the ratio of the baryon rest-mass of each star to the maximum allowed mass for a spherical star. Products we found when we stopped simulations are shown in the last column. All quantities are normalized by κ appropriately to be dimensionless. Here, $M_{*\,\max}^{\rm sph}$ denotes the maximum allowed mass of a spherical star ($\bar{M}_{*\,\max}^{\rm sph} \approx 0.162$ at $\bar{\rho}_{\max} \approx 0.524$ for $\Gamma = 2.25$). BH and NS denote black hole and neutron star.

timesteps using FACOM VPP5000 in the data processing center of National Astronomical Observatory of Japan (NAOJ).

In Figs. 1 and 2, we display the density contour lines and velocity vectors for ρ and v^i at selected timesteps for simulations of models (A) and (C). In the merger for model (A), the product we found when we stopped the simulation is a massive neutron star. However, the merged object does not promptly settle down to an axisymmetric rotating neutron star. Soon after the onset of the merger, the merged object develops spiral arms (see fourth panel of Fig. 1), because it is rapidly rotating and hence appears to be dynamically unstable to formation of bar and spiral arms. The spiral arms do not widely spread outward but wind about the central core (see fifth panel). After the winding, the spiral arms develop again (see sixth panel) and subsequently wind about the core soon (see seventh panel). After repeating this process for several times, the merged object gradually settles down to an ellipsoidal rotating star (see eighth and ninth panels). During these oscillations, gravitational waves of fairly large amplitude are excited, and the emission will eventually trigger the collapse of this massive neutron star to a black hole (see below).

The total baryon rest-mass of the binary for model (A) is $\approx 16\%$ larger than the maximum allowed value of a spherical star in isolation. Even with such a large mass, the merged object does not collapse to a black hole within a couple of dynamical timescales. Since the merger proceeds very mildly, the shock heating does not appear to play an important role in merging. Indeed, a function, $P/(\kappa\rho^\Gamma)$, which indicates the change of the entropy, increases by at most 10% around the central region of the merged object. This implies that the rotational centrifugal force plays an important role for supporting the self-gravity of a massive neutron star. In Fig. 3, the angular velocity along x

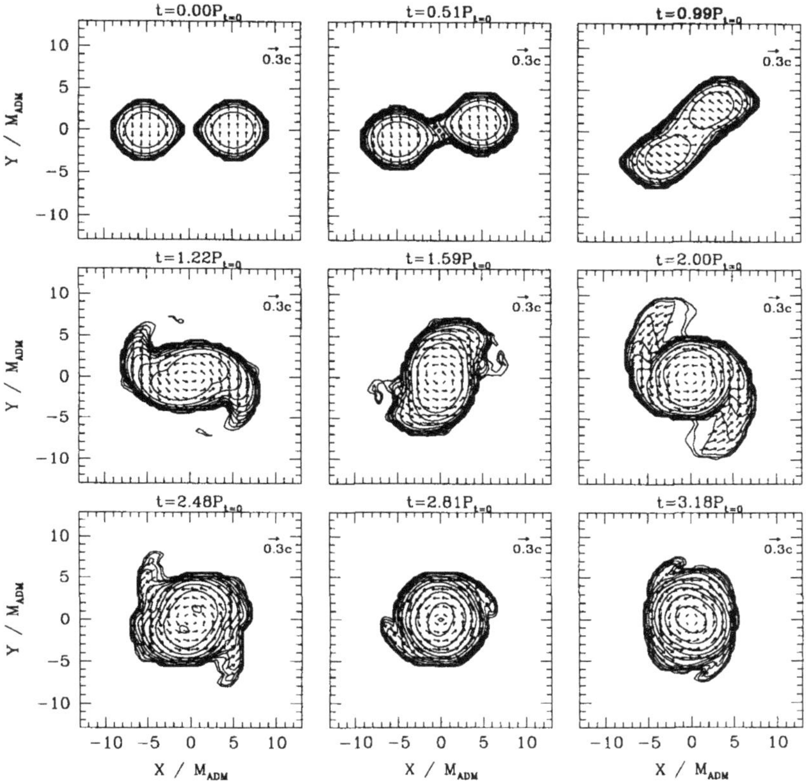

Figure 1. Snapshots of the density contours for ρ in the equatorial plane for model (A). The contour lines are drawn for $\rho/\rho_0 = 10^{-0.3j}$, where ρ_0 denotes the maximum value of ρ_0 at $t = 0$ (here $\rho_0 = 0.139\kappa^{-n}$), for $j = 0, 1, 2, \cdots, 10$. Vectors indicate the local velocity field (v^x, v^y) and the scale is shown in the top left-hand frame. $P_{t=0}$ denotes the orbital period of the quasiequilibrium configuration given at $t = 0$. The length scale is shown in units of $GM_{\rm ADM0}/c^2$ where $M_{\rm ADM0}$ is the gravitational mass at $t = 0$. Note that $t = 3.18P_{t=0} \approx 740M_{\rm ADM0}$.

and y-axes ($|v^y/x|$ and $|v^x/y|$) are displayed. It shows that the formed massive neutron star is differentially rotating, and the magnitude of the angular velocity is of order of the Kepler velocity, i.e., $\Omega M_{\rm ADM}(R/M_{\rm ADM})^{3/2} = O(1)$ where R denotes a characteristic radius of the merged object ($\approx 5M_{\rm ADM}$).

For models (B) and (C) in which $(M/R)_\infty \geq 0.14$, a black hole is formed within $t \lesssim 3P_{t=0}$. In both cases, we were able to locate apparent horizons at the last stage of simulations using our apparent horizon finder [44]. However, the formation process is slightly different among these models: For model (C),

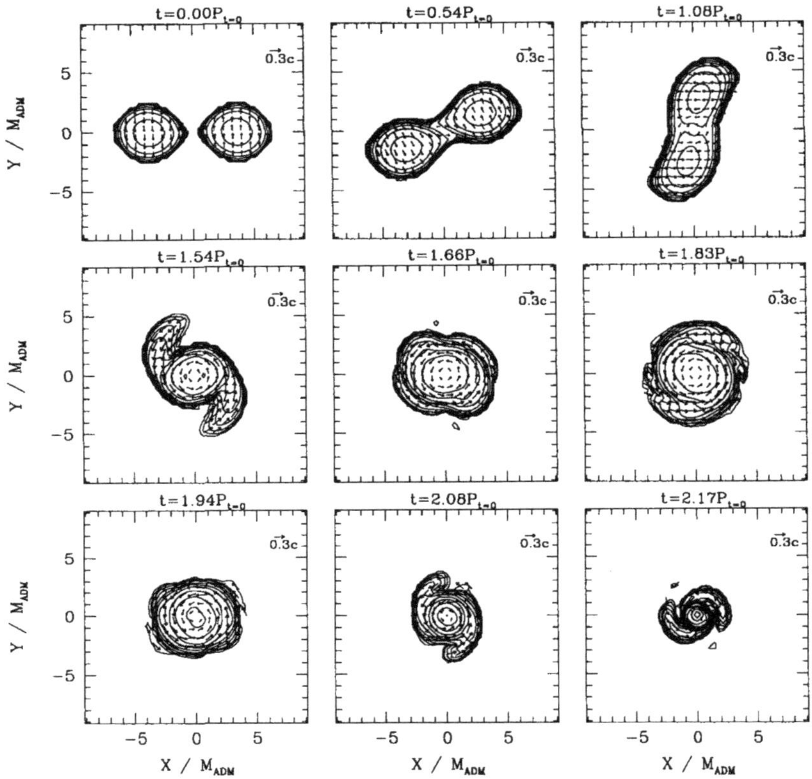

Figure 2. The same as Fig. 1, but for model (C). The contour lines are drawn for $\rho/\rho_0 = 10^{-0.3j}$ for $j = 0, 1, 2, \cdots, 10$. $\rho_0 = 0.52\kappa^{-n}$ which is the critical density of stability for spherical stars against gravitational collapse. The thick solid circle in the final panel denotes the apparent horizon.

the formation timescale is short. After the merged object once experiences a bounce for a short time (see fifth and sixth panels in Fig. 2) it soon collapses to a black hole. For model (B), the formation timescale to a black hole is much longer. In this case, the merged object quasi-radially oscillates for a couple of times after the first contact. The collapse toward a black hole likely sets in after dissipating the angular momentum by gravitational radiation. (Indeed, the angular momentum decreases by $\sim 10\%$ from the late inspiral to the merger stages. Also see Eq. (8).) The difference with regard to the formation process of black holes is significantly reflected in the waveform of gravitational waves (*cf.* Fig. 4) and the Fourier spectrum (*cf.* Fig. 6).

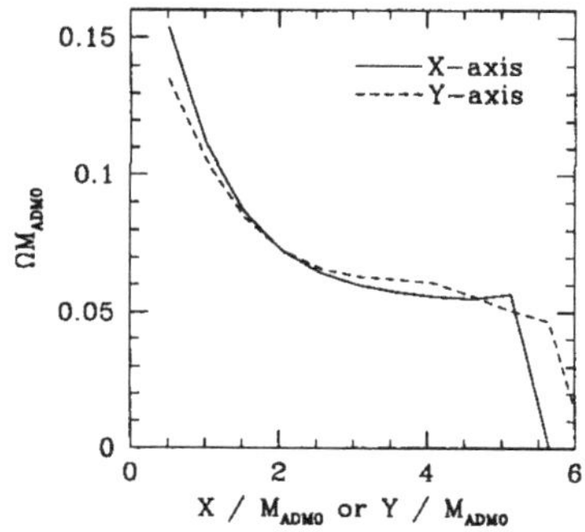

Figure 3. The angular velocity in units of M_{ADM0}^{-1} along x and y axes at $t/P_{t=0} = 3.18$ for model (A).

A noteworthy result in the late stage for models (B) and (C) is that the baryon rest-mass fraction outside the apparent horizon is less than 1% of the total rest-mass. This implies that the fraction of the baryon rest-mass in the disk around the black hole is very small. The main reason is that the specific angular momentum of all fluid elements before merger sets in is too small to form a disk [25]. Almost all fluid elements do not have a large angular momentum to avoid being swallowed by the central black hole. Similar conclusion has been drawn by post-Newtonian [10] and semi-relativistic simulations [12], recently.

In Fig. 4, we show $\bar{h}_+$ and $\bar{h}_\times$ as a function of a retarded time $(t - z_{\mathrm{obs}})/P_{\mathrm{t=0}}$ for models (A), (B) and (C). The waveforms are quite similar among three cases in the early merger stage $t - z_{\mathrm{obs}} \lesssim P_{t=0}$, but they vary as the merger proceeds toward final outcomes.

In the case of massive neutron star formation, quasi-periodic gravitational waves of a fairly large amplitude, which are excited due to the nonaxisymmetric deformation and oscillation of the merged object, are emitted for a long timescale after the merger. The wavelengths of these quasi-periodic waves are likely associated with several fundamental oscillation modes of the merged object, because the quasi-periodic oscillation is not a simple sine-cosine curve but has a modulation. Since the radiation reaction timescale is much longer than the dynamical (rotational) timescale of the massive neutron star, the quasi-periodic waves will be emitted for many rotational cycles.

Even for the case of black hole formation, quasi-periodic gravitational waves are excited due to the nonaxisymmetric oscillation of a transient merged object before collapsing to a black hole. Since the formation timescale of the

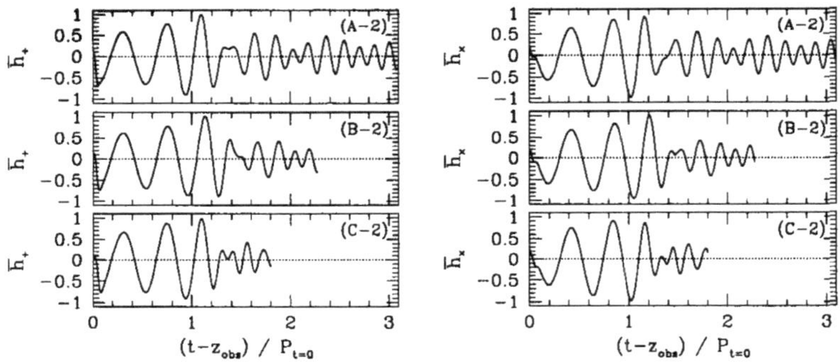

Figure 4. $\bar{h}_+$ and $\bar{h}_\times$ as a function of retarded time for models (A), (B) and (C).

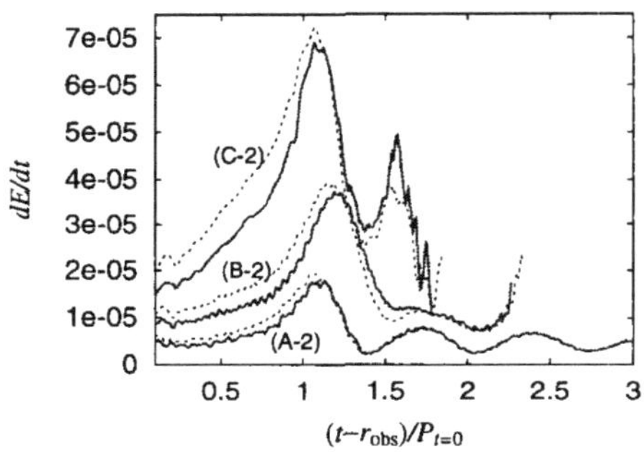

Figure 5. Luminosity of gravitational waves as a function of retarded time for models (A), (B) and (C). Solid and dashed lines denote the luminosity computed at $r_{\rm obs} \approx L$ and $0.77L$, respectively. $r_{\rm obs}$ denotes the radius at which the gauge invariant variables are extracted. The unit of the luminosity is $c^5/G \approx 3.63 \times 10^{59}$erg/sec. It is difficult to obtain a smooth curve of dE/dt because computing the energy luminosity requires to take time derivative of the gauge invariant quantity, which introduces a fairly large numerical noise.

black hole is different between models (B) and (C) depending on initial compactness of neutron stars, duration for emission of the quasi-periodic waves is also different. Because the computation crashed soon after the formation of the apparent horizon, we cannot draw definite conclusion from the present simulation with regard to gravitational waves in the last stage. However, we can expect that after the formation of black holes, their quasi normal modes (QNMs) are excited and gravitational waves will damp eventually. To complete waveforms up to the ringing tail of QNMs, we need either to develop an implementation such as the adaptive mesh refinement [46] and/or a horizon excision or to use a perturbative extraction technique developed by Baker et al.[47]

The properties mentioned above are also reflected in the energy luminosity, as shown in Fig. 5. Here, we display the luminosity for $l = |m| = 2$ modes as a function of the retarded time for (A), (B) and (C). Typical magnitude of the total radiated energy $\Delta E/M_{\rm ADM}$ is less than 1%. On the other hand, typical magnitude of $\Delta J/J_0$ is $\sim 10\%$, which is consistent with the decrease of the angular momentum J. For model (C) in which a black hole is quickly formed, a sharp peak is found before the computation crashes. This peak is likely associated with an oscillation for a short-lived transient merged object. The peaks associated with QNMs is expected to appear after this peak, and the luminosity damps eventually. For model (B), a long-lived transient object emits quasi-periodic gravitational waves for a certain duration. Consequently, the luminosity remains fairly large for a longer timescale than for model (C).

For model (A), the evolution of the luminosity is very different from that of other models after the first peak is reached. In this case, the luminosity oscillates with average amplitude $dE/dt \sim 5 \times 10^{-6}$ for a long time. Assuming that the energy luminosity remains to be $\sim 10^{-6}$ as indicated in Fig. 5 and that the emission timescale is approximately derived by dividing the kinetic energy T by dE/dt, we obtain

$$t_{\rm GW} \sim \frac{T}{dE/dt} \sim 0.1{\rm sec}\left(\frac{M}{3M_{\odot}}\right)\left(\frac{5M}{R}\right)\left(\frac{10^{-6}}{dE/dt}\right). \tag{8}$$

where we assume that T is about 0.1–0.2 times of the binding energy which could be approximated by $\sim GM^2/R$. Thus, within ~ 0.1 second, the massive neutron star will collapse to a black hole. As argued in Ref. 44, there are many processes of dissipation or transportation of the angular momentum inside the massive neutron star, besides the emission of gravitational waves. However, characteristic timescale for such processes is much longer than 1 second. Thus, the emission timescale of gravitational waves would be shortest in this case. On the other hand, if $t_{\rm GW}$ were longer than 10 seconds, other processes such as magnetic braking effect could be more important than the effect of gravitational radiation [45].

In Fig. 6, the spectrum of gravitational waves in the merging stage is displayed. Since the computation crashed soon after formation of the black hole, the spectrum of models (B) and (C) are incomplete for high frequency side, $f_{\rm GW}/f_{\rm QE} \gtrsim 4$, where $f_{\rm QE}$ is given in Eq. (7). In these cases, QNMs of black holes will be excited at the stage of black hole formation. However, the frequency would be very high as $f_{\rm GW} > 5{\rm kHz} \sim 5 f_{\rm QE}$ [48]. Irrespective of models, we should also point out that in the real spectrum of the coalescing binaries, $\bar{h}_{+,\times} \propto f_{\rm GW}^{-7/6}$ for $f_{\rm GW} \lesssim f_{\rm QE}$ [49], which is not taken into account in the present results. Here we focus only on the spectrum for quasi-periodic

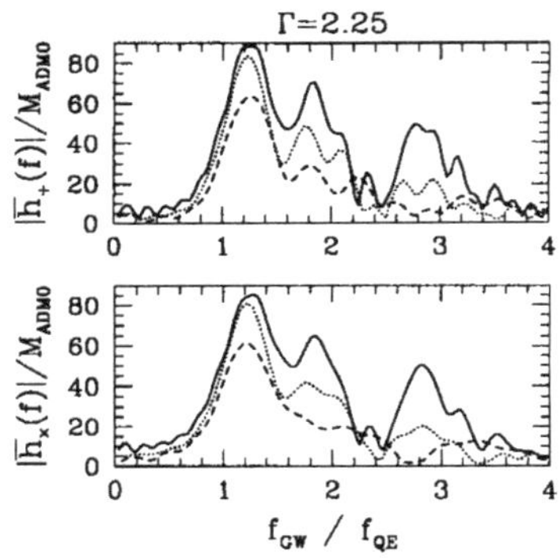

Figure 6. Fourier spectrum of $\bar{h}_+$ and $\bar{h}_\times$ for models (A) (solid line), (B) (dotted line) and (C) (dashed line) (left).

oscillations of merged objects for which $f_{\rm GW} \sim (2-3)$kHz. For the spectrum of model (A), one should keep in mind that the peak associated with the quasi-periodic oscillations is underestimated in Fig. 6 because the quasi-periodic waves will be emitted for a long time to $t \gg P_{t=0}$ as indicated in Eq. (8).

From Fig. 6, we find two typical frequencies of the quasi-periodic oscillation which are $\sim 1.8 f_{\rm QE}$ and $\sim 2.8 f_{\rm QE}$ for $\Gamma = 2.25$ [43]. As we explained above, the peaks of the Fourier spectrum which are associated with the quasi-periodic oscillations are determined by its accumulated cycles. Namely, if we would not find the large peak, we could judge that a black hole is formed in a short timescale after merger. This implies that even without detecting QNMs or other signals of the black hole, we will be able to obtain an evidence for black hole formation. On the other hand, if we would find a large peak, we could conclude that a massive neutron star is formed at least temporarily.

Since $f_{\rm QPO}$ is rather high, it may be difficult to detect the quasi-periodic oscillation by first generation, kilo-meter-size laser interferometers such as LIGO I. However, the resonant-mass detectors and/or specially designed advanced interferometers such as LIGO II may be available to detect such high frequency gravitational waves in future. These detectors will provide us a wide variety of information on neutron star physics.

5 Summary

As demonstrated in this article, fully GR simulations for binary neutron star merger are feasible stably and fairly accurately to clarify the final products (black hole or neutron star), to estimate disk mass around the final objects, and to compute gravitational waveforms.

However, there remains many issues to be performed and many technical problems to be improved or solved in future. The following is a list for them.
(1) So far, we have used quasiequilibrium states of *conformally flat three-metric* as initial conditions for simplicity. The conformal flatness approximation becomes a source of a certain systematic error for obtaining realistic quasiequilibrium states. As a result, this approximation would introduce a systematic error on initial condition and subsequent merger simulation. Since magnitude of the neglected terms in the initial conditions obtained by the conformal flatness approximation seems to be small, we expect that this effect is not very serious. However, it is not obvious. To exclude such question, it is necessary to perform simulations using more realistic quasiequilibrium states of generic three geometry as initial conditions [50].
(2) So far, numerical simulations have been performed for binaries of equal mass neutron star. For unequal mass, merger process is likely modified even qualitatively. In this case, a neutron star of smaller mass could be tidally disrupted before the orbital separation becomes sufficiently small. Then, the tidally disrupted object likely forms a disk around the massive neutron star while the massive one collapses to a black hole after accreting a fraction of matter from the tidal debris. With this change of the merger process, gravitational waveforms could be modified significantly.
(3) Simulations have been performed using simple polytropic equations of state. Gravitational waveforms and their characteristic frequency depend on the stiffness of equations of state as shown by Newtonian and post-Newtonian simulation [5,10,12] as well as by GR simulations [43]. Obviously, it is necessary to adopt realistic equations of state to compute more realistic gravitational waveforms to prepare for theoretical templates.
(4) Even using present supercomputers with about several hundred GBytes memory, we can take at most $(10^3, 10^3, 10^3)$ for the grid number. This implies that if we use a uniform grid, outer boundaries are located in the local wavezone or in the distant nearzone. As a result, gravitational waveforms are contaminated by unphysical boundary conditions imposed at the local zone. To avoid such numerical effect, it seems to be necessary to develop certain numerical technique such as the nested grid technique to enlarge the computational region. Also, the grid spacing would not be small enough to resolve the black hole forming region accurately. To improve the resolution, the adaptive mesh refinement technique would be necessary [46].

Acknowledgments

M. S. would like to thank scientific organizers of the 16th international conference on General Relativity and Gravitation for invitation. Numerical computations were performed on the FACOM VPP 5000 machines in the data processing center of NAOJ. This work was in part supported by a Japanese Monbu-Kagaku-sho Grant (No. 13740143) and NSF Grant PHY00-71044.

References

1. R. A. Hulse and J. H. Taylor, Astrophys. J. **201** (1975), L55: J. H. Taylor and J. M. Weisberg, Astrophys. J. **345** (1989), 434.
2. V. Kalogera, R. Narayan, D. N. Spergel and J. H. Taylor, astro-ph/0012038. See also E. S. Phinney, Astrophys. J. **380**, L17 (1991); R. Narayan, T. Piran and A. Shemi, Astrophys. J. **379**, L17 (1991).
3. For example, K. S. Thorne, in *Proceeding of Snowmass 95 Summer Study on Particle and Nuclear Astrophysics and Cosmology*, eds. E. W. Kolb and R. Peccei (World Scientific, Singapore, 1995), p. 398, and references therein: M. Ando et al., (the TAMA collaboration), Phys. Rev. Lett. **86** (2001), 3950.
4. L. Blanchet, this volume: L. Blanchet, G. Faye, B. R. Iyer and B. Joguet, gr-qc/0105098 & 0105099, and references cited therein.
5. X. Zhuge, J. M. Centrella, and S. L. W. McMillan, Phys. Rev. D **50**, 6247 (1994); **54**, 7261 (1996).
6. C. Cutler et al., Phys. Rev. Lett. **70**, 1984 (1993).
7. F. Rasio and S. L. Shapiro, Astrophys. J. **401** (1992), 226; **432** (1994), 242.
8. K. Oohara, T. Nakamura and M. Shibata, Prog. Theor. Phys. Supple. **128** (1997), 183, and references cited therein.
9. M. Ruffert, H.-Th. Janka and G. Schäfer, Astron. Astrophys. **311** (1996), 532: Their most recent work is, M. Ruffert and H.-Th. Janka, astro-ph/0106229 and see also references cited therein.
10. J. A. Faber and F. A. Rasio, Phys. Rev. D **62** (2000), 064012; **63** (2001), 044012.
11. S. Ayal, T. Piran, R. Oechslin, M. B. Davies and S. Rosswog, Astrophys. J. **550** (2001), 846.
12. R. Oechslin, S. Rosswog, and F. Thielemann, gr-qc/0111005.
13. M. Shibata and K. Uryū, Phys. Rev. D **61** (2000), 064001: in proceedings of Gravitational Waves: A Challenge to Theoretical Astrophysics, edited by V. Ferrari, J. C. Miller and L. Rezzolla (The Abdus Salam ICTP Publications, 2001), 137 : in proceedings of 20th Texas Symposium on Relativistic Astrophysics, edited by J. C. Wheeler and H. Martel, to be published (astro-ph/0104409).
14. A. Buonanno and Y. Chen, personal communication.
15. T. Piran, this volume for review.
16. M. Shibata, Phys. Rev. D **60** (1999), 104502.
17. K. Oohara and T. Nakamura, Prog. Theor. Phys. **136** (1999), 270.
18. J. Font et al., gr-qc/0110047 and references cited therein.
19. e.g., S. L. Shapiro and S. A. Teukolsky, *Black Holes, White Dwarfs, and Neutron Stars*, Wiley Interscience (New York, 1983).
20. C. S. Kochanek, Astrophys. J. **398**, 234 (1992): L. Bildsten and C. Cutler, Astrophys. J. **400**, 175 (1992).

21. M. Shibata, Phys. Rev. D **58** (1998), 024012: S. A. Teukolsky, Astrophys. J. **504** (1998), 442: See also, S. Bonazzola, E. Gourgoulhon and J.-A. Marck, Phys. Rev. D **56** (1997), 7740: H. Asada, Phys. Rev. D **57** (1998), 7292.
22. J. Isenberg and J. Nester, in *General Relativity and Gravitation* Vol.1, ed. A. Held, (Plenum Press, New York 1980): J.R. Wilson and G.J. Mathews, in *Frontiers in Numerical Relativity*, ed. C.R. Evans, L.S. Finn, and D.W. Hobill, (Cambridge University Press, Cambridge, England, 1989).
23. K. Uryū and Y. Eriguchi, Phys. Rev. D **61** (2000), 124023.
24. E. Gourgoulhon et al., Phys. Rev. D **63** (2001), 064029. See also references cited therein.
25. K. Uryū, M. Shibata and Y. Eriguchi, Phys. Rev. D **62** (2000), 104015.
26. J. F. Friedman, K. Uryū, and M. Shibata, Phys. Rev. D to be published (gr-qc/0108070).
27. M. D. Duez, T. W. Baumgarte and S. L. Shapiro, Phys. Rev. D **63** (2001), 084030: M. D. Duez et al., Phys. Rev. D to be published (gr-qc/0110006).
28. M. Shibata and K. Uryū, Phys. Rev. D **64** (2001), 104017.
29. N. Bishop et al., Phys. Rev. Lett. **76** (1996), 4303: A. M. Abrahams et al., Phys. Rev. Lett. **80** (1998), 1812: L. Rezzolla et al., Phys. Rev. D **59** (1999), 064001.
30. For a review on initial data for numerical relativity, see e.g., Cook, G., Living Rev. Relativity 3, (2000), 5. [Online article]: cited on 15 Aug 2001 http://www.livingreviews.org/Articles/Volume3/2000-5cook/
31. R. Arnowitt, S. Deser, and C. W. Misner, in *Gravitation: An Introduction to Current Research*, ed. L. Witten (New York: Wiley, 1962): J. W. York, Jr., in *Sources of Gravitational Radiation*, edited by L. Smarr (Cambridge University Press, 1979), 83.
32. L. Lehner, this volume for review.
33. T. Nakamura et al., Prog. Theor. Phys. Supple. **90**, chapter 1.
34. M. Shibata and T. Nakamura, Phys. Rev. D **52** (1995), 5428.
35. T. W. Baumgarte and S. L. Shapiro, Phys. Rev. D 59 (1999), 024007.
36. M. Alcubierre et al., Phys. Rev. D **61** (2000), 041501; **62** (2000), 044034.
37. M. Alcubierre et al., Phys. Rev. D **64** (2001), 064501.
38. E. Seidel, in proceedings of *Gravitational Waves: A challenge to theoretical astrophysics* eds. V. Ferrari, J. C. Miller and L. Rezzolla (The Abdus Salam ICTP Publications, 2001), p.113.
39. M. Shibata, T. W. Baumgarte, and S. L. Shapiro Phy. Rev. D **61** (2000), 044012.
40. L. Smarr and J. W. York, Phys. Rev. D **17** (1978), 1945.
41. M. Shibata, Prog. Theor. Phys. **101** (1999), 1199.
42. B. Brügmann, personal communication.
43. M. Shibata and K. Uryū, Prog. Theor. Phys. **107** (2002).
44. M. Shibata, Phys. Rev. D **55** (1997), 2002: M. Shibata and K. Uryū, Phys. Rev. D **62** (2000), 087501.
45. T. W. Baumgarte, S. L. Shapiro, and M. Shibata, Astrophys. J. Lett. **528** (2000), L29.
46. M. Berger and J. Oliger, J. Comp. Phys. **53** (1984), 484.
47. J. Baker, M. Campanelli and C. Lousto, gr-qc/0104063: J. Baker et al., Phys. Rev. Lett. **87**, 121103 (2001).
48. E. W. Leaver, Proc. R. Soc. London **A402** (1985), 285.
49. C. Cutler and E. E. Flanagan, Phys. Rev. D **49** (1994), 2658.
50. K. Uryū and M. Shibata, in preparation.

Public Lecture

THE LIGHTER SIDE OF GRAVITY

JAYANT V. NARLIKAR
Inter-University Centre for Astronomy and Astrophysics, Post Bag 4, Ganeshkhind, Pune 411 007, INDIA

This talk presents a light-hearted look at the phenomenon of gravity, the most enigmatic of all known natural interactions. The fact that a major international society is wholly devoted towards understanding this basic interaction of nature is ample testimony to its intellectual challenges. The GR-meetings are held once every three years to bring together workers in the field to share their ignorance as well as expertise in the field of gravity. It would be presumptuous on my part to attempt to tell anything new to the experts gathered here. This evening I will stay away from the complexities and take a light hearted look at some ideas and phenomena that make gravity so peculiar.

1 A (pseudo-)Historical Interlude

Perhaps, I should begin at the beginning. But where was the beginning? Much has been said and debated about the apple that fell on Newton's head, starting thereby a chain of thoughts that led him to the genesis of the law of gravitation. Although some argue that that particular event never happened, it certainly has added to the folklore on gravity.

However, a few years ago, one of our distinguished colleagues from India named Channapatna Venkataramaya Vishweshwara, whom the relativists not surprisingly prefer to address as 'Vishu', made a thought-discovery of historical importance (–if thought experiments are allowed in science, why not thought-discoveries in the history of science?–). He made this discovery at a gravity conference in the Indian coastal resort of Goa, and I can do no better than quote him verbatim on this topic:

"*...Many were the reasons for selecting Goa as the venue for a conference on gravitation. For instance, the natural beauty of the land with its emerald sea, the azure sky and the vast stretches of the golden sand. The warm hospitality and open friendliness of the people. A fascinating culture in which the East and the West have mingled together. In addition to all these, there was a historical reason as well.*

"*Legend has it that a sage belonging to this region discovered the universal law of gravitation some three hundred years before Isaac Newton. It so happens that there are hardly any apple trees in Goa, but one can find coconut groves all around. Consequently the discovery of the law of gravitation by our sage was occasioned by the fall of a coconut, as he sat in a contemplative mood. Needless to add, the world remained ignorant of his finding. This was indeed the first authentic case in unrecorded history of perishing without publishing.*"

This episode of unconfirmed historical authenticity will set the tone of my

Figure 1. The episode leading to the 'discovery' of the law of gravitation depicted by C.V. Vishveshwara.

talk, which will end with another episode whose authenticity is also questioned by some historians.

2 The Experiment of Tweedledum and Tweedledee

Let us continue with another interesting thought experiment, this time in science, by two relativists, Hermann Bondi and William McCrea. The experiment concerns two remarkable creatures called Tweedledum and Tweedledee, so named after the famous characters in Lewis Carroll's Alice novel, *Through the Looking Glass*. In the Bondi-McCrea experiment, these characters are made of pliable material that allows them to change shape. Originally, they had identical spherical shapes and were set moving around each other in highly eccentric orbits under their mutual gravitational force. They were given an order that they must always move along these orbits. Let us assume further that both Tweedledum and Tweedledee always have shapes symmetrical about axes perpendicular to the plane of their orbits.

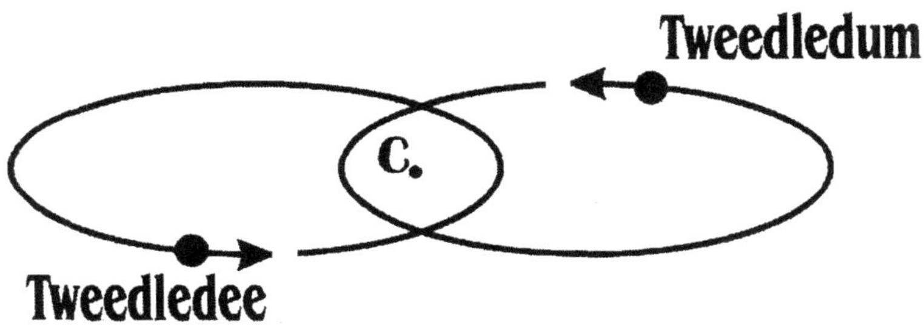

Figure 2. The highly eccentric elliptical orbits of Tweedledum and Tweedledee.

The orbits are identical, and the two move along their respective orbits in such a way that the point C midway between their centres is fixed. This is characteristic of the way any two objects move in each other's gravitational attraction. Even in the case of the Earth and the Sun, although we usually talk of the motion of the Earth only, the Sun should in principle also move under the Earth's gravitational force. However, in the Earth-Sun system, because the Sun's mass greatly exceeds that of the Earth, the Sun's motion is negligible. The centre of mass of the Earth and the Sun in fact lies within the Sun.

But to return to Tweedledum and Tweedledee: they would have no difficulty obeying the command to move along their assigned orbits if they were rigid creatures. Since they are pliable, their shapes get distorted by each other's tidal forces. Let us understand the tidal effect first.

In Figure 3, we have two initially spherical objects A and B. What happens to B if it is subject to A's gravitational force? Because of this force, in relation to B's centre, its part nearest to A will bulge in the direction of A, while its part away from A will bulge away from A. In fact it is because of this bulging tendency that the oceans rise and undergo tides under the

Moon's gravitational attraction on the Earth. Now, if B is also constrained to be symmetrical about an axis perpendicular to this direction (as Tweedledum and Tweedledee are), it will bulge all along its equator, as shown in Figure 3. The same happens to A under the tidal force of B.

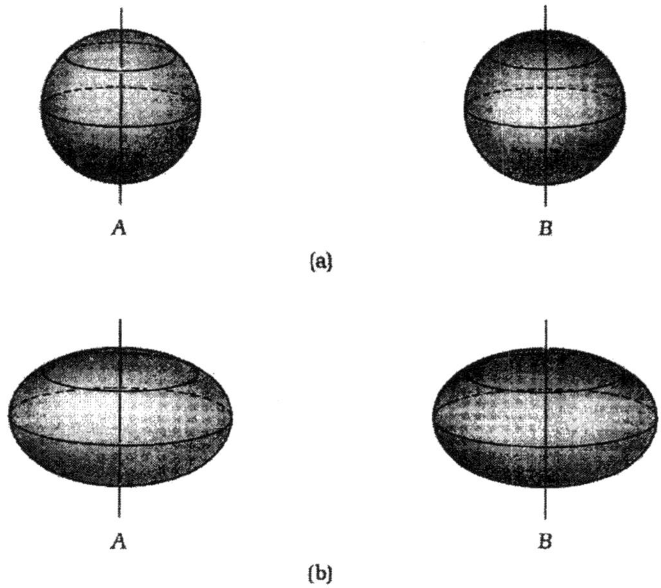

Figure 3. Initially, as in (a) the shapes of A and B are spherical. Because of their gravitational interaction both bulge out as in (b), attaining oblate spheroidal shapes.

When a sphere bulges along its equator as in Figure 3, it becomes an *oblate spheroid*. Had B instead been elongated at the poles, it would have become a *prolate spheroid*, as shown in Figure 4. An oblate spheroid is bun-shaped, and a prolate spheroid is egg-shaped, more like a rugby ball.

In the situation described in Figure 3, when the two spheres become oblate spheroids their mutual gravitational attraction *increases*. This is because A attracts B more powerfully as an oblate spheroid than as a sphere. This is the reason why the job of Tweedledum and Tweedledee is made more difficult - as they go around each other, their shapes tend to become oblate and the force between them increases, an effect which in turn tends to distort their orbits. However, remember that they are under strict instructions not to allow their

Figure 4. Typical oblate and prolate spheroids.

orbits to change. To overcome this difficulty, one course is open to them. If one of them becomes oblate, the other must become prolate, because the prolate spheroid exerts less force on its companion than a sphere. Thus, they can manage to keep their mutual force of attraction unchanged and thereby keep their orbits unchanged.

Now Tweedledum is clever and unscrupulous whereas Tweedledee is naive. Tweedledum uses his intelligence to capitalize on the situation in the following way. He makes Tweedledee sign an unfair agreement stating that Tweedledum will take the initiative in this shape-change operation: whenever Tweedledum becomes oblate, Tweedledee must become prolate, and vice versa. The exact shape that Tweedledee must acquire in order to keep their mutual gravitational foci unchanged, will be calculated and communicated to him by Tweedledum. The bottom line of this calculation is the criterion that their orbits must not change. Not knowing how to make this calculation himself, Tweedledee simply signs along the dotted line. This is a mistake for which he will have to pay dearly, as we will now see.

Moving in highly eccentric orbits, Tweedledum and Tweedledee alternately come close together and move farther apart. In Figure 5a, they are shown near each other. At this stage, there are huge tidal forces between them, and here Tweedledum allows himself to become oblate. To achieve, this, he of course does not have to do any work; as discussed earlier, the tidal force does the work of changing his shape. *And this work done by an out-*

side force increases Tweedledum's energy reservoir. What about Tweedledee? He must go into an appropriate prolate shape as per the agreement he has signed. To do this, he has to work against the tidal force, and spend energy. Consequently, his energy reservoir gets depleted.

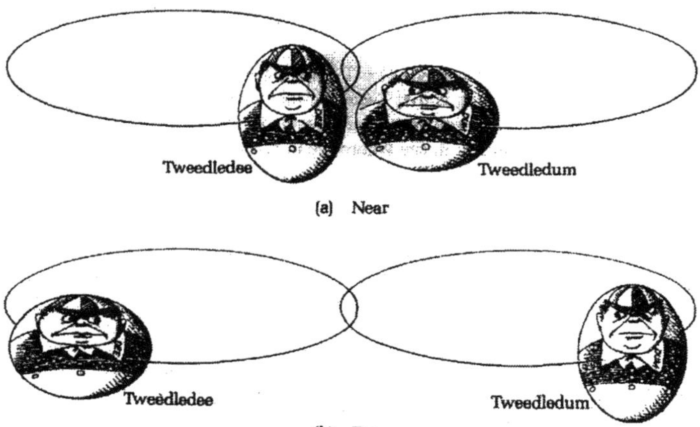

Figure 5. The shape changes produced by Tweedledum and Tweedledee when they come close, as in (a) and move apart, as in (b).

Meanwhile, the two have started moving apart. When their separation is close to the maximum, as Figure 5b, Tweedledum changes himself to a prolate shape, thereby making Tweedledee go oblate. However, when the two are far apart, the tidal force between them is weak. Hence, even though Tweedledum does some work to become prolate, he has to spend very little energy doing so. Correspondingly, although Tweedledee gains some energy because this time the tidal forces *help* him change his shape, the energy gained by him is small.

We now see why this agreement is unfair. During each orbit, Tweedledum alternately gains a lot of energy and spends very little, while Tweedledee alternately loses a lot of energy and gains very little. What is in fact is happening, is that Tweedledum is cleverly using the tidal force to extort energy from poor Tweedledee, although the latter cannot, on the face of it, see where the unfairness of the agreement lies. You will readily see a parallel between this scenario and the occasional unfair trade agreements signed between countries!

3 Tidal Disruption

The crucial property of the tidal force that leads to the remarkable effects of the Tweedledum-Tweedledee experiment is that its strength diminishes with distance very rapidly. In fact, its strength diminishes in inverse proportion to the *cube* of the distance. Thus, if the farthest distance of separation between Tweedledum and Tweedledee is *ten* times the distance of their closest separation, the tidal force in the former case is *one-thousandth* of what it would be in the latter case. This is why Tweedledee had to work so much harder than Tweedledum in changing from oblate to prolate form.

The name 'tidal force' comes from its manifestation in the Earth-Moon system, where the Moon's gravity causes the Earth to bulge by the seas rising, especially at the Full and new Moon, when the effect of the Sun's gravity also contributes in the same direction as the Moon's. However, as mentioned before, the tidal force falls as the inverse third power of distance from the gravitating object, which is why the Moon, despite its lower mass is far more effective than the Sun in causing tides on the Earth, because it happens to be much closer to the Earth. We therefore expect the tidal force to be very powerful in circumstances where two astronomical objects are close to each other. How close can a planet be to its parent star? How close can a satellite come to the planet around which it revolves? The answers to these questions invariably involve calculations of the tidal force. If a satellite comes too close to a planet, the tidal force may become so enormous that it tears the satellite apart. The same consideration applies to a planet near its parent star. The limit beyond which the tidal forces become highly disruptive is known as the *Roche limit*, so named after the work of Eduard Roche. The rings around Saturn indicate a 'failed' satellite because of the strong tidal force of the planet.

Another circumstance in which the tidal force becomes significant is that of a binary star system. Like Tweedledum and Tweedledee, two stars in a binary system go around each other in elliptical orbits. A typical binary system is shown in Figure 6a. The stars A and B (which need not be and usually are not of identical mass) orbit around their common centre of mass. The dashed line shows the so-called *Roche lobe*. If either star becomes large enough to fill and overflow its Roche lobe, the tidal force of its companion will begin to make its disruptive presence felt.

Now, astrophysicists find that as a star like the Sun evolves, in due course it passes through the red-giant phase, when its physical size becomes very large. If one of the stars in our binary system becomes a red giant, its outer surface begins to spill over its Roche lobe, and the situation shown in Figure

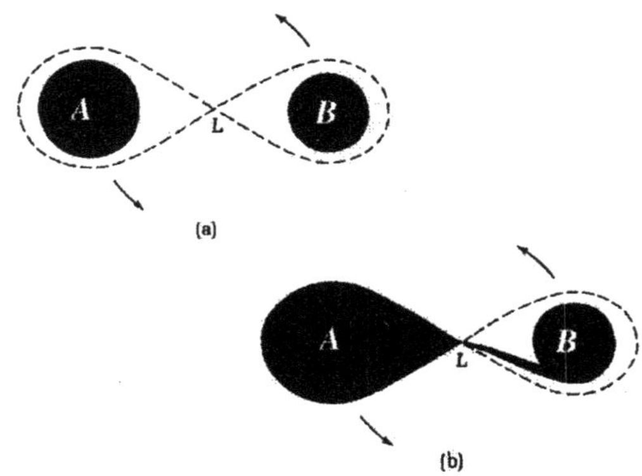

Figure 6. The dashed curve indicates the Roche Lobe of the binary system. At the Lagrangian point L the gravitational pulls of A and B balance each other. In (b) we see how star A expands and loses surface material to B which flows across the Roche Lobe towards B.

6b will occur. The companion star B now starts exerting a tidal force on A in such a way as to pull material from the surface of A that is closest to it. This surface, therefore, becomes distorted, and the material 'falls' towards B.

On an even grander scale, astronomers are now encountering examples of tidal interaction between galaxies, which leads to material being pulled out of one galaxy by another. This kind of interaction can lead to several interesting results like the birth of new stars, distortion of galactic shapes, emission of extra radiation by disturbed gas, etc. T he 'toadstool' shown in Figure 7 is an example of two-galaxy interaction.

Having described the gravitational interaction between matter and matter, let us now see whether the interaction makes matter attract light.

4 The Bending of Light

While Issac Newton was working on issues related to light, he had conjectured as to whether light is attached towards matter by the latter's gravitational force. He asked the following important question :

Do not Bodies act upon Light at a distance, and by their action bend its Rays; and is not this action (caeteris paribus) strongest at the least distance?

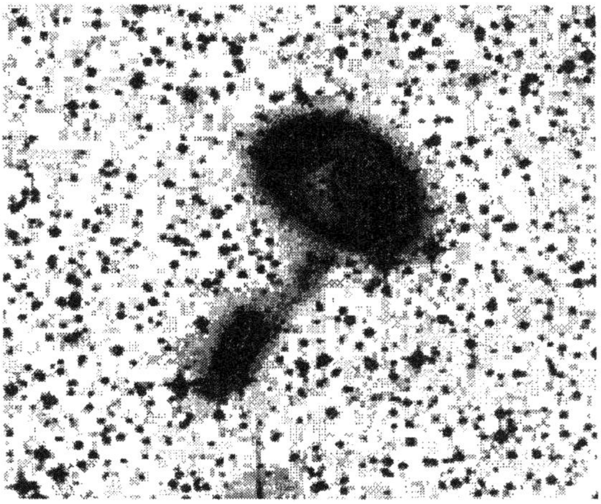

Figure 7. The 'toadstool' showing how one galaxy pulls material out of another when they are close enough.

— *Opticks, Query 1*

That the possibility of light being bent by gravity occurred to Newton was not surprising considering his genius, and also considering his belief that light is made of particles (which he called *corpuscles*). However, he had no experimental or observational means to settle his conjecture and so he left the issue unsettled.

4.1 A 'Newtonian' calculation

We can, however, use Newtonian ideas to work out how much, if any, bending light would suffer if it passed close to a massive body. Figure 8 illustrates the situation.

It shows the track of a typical particle coming from a great distance, with the speed of light. As it approaches the massive sphere, two possibilities arise. (i) It is immune to the force of gravity of the sphere and so continues along its original straight track. (ii) It *is* attracted by the spherical mass and bends round it and to emerge in another direction. The two possibilities are shown by (i) and (ii) in Figure 8.

We may consider (i) or (ii) as the possible outcome of Newton's conjecture about light. In the second alternative the bending can be calculated as the

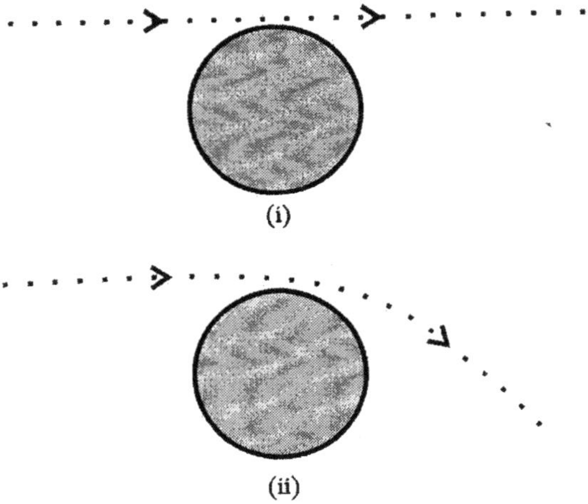

Figure 8. In (i) light is unaffected by the spherical mass, whereas in (ii) its direction is 'bent' by the gravitational pull of the mass. Which alternative represents reality?

angle between the two emerging directions (i) and (ii). The answer is the following : Multiply twice the value of Newton's gravitational constant by the mass and divide by the distance of closest approach by the ray to the centre of the sphere and also by the speed of light multiplied to itself. This gives the bending angle in radians.To express the angle in the conventionally better known unit of degrees, remember that π radians make 180 degrees. As we will shortly see, either the degree or the radian is too large a unit to measure bending of light. One needs to use the unit of an *arcsecond* which is the 3600th part of a degree. To convert radians into arcseconds a rough and ready rule is to multiply by 200,000.

If we work this out for a light ray grazing the surface of the Sun the answer is an angle as small as 0.87 seconds of arc. The smallness of this value is indicative of the rather weak effect gravity has on light, at least in the environment of our Solar System wherein the Sun is by far the strongest gravitating body.

I should also mention that even this effect has been obtained by extrapolating Newton's ideas somewhat. Thus if we take the modern concept of light as made of packets of energy, called *photons*, and assume that each photon is attracted by any mass as per Newton's law of gravitation, then we arrive at the result just described.

4.2 The bending of light in general relativity

How is the path of light affected by gravity according to Einstein's general theory of relativity?

Recall that in relativity gravitation is not looked upon as a force in the Newtonian sense. Rather, it is identified as an effect on spacetime geometry. So to solve the problem analogous to the Newtonian one which we just looked at, we first find out the non-Euclidean geometry in the vicinity of the gravitating sphere and then work out the path of a light ray in such a spacetime.

It may be worth mentioning for the benefit of the 'uninitiated' that the geometry we learn at school is that systematized by the Greek mathematician Euclid (*ca* 300 BC). If we change the basic assumptions of Euclid's geometry, we may get different *new* geometries, which as mathematical disciplines are every bit as respectable as the Euclidean geometry. But do they have any applications? Here Einstein had made the highly original suggestion that in the presence of gravitational influence, the geometry of space and time is different from Euclid's. His basic equations of general relativity tell us how to determine this non-Euclidean geometry in the presence of gravitating matter.

This was the problem solved by Karl Schwarzschild in 1916 when he showed how to determine the spacetime geometry outside a massive sphere. The second part of the problem is then straightforward. Notice that unlike the Newtonian case, there is no ambiguity here. We continue to assume that light travels in a straight line. However, travelling in a spacetime with geometry determined by gravity, *light has to modify its trajectory* from what it would have been if the geometry had been Euclidean. If we continue to use the Euclidean yardstick for 'straightness', then we have to conclude that light rays as per the non-Euclidean geometry determined by Schwarzschild, will appear to bend. A more correct statement is that given the prevalent non-Euclidean geometry, the straight line path followed by light as per its rules, will be different from the Euclidean path.

However, since we are comparing the relativistic result with Newton's alternative (ii) we will follow the rather loose but more commonly used phrase 'bending of light'.

How much does light bend? Having seen the Newtonian answer, the relativistic one is easy to state : *It is exactly double the Newtonian value.* In other words, the bending of light passing close to the Sun, will be by an angle 1.74 seconds of arc. Although the exact details of the spacetime geometry around a spherical mass became available in 1916 after the work of Schwarzschild, Einstein himself had worked out the relativistic bending of light in 1915, soon after formulating his equations of gravitation. In those early days very few scientists really understood what general relativity was all about. Most had found the notion of non-Euclidean geometry, actually applying to space and time, very bizarre and counter-intuitive.

A.S. Eddington, however, was one of those few who did grasp the essence of general relativity. Being an astronomer who had gone through the rigors of the Cambridge Mathematical Tripos to emerge as a Senior Wrangler, Eddington could not only appreciate the mathematical elegance of general relativity but also think of an astronomical test of the bending of light.

4.3 The 1919 eclipse expedition

Figure 9 is an adaptation of the second possibility of Figure 8. It shows a star A whose light grazes the surface of the Sun before coming towards the observer. The observer O therefore sees the star image at A' instead of at A; that is the star's image jumps from its regular position *if it happens to be just behind the eclipsed Sun.*

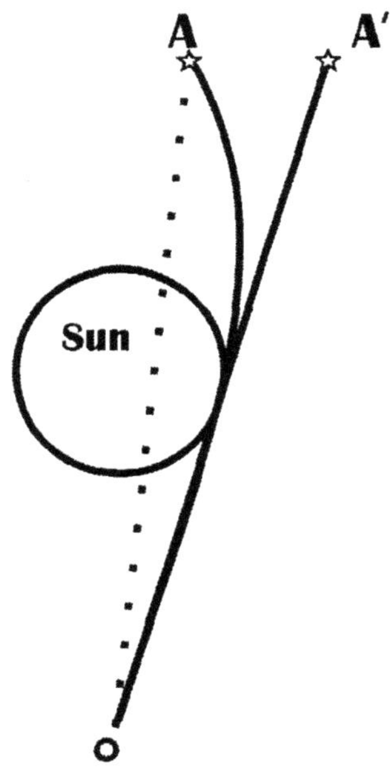

Figure 9. The 'bending' of light scenario. See text for details.

The expected shift in the direction of the star is no more than 1.74 seconds of arc. However, even to observe this shift we need a total solar eclipse. Otherwise, how can we see the star with the dazzlingly bright Sun in the forefront? Realizing this, Eddington proposed a measurement of this phenomenon at the time of the total solar eclipse of 29 May, 1919. Two teams set out, one consisting of Eddington himself and E.T. Cottingham to the island of Principe, in the Gulf of Guinea and the other of C.R. Davidson and A.C.D. Crommelin to Sobral in Brazil. A grant of £1000 obtained by the Astronomer Royal Sir Frank Dyson made this project possible.

In the end, (for the solar eclipse observation can be a chancy affair!) both teams were rewarded with adequate seeing conditions and measurements could be carried out.

The results of the observations were announced by the Astronomer Royal, Sir Frank Dyson on November 6, 1919, at a crowded joint meeting of the Royal Society and the Royal Astronomical Society. There had been great excitement as to what the findings would be. Would light show any bending at all? Or would it bend as calculated by the Newtonian methods? Or, would the answer favour relativity?

The results did favour general relativity. Within the estimated error bars, the bending of light was closer to the value 1.74 seconds of arc, rather than to half the value obtained from Newtonian gravity.

The success of the eclipse expedition made Einstein an instant celebrity. Although the concept of curved spacetime still remained beyond the common layperson, the results confirmed that nature did seem to follow these apparently crazy ideas.

And, of course, it was the first indication to the astronomer that because of bending of light by en-route masses, the observed positions of images in the sky may not represent reality.

But several decades had to pass for this indication to sink in.

4.4 A diversion

But there was another side to that historical expedition! The 1919 Solar Eclipse expedition that provided the first observational evidence for the bending of light by gravity also had a by-product : a problem in probability theory posed by Eddington.

Recall that four observers had been deputed to make the measurements : Davidson and Crommelin went to observe in Sobral in Brazil while Cottingham and Eddington went to the island of Principe in the Gulf of Guinea. A great deal depended on their results. Would light show any bending by gravity? Would the effect be as predicted by the (hybrid version of) Newtonian gravity or double that value as predicted by Einstein?

In an amusing after-dinner speech before they set out, Crommelin referred to the four observers C, C', D and E and hinted that a problem might arise as to the veracity of their claims with so much at stake and given the probability that each of them may be tempted to lie once in a while! The question was subsequently restated by Eddington in the following form :

A, B, C and D each speak the truth once in three times (independently). D makes a statement and A affirms that, B denies that, C declares that D is lying. What is the probability that D was speaking the truth?

This convoluted problem is solved by the use of arguments based on conditional probability. Try to solve it, if you like! The answer : the probability that D was telling the truth is 25/71. [For a solution see Eddington's article in the *Mathematical Gazette*, **19** ,256,(1935)]

5 Gravitational Lensing

Figure 10 shows an ordinary lens, the type of which is used in magnifying letters. The ray-diagram shows how the magnification occurs. The rays emerging from the object AB, appear to come from a much larger source $A'B'$ which is the 'virtual' image of AB.

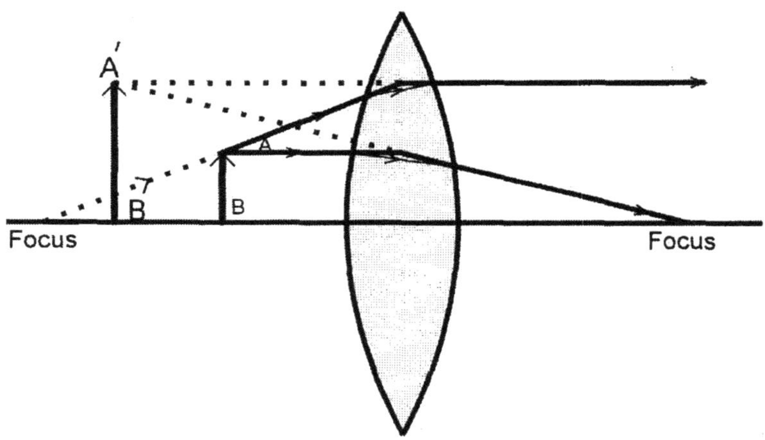

Figure 10. Image formation by a convex lens. The rays starting from A appear to emerge from A' when viewed from the other side and the size AB of the source appears magnified in the image $A'B'$.

Lenses are of various kinds. The one in Figure 10 has both its surfaces *convex*. There are others with both surfaces *concave* or with one surface convex, the other concave. All of them form images of real objects by bending light rays suitably. The cause of bending here is of course, *refraction* of light. It is the same phenomenon that leads to mirages in deserts.

Since gravitation can also bend rays of light, can we likewise encounter situations where the phenomenon produces lenses? This issue was first raised in the 1930s by the Caltech astronomer Fritz Zwicky. He wrote in 1937 :

Last summer Dr. V.K. Zworykin (to whom the same idea had been suggested by Mr. Mandl) mentioned to me the possibility of an image formation through the action of gravitational fields. As a consequence I made some calculations which show that extragalactic nebulae offer a much better chance than stars for the observation of gravitational lens effects.

There have been occasions, when a highly perceptive idea remains neglected because the scientific community is not ready for it. Zwicky's ideas were three to four decades ahead of their time and were forgotten. They were recalled and appreciated much later, when he himself was no more. Likewise the results obtained by S. Refsdal and Jeno Barmothy in the mid–1960s wherein working independently, these scientists explored the possibility of lensing by galactic masses and their effect on images of quasars, were considered interesting but hardly relevant to actual observations.

Then in 1978-79 the subject erupted all of a sudden, catapulting gravitational lensing onto the centre stage of astronomy. Zwicky's prophesy expressed in 1937 and reproduced below became realized :

Provided that our present estimates of the masses of cluster nebulae are correct, the probability that nebulae which act as gravitational lenses will be found becomes practically a certainty.

6 Cosmic Illusions

The example of a star's image shifting because of the bending of its light by the Sun's gravitation conjures up many more interesting possibilities.

Imagine, for example, a source of light and an observer on the Earth, with a massive object like a galaxy lying in between them. Figure 11 shows a possibility where there are *two* possible and different routes for light from the source to arrive at the observer. The light path is 'bent' by the galaxy in both cases, but the interesting result is that both alternative routes are made possible by the non-Euclidean geometry in the neighbourhood of the bending galaxy.

Why is it interesting? Because the observer will now see not one but two

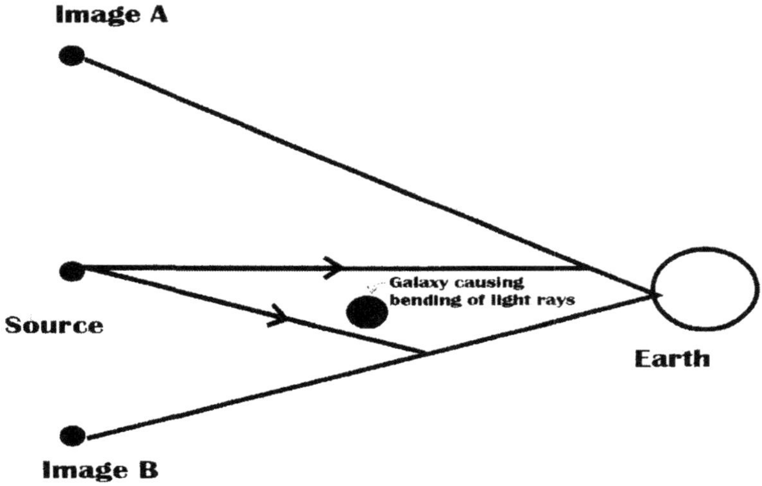

Figure 11. A single source is lensed and is seen as two images at A, B because of the bending of light rays by an intervening galaxy, as they reach the observer on the Earth.

displaced images shown by A and B in the figure)! And, of course, with more alternative light paths possible we might see even more images. In an idealized situation where the source, the gravitating mass, and the observer are exactly aligned, and the geometry has a symmetry about this axis, we should see *infinitely* many such images all distributed along a ring perpendicular to the axis. This ring is sometimes called the *Einstein ring.*

In 1979, a group of astronomers, D. Walsh, R.F. Carswell and R.J. Weymann, came up with a startling case precisely of this kind. The case concerned a pair of quasars with the catalogue names 0957+561 A & B.

Quasars are extremely powerful but very compact sources of light, so much so that before their significance was appreciated in the 1960s, a few bright quasars were mistaken for stars. Closer studies, however, revealed them to be much more powerful than typical stars and also considerably more distant from us than stars in our Milky Way Galaxy.

The quasars 0957+561 A & B were known to be such powerful and distant sources, but with the peculiarity that their directions as seen from here were very close to each other. Only a small angle of 6 arcseconds separated A from B. However, when Walsh, Carswell, and Weymann studied them more closely they found considerable similarity in their spectra and other features. While there is nothing to prevent two very similar looking objects from appearing near each other, like human twins, one expects such quasar twins to be rather rare.

This is where the above gravitational lensing idea begins to assert itself. One could ask, Instead of there being two identical looking quasars in reality, are we seeing two images of only one real source? If so, where is the gravitating galaxy of Figure 11? Walsh et al. looked and found a galaxy that just about did the trick. Subsequently, it was noted that this galaxy belongs to a cluster which collectively also contributes to the bending of light, just as Zwicky had predicted.

The discovery demonstrated that gravitational bending of light can very subtly distort the astronomer's perception of reality. After 0957+561 A & B there were other cases of multiple quasars identified as gravitationally lensed images. In some cases three and even four images have been seen. One may see a single image, but in a distorted form. For example, in Figure 12 we see an arc extending across several galaxies. This may well be a distorted vision (like our own images seen reflected by curved mirrors!) produced by lensing. Another characteristic of lensing is that it may brighten one of the images by a large factor by concentrating a number of light rays together.

Figure 12. The arc in Abell Cluster 370 is a distorted image due to gravitational lensing.

Is there any way by which an astronomer can check whether two closely situated quasars are distinct real entities or images of one source? If the light from the quasars fluctuates with time then the answer is, 'yes, possibly'. For, even though the images are nearly identical, the lengths of their light paths are in general different. Thus, in Figure 11, the path which gave the image A may be longer than the path for the other image, in which case we are seeing in A an image of the source at an *earlier* epoch than in B. Thus, the light variations seen in A should be repeated in B after a specific gap, like a few weeks or a few moths or so. Hence, if we continuously monitor both A and B, we may be able to match their light fluctuation patterns by imposing a suitable time gap between the two.

Attempts like this are being made to discover whether the light variations of the two images 0957+561 A & B can be so matched. There are some indications that such matching can be done but it is still too early to say that the result is confirmed.

Whatever the outcome of these investigations, the mere possibility of gravitational lensing has cautioned the astronomer against the old adage 'Seeing is believing'.

7 The Black Hole of Calcutta

The bending of light is of course great in a very strong gravitational field. Can it be so much that in some cases, light is bent so much that it cannot escape the vicinity of the object at all? Such an object would remain invisible, even though it exerts gravitational influence on the surroundings. The notion of a *black hole* arose from this extreme limit. Astrophysicists believe that black holes form when a massive object finds itself unable to hold itself in equilibrium under its own force of gravity. Such an object would continue to shrink in size until it reaches a stage that the force of gravity on its surface becomes so strong as to hold in everything including light. Imagine an object ten times the mass of the Sun. When it becomes a black hole its diameter would be a mere sixty kilometres! Even the great Eddington found the notion of a black hole too bizarre to be realized in nature. However, today, despite its weird properties, the black hole has proved to be a very useful concept in astrophysics.

However, going back to history, if one were to ask the following quiz question:

Who talked of black holes first?

which of the four alternative answers would be the correct one?:
(A) John Wheeler in 1968
(B)·Pierre-Simon de Laplace in 1799

(C) John Michell in 1784
(D) British Historians around 1757?

The correct answer is (D)!

The oldest mention of a black hole is found not in books of physics or astronomy but in books of history. In the summer of the year 1757, Nawab Siraj-Uddaula, the ruler of Bengal in eastern India, marched on Calcutta to settle a feud with the British East India Company. The small garrison stationed in Fort William at Calcutta was hardly a match for the Nawab's army of 50,000. In the four-day battle that ensued, the East India Company lost many lives, and a good many, including the company's governor, simply deserted. The survivors had to face the macabre incident now known as the *Black Hole of Calcutta.*

The infuriated Nawab, whose army had lost thousands of lives in the battle, ordered the survivors to be imprisoned in what came to be known as the Black Hole, a prison cell in Fort William. In a room 18 feet by 14 feet, normally used for housing three or four drunken soldiers, the 146 unfortunate survivors were imprisoned. The room had only two small windows. During the 10 hours of imprisonment, from 8 p.m. on 20 June to 6 a.m. on 21 June in the hottest part of the year, 123 prisoners died. Only 22 men and 1 woman lived to tell the tale.

Apart from its macabre aspect, the Black Hole of Calcutta did bear some similarity to its astronomical counterpart, involving as it did a large concentration of matter in a small space from which there was no escape. Indeed the incident is so macabre that most Indian historians discount it as fictitious, arguing that it was blown up by the British East India Company to find a reason to engage the Nawab in a battle and annex his state.

Whatever the real facts, I should mention that on the location of the notorious Black Hole of Calcutta, today there stands the General Post Office...which may perhaps explain the long postal delays!

I started on this light-hearted tour of gravity with a pseudo-historical interlude...I end the tour with this debatable historical incident.

Workshop Reports

A1: EXACT SOLUTIONS AND THEIR INTERPRETATION

JIŘÍ BIČÁK

Institute of Theoretical Physics, Faculty of Math and Physics, Charles University
V Holešovičkách 2, 180 00 Prague 8, Czech Republic
E-mail: bicak@mbox.troja.mff.cuni.cz

This is the account of the workshop *Exact solutions and their interpretation* at the 16th International Conference on General Relativity and Gravitation held in Durban, July 15-21, 2001. Work reported in 32 oral contributions spanned a wide variety of topics, ranging from exact radiative spacetimes to cosmological solutions. Two invited review talks, on the role of exact solutions in string theory and in cosmology, are also described.

1 Introduction

In accord with past tradition, the workshop A.1 has once again received the highest number of papers (73) in GR16, thus providing ample proof that the interest in discovering and understanding solutions to the Einstein equations did not fade away with the turn of the millennium. In comparison, a closely related workshop on mathematical studies of the field equations received 48, three workshops devoted to quantum issues totalled 52, mathematical cosmology received 38 and relativistic astrophysics 31, etc. Obviously, quantity does not necessarily mean quality. Also, not all the contributors could have been expected to come to Durban. Despite this, I asked the organizers to allot extra 3 hours to the A.1 workshop, with the welcome side effect that no need of a poster session arose. During 9 hours, 32 original contributions (15 minutes each) were presented. In addition, I turned to Roberto Emparan and Malcolm MacCallum to give reviews (30 minutes each) on exact solutions in string theories and on cosmological models. Their reviews also attracted a great number of GR16 participants, who normally would not have attended talks at the A.1 workshop.

In the Introductory remarks, I very briefly recalled the role of exact – sufficiently explicit – solutions for the development and understanding of general relativity. Despite significant recent advances in demonstrating the *existence* of some quite general classes of solutions which, of course, are "no less exact", we still rightly welcome special *explicit* formulae. Such questions as location and character of singularities and the global properties of a corresponding spacetime, are best addressed if an explicit solution is available. A special explicit solution frequently stimulates questions relevant to more general situations. For example, the presence of the Cauchy horizon in the Reissner-Nordström solution inspired a great amount of work on the instabilities of the Cauchy horizons of other types.

The role of the most important exact vacuum solutions such as black-hole solutions, stationary axisymmetric fields, and several classes of exact radiative spacetimes and vacuum cosmological models, in the development of general relativity and relativistic astrophysics, was reviewed recently [1].

In the following, the contributions are arranged in essentially the same order as they were given during the workshop. Although the topics covered were quite diverse, wherever possible, the contributions were grouped together: on exact radiative spacetimes, on the mathematical techniques used in analyzing exact solutions, on cosmological solutions, on the black-hole and other stationary solutions, and three contributions are covered by "Miscellanea". The two invited reviews are summarized at the end of the article.

With minor exceptions the following text is the combination of the material written by those who gave the talks (with their names in italics), which I have edited, modified in a number of cases, and in all cases had to shorten because of length constraints. I thank all the authors for supplying me with this material and, in particular, for adding the references which make the report more useful. I also thank T. Ledvinka and M. Varadarajan for help.

2 Exact radiative spacetimes

Standing gravitational waves (*H. Stephani*)
H. Stephani proposed an intuitive definition of standing gravitational waves: (i) the constitutive parts of the metric functions should depend on the timelike coordinate only through a periodic factor, and they should also depend on spacelike coordinates ("product structure"); (ii) the time average of some of the metric functions should vanish; in particular, the analogue of the Poynting vector (if there is any) should be divergence-free and the time average of its spatial components should be zero. He then searched for standing gravitational waves among the main classes of exact vacuum solutions – in most cases with a negative result. There are none among the plane waves and the algebraically special solutions. He can find them only as simple subcases of the (generalized) Einstein-Rosen waves. But in particular for the Gowdy universes, the criterion "product structure" gives contradicting results, depending on the choice of the coordinate system. The criterion "Poynting vector" again turns out to be ambiguous. The most probable choice is that connected with the so-called "C-energy" since the energy-momentum pseudotensor gives unconvincing answers, and

for the other forms based on a Lagrangian the timelike component of the Poynting vector turns out to be negative. Although Stephani could find Einstein-Rosen waves which meet his two conditions, the coordinate dependence of the criteria remains disturbing. One may expect more standing wave solutions in the classes with only one or none spacelike Killing vector.

Cylindrical spacetimes endowed with angular momentum
(*G. A. Mena Marugán*)
In this work the general solutions to the vacuum Einstein equations with cylindrical symmetry were studied with the symmetry axis being allowed to be singular – it could be endowed with spin and a linear energy density. First a gauge-fixing procedure adapted to cylindrical symmetry that removes all the constraints of canonical gravity was given. The corresponding reduced system has two field-like degrees of freedom which depend only on the radial and time coordinate. The gauge-fixed metric is expressed in terms of these fields. The solution contains also two constant parameters that describe the deficit angle outside the axis and the density of the angular momentum in the axis direction. The symmetry axis is regular if and only if the two constants vanish. It is proved that the gauge-fixed metric is rigorously defined and there exists a Hamiltonian that generates the reduced evolution if one imposes suitable boundary conditions at spatial infinity and on the axis on the two metric fields and, in addition, introduces a certain restriction on the initial data if the angular momentum does not vanish. The explicit form of the reduced Hamiltonian which provides the energy density in the axis direction is also found. This density is preserved in the evolution and turns out to be bounded both from below and above. The considered spacetimes describe cylindrical gravitational waves surrounding a spinning string. Moreover, the deficit angle detected at spatial infinity is equal or greater than the deficit seen close to the axis. Gravitational waves have thus a positive contribution to the energy density. For details, see Ref. [[2]].

On Killing vectors and Bondi expansions (*J. A. Valiente Kroon*)
Asymptotically flat radiative spacetimes have frequently been studied by solving a characteristic initial value problem by means of expansions in inverse powers of a radial parameter (Bondi expansions). If one assumes that the spacetime has a Killing vector field, then by performing similar expansions on the Killing equations it is possible to deduce the asymptotic form of the Killing vectors compatible with asymptotic flatness.[3] Furthermore, one can deduce so-called "constraint equations" for the news function, the mass aspect and the Newman-Penrose constants. The particular form of these

quantities is thus restrained. By resorting to the analysis of the orbits of the Killing vectors at null infinity, it is possible to give the solution to the constraint equations. For Killing fields that are non-supertranslational the characteristics of the constraint equations are the orbits of the restriction of the Killing field to null infinity. The possible classes of Killing fields are discussed by analysing their orbits at null infinity. If axial symmetry is imposed, and the spacetime is assumed to be radiative, then one recovers known results on boost-rotation symmetry as the only other admissible symmetry (see Ref. [4] and references therein). As an application, the formalism was used to study the Einstein-Maxwell boost-rotation symmetric Petrov type D spacetimes.[5] Valiente Kroon also studied the late time behaviour of the polarisation states, of the Newman-Penrose constants and the Bondi mass. In particular, one finds that the leading term behaviour of the Bondi mass loss is proportional to the square of the norm of the only non-vanishing Newman-Penrose constant. Since it is of quadrupolar nature, this result resembles, at least formally, Einstein's quadrupole formula.

Boost-rotation symmetric radiative spacetimes and fields: post-GR15 developments (*J. Bičák*, P. Krtouš, V. Pravda and A. Pravdová)
These spacetimes, representing uniformly accelerated objects (e.g. black holes), are significant since in rotationally symmetric, locally asymptotically flat at null infinity electrovacuum spacetimes the only additional allowable symmetry that does not exclude radiation is the boost symmetry (for latest formulations and proofs, see Refs. [4,3] and references therein.) These solutions are the only explicitly known spacetimes in which strong initial data can be chosen on a hyperboloidal initial hypersurface which lead to complete, asymptotically radiative spacetimes in future. Their properties and their use as test-beds in approximation methods, numerical relativity and quantum gravity models have been reviewed recently.[1,6]

The spinning C-metric is the only known example of such spacetime with Killing vectors not hypersurface orthogonal. It was brought into the canonical form of the boost-rotation symmetric spacetimes, and its interpretation as a metric representing uniformly accelerated, rotating black holes was analyzed.[7] The geodesics in the standard C-metric were also studied.[8]

Most recently, scalar and electromagnetic fields produced by uniformly accelerated (test) charges in de Sitter space were analyzed.[9] New effects arise because de Sitter space has spacelike conformal infinities. The radiative properties of the fields depend on the way in which a given point of $\mathcal{J}$ is approached. With particle horizons, purely retarded fields become singular or even cannot be constructed at the "creation light cone" of the "point" at which a source "enters" the universe. Smooth fields involving both retarded and advanced effects were constructed. In particular, the classical Born's solution for the field of two uniformly accelerated charges in Minkowski space were generalized to the de Sitter universe.[9]

Geodesics in expanding impulsive gravitational waves (J. Podolský and *R. Steinbauer*)
Introduced in a classical paper of Penrose by using his "scissors and paste" method, the explicit solution for spherical impulsive gravitational waves in continuous coordinates was later on given by Penrose and Nutku, and Hogan. It was only recently related explicitly to the impulsive limit of the Robinson-Trautman type N solutions.[10] The latter however has to be considered as formal since the metric tensor contains terms proportional to the square of δ-function. The transformation relating the distributional coordinate system to the continuous one is discontinuous. It is analogous to the one relating the distributional to the continuous form of the metric tensor for impulsive pp-waves which has been analyzed rigorously.[11]

The authors investigated the geodesics in spacetimes with spherical impulsive waves. In particular, employing the continuous form of the metric they found a large class of privileged and simple geodesics which can be related to the geodesics in the distributional form of the metric "in front" and "behind" the spherical impulse. This provides foundations for a rigorous (distributional) treatment of impulsive Robinson-Trautman solutions of type N.

3 Mathematical methods

Killing vectors, homothetic vectors, and conformal Killing vectors in the Newman-Penrose formalism (*G. Ludwig* and B. Edgar)
Finding Killing vectors (KV) or a homothetic vector (HV) of a given metric can be a formidable task. The authors present a new and simpler approach within the Newman-Penrose formalism. Leaning on results[12,13] previously derived in the Geroch-Held-Penrose formalism they show that the Killing or homothetic equations can be replaced by the NP-Lie equations, a set of equations involving the commutators of the Lie derivative with the four NP differential operators applied to the four coordinates. Provided that these operators refer to a preferred tetrad relative to the KV or HV, the equations can be solved for the Lie derivative of the coordinates, i.e., for the components of the KV or HV. If part of the tetrad (null directions and gauge) can be defined intrinsically, then that part is generally preferred relative to any KV or HV or conformal Killing vector. In case when part

of the tetrad cannot be defined intrinsically, the tetrad used is defined only up to one null rotation parameter and/or a gauge factor, and the NP-Lie equations become more involved. The general method remains the same and is still more efficient than conventional methods. See Ref. [14] for details. The following contribution is devoted to a closely related problem.

Tetrads and symmetries (*B. Edgar* and G. Ludwig)
Two of the main tools in the search for exact solutions of Einstein's equations are tetrad formalisms and use of symmetry. The authors propose a method which analyzes efficiently the Killing vector structure of metrics as they are calculated in tetrad formalisms; this method builds on the following results:[15] a vector field $\boldsymbol{\xi}$ is a Killing vector field iff there exists a tetrad $\hat{Z}^{\mu}_{m}$ $(m = 1, 2, 3, 4)$, which is Lie derived with respect to this field, $\mathcal{L}_{\xi}(\hat{Z}^{\mu}_{m}) = 0$. If we are given, in some ${Z_m}^{\mu}$, a spacetime containing at least one KV $\boldsymbol{\xi}$, there exists a tetrad ${\hat{Z}_m}{}^{\mu}$ which is Lie derived by that KV and which is given by a transformation linking ${\hat{Z}_m}{}^{\mu}$ to ${Z_m}^{\mu}$ for some values of the six Lorentz parameters. A direct procedure to find KVs is to substitute ${Z_m}^{\mu}$ via ${\hat{Z}_m}{}^{\mu}$ and six arbitrary parameters into $\mathcal{L}_{\xi}({\hat{Z}_m}{}^{\mu}) = 0$, and integrate this system for $\boldsymbol{\xi}$ and for all the parameters simultaneously. The explicit KVs and the parameters giving ${\hat{Z}_m}{}^{\mu}$ are then obtained. Rather than integrate $\mathcal{L}_{\xi}({\hat{Z}_m}{}^{\mu}) = 0$, it is efficient[12] to deal with the equivalent commutator equations for $\mathcal{L}_{\xi}$ and $\hat{\nabla}_m = {\hat{Z}_m}{}^{\mu}\nabla_{\mu}$.

Conformally decomposable spacetimes (*J. Carot* and B. O. J. Tupper)
Here spacetime (M, g) is studied whose metric is conformally related to the metric of a 2+2 locally decomposable spacetime $(M, \tilde{g})$; i.e., it is a product of two 2-dimensional spaces. Starting from the fact that the underlying decomposable spacetime cannot posses proper conformal Killing vectors (CKV) (unless it is conformally flat),[16] the authors show that the CKV of (M, g) must be the Killing and homothetic vector fields of $(M, \tilde{g})$ and carry out a complete classification of all the different possibilities for the conformal Lie algebra of (M, g) deriving canonical forms for both the metric and the CKV. A number of results on KV and HV fields are reviewed and new ones derived, especially those concerning homotheties and isometries with fixed points. The authors also obtain some classes of physically relevant examples, such as perfect fluids and electromagnetic fields.

Equations for complex-valued, twisting, type-N vacuum solutions, with one or two Killing/homothetic vectors (*D. Finley*)
The goal of understanding solutions of Petrov type N with non-zero twist

is still not realized. Although they can hardly represent radiative fields of bounded sources,[17] they attract researches because of a rich mathematical structure. One way to understand them is the use of HH spaces, i.e., complex-valued spacetimes, of Petrov type $N \times N$. They are determined by three PDEs for two functions, λ and a, of three independent variables (and also two gauge functions which may be chosen to be two of the independent variables). As in integrable systems, these form a second order, linear system for λ; however, here the integrability conditions, involving a, are complicated. Therefore, with the hope of finding new solutions, Finley assumed that these spacetimes admit both one and two homothetic or Killing vectors. The case with one Killing and one homothetic vector reduces equations to two ODEs for two unknown functions of the one remaining variable. Finley described the explicit forms of the metric, tetrad, connections, and curvature for twisting HH spaces of Petrov type $N \times N$, modulo the determining equations. The details can be found in recent Ref. [18].

Algebraic approach to the study of spacetimes with an isometry (*C. F. Sopuerta* and F. Fayos)
In this contribution a new formalism for the study of spacetimes possessing a non-null Killing vector $\boldsymbol{\xi}$ is presented. The main quantity in this approach is 2-form $\boldsymbol{F} = \boldsymbol{d\xi}$ satisfying Maxwell's equations. Considering $(\xi^\alpha, F_{\alpha\beta})$, the equations $\nabla_\alpha \xi_\beta = \frac{1}{2} F_{\alpha\beta}$, $\nabla_{[\alpha} F_{\beta\gamma]} = 0$, $\nabla_\beta F^{\alpha\beta} = J^\alpha = 2R^\alpha{}_\beta \xi^\beta$, and studying the integrability conditions, the authors find that all the components of the Weyl tensor can be expressed in terms of $(g_{\alpha\beta}, \xi^\alpha, F_{\alpha\beta}, T_{\alpha\beta})$, $T_{\alpha\beta}$ being the energy-momentum tensor. An extension of the Newman-Penrose formalism is set up in which a null tetrad adapted to $F_{\alpha\beta}$ is chosen, and the tetrad components of ξ^α and $F_{\alpha\beta}$ are used. The dependence of $C_{\alpha\beta\gamma\delta}$ on $(g_{\alpha\beta}, \xi^\alpha, F_{\alpha\beta}, T_{\alpha\beta})$ is algebraic. Substituting $C_{\alpha\beta\gamma\delta}$ into the second Bianchi identities one obtains first-order equations for the spin coefficients. Therefore, all the compatibility conditions reduce to compatibility conditions for the first-order equations for the spin coefficients. This formalism naturally provides a refinement of the Petrov classification: apart from the algebraic structure of $C_{\alpha\beta\gamma\delta}$ one has the structure of $F_{\alpha\beta}$ (see Ref. [19]).

Painlevé analysis in general relativity (*R. Halburd*)
This talk illustrated the use of complex-analytic methods as "detectors" of solvable models. The problem of determining the metric for a non-static shear-free spherically symmetric fluid (charged or neutral) reduces to the problem of determining a one-parameter family of solutions to $d^2y/dx^2 = f(x)y^2 + g(x)y^3$, where f and g are arbitrary. Most of the

solutions are special cases of the family given by Sussman. It was shown[20] that Sussman's solutions correspond to special cases such that the general solution has a simple singularity structure in the complex domain (the Painlevé property). New solutions were also given in terms of Airy functions and Painlevé transcendents. Halburd described all choices for f and g for which the equation above admits a one-parameter family of solutions such that all movable singularities are poles. This is weaker than the Painlevé property and yields a number of solutions described by Riccati equations besides those given in Ref. [20]. Conditions for matching these to an external metric were determined.

4 Cosmological Solutions

Hypersurface homogeneous rotating dust models (*A. Krasiński*)
For matter that moves geodesically and with nonzero rotation, Plebański coordinates (P.c.) (t, x, y, z) can be chosen so that the velocity field, the metric and the rotation vector field have the forms $u^\alpha = \delta^\alpha_0$, $g_{0\alpha} = \delta^0_\alpha + y\delta^1_\alpha$, $w^\alpha = n\delta^\alpha_3$, where n is the number density of the particles of the matter. If any symmetries exist, then every Killing field, expressed in P.c., must have the form $k^\alpha = (C - \phi - y\phi_{,y})\delta^\alpha_0 + \phi_{,y}\delta^\alpha_1 - \phi_{,x}\delta^\alpha_2 + \lambda\delta^\alpha_3$, where $\phi(x, y)$ and $\lambda(x, y)$ are arbitrary functions and C constant. If ϕ = constant, then $k^\alpha = Cu^\alpha + \lambda w^\alpha$, and this property is invariant under all the allowed transformations of the P.c. If $\phi \neq$ constant, coordinates can be adapted to this Killing field so that $k^\alpha = \delta^\alpha_1$. When 3 Killing fields exist, a classification of all the metric forms can be provided. The corresponding symmetry algebras can all be assigned to the Bianchi classes; the orbits can be spacelike as well as timelike or null. Apart from a few classes in which the orbit has to be timelike, most of the 25 classes contain a free parameter that measures the tilt of the velocity field with respect to the symmetry orbits, and allows one to consider all possible positions of the orbits relative to velocity, including the nonrotating, isotropic and spatially homogeneous Friedmann limit. The results have all been published by A. Krasiński in J. Math. Physics (see Ref. [21] and citations therein).

Spherical inhomogeneous cosmologies with pressure (*J. R. Gair*)
A generalization to the Lemaître-Tolman-Bondi solution may be obtained describing a spherically symmetric universe filled with dust which has angular momentum. The dust is labeled by a comoving radial coordinate, R, such that each dust particle remains on a surface R = constant. The dust particles within each of these shells move in every tangential direction with a certain

angular momentum about the symmetry centre. This angular momentum is conserved and so it is a function of shell label only. It produces an effective tangential pressure, the radial pressure vanishes. Einstein derived a static solution of this form, Datta and Bondi generalized to the dynamic case, but did not obtain the metric, Magli derived the metric in mass-area coordinates. Now Gair has obtained a new solution[22] in terms of more physical coordinates - the shell label, R, and the proper time, τ, experienced by the dust particles. The equation of motion for the areal radius, r, is of the form $(\partial r/\partial \tau) = \sqrt{E(R) - V(r, R)}$. The potential $V(r, R)$ for a given shell is the potential of a test particle moving in a Schwarzschild-de Sitter metric. The solution contains four arbitrary functions of R – the angular momentum $L(R)$, the mass $M(R)$, the 'energy' $E(R)$ and the initial position of the shells, $\tau_0(R)$. The shape of the potential permits seven different types of evolution for each shell: expansion and recollapse, collapse and bounce, unbounded expansion/collapse, oscillations, static, coasting expansion/collapse and hesitation. Self-similar solutions may be obtained by choosing $M \propto R$, $L \propto R$ and $E =$ constant, and a solution with a null fluid source by letting $L \to \infty$.

On some perfect fluid solutions of Stephani (*A. Barnes*)
In 1987 H. Stephani derived several solutions for a geodesic perfect fluid flow with zero pressure (dust). Barnes generalized[23] these solutions with non-zero rotation to include a non-zero cosmological constant Λ. All solutions are of Petrov type D and the magnetic part of the Weyl tensor (relative to the fluid flow) vanishes. In general the solutions admit no Killing vectors. The fluid flow is shearing, twisting and expanding. In the limit when the energy density vanishes the solutions reduce to spacetimes of constant curvature. The solutions can be matched across timelike hypersurface to a de Sitter, anti de Sitter or Minkowski spacetime, depending on Λ.

Locally isotropic spacetimes (M. A. H. MacCallum and *F. C. Mena*)
The authors consider spacetimes which locally (at every point in a neighbourhood) have a non-trivial isotropy group. The possible cases are defined in terms of symmetry of the Riemann tensor and its derivatives, following the method introduced by Cartan and Karlhede (see Ref. [24]) which gives a unique local characterization of a spacetime. They review and sharpen previous results for the cases with continuous isotropy groups. For the discrete case they analyze the possible maximal isotropy groups allowed by the algebraic structure of the Riemann tensor. The aim is to prove results analogous to those of Schmidt from 1969, who showed that a perfect fluid spacetime which locally admits the reflections in three perpendicular

spatial axes as isotropies must admit a 3-parameter group of isometries. The authors obtain several results of this sort for groups of spatial reflections. For example, they show that for a spacetime with a local discrete isotropy group L generated by reflection of a plane, whose energy-momentum tensor is that of a vacuum or perfect or imperfect fluid, if $R_{\alpha\beta\gamma\delta}$, $R_{\alpha\beta\gamma\delta;\mu\nu}$ and $R_{\alpha\beta\gamma\delta;\mu\nu\rho}$ are invariant under L, then there is a local group G_2 of isometries acting transitively on the surfaces to which the reflection plane is tangent.

On stationary, spherically symmetric inhomogeneous spacetime with matter content obeying $p = \alpha\rho$ (*S. M. Wagh*, D. W. Deshkar and P. S. Muktibodh)
The authors consider the shear-free, spherically symmetric, inhomogeneous spacetimes admitting the separable metric of the form[25]

$$ds^2 = -y^2 A^2 dt^2 + 2(y')^2 R^2 dr^2 + y^2 R^2 \left(d\theta^2 + \sin^2\theta\, d\phi^2\right)$$

where $y = y(r)$, $A = A(t)$, $R = R(t)$. The coordinates are comoving and the spacetime has non-vanishing energy flux. One nontrivial field equation and the equation of state determine $R(t)$. For example, the imperfect matter may obey $p = \alpha\rho$, $\alpha =$ constant. The radial distribution of ρ is unspecified since the field equations do not determine the metric function $y(r)$ and $\rho \propto 1/y^2$. A timelike Killing vector exists when $\dot{R}/A =$ constant. The stationary solution has an application for an inhomogeneous steady state cosmology. In the absence of a timelike KV, the metric can describe an inhomogeneous big bang cosmology.

5 Stationary systems and black holes

On electromagnetic Thirring problems (M. King and *H. Pfister*)
The authors consider models consisting of two spherical shells of radii a and $R \geq a$, the first one carrying a charge q but no rest mass, the second one mass M but zero charge (the mass shell fulfils the weak energy conditions). To these shells small angular velocities are applied. The coupled Einstein–Maxwell equations were solved exactly in M/R and q/R, and to first order in the angular velocities. The dragging effects and the induced magnetic fields were calculated.[26] The authors confirm some results of Cohen, and Ehlers and Rindler; in contrast to the latter they argue that the results completely fulfil Machian expectations. The collapse limit of the system was also discussed.

Then only one slowly rotating shell carrying both mass and charge was considered. For M/R and q/R for which the weak energy condition is violated, the dragging constant can become negative (antidragging!). Interesting is the behavior of the gyromagnetic ratio: it is very near to $G = 2$

in most of the parameter space, e.g., $1.92 < G < 2.05$ for all $M/2R > 1$ and q/R that the weak energy condition is fulfilled. The authors argue that the "robustness" of the value $G = 2$ and its "coincidence" with $G \approx 2$ for the simplest rotating, charged particles (electron and muon) hints to a deep connection between general relativity and quantum theory.

Exact relativistic treatment of stationary counterrotating dust disks (J. Frauendiener and *C. Klein*)
Relativistic dust disks with counter-rotating matter have been discussed since the work of Morgan and Morgan. The rigidly rotating single component dust disk was treated numerically and perturbatively by Bardeen and Wagoner and solved analytically by Neugebauer and Meinel in terms of Korotkin's finite gap solutions. The authors discuss a class of stationary counter-rotating dust disks which interpolates continuously between the rigidly rotating disk and a static disk. The whole metric is given explicitly in terms of theta functions on a hyperelliptic Riemann surface. The metric functions are analyzed analytically and the hyperelliptic functions evaluated numerically by using spectral methods. Physically interesting limiting cases as the Newtonian, the static, and the ultrarelativistic limit, where the central redshift diverges, are studied. Explicit expressions for the mass and the angular momentum are given. Following Bičák and Ledvinka who studied counter-rotating disks of infinite extension as sources for the Kerr metric the authors show that the matter in the disk can be interpreted as two streams of pressureless matter which move on geodesics of the inner geometry of the disk. The occurence of ergospheres and the dragging of the inertial frames is also discussed. See Ref. [27] for details and references.

Generalized Kerr-Schild metrics and the gravitational field of a massless particle on the horizon (*H. Balasin*)
Following the work of Aichelburg and Sexl, and 't Hooft and Dray on the gravitational field of a massless particle in Minkowski space, and on the horizon of a Schwarzschild black hole, respectively, Balasin investigated the change in the spacetime structure generated by a massless particle moving along the horizon $\mathcal{H}$ of an arbitrary stationary black hole. The generalized Kerr-Schild class characterized by $\tilde{g}_{\alpha\beta} = g_{\alpha\beta} + fk_\alpha k_\beta$, $k^\alpha k^\beta g_{\alpha\beta} = k^\alpha k^\beta \tilde{g}_{\alpha\beta} = 0$, $(k\nabla)k^\alpha = (k\tilde{\nabla})k^\alpha = 0$ gives a framework well adapted to the problem. Identifying the null generators of the horizon with the geodetic null vector field k^a and taking the scalar function f to be concentrated on the horizon, i.e. $f = \delta_{\mathcal{H}}\tilde{f}$, allows one to investigate the situation in terms of properties of supporting null hypersurface (the horizon)

and the parent geometry represented by $g_{\alpha\beta}$. Balasin obtained a generalized 't Hooft-Dray equation for the reduced "profile" $\tilde{f}$, whose coefficients are determined by optical scalars of the congruence forming $\mathcal{H}$ and curvature quantities thereon. See Ref. [28] and references therein.

The generalized Wahlquist-Newman solution(*M. Mars*)
The Kerr metric and the Wahlquist metric share the property that their Simon tensor vanishes. Recent work has clarified the geometrical meaning of the vanishing of the Simon tensor in terms of a natural relationship between the Weyl tensor and the covariant derivative of the stationary Killing vector (the so-called Papapetrou form). The Kerr-Newman-de Sitter spacetime satisfies the same relationship.[29] The electromagnetic field and the Papapetrou form are also closely connected in this metric. This leads to the likely existence of a charged generalization of the Wahlquist metric – Wahlquist-Newman metric. This was explicitly found in Ref. [29]. It was originally given by García using different hypotheses. This family contains eight essential parameters and includes, besides the Wahlquist and Kerr-Newman-de Sitter metrics, the whole Plebański limit of the rotating C-metric. A weakening of the relationship between the Weyl tensor and the Papapetrou form leads to a larger family containing one arbitrary function of one coordinate. The family is generically of Petrov type *II* and possesses only one Killing vector. These metrics admit a well-defined "infinity", with topology $\mathbb{R} \times S$, where S is a two-dimensional manifold, which may be compact or not. The metrics are in general not asymptotically flat but have a quite rich asymptotic behaviour.

General vacuum solution for axially symmetric stationary space-time (Z. Ya. Turakulov and *N. Dadhich*)
The authors have obtained the most general vacuum solution for the axially symmetric stationary metric in which both the Hamilton-Jacobi equation for particle motion and the Klein-Gordon equation are separable. It can be transformed into the Kerr-NUT solution. Like the Kerr solution, the Kerr-NUT solution is thus also unique under these assumptions. It is the most general axially symmetric stationary asymptotically non-flat vacuum spacetime admitting regular horizon. See Ref. [30] for details.

On the integrability of the exact and perturbative Einstein's equations for axistationary perfect fluids (*M. Bradley*, G. Fodor, M. Marklund and Z. Perjés)
In this contribution the integrability of stationary axisymmetric perfect

fluid spacetimes was investigated using a tetrad approach with the Riemann tensor, the Ricci rotation coefficients and the tetrad vector components as variables.[31,32] The system of equations is integrable in the generic case according to the Cauchy-Kowalewski theorem. The equation of state is barotropic if and only if $\omega_1\sigma_2 = \omega_2\sigma_1$ holds, where ω_i is the vorticity and σ_i the shear (with tetrad e_1 and e_2 in the "r, θ"-plane). A physically realistic rotating incompressible perfect fluid must be of Petrov type I. If the equation of state is linear, the Petrov type cannot be D. The authors linearize the equations around the interior Schwarzschild solution. The resulting solution of the linearized equations is known to be a first order approximation of a solution to the full non-linear system because of the integrability of the system. An example of the integration was given.

Plane symmetric analogue of NUT space (*M. Nouri-Zonoz* and A. R. Tavanfar)
The authors introduce a new definition of spacetime symmetry which is in accordance with the symmetry of its curvature invariants and its gravoelectromagnetic fields. Applying this definition, they investigate exact vacuum solutions corresponding to both static and stationary plane symmetric spacetimes using concepts of the (1+3)–decomposition or threading formalism. Demanding the presence of a plane symmetric gravomagnetic field, they find a family of two parameter (m and l) solutions, every member of which is the plane analogue of NUT space. See Ref. [[33]] for details.

New exact solution to the static balance problem (*R. B. Mann* and T. Ohta)
The static balance problem of finding equilibrium solutions in which the gravitational attraction of two bodies is balanced by their electromagnetic repulsion was solved by Majumdar and Papapetrou. These solutions are extremal in the sense that the mass of each body is equal to its charge in gravitational units, $e_i = \pm\sqrt{4\pi G}m_i$. No one has yet found equilibrium states in which this condition does not hold for both bodies. Within $(1+1)$ dimensional ("lineal") gravity the authors considered the motion of two charged gravitating bodies. They solved for the Hamiltonian as a function of the degrees of freedom of the system, which in this case are the centre-of-inertia momentum and the relative proper separation of the bodies. Demanding that the centre-of-inertia momentum p had vanishing time derivative they found the equation[34] $\kappa/2\left(\sqrt{p^2+m_1^2}-p\right)\left(\sqrt{p^2+m_2^2}-p\right) - e_1e_2 = 0$. This condition is the generalization of the Majumdar-Papapetrou condition. It

depends on the momentum and can be satisfied for some fixed momentum p_c whilst $e_i \neq \pm\sqrt{4\pi G}m_i$.

Kóta-Perjés metrics and their Kerr-Schild class (*M. Vasúth*)
In this work properties of Kerr-Schild metrics and Kóta-Perjés spacetimes were discussed. Gergely and Perjés have examined under which conditions the Kerr-Schild pencil of a vacuum metric generates another vacuum spacetime. Using parameter η, which governs the shear of the null congruence, one can classify the possible solutions of the problem. When η is arbitrary, the solution is either Kasner metric or another Kóta-Perjés metric depending on the curl of the null congruence. In the original paper there are three Kóta-Perjés metrics. In the above classification only two of them were listed. The author shows that the third metric is a mate of the other Kóta-Perjés metric in the class determined by the same special value, $\sin\eta = \pm 1/\sqrt{2}$, of the parameter. Some global properties of this third metric were analyzed.

De Sitter-Schwarzschild geometry (*I. G. Dymnikova*)
Under spherical symmetry, the weak energy condition, the regularity at the center, the asymptotic flatness and the finiteness of the ADM mass define the family of regular solutions which include the class of metrics asymptotically de Sitter as $r \to 0$ and asymptotically Schwarzschild as $r \to \infty$. The source is anisotropic perfect fluid which can be interpreted as a vacuum invariant under boosts in the radial direction.[35] According to the author, it connects the de Sitter vacuum in the origin with the Minkowski vacuum at infinity and corresponds to an extension of $\Lambda g_{\mu\nu}$ to the cosmological tensor $\Lambda_{\mu\nu}$ which allows Λ to become evolving and clustering.[36]

With masses $M > M_{crit}$, the spacetime contains an infinite sequence of vacuum non-singular black and white holes whose singularities are replaced with regular cores asymptotically de Sitter as $r \to 0$. A white hole core can model initial stages of the universe starting from nonsingular big bang, followed by the Kasner-type expansion. The author suggests that the instability of de Sitter vacuum near the regular surfaces $r = 0$ results in a possibility of multiple quantum birth of baby universes inside.[36] A gravitating particle-like structure without horizons with de Sitter vacuum in the origin was applied to estimate the lower limits on sizes of fundamental particles.

Hairy black holes, horizon mass, and solitons (*D. Sudarsky*, A. Ashtekar and A. Corichi)
In this work colored static black hole solutions to the Einstein-Yang-Mills (EYM) equations and hairy black holes in other theories were considered

within the isolated horizons framework.[37] Hairy black holes may be regarded as 'bound states' of ordinary black holes without hair and colored solitons. This model is used to predict the qualitative behavior of the horizon properties of hairy black holes, to provide a physical 'explanation' of their instability and to put qualitative constraints on the end point configurations that result from this instability. The model explains qualitative aspects of the behavior of surface gravity of the colored black holes. Predictions for the value of the ADM mass of the soliton configurations found often in association with the hairy black holes in, for example, Einstein-Yang-Mills-Higgs theory can be made. These predictions allow one to evaluate the difference in the values of the mass of these solitons in terms of quantities associated solely with the corresponding black holes and the values of the ADM masses of the various Bartnik-McKinnon solitons. The results have been reported in Ref. [38], numerical calculations in Ref. [39].

Isolated horizons in 2+1 dimensions
(*J. Wiśniewski*, A. Ashtekar and O. Dreyer)
The framework of isolated horizons[37] is here extended to 2+1 dimensional general relativity with negative cosmological constant. An analog of the orthonormal null tetrad is introduced, where l^μ, n^ν are null, l^μ is tangent to the horizon and m^ρ is spacelike. One can then construct the analog of the Newman-Penrose framework.[40] A (weakly) isolated horizon is a pair $(\Delta, [l])$, where Δ is a null surface and two l^μ's are equivalent if they differ by a multiplicative constant. Expansion of Δ vanishes and the intrinsic connection 1-form on Δ is Lie-dragged by l^μ. Equations of motion are imposed on Δ and an energy condition on stress-energy tensor is assumed. The consequences of this definition are: intrinsic metric on Δ is Lie-dragged by l^μ, the 0-th law of black hole mechanics holds, and one can always choose a gauge in which electrostatic potential is constant on Δ. Using Hamiltonian methods as in 3+1, the quasi-local definitions of mass and angular momentum can be introduced. The first law of thermodynamics is equivalent to the requirement that the time evolution vector field t^μ gives rise to a Hamiltonian vector field on the phase-space of isolated horizons. However, in presence of non-zero total charge, the appropriate boundary conditions turn out to be more subtle than in 3+1 dimensions. Energy can only be uniquely defined up to an additive function of the electric charge. The concepts introduced were used to analyze the charged generalization of BTZ solution.

Asymptotic behaviour of the proper length and volume of the Schwarzschild singularity (*A. Qadir* and A. A. Siddiqui)

It has been proved that the proper length of the Qadir-Wheeler suture model, constructed from two sections of two closed FLRW models joined by a Schwarzschild region, goes to infinity as the singularity is approached, while its proper volume shrinks to zero.[41] Asymptotically, the proper length goes as $K^{1/3}$ and the proper volume as $K^{-2/3}$, K is the mean extrinsic curvature. A similar analysis could provide the asymptotic behaviour of the Schwarzschild black hole near the singularity – with K taken as a time parameter. Trivial foliations, as by constant Kruskal time hypersurfaces, or by hypersurfaces of constant values of the compactified Kruskal time, do show that the proper length approaches zero and the volume approaches infinity. The relevant K-slicing for the Schwarzschild geometry is not so simple, the difference from the suture model being that it has a boundary which limits the part of the region near the singularity whose proper length is to be measured. The authors prove that, as for the suture model, the length does indeed go as $K^{1/3}$ and the volume as $K^{-2/3}$.

6 Miscellanea

The problem of coordinates in general relativity (*A. Chamorro*)
The author defines quasi-Minkowskian locally inertial coordinates (QMLIC) relative to an observer in free fall that is instantaneously at rest at an event by usual requirements: (i) the world line of the observer (WLO) is described in the QMLIC by zero spatial coordinates, time is observer's proper time; (ii) on the WLO the metric coincides with Minkowski metric and its derivatives vanish; (iii) the coordinate lines of the QMLIC have as unit tangent vectors, an orthonormal tetrad which is parallelly transported along the WLO, one vector being WLO's tangent; (iv) if the tangent vectors have components which are the derivatives of the original coordinates with respect to each of the QMLIC, then, in a neighborhood of the event, their absolute differentials differ from zero by a curvature dependent term. The author shows that in a weak gravitational field there are in general infinitely many different QMLICs all of which reduce to the unique Minkowskian coordinates when the curvature vanishes. Once the curvature term and the tetrad are fixed, so are the corresponding QMLIC.

Coframe teleparallel models of gravity. Exact solutions (*Y. Itin*)
Itin studies the coframe teleparallel theory of gravity with the most general quadratic Lagrangian.[42] The coframe field on a differentiable manifold is a basic dynamical variable. A metric tensor as well as a metric compatible connection are generated by the coframe in a unique manner. For a special

choice this theory gives an alternative description of Einsteinian gravity - teleparallel equivalent of GR. The field equations of the theory were studied by a "diagonal" coframe Ansatz.[43] Itin obtained the explicit form of all spherically symmetric static solutions of the "diagonal" type to the field equations for an arbitrary choice of free parameters and proved that the unique asymptotically flat solution with Newtonian limit is the Schwarzschild solution that holds for a subclass of teleparallel models.

Matching preserving the symmetry (*R. Vera*)
The matching of spacetimes is usually implicitly assumed to preserve some of the symmetries of the problem. Some previous work performed matchings by taking into account the orbits of the preserved groups at both sides of the matching hypersurface. In the work of Vera,[44] a general definition for such a kind of matching is given. The matching hypersurface is restricted to be 'tangent' to the orbits of a desired group of isometries admitted at both sides. The definition implies that the matching hypersurface inherits the preserved symmetry and its algebraic type. When matching is preserving a 2-dimensional group of isometries, two functions constructed from the exterior product of the two Killing forms and the differential of one of the Killing forms at one side, and the analogous two functions at the other side, have to coincide at the matching hypersurface. The orbits of the group generate orthogonal surfaces if and only if these two functions vanish. This is related to the 'circularity condition' in stationary, axisymmetric interior problems. This implies the absence of convective motions in fluids without flux. The functions vanish in the vacuum exteriors and, therefore, have to vanish also on the matching hypersurface.

7 Invited reviews

The role of exact solutions in string theory (*R. Emparan*)
There are at least two levels at which one can discuss classical exact solutions within the context of string theory: (i) classical solutions of full string theory, and (ii) classical solutions of string theory at low energies. The latter are approximations to the former, where terms in the equations that could give rise to corrections at curvature scales of the order of the string length are neglected. Solutions in (ii) are typically exact solutions of GR, possibly coupled to a variety of scalar and p-form gauge fields. Only a few among these belong to the more stringent class (i).

The techniques for constructing exact solutions to the full classical string theory are rather different from those familiar in GR. Conformal sigma models provide automatically solutions of this sort (see, e.g., Ref. [[45]]). Besides, supersymmetry is often crucial in establishing the absence of stringy corrections to some solutions of the low energy effective action, such as the extremal Reissner-Nordstrom black holes, pp-waves, extremal p-branes, AdS spacetimes etc.[46] Supersymmetry is also of help in solving the field equations, since it reduces them to first order partial differential equations.

Many of the techniques developed for GR can be used in low energy string theory. String theory provides additional solution-generating techniques, based on the existence of dualities (S and T) which relate solutions to different low energy limits of string/M-theory. Other techniques make use of additional, compactified dimensions: for example, performing boosts along them give rise to electric Kaluza-Klein charges.

Exact gravitating solutions were far from the center stage in string theory until around the late '80s. Since '96–'97 they have come to play a crucial role, following the successful microscopic description of black holes using D-branes,[47] and the formulation of the AdS/CFT correspondence, reviewed in Maldacena's contribution to this conference. Besides this, some old and well-known exact solutions of GR have received new clothes through their use in string theory and related higher-dimensional scenarios. For example, the Taub-NUT solution, whose interpretation in conventional GR is obscure, has undergone a transformation (via Euclidean rotation, and the addition of extra dimensions) into a solution to 11D supergravity, which provides the lift to M-theory of a D6-brane.[48] Another interesting example is the use of the C-metric, to describe black hole pair creation and exact solutions for black holes on branes.[49] Much work extends the techniques of GR to cover Kaluza-Klein reductions, couplings to higher-dimensional p-form gauge fields etc, but it would be desirable to unravel the structure of the simplest situation – *pure* gravity ($R_{\mu\nu} = 0$) in dimensions $d > 4$. Along this line, some solutions with qualitatively new features, such as higher-dimensional rotating black holes,[50] and the instabilities of black branes,[51] have been discovered. A more systematic approach to the construction and study of such solutions appears is still lacking.

Cosmological solutions: selected themes (*M. A. H. MacCallum*)

In this last talk of the workshop, MacCallum first discussed where exact solutions impinge on the cosmological theory. The "standard model" nowadays is the Friedmann-Lemaitre-Robertson-Walker (FLRW) model with $k = 0$, with cosmological constant Λ, and with inflation. Its predictions fit the cosmic microwave background and the large-scale structure, but worries remain: (i) because of the large number of "initial" assumptions, such as what is the inflaton, what is the initial quantum state etc; (ii) there are uncertainties in its fit to galactic structures; (iii) no convincing reasons are available for the value of Λ. The role of exact cosmological solutions is that they capture full non-linearity of the Einstein equations and provide tests of the claims of the standard model. For example, phase-plane arguments show that there are anisotropic solutions staying close to the FLRW models for as long as one wishes.[52] Further, there are models from which it is evident that inflation does not necessarily remove anisotropy. It is not clear how generic these models are, not even what "generic" means. Exact solutions may be useful for observational data analysis. For example, number counts can be explained by the inhomogeneous Lemaitre-Tolman-Bondi models without an evolution of the sources equally well as by the standard FLRW models with sources evolving. By using exact cosmological solutions one can study (past and future) asymptotics, gravitational waves in a non-linear regime (solutions with G_2 symmetry groups) and see, within a more general framework, what is special about the FLRW models.

In the second part, MacCallum mentioned some progress made in the last decade or so in the field of exact cosmological models. Most of the results are summarized in Refs. [52,53] where a number of other citations can be found (for the references on vacuum cosmological models, see also Ref. [1]). New fluid solutions with a hypersurface homogeneity were found, mainly as fixed points of dynamical systems. Other solutions were obtained by generating mechanisms (in particular those with $p = \rho$), by separation Ansätze, or by use of homotheties. More attention has been paid to solutions with Λ. Methods using dynamical systems techniques, symmetries of Hamiltonians, etc are summarized in Ref. [52] (see also the contribution of Uggla to the workshop A.3 in these Proceedings). Current efforts are extending to tilted models and cases where self-similar asymptotics does not work. Interesting open questions arise as to whether these new models can be used to test averaging, or to study structure formation.

Finally, MacCallum mentioned the problem of "local" objects in cosmology, in particular black holes in spacetimes which are asymptotically FLRW. A simple example is given by joining the Schwarzschild solution to a Whittaker solution in the Einstein universe.[54]

Most of the papers quoted bellow are also contained on the qr-qc archives. Here we give the archive number only if the journal citation is not available.

References

1. J. Bičák *Selected solutions of Einstein's field equations: their role in general relativity and astrophysics*, in *Einstein's Field Equations and Their Physical Meaning*, ed. B. G. Schmidt, Lecture Notes in Physics **540** (Springer Verlag, Berlin–New York, 2000), pp. 1–126.
2. G. A. Mena Marugán, *Phys. Rev.* D **63**, 024005 (2000); N. Manojlović and G. A. Mena Marugán, *Class. Quant. Grav.* **18**, 2065 (2001).
3. J. A. Valiente Kroon, *J. Math. Phys.* **41**, 898 (2000).
4. J. Bičák and A. Pravdová, *J. Math. Phys.* **39**, 6011 (1998); *Class. Quant. Grav.* **16**, 2023 (1999).
5. R. Lazkoz and J. A. Valiente Kroon, *Phys. Rev.* D **62**, 084033 (2000).
6. V. Pravda and A. Pravdová, *Czech. J. Phys* **50**, 323 (2000).
7. J. Bičák and V. Pravda, *Phys. Rev.* D **60**, 044004 (1999).
8. V. Pravda and A. Pravdová, *Class. Quant. Grav.* **18**, 1205 (2001).
9. J. Bičák and P. Krtouš, *Phys. Rev.* D **64**, 124020 (2001); and submitted.
10. J. Podolský and J. B. Griffiths, *Class. Quant. Grav.* **16**, 2937 (1999).
11. M. Kunzinger and R. Steinbauer, *Class. Quant. Grav.* **16**, 1255 (1999).
12. S. B. Edgar and G. Ludwig, *Gen. Rel. Grav.* **32**, 637 (2000).
13. G. Ludwig and S. B. Edgar, *Class. Quant. Grav.* **17**, 1683 (2000).
14. G. Ludwig and S. B. Edgar, *gr-qc/0111040*.
15. C. D. Collinson, *Gen. Rel. Grav.* **22**, 1163 (1990).
16. A. A. Coley and B. O. J. Tupper, *J. Math. Phys.* **33**, 1754 (1992).
17. J. Bičák, *Ann. Phys. (Leipzig)* **9**, 207 (2000).
18. D. Finley, *gr-qc/0108055*.
19. F. Fayos and C. F. Sopuerta, *Class. Quant. Grav.* **16**, 2965 (1999); **18** 353 (2001).
20. R. Halburd, *Class. Quant. Grav.* **18**, 11 (2001).
21. A. Krasiński, *J. Math. Phys.* **42**, 355 (2001); **42**, 3628 (2001).
22. J. R. Gair, *Class. Quant. Grav.* **18**, 4897 (2001)
23. A. Barnes, *Class. Quant. Grav.* **16**, 919 (1999))
24. M. A. H. MacCallum *et al*, in *Algebraic computing in General Relativity*, Eds. M. Rebouças & W. Roque (Oxford University Press, Oxford, 1994).
25. S. M. Wagh *et al*, *Class. Quant. Grav.* **18**, 2147 (2001).
26. M. King and H. Pfister, *Phys. Rev.* D **63**, 104004 (2001); and to appear.
27. J. Frauendiener and C. Klein, *Phys. Rev.* D **63**, 084025 (2001).
28. H. Balasin, *Class. Quant. Grav.* **17**, 1913 (2000).
29. M. Mars, *Phys. Rev.* D **63**, 064022 (2001).
30. Z. Y. Turakulov and N. Dadhich, *gr-qc/0104027*.
31. M. Bradley and M. Marklund, *Class. Quant. Grav.* **13**, 3021 (1996).

32. G. Fodor *et al*, *Class. Quant. Grav.* **16**, 453 (1999).
33. M. Nouri-Zonoz and A. Tavanfar, *Class. Quant. Grav.* **18**, 4293 (2001).
34. R. B. Mann and T. Ohta, *Class. Quant. Grav.* **17**, 4059 (2000).
35. I. G. Dymnikova, *Int. J. Mod. Phys.* **D5**, 529 (1996).
36. I. G. Dymnikova *et al*, *Phys. Lett.* **B472**, 33 (2000); **B506**, 351 (2001).
37. A. Ashtekar *et al*, *Phys. Rev. Lett.* **85**, 3564 (2000).
38. A. Ashtekar *et al*, *Class. Quant. Grav.* **18**, 919 (2001).
39. A. Corichi *et al*, *Phys. Rev.* D **64**, 107501 (2001).
40. A. Ashtekar *et al*, in preparation
41. A. Hussain and A. Qadir, *Phys. Rev.* D **63**, 083502 (2001)
42. F. W. Hehl *et al*, *Physics Reports* **258**, 1–171 (1995).
43. Y. Itin, *Int. Journal of Modern Phys.* **10**, 547 (2001).
44. R. Vera, *Ph.D. thesis*, Univ. de Barcelona (1994); and in preparation.
45. E. Witten, *Phys. Rev.* D **44**, 314 (1991).
46. K. S. Stelle, *hep-th/9803116.*
47. A. Strominger and C. Vafa, *Phys. Lett. B* **379**, 99 (1996).
48. P. K. Townsend, *Phys. Lett. B* **350**, 184 (1995).
49. R. Emparan *et al*, JHEP **0001**, 007 (2000).
50. R. C. Myers and M. J. Perry *Ann. Phys. (N.Y.)* **172**, 304 (1986).
51. R. Gregory and R. Laflamme, *Phys. Rev. Lett* **70**, 2837 (1993); G. T. Horowitz and K. Maeda *hep-th/0105111.*
52. J. Wainwright and G. F. R. Ellis eds., *Dynamical Systems in Cosmology* (Cambridge University Press, Cambridge, 1997).
53. A. Krasiński, *Inhomogeneous Cosmological Models* (Cambridge University Press, Cambridge, 1997).
54. K. R. Nayak *et al*, *Phys. Rev.* D **63**, 024020 (2001).

A2: COMPLEX METHODS, TWISTORS AND CONNECTION VARIABLES

K.P.TOD
Mathematical Institute
Oxford OX1 3LB
England
E-mail: tod@maths.ox.ac.uk

Section a2 was presented in two 2-hour sessions, each with five talks. Some of the selected speakers were eventually unable to attend the conference but we were able to present all of the material originally chosen. In summarising the talks below I have given archive references even for material published in refereed journals, on the basis that archives are easier for many people to find.

The first session was scheduled to begin with two coordinated talks on related abstracts, *Geometry of isolated horizons* (J.Lewandowski, A.Ashtekar and C.Beetle) to be given by Lewandowski and *Isolated horizon mechanics* (C.Beetle, A.Ashtekar, S.Fairhurst, B.Krishnan and J.Lewandowski) to be given by Ashtekar. Since Lewandowski was unable to attend the conference, Ashtekar gave a single talk covering the material in both (for which see gr-qc/0006006 and subsequent e-prints).

An isolated horizon (or IH) is a null hypersurface with certain geometrical properties, which is intended to play the role of the inner boundary of a region of space-time whose outer boundary is future-null-infinity, $\mathcal{I}^+$. The definition of an IH is modelled on local properties of a Killing horizon, and a space-time with an IH may be thought of as an approximation to a space-time with an event horizon. Unlike a Killing horizon, an IH is compatible with the presence of (outgoing) gravitational radiation and, unlike an event horizon, the definition of an IH is local.

The definition of an IH proceeds in two stages. One first defines a *non-expanding horizon* as a null hypersurface N, topologically of the form $S^2 \times \mathbf{R}$ with spherical space-like cross-sections and with vanishing expansion. Now suppose that ℓ is a choice of vector field tangent to the null generators of N, and impose the condition that $-T_{ab}\ell^b$ be time-like and future-pointing. This is implied by the dominant energy condition (the sign is due to choice of conventions), and implies that N is necessarily shear-free. This condition means that the degenerate metric of N is Lie-dragged by ℓ, that $T_{ab}\ell^b$ actually vanishes at N, and that the components of the Weyl tensor corresponding to radiation across N vanish. For the second stage, one notes that there is a unique covariant derivative D on vectors tangent to N defined by the 4-

dimensional metric covariant derivative. Now a non-expanding horizon N is *isolated* if D commutes with Lie-derivative along ℓ for some choice of ℓ (which is then unique up to constant rescaling).

A generalisation of the zero-th law of black hole mechanics follows immediately from the definitions. The first law requires definitions of mass and angular momentum of isolated horizons. These are introduced using Hamiltonian methods. A key new result is that the first law is a necessary and sufficient condition for time evolution to be Hamiltonian. Thus, not only does the framework generalize the first law to non-stationary contexts, but it sheds new light on its origin (gr-qc/0005083, gr-qc/0103026).

The framework has also proved to be useful in other contexts, ranging from numerical relativity to background independent quantum gravity: i) it plays a key role in an ongoing program for extracting physics from numerical simulations of black hole mergers (gr-qc/0006006); ii) it has led to the introduction of a physical model of hairy black holes, systematizing a large body of results obtained from a mixture of analytical and numerical investigations (gr-qc/0011081); and, iii) it serves as a point of departure for statistical mechanical entropy calculations in which all non-rotating black holes (extremal or not) *and* cosmological horizons are incorporated at a single stroke (gr-qc/9905089).

The next talk was *Canonical formulation of N=3 supergravity in terms of Ashtekar variables* by M.Tsuda. He described the canonical formulation of $N = 3$ supergravity in terms of Ashtekar variables, starting from the chiral Lagrangian (gr-qc/0011046). This is the lowest value of N at which spin-1/2 fields appear in the theory. No auxiliary fields are needed, unlike the case of 2-form gravity, and the constraints can be multiplied by a single non-polynomial factor to make them polynomial (as is the case with $N = 2$). The constraint algebra is still under investigation.

The abstracts chosen for the last two talks in the first session were *Conformal connection and equivalence problem for third-order ODEs* of M.Godliński, P.Nurowski and J.Korbicz and *Anti-Kähler manifolds and a new class of Einstein metrics* of A.Borowiec. In the event, neither Nurowski nor Borowiec was able to attend the conference so I presented these abstracts myself.

The abstract of Godliński et al considers the equivalence problem for third-order ODEs

$$\frac{d^3y}{dx^3} = F(x, y, \frac{dy}{dx}, \frac{d^2y}{dx^2})$$

under point transformations (a problem solved by Cartan) and under contact transformations (a problem solved by Chern). The revival of interest in this problem arises from the fact that the classification of equivalence classes can be described in terms of geometrical structures on the 3-dimensional 'moduli space' of solutions of the ODE. If the Wünschmann invariant, a contact invariant constructed from F, vanishes then the moduli space admits a conformal metric and conversely every conformal metric arises in this way. The remaining contact invariants of the ODE can be expressed in terms of the Cotton-York tensor of the conformal metric. An interesting outstanding question is how to relate those conformal metrics which admit Einstein-Weyl structures to this classification.

The abstract of Borowiec was based on joint work with Francaviglia and Volovich (math-ph/9906012) and with Ferraris, Francaviglia and Volovich (dg-ga/9612009) on what these authors call anti-Kähler manifolds. To be anti-Kähler, a manifold needs a metric g (non-degenerate, symmetric but not necessarily positive-definite), and a complex structure J which is covariant constant in the metric covariant derivative ∇^g (and therefore in particular is integrable) and is compatible with g in the sense

$$g(JX, JY) = -g(X, Y)$$

Note that the sign here is different from that in the usual compatibility condition, so that the tensor h defined by

$$h(X, Y) = g(JX, Y)$$

is symmetric rather than antisymmetric (with the other, more familiar, choice of sign, h would be the Kähler form). The tensor h is also non-degenerate, and covariant constant in the metric covariant derivative ∇^g. Since ∇^h is the unique torsion-free connection annihilating h, it follows that $\nabla^g = \nabla^h$. Also the signatures of g and of h are now both necessarily zero. The key observation is that the combination $\hat{g} = g - ih$ is now *holomorphic* and conversely every anti-Kähler structure arises as the real part of a holomorphic metric. Various restrictions on the underlying manifold follow from the requirement that it admit a holomorphic metric. The new class of Einstein metrics of the title are obtained as follows: any (real, Riemannian) solution of the Einstein equations, say of dimension n, is analytic and so may be complexified; the result is a holomorphic metric whose real and imaginary parts are anti-Kähler Einstein metrics of dimension $2n$. The authors also note that these anti-Kähler Einstein manifolds can be obtained from a Palatini variational principle with a suitable Lagrangian.

The second session began with two related talks on canonical analysis in the 2+2-formalism. Recall that the 2+2-formalism, named by analogy with the 3+1-formalism, is based on a decomposition of space-time into a two-parameter family of (usually) space-like 2-surfaces; the two coordinates labelling the 2-surfaces may both be null or one null and one time-like (for a discussion see e.g. *http://www.maths.soton.ac.uk/~jav/2+2.html).*

First P.Lambert spoke on *2+2-Hamiltonian using SO(3) connection variables*, the abstract of Lambert, Vickers and d'Inverno. These authors use a double-null formalism and a form of the action using self-dual 2-forms and $SO(3)$-connection variables. The Hamiltonian is expressed in terms of these variables and the authors carry through the constraint analysis. There are 26 or 32 primary constraints, depending on the form chosen for the Hamiltonian, and 14 secondary constraints.

Next Jong Hyuk Yoon spoke on *Constraint-free Hamiltonian dynamics and quasi-local conservation equations in the (2,2)-formalism* (see e.g. gr-qc/0004074). He uses a null+time-like foliation; data are thought of as given on the null hypersurfaces and evolved in the time-like direction. Now the constraints on the null hypersurfaces take the form of conservation laws for quasi-local quantities associated with the space-like 2-surfaces. In various limits, these quasi-local quantities become familiar notions of mass, momentum and angular momentum.

L.Szabados spoke third on *The Chern-Simons functional and the dynamics of GR.* He observes that there are a variety of Chern-Simons functionals associated with different spinor bundles on a space-like hypersurface Σ (gr-qc/9909052). Among these he chooses to study $Y = Y_{(1,0)}$ defined on the bundle of unprimed, valence-1 spinors. Under conformal rescaling, the real part of this functional is always unchanged, and the imaginary part is also unchanged *to first order* iff the Einstein vacuum equations hold at Σ. As functions on the (ADM) phase space, the functional $Re(Y)$ has vanishing Poisson bracket with Misner time, while the Poisson bracket of $Im(Y)$ with Misner time is the Hamiltonian for the vacuum equations. The interest of this is that both Misner time and $Im(Y)$ have been suggested as 'natural time variables'.

The next presentation was J.Frauendiener *On the Penrose inequality* (see gr-qc/0105093). By the Penrose inequality here is meant the inequality

$$16\pi M^2 \geq A$$

where M is the ADM mass of an asymptotically-flat hypersurface Σ and A the area of a marginally-trapped 2-surface (or MTS) on Σ, assuming there is one. This inequality is a prediction of the Cosmic Censorship Hypothesis. Other interpretations of the Penrose inequality follow from other interpretations of the mass M and the surface A under consideration: for example A could be the area of the intersection of an event-horizon with Σ or the total area of all outermost MTSs on Σ; Σ could be asymptotically null with M the Bondi mass etc.

If Σ is umbilic (or time-symmetric) then the Penrose inequality relates quantities defined just in terms of the (intrinsic) Riemannian geometry of Σ. This case has been proved by Huisken and Illmanen (Internat. Math. Res. Notices **20** (1997) pp1045-1058) by proving the existence of an inverse-mean-curvature flow, first suggested by Geroch, which flows the MTS out to infinity on Σ, while increasing the Hawking mass, a quantity which gives equality in the formula above and tends to M at infinity. Frauendiener proposes a flow which generalises this inverse-mean-curvature flow and drops the assumption that there is a time-symmetric Σ on which the MTS moves. His flow again increases the Hawking mass, so that this version of the Penrose inequality would be proved if the flow was known to exist and to move the MTS out to infinity.

Finally O.Richter gave the only explicitly twistor theory talk in the session, on *Algebro-geometric solutions to the Ernst equations and twistor theory*, describing work with C.Klein (gr-qc/9806051, 9811039, 9908042). The stationary, axisymmetric Einstein vacuum equations are known to be equivalent to a symmetry reduction of the self-dual Yang-Mills equations. Thus they can be solved by a twistor method - there is a one-to-one correspondence between solutions and certain rank-2 holomorphic vector bundles over 'reduced twistor space', a non-Hausdorff Riemann surface. To obtain the space-time, one needs to solve a matrix Riemann-Hilbert problem on a family of 2-spheres. In a particular case, this can be reduced to a scalar Riemann-Hilbert problem on a hyperelliptic Riemann surface and this can be solved in terms of θ-functions. These give the algebro-geometric solutions of the title and for them the authors can give the metric and vector-bundle explicitly. The genus-2 sub-case includes the metric of a counter-rotating disk of dust.

In summary, these were two interesting sessions. This Section continues to represent a region of active research with a distinctive character within the wider world of GR.

Acknowledgments

I am grateful to the organisers of GR16 for local expenses, to St John's College and the Mathematical Institute, Oxford for travel expenses, and to Stephen Siklos and the authors of the abstracts presented for comments on this report.

A3: MATHEMATICAL STUDIES OF THE FIELD EQUATIONS

LARS ANDERSSON
Department of Mathematics
University of Miami
Email: larsa@math.miami.edu

Due to length constraints, some of the topics covered at the session will be very briefly mentioned or left out altogether. The summary is mainly based on extended abstracts submitted by the speakers. I would like to thank all participants for their contributions and help with this summary. An extended summary can be found at `http://xxx.lanl.gov/abs/gr-qc/0204087`.

1 The Horowitz-Myers Soliton

According to a conjecture by Maldacena[1], supported by a growing body of evidence, there is a correspondence between string theory in Anti-deSitter (AdS) spacetime and a conformal field theory (CFT) on the boundary of AdS. Stability considerations motivated by the AdS/CFT conjecture led to the conjecture that the boundary of asymptotically AdS space-times should be connected. The Riemannian version of this was proved by Witten and Yau[2] and generalized by Cai and Galloway[3]. The non-supersymmetric version of the AdS/CFT correspondence, leads to consideration of asymptotically locally AdS spacetimes with conformal boundary such that one spatial direction is compactified on a circle. If we allow nontrivial topology at infinity, the positive mass theorem for general relativity is not valid in general[4]. In case of negative cosmological constant, $\Lambda < 0$, a partial Cauchy surface in an asymptotically locally AdS $n+1$-dimensional space-time may have any orientable $n-1$-manifold as boundary at infinity. The Kottler and Nariai metrics give explicit examples of static AdS metrics on manifolds with nontrivial topology at infinity[5]. These examples all have horizons.

Horowitz and Myers[4] have found a static solution of the Einstein equations with negative cosmological constant, $\Lambda < 0$, which is asymptotically locally AdS and contains no horizon. I will refer to this spacetime as the HM soliton. The HM soliton in $n+1$ dimensions is given by the metric

$$ds^2 = -r^2 dt^2 + \frac{1}{V(r)} dr^2 + V(r) d\phi^2 + r^2 \sum_{i=1}^{n-2} (dy^i)^2 \qquad (1)$$

where $V(r) = \frac{r^2}{\ell^2}\left(1 - \frac{r_0^n}{r^n}\right)$, $\ell^2 = -\frac{n(n-1)}{2\Lambda}$, and r_0 is a constant. In order to avoid an orbifold singularity, ϕ must be identified with period $\beta_0 = \frac{4\pi\ell^2}{nr_0}$. After making the y^i coordinates periodic as well (with arbitrary period), the resulting spacetime has conformally compact Cauchy surfaces with flat torus conformal boundary.

Computing the mass of the soliton using a standard method gives the negative answer $E = -\frac{\beta r_0^2}{4G_5\ell}$. The mass at infinity of metrics asymptotic to the soliton is related via the AdS/CFT correspondence to the negative Casimir energy of a non–supersymmetric gauge theory on the boundary. If a non–supersymmetric version of the AdS–CFT conjecture holds, then stability considerations imply that the soliton must be the lowest energy state with the given asymptotic conditions. This leads to a new positive mass conjecture,

Conjecture 1 (Horowitz and Myers) *Spacetimes satisfying a dominant energy condition with $\Lambda < 0$, and with the asymptotic behavior of the soliton (1), have mass bounded from below by E. The soliton is the unique lowest mass solution for all spacetimes in its class.*

This conjecture is supported by perturbation calculations up to second order[4,6]. A further piece of supporting evidence was provided by the recent work of Galloway, Surya and Woolgar[7], described by **Greg Galloway** in his talk at session A.3. Galloway et al. proved that the HM soliton is the unique static metric in a class of metrics with the given asymptotic conditions. In order to formulate the main result of Galloway et. al., let (Σ, h, N) be a solution of the static vacuum field equations $R_{ab} = N^{-1}\nabla_a\nabla_b N + \frac{2\Lambda}{n-1}h_{ab}, \Delta N = -\frac{2\Lambda}{n-1}N$, which is conformal to a compact manifold with boundary $(\tilde{\Sigma}, \tilde{h})$ with conformal factor $\tilde{N} = 1/N$ (with $\tilde{N} = 0$ on $\partial\tilde{\Sigma}$. Then (Σ, h, N) is said to satisfy condition (S) if the level sets $\tilde{N} = c$ near infinity are weakly convex in the sense that the principal curvatures are either all positive or all negative. The mass is computed as an integral over the conformal boundary of the mass aspect of (Σ, h, N), defined as (specializing the Ashtekar–Magnon conformal mass to the static case) $\mu = -\partial^{n-2}\tilde{R}/\partial x^{n-2}|_{x=0}$ (where $x = \tilde{N}$ and $\tilde{R}$ is the scalar curvature function of $(\tilde{\Sigma}, \tilde{h})$).

Theorem 1.1 *Let (Σ, h, N) be a static spacetime as above, satisfying condition (S). Suppose in addition, that (a) The conformal boundary of (Σ, h, N) is the same as that of the H&M soliton, (b) The mass aspect μ of (Σ, h, N) is pointwise negative, and (c) The kernel of the inclusion map $i_* : \Pi_1(\partial\tilde{\Sigma}) \to \Pi_1(\tilde{\Sigma})$ is generated by the S^1 factor. Then the spacetime $M^{n+1} = \mathbb{R} \times \Sigma, \quad g = -N^2dt^2 \oplus h$. determined by (Σ, h, N) is isometric to the AdS soliton (1).*

The proof uses complete null lines constructed using the convexity of the boundary together with a null splitting theorem of Galloway[8] to prove that the universal cover of the conformal compactification $(\tilde{\Sigma}, \tilde{h})$ splits as a metric product $\mathbb{R}^{n-2} \times W$ with the metric on W determined uniquely by the field equations.

2 Dynamical systems in general relativity

Claes Uggla reviewed some applications of dynamical systems methods in general relativity. Dynamical systems formulations arise naturally in the study of Einstein's field equations for special classes of spacetimes, such as hypersurface–homogenous and hypersurface-self-similar spacetimes[9]. Further examples are provided by the geodesic equations of certain classes of spacetimes, where dynamical systems techniques can be used to e.g., calculate the null geodesics and the corresponding cosmic microwave background radiation pattern in a cosmological model[10,11], and the density perturbations of spatially homogenous models. The classical example of application of dynamical systems methods to the Einstein equations is the study of spatially homogenous (Bianchi) space–times, with or without matter. The Bianchi family of spacetimes contains the Friedman spacetimes as a special case, and gives a rich set of examples of the influence of matter and geometry on the early and late time evolution of space–times. It is a natural problem to analyze completely the behavior of the resulting family of dynamical systems.

The book on dynamical systems in cosmology edited by Ellis and Wainwright[12] gives a good overview of the state of the art of this approach, some years ago. As an outgrowth of that work, substantial progress has been made towards understanding the Bianchi systems both near the initial singularity and in the asymptotically expanding direction. **Hans Ringström**[13,14] proved that Bianchi VIII and IX have oscillating singularity, which is an important step in understanding the full BKL picture for these models. An important step in the analysis was to prove that Bianchi II is an attractor for Bianchi IX. This sort of hierarchical behavior in the Bianchi models, is an important organizing principle also for further work on systems with lower dimensional isometry groups. Ringström has also, as he described in his talk, been able to analyze in detail[15] the asymptotic behavior of the expanding vacuum Bianchi class A spacetimes (this excludes Bianchi IX). For the most complicated of these, Bianchi VIII, the result is that the geometry is asymptotically Taub, with precise control on the asymptotic behavior of the commutator variables. For Bianchi VII_0, the system approaches (in terms of scale free variables), a flat VII_0 spacetime. The Cauchy surfaces of these spacetimes are all Seifert fibered. The results show that the scale free geometry of the spatial sections in these spacetimes collapse along the fiber direction. The result is consistent with the picture for the limiting behavior in the asymptotic expanding direction developed by M. T. Anderson[16] Analyzing the hierarchical structure of systems with small symmetry group, such as Gowdy models, or without symmetry, is likely to be a crucial step in understanding the global dynamics of the Einstein equations.

3 Well-posed forms of the 3+1 trace-decoupled Einstein equations

The standard ADM form of the Einstein evolution equations fails to be well posed both from the mathematical and numerical points of view. Due to the demand for stable numerical procedures for solving the Einstein evolution equations, a great deal of interest has recently been focussed on well posed hyperbolic formulations of the Einstein equations. This is of particular importance in solving boundary value problems. **Oscar Reula** discussed various well-posed reformulations of the Einstein equations. It is possible to derive a well-posed form of the standard 3+1 Einstein equations[17] by reducing to first order form, densitizing the lapse function and adding combinations of the constraints to the evolution equations[18,19]. An alternative approach, which recently has been shown to possess striking computational advantages over the standard form[20], is to decompose the fundamental variables in order to extract the trace of the extrinsic curvature and the determinant of the 3-metric with the purpose of evolving them independently from the remaining components. This trace-decoupled form of the Einstein equation turns out, oddly enough, to be ill-posed. Reula and Frittelli have used techniques similar to those used for the standard evolution equations, to obtain versions of the 3+1 Einstein equations which are both trace-decoupled and well posed. This well posed version propagates the constraints in a stable manner, which is relevant to unconstrained evolution. It uses essentially the same variables as the original system, but requires a gauge condition, namely that the lapse to be proportional to the determinant of the intrinsic geometry of the surfaces.

4 Critical phenomena

Critical phenomena in general relativity have been extensively studied since the discovery by Choptuik of universal behavior in the collapse of a self-gravitating scalar field. Since then a number of matter models, such as charged self-gravitating scalar fields, Yang–Mills and dilaton fields have been studied and a systematic understanding of these phenomena has been gained, in particular through the study of the stability of models with continuous or discrete self–similarity. We now have a good qualitative understanding of criticality in gravitational collapse: there is an intermediate (codimension-1) attractor of the evolution flow in Phase Space, sitting on the threshold of black hole formation. The symmetries of that attractive solution determine the type of criticality: 'type I' if the solution is static, 'type II' when it is self-similar.

The Einstein–Vlasov system describes the evolution of a statistical ensemble of non-interacting particles coupled to gravity through their average properties. **Jose M. Martin–Garcia** described joint work with Carsten Gundlach on critical phenomena in the Einstein–Vlasov system, which is important in Critical Phenomena Theory because it is the only one where type II critical phenomena have not yet been found.

There are two different ways of checking for the existence of criticality. On the one hand, it is possible to evolve numerically families of initial conditions and study the collapse threshold, looking for intermediate attractors and universality. On the other hand, one can construct static or self-similar solutions and then investigate whether they are codimension-1 linearly stable.

There are two relevant works in this problem, both using the method of evolution of initial conditions: Rendall, Rein and Schaeffer[21] have not found criticality and suggest that, if there is any, it must be type I. Olabarrieta and Choptuik[22] confirm the apparent nonexistence of type II phenomena and have found some signs of type I criticality, though not conclusive because there is not universality in the critical exponents. Gundlach and Martin–Garcia look for type II criticality in the massless Einstein-Vlasov system using the second method.

In his talk Martin–Garcia reported on the first step: the numerical construction of self-similar solutions, which is an interesting problem in itself. The situation resembles what is already known analytically for static solutions of the same system[23]. There is a free function of two variables, representing different distributions of the energy-momentum among the particles of the system, but all giving rise to the same gravitational field. Therefore a candidate critical solution can not be isolated. It is likely that a similar result in the massive case could explain the observed non-universality.

Recent numerical investigations[24,25] of near-critical scalar field collapse in 2+1 dimensional AdS spacetime show a continuous self-similar (CSS) behaviour near the central singularity and power law scaling for the black hole mass. Garfinkle[26] found that this behaviour is well approximated, at intermediate times, by a member of a one-parameter family of exact CSS solutions to the equations with vanishing cosmological constant Λ. However the extension of these solutions to $\Lambda \neq 0$, necessary to explain black hole formation (as vacuum black holes exist only for $\Lambda < 0$) is a hard problem which has not been solved up to now.

Gérard Clément in his talk discussed joint work with Alessandro Fabbri[27] on quasi-CSS 2+1 dimensional scalar field spacetimes. They have derived by a limiting process a new class of $\Lambda = 0$ CSS solutions

$$\begin{aligned} ds^2 &= dudv - (-u)^{2/(1+c^2)} d\theta^2 , \\ \phi &= -\frac{c}{1+c^2} \ln(-u) \end{aligned} \tag{2}$$

These present for $c^2 \leq 1$ a point singularity ($u = 0$, $v = +\infty$), and for $c^2 < 1$ a null line singularity $u = 0$. They can be extended to exact solutions of the $\Lambda = -l^{-2} < 0$ equations by means of the ansatz

$$ds^2 = e^{2\sigma(x)} dudv - (-u)^{\frac{2}{c^2+1}} \rho^2(x) d\theta^2 , \quad \phi = -\frac{c}{c^2+1} \ln|u| + \psi(x) , \tag{3}$$

where $x = uv$, with the initial conditions $\rho(0) = 1$, $\sigma(0) = \psi(0) = 0$. These extended solutions inherit the CSS behaviour close to the central singularity, and have the correct AdS behaviour at spatial infinity. To show that these quasi-CSS solutions are indeed threshold solutions for black hole formation, one studies their linear perturbations. The case $c = 0$ is at the same time simple and interesting. In this case the scalar field decouples, $\phi = 0$, and the corresponding quasi-CSS solution is the vacuum BTZ[28] solution. The linear perturbation equations with appropriate boundary conditions lead to a unique solution, with eigenvalue $k = 2$. This solution is found to be an exact linearization in M of the BTZ black hole. In the case of genuine scalar perturbations, one finds two possible modes $k_a = c^2 + 3/2$ and $k_b = c^2 + 2$. For the a mode the singularity and the apparent horizon appear simultaneously on the null line $u = 0$ at the time $v = 0$ and evolve in the region $v > 0$ in a physically meaningful way, while for the b mode the singularity still appears for $v = 0$, but the apparent horizon seems to be eternal, as in the case of the static BTZ black hole. The b mode, which cannot describe actual gravitational collapse with regular initial conditions, appears to be unphysical. The critical exponent is related to the eigenvalue k by $\gamma = 1/k$. In the BTZ case $c = 0$ Clément and Fabbri obtain $\gamma = 1/2$. In the scalar field case, choosing the critical value $c^2 = 1$ they obtain for the a mode $\gamma = 0.4$. This value differs signicantly from those found in the numerical analysis[24,25], i.e. $\gamma \sim 1.2$ and $\gamma \sim 0.81$, showing that further work is needed in order to clarify this issue.

5 Isolated systems

It is an open problem to construct a nonflat vacuum spacetime, which has a regular conformal infinity in the sense of Penrose. The conformal compactification of a nonflat spacetime must have a singularity at spatial infinity i_0, due to the slow falloff of the Weyl tensor at i_0. **Sergio Dain** described joint work with Helmut Friedrich[29], which constructs a large class of asymptotically flat initial data with non-vanishing mass and angular momentum for which the metric and the extrinsic curvature have asymptotic expansions at space-like infinity in terms of powers of a radial coordinate. These asymptotic expansions are of the form $\tilde{h}_{ij} \sim (1 + \frac{2m}{\tilde{r}})\delta_{ij} + \sum_{k\geq 2} \frac{\tilde{h}^k_{ij}}{\tilde{r}^k}, \tilde{\Psi}_{ij} \sim \sum_{k\geq 2} \frac{\tilde{\Psi}^k_{ij}}{\tilde{r}^k}$, where $\tilde{h}^k_{ij}$ and $\tilde{\Psi}^k_{ij}$ are smooth function on the unit 2-sphere (thought as being pulled back to the spheres $\tilde{r} = const.$ under the map $\tilde{x}^j \to \tilde{x}^j/\tilde{r}$). In earlier work[30,31], stationary data were shown to admit expansions of this type.

In the study of isolated system, an important and difficult problem is to give an unambiguous definition of angular momentum. The difficulty lies in the fact that without introducing extra structure, there is no unambiguously defined Poincaré subgroup of the asymptotic isometries of an isolated system. This is known as the supertranslation ambiguity. This ambiguity is known as supertranslation ambiguity. In order to provide with an unambiguous notion of angular momentum in radiative spacetimes, it is essential to make use of a reference frame system, which embodies the notion of rest frame[32]. **Osvaldo Moreschi** presented a definition of angular momentum for radiative spacetimes which does not suffer from supertranslation ambiguity. The definition uses the construction of so called 'nice sections' at $\mathcal{I}^+$. The defining equations for these have recently been proved[33,34] to have solutions in terms of a 4-parameter family of translations. The solutions have the expected physical properties of a rest frame. The nice section construction singles out a Poincaré structure from the infinite dimensional BMS group. In particular, given a fixed observational point p at $\mathcal{I}^+$, there is precisely a 3-degree of freedom of spacelike translations which generate all the nice sections that contain p. In contrast, without this constructions there is an infinite dimensional family of general sections that contain p, one for each supertranslation. The condition that the spatial part of the Bondi 4–momentum is zero singles out a one parameter family of nonintersecting[34] sections S_{cm} of $\mathcal{I}^+$, which define a center of mass frame and can be used to describe the detailed asymptotic structure of the spacetime. The angular momentum is defined in terms of a charge integral over the S_{cm}. Detectors of gravitational waves can be considered as observers at future null infinity. The construction of center of mass sections, provides for each point p at scri a prescription that singles out the unique center of mass section containing p, together with the appropriate rest frame to calculate intrinsic quantities like angular momentum and multipole moments, which are important for the description of the gravitational waves that one wishes to detect.

John L. Friedman described described in his talk joint work with Kōji Uryū and Masaru Shibata. They consider compact binary systems, modeled as vacuum or perfect-fluid spacetimes with a helical Killing vector k^a, heuristically, the generator of time-translations in a corotating frame. Systems that are stationary in this sense are not asymptotically flat, but have asymptotic behavior corresponding to equal constant fluxes of ingoing and outgoing radiation. For black-hole binaries, a rigidity theorem implies that the Killing vector lies along the horizon's generators, and from this one can deduce the zeroth law (constant surface gravity of the horizon). Remarkably, although the mass and angular momentum of such a system are not defined, a finite Noether charge is defined on any sphere enclosing the matter and black holes, its value the same on each such sphere. An exact first law relates the change in this charge to the changes in the vorticity, baryon mass and entropy of the fluid and in the area of black holes. Modeling a binary system by a spacetime with a helical Killing vector is accurate if the energy emitted in a dynamical time is small compared to the kinetic energy of the system. Other approaches use a post–Newtonian approximation, or conformally flat spacelike slices that satisfy a truncated set of field equations. Like the post-Newtonian spacetimes, these conformally flat spacetimes are nonradiative and asymptotically flat. In the isolated horizon framework, for a horizon with a single Killing vector, one shows the existence of a charge E defined on an isolated horizon for which $\delta E = \kappa \delta A$. The first law for black holes with helical symmetry, in contrast, relates this change in the black-hole charges to the changes in the Noether charge of a sphere surrounding all black holes and all matter and to the changes in the entropy, baryon number, and circulation of the fluid. The existence of such a first law depends precisely on what is *not* assumed in the isolated horizon framework: a globally defined Killing vector.

6 Variational formulation

Naresh Dadhich and **Niall Ó Murchadha** gave talks relating to the variational formulation of general relativity. Dadhich considered the conditions under which one can deduce the field equations from the equations of motion for particles. In the context of electomagnetic field, this question was, according to Dyson [35], first considered by Feynman. By considering commutation relations between position and velocity, he could obtain the Bianchi set of the equations. This gave rise to considerable activity in this direction involving rederivations, generalizations and extensions. However the attention centred on the Bianchi set. Singh and Dadhich[36] derived the full set of the Maxwell equations by demanding the law of motion to be linear in velocity and derivable from a Lagrangian. A similar procedure has been applied to the Einstein equations[37]. In this case, the fact that the particle law in general relativity is quadratic in the 4–velocity leads to the Einstein equations. This is a novel way of deducing the field equations from the particle law of motion.

Niall Ó Murchadha described joint work with Julian Barbour and Brendan Foster[38], where a derivation of General Relativity was given which does not depend on the relativity principle. The fundemental idea is to choose superspace (the space of all riemannian three geometries) as the configuration space and to consider a parametrised curve in this configuration space, with parameter λ. Ó Murchadha et al. consider a generalized Jacobi-type action of the form $\int d\lambda \int dx^3 \sqrt{gPT}$ on such curves where one first integrates on each given three manifold and then integrates along the curve. 'P' is the potential which is a scalar on the three manifold but otherwise arbitrary, while 'T' is the generalized kinetic energy which means that it is a quadratic function of $\partial g_{ij}/\partial\lambda$. It turns out that in order for the constraints of the theory to be propagated, it is necessary to choose the potential energy term to be the three scalar curvature and the kinetic energy to be defined with the DeWitt supermetric. This recovers the Baierlein, Sharp, Wheeler parametrised action for G.R. as the unique self-consistent Jacobi-type action on superspace.

References

1. J. Maldacena. Plenary talk at GR16. In *GR16 Proceedings*, 2001.
2. Edward Witten and S.-T. Yau. Connectedness of the boundary in the AdS/CFT correspondence. *Adv. Theor. Math. Phys.*, 3(6):1635–1655 (2000), 1999.
3. Mingliang Cai and Gregory J. Galloway. Boundaries of zero scalar curvature in the AdS/CFT correspondence. *Adv. Theor. Math. Phys.*, 3(6):1769–1783 (2000), 1999.
4. Gary T. Horowitz and Robert C. Myers. AdS-CFT correspondence and a new positive energy conjecture for general relativity. *Phys. Rev. D (3)*, 59(2):026005, 12, 1999.
5. Piotr T. Chruściel and Walter Simon. Towards the classification of static vacuum spacetimes with negative cosmological constant. *J. Math. Phys.*, 42(4):1779–1817, 2001.
6. Neil R. Constable and Robert C. Myers. Spin-two glueballs, positive energy theorems and the AdS/CFT correspondence. *J. High Energy Phys.*, (10):Paper 37, 30 pp. (electronic), 1999.
7. G.J. Galloway, S. Surya, and E. Woolgar. A uniqueness theorem for the

ads soliton, 2001. hep-th/0108170.

8. Gregory J. Galloway. Maximum principles for null hypersurfaces and null splitting theorems. *Ann. Henri Poincaré*, 1(3):543–567, 2000.
9. Ulf S. Nilsson and Claes Uggla. Hypersurface homogeneous and hypersurface self-similar models. *Classical Quantum Gravity*, 14(7):1965–1980, 1997.
10. Ulf S. Nilsson, Claes Uggla, and John Wainwright. A dynamical systems approach to geodesics in Bianchi cosmologies. *Gen. Relativity Gravitation*, 32(10):1981–2005, 2000.
11. U S. Nilsson, Uggla C., J. Wainwright, and W. C. Lim. An almost isotropic microwave temperature does not imply an almost isotropic universe. *The Astrophysical Journal Let.*, page 521, 1999.
12. J. Wainwright and G. F. R. Ellis, editors. *Dynamical systems in cosmology*. Cambridge University Press, Cambridge, 1997. Papers from the workshop held in Cape Town, June 27–July 2, 1994.
13. H. Ringström. Curvature blow up in Bianchi VIII and IX vacuum spacetimes. *Classical Quantum Gravity*, 17(4):713–731, 2000.
14. Hans Ringström. The Bianchi IX attractor. *Ann. Henri Poincaré*, 2(3):405–500, 2001.
15. H. Ringström. The future asymptotics of Bianchi VIII vacuum solutions. *Classical Quantum Gravity*, 18(18):3791–3824, 2001.
16. M. T. Anderson. On long-time evolution in general relativity and geometrization of 3-manifolds. *Comm. Math. Phys.*, 222:533–567, 2001.
17. Jr. James W. York. Kinematics and dynamics of general relativity. In Larry L. Smarr, editor, *Sources of gravitational radiation*, pages 83–126, Cambridge-New York, 1979. Cambridge University Press.
18. Simonetta Frittelli and Oscar Reula. First-order symmetric hyperbolic einstein equations with arbitrary fixed gauge. *Phys. Rev. Lett.*, 76:4667–4670, 1996.
19. Simonetta Frittelli and Oscar A. Reula. Well-posed forms of the 3 + 1 conformally-decomposed Einstein equations. *J. Math. Phys.*, 40(10):5143–5156, 1999.
20. Thomas W. Baumgarte and Stuart L. Shapiro. Numerical integration of Einstein's field equations. *Phys. Rev. D (3)*, 59(2):024007, 7, 1999.
21. Gerhard Rein, Alan D. Rendall, and Jack Schaeffer. Critical collapse of collisionless matter - a numerical investigation. *Phys. Rev. D*, 58:044007, 1998.
22. I. Olabarrieta and M. Choptuik. Critical phenomena at the threshold of black hole formation for collisionless matter in spherical symmetry. gr-qc/0107076.

23. Gerhard Rein. Static solutions of the spherically symmetric Vlasov-Einstein system. *Math. Proc. Cambridge Philos. Soc.*, 115(3):559–570, 1994.
24. Frans Pretorius and Matthew W. Choptuik. Gravitational collapse in $2+1$ dimensional AdS spacetime. *Phys. Rev. D (3)*, 62(12):124012, 15, 2000.
25. Viqar Husain and Michel Olivier. Scalar field collpase in three-dimensional AdS spacetime. *Classical Quantum Gravity*, 18(2):L1–L9, 2001.
26. David Garfinkle. Exact solution for (2+1)-dimensional critical collapse. *Phys. Rev. D (3)*, 63:044007, 5, 2001.
27. G. Clément and A. Fabbri. Critical collapse in 2+1 dimensional ads spacetime: quasi-css solutions and linear perturbations. gr-qc/0109002., 2001.
28. Máximo Bañados, Claudio Teitelboim, and Jorge Zanelli. Black hole in three-dimensional spacetime. *Phys. Rev. Lett.*, 69(13):1849–1851, 1992.
29. Sergio Dain and Helmut Friedrich. Asymptotically flat initial data with prescribed regularity. *Comm. Math. Phys.*, 222(3):569–609, 2001. gr-qc/0102047.
30. R. Beig and W. Simon. Proof of a multipole conjecture due to Geroch. *Comm. Math. Phys.*, 78(1):75–82, 1980/81.
31. Sergio Dain. Initial data for stationary space-time near space-like infinity. *Class. Quantum Grav.*, 18(20):4329–4338, 2001. gr-qc/0107018.
32. Jeffrey Winicour. Angular momentum in general relativity. In *General relativity and gravitation, Vol. 2*, pages 71–96. Plenum, New York, 1980.
33. Osvaldo M. Moreschi and Sergio Dain. Rest frame system for asymptotically flat space-times. *J. Math. Phys.*, 39(12):6631–6650, 1998.
34. Sergio Dain and Osvaldo M. Moreschi. General existence proof for rest frame systems in asymptotically flat spacetime. *Classical Quantum Gravity*, 17(18):3663–3672, 2000.
35. Freeman J. Dyson. Feynman's proof of the Maxwell equations. *Amer. J. Phys.*, 58(3):209–211, 1990.
36. Parampreet Singh and Naresh Dadhich. The field equation from Newton's law of motion and the absence of magnetic monopole. *Int. J. Mod. Phys. A*, 16:1237–1247, 2001.
37. Parampreet Singh and Naresh Dadhich. Field theories from the relativistic law of motion. *Modern Phys. Lett. A*, 16(2):83–90, 2001.
38. J. Barbour, B. Foster, and N. Ó Murchadha. Relativity without relativity. gr-qc/0012089.

A4: APPROXIMATION METHODS

C. M. WILL
Department of Physics and McDonnell Center for the Space Sciences
Washington University, St. Louis MO 63130 USA
E-mail: cmw@wuphys.wustl.edu

The Workshop A4 on Approximation Methods was divided into three sessions. The first, on post-Newtonian methods, focussed on efforts to calculate the equations of motion and gravitational waveform of inspiralling compact binaries to very high order in a post-Newtonian approximation (essentially in powers of $(v/c) \sim (Gm/rc^2)^{1/2}$). Progress was reported on the effort to reach third post-Newtonian order and beyond. The second session focussed on perturbations of black holes and stars, with particular interest in linking black hole perturbations with numerical relativity, and in the unstable r-modes of relativistic stars. The final session was devoted to miscellaneous perturbation methods.

1 Post-Newtonian Methods

It has been realized for some time that, in order to maximize the usefulness of matched filtering for extracting information from gravitational waves from inspiralling compact binaries detected by laser interferometers, template waveforms will be needed that are accurate to at least third post-Newtonian order (3PN). Waveforms accurate to 2PN order were derived by Blanchet, Damour and Iyer[1], and Will and Wiseman[2], and reported jointly.[3] The formidable task of reaching 3PN and 3.5PN order has been tackled by three independent groups, Blanchet and Iyer (and their students and collaborators), Schäfer, Jaranowski and Damour, and by Will and his collaborators. Two of these groups reported on progress at the workshop.

The team of Blanchet *et al.* reported on reaching the 3PN level for equations of motion and 3.5PN for the gravitational waveform. Faye (with de Andrade and Blanchet) discussed development of a Lagrangian from which one can derive the third post-Newtonian (3PN) equations of motion of compact binaries (neglecting the radiation reaction damping)[4]. The 3PN equations of motion were computed previously by Blanchet and Faye in harmonic coordinates[5,6]. The Lagrangian derived depends on the harmonic-coordinate positions, velocities and accelerations of the two bodies. At the 3PN order, one undetermined physical parameter λ appeared, which reflects the incompleteness of the point-mass regularization used when deriving the equations of motion. In addition the Lagrangian involves two unphysical (gauge-dependent) constants r'_1 and r'_2 parametrizing some logarithmic terms. Faye showed expressions for the ten Noetherian conserved quantities, associated with the invariance of the Lagrangian under the Poincaré group. By performing an infinitesimal "contact" transformation of the motion, he showed that the 3PN harmonic-coordinate Lagrangian is physically equivalent to the 3PN Arnowitt-Deser-Misner Hamiltonian obtained recently by Damour, Jaranowski and Schäfer.[7,8,9]

Bala Iyer (with Blanchet, Faye, and Joguet) applied the multipolar-post-Minkowskian approach to the problem of the generation of waves by compact binary inspiral. They computed the 3PN contribution to the total energy flux. New inputs are the 3PN mass quadrupole moment and the 2PN current quadrupole and 2PN mass octupole. Wave tails and tails of tails in the far zone were included to 3.5PN order. The recently derived 3PN equations of motion were then used to compute the time-derivatives of the moments. There was perfect agreement to the 3.5PN order with perturbation calculations of black holes in the test mass limit for one body. Technical inputs in the computation included a model of point particles for describing the compact objects and a Hadamard-type self-field regularisation. Because of a physical incompleteness of the Hadamard regularisation at the 3PN order the energy flux depends on one unknown physical parameter, a combination of the parameter λ in the equation of motion and a parameter θ coming from the mass quadrupole moment. The orbital phase whose prior knowledge is crucial for searching and analyzing the inspiral signal was finally computed through 3.5PN order beyond the quadrupole radiation by combining the 3.5PN-accurate binary's center-of-mass energy and emitted gravitational flux, and by using the standard energy balance argument.[10]

Poujade (with Blanchet) discussed the issue of algorithmically iterating the PN expansion to all orders, based on a general algorithm for computing, by means of a multipolar post-Minkowskian expansion, the gravitational field in the region exterior to an isolated material source, developed by Blanchet and Damour.[11] In that method, the exterior field is parametrized by some time-varying multipole moments (of any order) describing the source. Subsequently the multipole moments, as explicit functionals of the pseudo-stress-energy tensor of the source, were obtained by means of a matching equation valid in the exterior near-zone of the source.[12] Poujade described a general systematic scheme[13], based on the matching equation, to calculate, formally up to any orders in $1/c$, the post-Newtonian expansion of the field in harmonic coordinates near and inside the material source. The general solution of the post-Newtonian development is written in the form of a particular Poisson-type solution plus some unknown "homogeneous" terms. The post-Minkowskian exterior field is then expanded in the near zone. By matching those expansions, in the intermediate zone where both developments are valid, one finds uniquely and iteratively those homogeneous terms which are associated with the reaction of the gravitational field on the source (including tails).

The second team of Will and collaborators reported on their progress toward 3.5PN order. This work uses a framework called Direct Integration of the Relaxed Einstein Equations (DIRE) to calculate equations of motion and gravitational radiation emission for isolated gravitating systems based on the post-Newtonian approximation.[14] This was a report of work in progress, and touched on results for radiation reaction at 3.5PN order in the equations of motion, and equations of motion for spinning bodies including radiation reaction. He also discussed the possible presence of structure-dependent terms in the equations of motion at 3PN order that emerge when one treats the bodies explicitly as fluid balls, instead of using formal *partie finie* and Hadamard regularization techniques.

2 Perturbations of Black Holes and Stars

Perturbation theory for stationary, relativistic stars and black holes has become a particularly active and rich subject, because of (a) the ability to carry out highly accurate numerical calculations as well as post-Newtonian-like calculations to extremely high (5.5PN) order, and thereby derive accurate gravitational waveforms; (b) the surprising concordance between certain results of black hole perturbation theory (especially when carried out to second-order) and those of numerical relativity for black hole mergers and inspirals; and (c) the discovery of the importance of a class of unstable perturbations of relativistic stars, known as r-modes. Some of these topics were addressed in this session.

Price (with Beetle, Landry and Whelan) discussed a method to address effects in the late stage of the inspiral of two black holes that may be entirely unrelated to gravitational radiation reaction. To understand these effects they are developing an approximation method in which radiation reaction is eliminated by specifying that the orbit and spacetime be precisely periodic. The computational problem is then converted from an evolution problem with initial data, to a boundary value problem with equal igoing and outgoing radiation. In addition to giving approximate answers for inspiral, a periodic solution can give initial data for numerical relativity evolution. Price reported results on simplified models and on progress in building 3D numerical solutions.[15]

Since its serendipitous discovery about three years ago, the gravitational-wave driven instability associated with the so-called r-modes in a relativistic star has attracted considerable attention. However, even though the initial discovery was associated with an attempt to study the r-modes in a fully relativistic context, most of the studies of this mechanism have been done in the context of Newtonian theory. Andersson described recent efforts to get to grips with the relativistic r-mode problem.[16] This requires a deeper understanding of the wider class of so-called inertial modes of which the r-modes form a relatively small subset. While the r-modes can be described by a purely axial perturbation, the typical inertial mode is a hybrid mixture of axial and polar quantities. In the case of relativistic barotropic stars, one can prove that no purely axial r-modes can exist — they are all hybrids. This makes the numerical calculation of the relativistic r-modes significantly harder than the corresponding Newtonian calculation. The case of non-barotropic stars is also rather complex, as one must deal with a singular eigenvalue problem. Andersson discussed the solution of this problem and presented a fully relativistic estimate of the characteristic growth-time of the r-mode instability.

Williams (with Andersson and D'Inverno) reported on the status of a recently initiated project to investigate the advantages of characteristic methods in studies of perturbed Kerr black holes. In preliminary work, numerical codes were developed for evolving the one-dimensional wave equation using Cauchy, double-null and null-timelike initial value formulations. Techniques for transforming between the three schemes were investigated. This work was extended to study the Regge-Wheeler equation governing black hole perturbations in Schwarzschild spacetimes and the Teukolsky equation governing black hole perturbations in Kerr spacetimes.[17] In the next phase, the quasinormal modes produced by reflection at a hard sphere[18] were studied in Cauchy and null formulations. Axisymmetric codes for evolving the three-dimensional wave equation were developed and comparisons made between numerical results and analytic mode calculations. Williams assessed the advantages of a characteristic approach over the more conventional approaches to mode calculations for rapidly rotating stars[19] and black holes.[20]

Shinkai and Hayward described a new approximation scheme in a dual-null decomposition of space-time, and reported numerical tests of its validity. The aim is to provide a computationally inexpensive estimate of the gravitational waveforms produced by a black-hole or neutron-star collision, given a full numerical simulation up to (or close to) coalescence, or an analytical model thereof. The scheme truncates the Einstein equations by removing second-order terms which would vanish in a spherically symmetric space-time. The scheme was numerically implemented, testing it against angular momentum by applying it to Kerr black holes. As error measures, they took the conformal strain and specific energy due to spurious gravitational radiation. The strain was found to be monotonic rather than wavelike. The specific energy was found to be at least an order of magnitude smaller than the 1% level expected from typical black-hole collisions, for angular momentum up to at least 70% of the maximum, for an initial surface as close as $r = 3m$.

Burston and Lun investigated first-order perturbations of the 3+1 Lun-Burston-Stark system of equations due to small changes of matter distribution on the initial hypersurface representing a single black hole. They discussed how linear wave type equations arise from the perturbation equations, and how they can be simplified by choosing appropriate gauge conditions. The wave equations represent gravitational radiation from a binary system composed of a single black hole and a companion governed by matter perturbations in the source terms. Finally they indicated how the perturbation method can be extended to include higher order terms.

3 Other Perturbation Techniques and Applications

The remaining session consisted of presentations that did not automatically fit into the theme of the other two sessions.

Moreschi discussed a method for expanding the vacuum asymptotic field equations around stationary spacetimes using the center of mass reference frame rather than a fixed Bondi system. The parameter used in the expansion is associated with the strength of gravitational radiation. He argued that this is the appropriate formalism to be used on systems where back reaction due to gravitational radiation is important.

In a series of shorter talks, Nakamura presented a study of the oscillatory behavior of an infinite gravitating Nambu-Goto string within a gauge-invariant, general-relativistic perturbation framework[21]; Bonnor used an approximation to investigate the stationary axisymmetric solution for two spinning mass particles[22]; Kim discussed the scattering of various waves in the traversable Lorentzian wormhole geometry[23]; Colistete described a method for finding orbits with small eccentricity in Schwarzschild and Kerr space-times in a fully relativistic scheme[24]; and Perjes presented a new method for finding metric perturbations in the Kerr geometry from Newman-Penrose-type quantities.

References

1. L. Blanchet, T. Damour and B. R. Iyer, *Phys. Rev. D* **51**, 5360 (1995).
2. C. M. Will and A. G. Wiseman, *Phys. Rev. D* **54**, 4813 (1996).
3. L. Blanchet, T. Damour, B. R. Iyer, C. M. Will and A. G. Wiseman, *Phys. Rev. Lett.* **74**, 3515 (1995).
4. V. C. de Andrade, L. Blanchet and G. Faye, *Class. Quantum Grav.* **18**, 753 (2001).
5. L. Blanchet and G. Faye, *Phys. Lett. A* **271**, 58 (2000).
6. L. Blanchet and G. Faye, *Phys. Rev. D* **63**, 062005 (2001).
7. P. Jaranowski and G. Schäfer, *Phys. Rev. D* **57**, 7274 (1998).
8. T. Damour, P. Jaranowski and G. Schäfer, *Phys. Rev. D* **62**, 021501(R) (2000); **63**, 029903 (2001), erratum.
9. T. Damour, P. Jaranowski and G. Schäfer, *Phys. Rev. D* **63**, 044021 (2001).
10. L. Blanchet, G. Faye, B. R. Iyer and B. Joguet, *Phys. Rev. D*, in press (gr-qc/0105099).
11. L. Blanchet and T. Damour, *Phil. Trans. R. Soc. Lond. A* **320**, 379-430 (1986).
12. L. Blanchet, *Class. Quantum Grav.* **15**, 1971 (1998).
13. O. Poujade and L. Blanchet in preparation.
14. M. E. Pati and C. M. Will, *Phys. Rev. D* **62**, 124015 (2000).
15. J. T. Whelan, C. Beetle, W. Landry, and R. H. Price, preprint (gr-qc/0110004).
16. K. Lockitch, N. Andersson and J. L. Friedman, *Phys. Rev D* **63**, 024019 (2000).
17. M. Campanelli, R. Gomez, S. Husa, J. Winicour and Y. Zlochower, *Phys. Rev. D* **63**, 124013 (2001).
18. B. P. Jensen, *Ann. de Physique* **14**, 117, (1989).
19. S. M. Morsink, N. Stergioulas, and S. R. Blattnig, *Astrophys. J.* **510**, 854, (1999).
20. E. W. Leaver, *Proc. R. Soc. London* **A402**, 285, (1985).
21. K. Nakamura, A. Ishibashi and H. Ishihara, *Phys. Rev. D* **62**, 101502(R) (2000); K. Nakamura and H. Ishihara, *Phys. Rev. D* **63**, 127501 (2001); K. Nakamura, preprint (gr-qc/0101065).
22. W. B. Bonnor, *Class. Quantum Grav.* **18**, 2853 (2001).
23. S.-W. Kim, preprint (gr-qc/0102078).
24. R. Kerner, J.W. van Holten, and R. Colistete, Jr., *Class. Quant. Grav.* **18**, 4725 (2001).

A5: COMPUTER METHODS

BEVERLY K. BERGER
Physics Division, National Science Foundation, Arlington, VA 22230 USA
E-mail: bberger@nsf.gov

Session A5 on numerical methods contained talks on colliding black holes, critical phenomena, investigation of singularities, and computer algebra.

1 Introduction

In this Chapter, I shall describe the current status of computer simulation and algebraic computing for general relativity and gravitation. Most of the effort, both as represented in this Workshop and world-wide, is focused on "numerical relativity," the name usually given to large scale simulations of isolated gravitational systems. Elsewhere in these proceedings is the contribution by Lehner based on his plenary talk on the status of the colliding black hole (CBH) problem. This problem is extremely difficult but urgently needs to be solved. The urgency arises because one desired output of CBH simulations is a set of gravitational wave forms that can be used by the ground based gravitational wave detectors (LIGO, VIRGO, GEO600, and TAMA) now coming into operation. The difficulties arise on many fronts. These include (1) the computational spacetime grid is not the physical spacetime grid; (2) CBH spacetimes are singular within the horizons of the black holes. Note that simulations cannot be followed (much) past the time at which NaN (not a number) or INF (infinity) is recorded as a value anywhere in the computational grid; (3) Setting up initial data is highly nontrivial. In the process of solving the constraints, the connection between the actual initial system and the desired physical initial state is lost. For example, the "no incoming radiation" boundary condition is problematic; (4) Existing 3D CBH codes which have made considerable progress in dealing with (1) and (2) are plagued by instabilities which cause them to crash long before any physically relevant results are obtained. Note that a code that runs long enough can overcome the problem of radiation in the initial data by evolving until after it dissipates; (5) Identification of outgoing gravitational radiation is also nontrivial because radiation is a small effect thus requiring accurate simulations for a credible extraction and because many computational grids must terminate before the radiation zone is reached. All of these issues are under intensive investigation and were discussed in this Workshop. Similar techniques have been developed to study the coalescence of neutron star binaries. Such simulations do not have to face the issues of singularity and horizon but do have the complication of matter.

The travails of the CBH simulations should not be taken to mean that all computer simulations in general relativity and gravitation have not yet been completely successful. In fact, rather the opposite is true. Most spectacular, of course, was the discovery by Choptuik of critical phenomena in the collapse of a spherically symmetric self-gravitating scalar field. Careful numerical techniques (including adaptive mesh refinement) allowed study of the critical system dividing scalar field configurations which eventually disperse from those which eventually form BHs. Scaling typical of that found in other phase transitions was observed. The critical solution itself (a zero mass naked singularity) displays periodic self-similarity — a phenomenon unknown outside gravitating systems. Choptuik's original discovery sparked a large number of further and ongoing numerical and analytic studies of critical phenomena in gravitational collapse. Some of these were discussed in this Workshop.

Another area where there has been some success has been the numerical investigation of cosmological singularities. While an actual singularity cannot be handled on the computer, the regular behavior in the approach to it may be studied. If, asymptotically as the singularity is approached, some types of terms in Einstein's equations dominate over others, it might be possible to characterize the approach to the singularity in a simple way. This was argued heuristically long ago by Belinskii, Lifshitz, and Khalatnikov (BKL) with their predictions of local Mixmaster dynamics (LMD) recently supported in the cosmological context with numerical experiment and, in some cases, mathematical proof. In fact, the synergy between the mathematical and numerical approaches in this area has been useful in understanding the relevant phenomenology.

Originally, this Workshop was designed to feature algebraic computing as well as computer simulation. However, the co-organizer, Jim Skea, was unable to attend. In the end, there was only a single talk on this subject where considerable progress was described toward the goal of automated classification of exact or perturbative solutions of Einstein's equations.

In the rest of this Chapter, the highlights of the talks in each area will be given. For more details please see the indicated references.

2 Colliding Black Holes and Neutron Stars

Lousto presented an overview of the Lazarus Project to combine 3D simulations and perturbation approaches in an ambitious attempt to follow a BH collision and extract waveforms.[1] This required the development of techniques to convert the coordinates suitable to the 3D simulation to those appropriate for the single black hole close-limit approximation. These simulations were checked by interchanging the use of 3D simulation and perturbation where both were valid. In addition, the method was applied to a single Kerr BH where computed spurious radiation indicates the degree of error. This approach has made it possible to compute the first complete waveforms covering the post-orbital dynamics of a binary black hole system.

A number of talks were given on various techniques needed for standard (i.e. $3+1$ formulation) 3D codes. It is currently believed that an important source of instability in 3D codes is constraint violating modes. Einstein's equations may be written in an infinite number of ways by adding arbitrary multiples of the constraints to the dynamical equations of motion. The goal is to identify a formulation which will drive the system to the constraint hypersurface and thus eliminate these instabilities. One such approach was discussed by Shinkai.[2]

Pollney discussed work (with AEI collaborators) on BH excision and gauge techniques.[3] They describe recent progress in an implementation of excision for 3D black holes along with a set of gauge conditions which respond naturally to spacetime dynamics. Through the combined use of these techniques, they are able to produce accurate and long-lived evolutions of highly distorted, rotating black hole spacetimes. Accurate waveforms can be extracted.

Diener presented the results of the head-on CBH code test for 3D simulations. Koppitz reported on preliminary work in the construction of initial data sets. The objective is to consider thin sandwich, Kerr-Schild, and post-Newtonian approximation initial data using various implementations.

Bondarescu described how one might visualize BH horizons as they evolve in CBH simulations.[4] The 3D codes allow one to locate and track the evolution of apparent and event horizons in the coordinates of the simulation. One typically has little information about their real geometry. Previous studies in axisymmetric spacetimes have visualized horizons via an embedding in flat space. A new method which goes beyond axisymmetry was discussed. The method correctly reproduces known results and, for the first time, has allowed construction of embeddings of non-axisymmetric, distorted black holes.

Ashtekar described a method to extract physics from strong field regimes in numerical simulations using the isolated horizon (IH) framework.[6] If the IH (but not the spacetime) is axisymmetric, e.g., then, knowing just the intrinsic metric and one Newman-Penrose spin coefficient, the IH framework enables one to calculate the angular momentum and mass of the final black hole. These quantities refer only to the black hole in equilibrium and do not include the angular momentum and mass in matter fields or radiation outside the IH and thus could be found numerically.

Grandclément described a new approach to binary black hole evolution prior to the plunge phase.[7] One important feature of this phase of the evolution is the location of the innermost stable circular orbit (ISCO). This has been calculated both using post-Newtonian approximations and numerically from quasi-static equilibria with different results. If two orbiting bodies are far apart, the time-scale associated with the gravitational radiation is much longer than the orbital period allowing one to assume that the two black holes are on exact circular orbits. Using multi-domain spectral methods, a sequence of two identical corotating black holes is computed. The ISCO appears as a turning point in the total energy and angular momentum curves. Its position agrees well (for the first time) with the post-Newtonian values.

Several talks were given on characteristic codes. For a review of this approach, see [[8]]. Bishop described the initial value problem for a 3D characteristic code. The code is robust and stable and has been extended to include matter,[9] although in a simple way that would not handle shocks. The code cannot evolve a whole spacetime, but requires data on an inner world tube. For this reason, it is important to identify initial data of the required type to represent a physical situation with strong gravitational fields.

D'Inverno and Pollney discussed lower dimensional code tests for a Cauchy characteristic matching (CCM) code. See (e.g.) [[10]]. The advantages of such a code are that it dispenses with an outer boundary condition and (since it uses a compactified coordinate) can yield global solutions in which gravitational waves can be identified unambiguously at future null infinity. Recently, a master vacuum axisymmetric CCM code has been completed. The main motivation for this work is to construct a three dimensional code possessing the characteristic, injection and extraction modules (now present in the 2D code) which can be attached to existing interior codes based on a finite grid.

A completely different approach involves a 3D code based on the conformal field equations developed by Friedrich. This work was presented by Husa. He discussed the evolution of linear and (mildly) nonlinear gravitational waves. This approach aims at the numerical computation of the global structure of generic isolated systems in GR using well defined and unambiguous algorithms even out to and beyond future null infinity. The basic ideas and algorithms were developed by Hübner and Weaver in recent years and implemented by Hübner. Hübner[11] obtained the complete future of (the physical part of) the initial slice, illustrating a theorem by Friedrich, which states that for sufficiently weak initial data a regular point i^+ exists. However Husa found that for higher amplitudes the gauge chosen in by Hübner results in code crashes which can be cured by some ad-hoc modification of the lapse. For still stronger (but still only mildly nonlinear) data, evolution up to i^+ would require the tuning of too many parameters to compensate for the lack of symmetry. Evolutions of axially symmetric initial data modeled in the spirit of "Brill waves", and evolutions of Minkowski space in a static gauge were discussed. While the former evolution is numerically stable and quite robust, the latter exhibit instabilities which seem to be rooted in exponentially growing constraint violating modes inherent in the analytical formulation.

Siebel presented the results of a study using the characteristic approach to identify gravitational radiation. The implementation follows the characteristic initial value formulation of general relativity based on Bondi's radiative metric. The study focuses on the evolution of neutron stars modeled by a $N = 1$ polytrope with fully relativistic hydrodynamics.[12] The code was used to study spherical relativistic stars with a scalar field to model the gravitational degrees of freedom. Depending on the specific neutron star model, the scalar field either induces oscillations of the star or makes it collapse to a black hole. The extracted frequencies of the oscillations agree very well with linear studies, except, as expected, at the threshold of black hole formation. The code was also applied to evolutions of neutron stars in axisymmetry and to their interaction with gravitational waves.

Stark (with Lun) considered an alternative $3 + 1$ scheme as applied to spherical perfect fluid collapse. The scheme utilizes the constraint equations contained in the Einstein field equations with those in the Bianchi identities to determine the 3-curvature tensor and the 3-covariant derivatives of the extrinsic curvature of the hypersurface. This yields a hyperbolic system of equations for the evolution of the gravitational potentials and hydrodynamical quantities. A numerical implementation of this system was used to model the Oppenheimer-Volkoff solution and the gravitational collapse of an initially homogeneous dust sphere to a BH.

Moreschi discussed modeling the collision of black holes with angular momentum with Robinson-Trautman (RT) spacetimes.[13] These spacetimes contain purely outgoing gravitational radiation and decay to leave a Schwarzschild-like horizon. These RT spacetimes are natural candidates to study systems settling to a single black hole and may be used as background spacetimes for a perturbation treatment. Previous studies of this type give excellent agreement with numerical relativity results for the head-on collision, zero spin case. They are then extended to include angular momentum.

Reula presented results using the collapse of spherically symmetric self-gravitating scalar fields as a model problem to study numerical instabilities.[14] There are instabilities present in this simple problem, and furthermore there are strong indications that, more generally, the instabilities are predominantly due to a longitudinal or Newtonian mode, namely the only one present in the spherically symmetric case. Free and constrained evolutions are compared. In [15] a symmetric hyperbolic formulation is used, and certain freedom, still available in that setting, is used to suppress the main instability found there, to allow a stable propagation. However, their discretization scheme has first order numerical dissipation and the boundary-value problem for the system is not known to be well posed. Reula discussed an improved system that overcomes these problems, preserves the constraints, and reproduces several known results.

Gentle applied Regge calculus methods to the evolution of Brill waves. This is the first significant test of Regge calculus in a highly dynamic setting. An axisymmetric lattice is obtain by triangulating and collapsing a hypercubic grid which is aligned with a cylindrical polar coordinate system (see [16] and references therein).

Neutron stars might act as a source of gravitational waves. Sperhake presented a new numerical approach to nonlinear oscillations in neutron stars.[17] This approach rewrites the equations to cancel out the background terms in the numerical evolution of the perturbations. A comparison of the resulting perturbative code with a "standard" non-perturbative method was made. The perturbative scheme reproduces the expected harmonic oscillations with high accuracy, while the non-perturbative scheme produced spurious behavior. The new method is then used to accurately evolve initial data of arbitrary amplitude to investigate non-linear coupling of eigenmodes for a simplified neutron star model.

Hawley (with Choptuik) explore the genericity of initial data for multi-scalar stars. They consider a class of general relativistic soliton-like solutions composed of multiple minimally coupled, massive, real scalar fields without self-interaction.

3 Numerical Investigation of Singularities

For reviews of critical phenomena see [18,19]. Gundlach described what happens to the critical behavior of the gravitational collapse of a perfect fluid when there are departures from spherical symmetry. Does the type II critical behavior in the gravitational collapse of a $p = k\rho$ perfect fluid persist beyond the restriction to spherical symmetry? It was found that for $1/9 < k < 0.49$, all non-spherical perturbations decay. For $0.49 < k < 1$, there are one or several growing modes in the polar perturbations. For $0 < k < 1/9$, there is exactly one non-spherical growing mode. It is an $l = 1$ axial mode, linked to infinitesimal rotation. BH mass and BH angular momentum become a function of 2-parameter families of initial data, chosen so that one of the parameters is linked to rotation (and thus can be tuned to zero by axisymmetry). At the BH threshold they depend on the initial data through one universal function of one variable,a combination of these two parameters. The formula obtained is completely analogous to the dependence of the magnetization of a ferromagnet as a function of temperature and external magnetic field near its critical point.

Aichelburg presented results (including those of Lechner) on a new transition between discrete and continuous self-similarity in collapsing $SU(2)$ σ-models parametrized by a dimensionless coupling constant.[20] In the intermediate range, a competition between CSS and DSS solutions gives rise to new phenomena in the dynamical evolution of a critical search: the appearance of episodes of approximate CSS, where repeatedly the evolution approaches and departs from CSS before leading to black hole formation or dispersion. This picture is supported by the explicit numerical construction of the CSS and DSS solutions and its comparison with critical time evolutions.

Szpak studied critical behavior in a model nonlinear wave problem.[21] The motivation was to study the existence of an universal intermediate attractor in the dynamics at the threshold of forming singularity.

Hobill (with Webster) presented studies of trapped surface formation in Brill wave evolution to see if a naked singularity could form. The oblateness or prolateness of the system is determined by measuring the ratio of polar circumference to equatorial circumference for a surface with constant logarithmic radial coordinate. It is found that for initial data with positive amplitude the systems are prolate and for negative amplitude the systems are oblate. It is seen that, as the Brill wave amplitude is increased (thereby decreasing the radius of the outermost minimal surface), the location of the outermost trapped surface eventually coincides with the minimal surface thereby preventing the formation of a naked singularity. This behavior is in agreement with recent results of Garfinkle and Duncan.[22]

I presented a summary of our (with D. Garfinkle, J. Isenberg, V. Moncrief, M. Weaver) program of investigation of collapsing spatially inhomogeneous cosmologies.[19] Significant recent progress has been made in rigorous studies of generic collapse in cases where one expects the approach to the singularity to be asymptotically velocity term dominated.[23] However, one expects that an even larger class of spacetimes exhibits local Mixmaster dynamics (LMD) in the vicinity of the singularity. While numerical evidence for LMD in spatially inhomogeneous spacetimes with T^2 and $U(1)$ symmetries is compelling, rigorous results for spatially inhomogeneous cosmological spacetimes with LMD do not exist. New mathematical methods to handle LMD are needed.

4 Algebraic Computing: Invariant Classification in Maple

D'Inverno provided a summary of the program to classify solutions to Einstein's equations automatically.[24] Recent developments are in collaboration with Pollney, Skea, Araújo, de Albuquerque, and Roveda. A short review of the equivalence problem in general relativity was presented which included a description of the computer database of exact solutions whose central site is located at http://www.jim.dft.uerj.br. The database contains over 200 solutions of Einstein's equations which have been classified with CLASSI, an extension of the general relativity system called SHEEP. In the next stage of this collaborative project attention has been focused on setting up a database which can easily be accessed and updated by the user community. The choice of platform is the general relativity package GRTensor which is an applications package of the computer algebra system Maple. The underlying algorithm is based on the Cartan-Karlhede invariant classification of geometries. In particular, new algorithms have been obtained for putting the Weyl spinor, Ricci spinor and general spinors into standard form and the derivative operators needed in the classification algorithm and the behavior of symmetric spinors under frame rotations have been investigated. More recent work was reported on a refinement of the JMS (Joly-MacCallum-Seixas) algorithm for completely determining the Segre type of a Ricci spinor. To date 30 of the 45 possible sub-cases have been completed and the algorithms have been implemented in GRTensor. In addition, the status of the work on obtaining the isometry group of a space-time (rather than simply its dimension as in the current database) was reported. It is currently possible to obtain the isometry group for spacetimes admitting a G2, G3 and some G4 symmetries. This may require the user to input limits on parameters and coordinates appearing in the line element, to overcome an ambiguity in determining the signature of a matrix occurring in the Bianchi classification. The goal is to complete the G4 case and extend the work to the G5, G6 and G7 cases. Work was also reported on the Maple package SPIDOR, which has been developed for obtaining spinor contractions. An example of its application is its ability to automatically generate the spinor curvature invariants built out of the Weyl spinor, its complex conjugate and the Ricci spinor up to the 20th order on a laptop. This package may be useful in relation to formalisms which have been developed to address particular issues in exact solutions.

References

1. J. Baker, B. Brügmann, M. Campanelli, C. O. Lousto, and R. Takahashi, *Phys. Rev. Lett.*, **87**, 121103 (2001).
2. H. Shinkai, G. Yoneda, gr-qc/0110008.
3. M. Alcubierre, B. Bruegmann, D. Pollney, E. Seidel, R. Takahashi, *Phys. Rev. D* **64** (2001) 061501.
4. M. Bondarescu, M. Alcubierre, E. Seidel, gr-qc/0109093.
5. M. Bondarescu, M. Alcubierre and E. Seidel, (2001), gr-qc/0109093.
6. A. Ashtekar, C. Beetle, O. Dreyer, S. Fairhurst, B. Krishnan, J. Lewandowski, J. Wisniewski, *Phys. Rev. Lett.* **85** (2000) 3564–3567.
7. P. Grandclément, E. Gourgoulhon and S. Bonazzola, gr-qc/0106016].
8. J. Winicour, *Living Reviews in Relativity* **4**.
9. N.T. Bishop, R. Gomez, L. Lehner, M. Maharaj, and J. Winicour, Phys. Rev. D **60**, 024005 (1999).
10. R.A. d'Inverno, M.R. Dubal, E.A. Sarkies, *Class. Quant. Grav.* **17** (2000) 3157-3170.
11. P. Hübner, *Class. Quant. Grav.* **18**, 1871-1884 (2001).
12. F. Siebel, J.A. Font, P. Papadopoulos, gr-qc/0108006.
13. O. M. Moreschi, A. Perez, *Class. Quant. Grav.* **18** (2001) 3701–3720.
14. M. S. Iriondo, O. A. Reula, gr-qc/0102027.
15. M. A. Scheel, et al, Phys. Rev. D, **58**, 044020 (1998).
16. A. P. Gentle et al, *Class. Quantum Grav.* **16**, 1979–1985 (1999).
17. U. Sperhake, P. Papadopoulos, N. Andersson, astro-ph/0110487.
18. C. Gundlach, *Living Reviews in Relativity* **2** (1999).
19. B. K. Berger, *Living Reviews in Relativity* **1** (1998).
20. J. Thornburg, Ch. Lechner, M. Purrer, P. C. Aichelburg, S. Husa, gr-qc/0012043.
21. N. Szpak, Master's Thesis, Jagellonian University, 1999
22. D. Garfinkle, and G.C. Duncan, *Phys. Rev. D* **63**, 044011 (2001).
23. L. Andersson, A.D. Rendall, *Commun. Math. Phys.* **218** (2001) 479.
24. D. Pollney, J. E. F. Skea, R. A. d'Inverno, *Class. Quantum Grav.* **17**, 643 (2000); 2267 (2000); 2885 (2000).

A6: ALTERNATIVE THEORIES OF GRAVITY

ROBERT MANN
Dept. of Physics, University of Waterloo, Waterloo, Ontario, Canada N2L 3G1
E-mail: mann@avatar.uwaterloo.ca

1 Introduction

The session on *Alternative Theories of Gravity* had a total of 26 submissions, of which two were referred to other parallel sessions. Of the remaining 24 submissions, 15 were accepted for oral presentation and 5 for poster presentation. Only 9 of the 15 oral presentations were actually given – the remaining presenters did not show up.

The topics presented were, suprisingly enough, relatively unconnected to alternative theories of gravity. Many of them were actually concerned with alternative ideas within the existing paradigm for gravitational theory, namely general relativity. This represents a significant departure from previous GR meetings, in which a considerable amount of attention was given to actual alteratives to general relativity itself.

1.1 Black Holes

There was one talk on black holes by T. Torii. He presented an examination of the no-hair conjecture in the context of asymptotically anti-de Sitter (AdS) spacetimes in general relativity [1]. By regarding a real scalar field as the matter field coupled to gravity (and assuming static spherical symmetry), he found that this scalar field must approach the extremum of its potential upon an examination of its asymptotic behaviour. From this it follows that there is no regular black hole solution when the scalar field is either massless or has a "convex" potential. Although there is a mode of the scalar field that grows near the local minimum of the potential, there is no corresponding growing mode around the local maximum. Rather the local maximum is a kind of "attractor" of the asymptotic scalar field.

Two examples were given of new black hole solutions with nontrivial scalar field configurations. These were obtained numerically in the symmetric/asymmteric double well potential models. In the double-well potential model the solution was shown to be stable over a certain range in parameter space. The subsequent interpretation is that the black hole no-hair conjecture is violated in asymptotically AdS spacetimes.

1.2 Extended Lagrangians

There were three talks on this topic. The first, by K. Shima and M. Tsuda, presented an approach for obtaining an alternative supersymmetric extension to the Einstein-Hilbert action of general relativity [2]. The idea is to consider extending the usual vierbein of general relativity by adding to it a term that is bilinear in a spin-3/2 field. This field can be realized as a Nambu-Goldstone fermion in a non-linear realization of supersymmetry, and is an extension of previous work done on the spin-1/2 case [3]. What is remarkable is that one can obtain a new consistent coupling of a spin-3/2 field to gravity without going through the usual SUGRA framework. The action is fully supersymmetric under a set of non-linear supersymmetry transformations. The authors propose that this be regarded as the framework for a new supersymmetric composite unified model for spacetime and matter, called a superon-graviton model or SGM. The model is based upon the SO(10) super-Poincare algebra. It may be the most economic model yet proposed that accomodates all observed particles in a single irreducible representation of a (semi)simple group. This is due (at least in part) to the non-linear realization of supersymmetry. In the SGM all particles that participate in the (super) Higgs mechanism – except for the graviton – are composites of Nambu-Goldstone fermions. The authors demonstrated the spacetime symmetries of their Einstein-Hilbert type SGM action, which requires a small non-zero cosmological constant. It was a refreshing approach toward incorporating supersymmetry into gravity.

The second talk, given by R. Colistete, was concerned with Lagrangians of the type $\mathcal{L} = \sqrt{-g}\,(R + \beta R^2)$ in the context of inflation. Such models, when coupled to a well-behaved perfect fluid (with $p > 0$, $\rho > 0$, etc.) can yield non-singular inflationary cosmological behaviour for Robertson-Walker metrics with closed spatial sections. Colistete was interested in determining which values of the parameter β could ensure this behaviour. He did this by inverting the problem, ie. by imposing the requisite scale factors so that the resultant perfect fluid was amenable to analysis. The possibility of graceful exit, from inflationary to radiation dominated era, is also investigated.

R. Zalaletdinov gave a talk on the problem of constructing a continuous (macroscopic) matter model for a given point-like (microscopic) matter distribution in general relativity. After reviewing existing approaches, he pointed out a physical analogy with a similar problem in classical macroscopic electrodynamics. He showed that a procedure of Szekeres in linearized general relativity for constructing a gravitational quadrupole polarization tensor had some inconsistencies: it only provided the terms linear in perturbations for the averaged field operator that do not contribute to the dynamics of the averaged field, and upon averaging the analogy between electromagnetism and gravitation breaks down. Extending to second-order perturbation theory, he wrote down a system of macroscopic field equations, which were Isaacson's equations with a source that incorporated (a) the quadrupole gravitational polarization tensor, (b) Isaacson's gravitational wave energy-momentum tensor and (c) the energy-momentum tensor of gravitational molecules. He proposed a suitable set of material relations that related all the tensors.

1.3 Geometric Methods

One talk was given on this subject, a nice review by P. Matteucci on the formalism of Gauge-Natural Bundles. This is a formalism developed over the past decade by M. Francaviglia and collaborators, and is quite useful in deriving a number of results for computing conserved charges both in general relativity and in its many extensions/alternatives. The particular extension of interest for this talk was Einstein (-Cartan) -Dirac Theory [4]. In order to properly describe the interaction between gravity and spinor fields, *spin-tetrads*, and not tetrads, are the appropriate objects to be considered [5,6], and *gauge-natural bundles* provide a suitable geometrical framework for doing so. Such bundles are fibre bundles associated to "abstract" principal bundles with an arbitrary structure group [7]. In this new geometric framework, if Σ denotes the spin bundle and Ξ the generator of a one-parameter subgroup of automorphisms of $\Sigma/\mathbb{Z}_2$, the superpotential along Ξ associated with the Hilbert Lagrangian $\mathcal{L}$ via the Poincaré-Cartan method [8] turns out to be

$$U(\mathcal{L},\Xi) := -\frac{1}{2\kappa}\check{\Xi}_{ab}\Sigma^{ab},$$

where $\check{\Xi}$ is the "vertical part" of Ξ w.r.t. the given spin connection, $\kappa := 8\pi G/c^4$ and $\Sigma^{ab} := {}^*(\theta^a \wedge \theta^b)$, θ being a spin-tetrad.

Despite its appealing geometric meaning, $\check{\Xi}$ is *not* in fact *determined* by the theory, thereby *requiring*, or *enabling* one—according to one's personal taste—, to *make a choice* (e.g., "Kosmann", "Witten lift") on the basis of the particular physical context under consideration.

1.4 Alternative Geometries

Four talks were given on this topic. The first, by M. Wanas, dealt with some recent work on a model that the author had worked on quite some time ago [9] called "Generalized Field Theory". The theory was originally constructed as an attempt to unify gravity with electromagnetism. The author re-derived these results in the context of a Parameterized Absolute Parallelism (PAP) geometry [10]. The idea of a PAP is to hypothesize a set of n vectors in an n-dimensional spacetime, each of which has a vanishing covariant derviative (the absolute parallelism condition). These vectors are then used as n-beins for an n-dimensional metric, and in turn form the foundation of an extended geometry with simultaneously non-vanishing curvature and torsion. There are a variety of ways to extend the connection beyond that given in standard Riemannian geometry, and one can parametrize each of the possiblities – hence the moniker 'PAP'. Since some features of these extended geometry have interesting implications for quantum physics, the present work may throw some light on the possibility of quantizing gravity [11]. For example, using the PAP-paths as trajectories of test particles, the quantum spin of a moving particle interacts non-trivially with the background spacetime torsion, with potential experimental implications [12]. The author presented more recent attempts to study possible interactions between the spin and the unified field.

H. Ellis presented an extended 5-dimensional geometry which is a hybrid of Kaluza's extra-dimensional geometry and Weyl's conformal geometry [13]. He concentrated on the geodesic equations of the extended metric

$$\hat{G} = e^{2\zeta}(dx^\mu \otimes g_{\mu\nu} dx^\nu) + \phi^2(A_\mu dx^\mu + d\zeta) \otimes (A_\nu dx^\nu + d\zeta),$$

which take the form

$$\frac{d(\mathring{m}u^\mu)}{d\tau} + (\mathring{m}u^\kappa)\Gamma_\kappa{}^\mu{}_\lambda u^\lambda = qF^\mu{}_\lambda u^\lambda - \mathring{m}A^\mu + e^{2\zeta}\frac{(q/\phi)^2}{\mathring{m}}\phi^{-1}\phi^{\cdot\mu}$$

and

$$\frac{dq}{d\tau} = \mathring{m} = e^\zeta[1 - (q/\phi)^2]^{\frac{1}{2}},$$

in which the rest mass $\mathring{m}$ and electric charge q of the test particle following the geodesic are variables defined kinematically on its path. The test particles exhibit quantum behavior: (1) they are born and die in full-blown motion at definite events; (2) all that share an event $\mathcal{E}$ of birth or death do so with the same charge magnitude $|q| = \phi(\mathcal{E})$; (3) conservation of 5-momentum at such an event entails conservation of electric charge in addition to conservation of 4-momentum, among the particles created and annihilated there; (4) at such events $e^{2\zeta} \to \infty$, causing the $\phi^{\cdot\mu}$ force to infinitely dominate the other three — this strongly biases the birth and death events to be concentrated deep in the discretely spaced potential wells of ϕ, and sparse elsewhere. He discussed progress obtained so far in numerically solving the equations, and claimed that the field equations for the extended metric $G_{\mu\nu}$ yielded something called a 'darkhole', an object that is geodesically complete, horizonless, and singularity free [14]. It is the analogue of a Schwarzchild black hole, and is comprised of two asymptotically euclidean regions connected by a throat that gravitationally attracts on the 'high' side of the potential and repels, more strongly, on the 'low' side.

Yakov Itin presented a talk in which he considered a general 3-parameter class of teleparallel models [15]. The field equations turn out to have a form completely similar to the Maxwell field equation $dF^a = T^a$. The source 3-form T^a can be shown to be connected with the diffeomorphism invariance of the Lagrangian via the Noether procedure. It can thus be interpreted as the total conserved energy-momentum current of the system. This energy-momentum tensor can be defined in a diffeomorphism invariant and translationally covariant way. The total energy-momentum current of a system is conserved, and so a redistribution of the energy-momentum current between the material and gravitational fields is possible at least in principle, in contrast to the situation in general relativity. The talk concluded by briefly showing calculations of the energy-momentum tensor for various teleparallel models, including a pure Yang-Mills type of model, and the generalized teleparallel equivalent of GR. The latter model furnishes a very close alternative general relativity.

Finally, Luis Ralay Tomas gave a talk – translated from Spanish by his daughter – on irreducible magnitudes in an alternative gravitational theory. He proposed the idea that the concept of Mass be considered as artificial and not representative of the geometric reality of space and time. This notion is to be replaced with something he called a "relation of surface" concept, abbreviated as RS. This is a dimensionless quantity and its value for the earth would be $RS_{earth} = 698977762.7$. A short demonstration as to how this quantity could be converted into the standard measurement of the earth's mass, of Newton's gravitational constant was given, and of the escape velocity from the earth was given. The details by which one would compute the RS for a given body were not clearly presented. However the motivation is that, by the removal of the concept of "mass" as a fundamental physical quantity, the irreducible magnitudes of physics become two (length and time) instead of three.

2 Summary

The number of submissions and participants in session A.6 was considerably lower at this conference than at previous meetings. Whether or not this is reflective of a general trend in the overall discipline remains to be seen. However the talks presented here were quite lively, providing us with both new perspectives on general relativity, and the ability to understand how nature might have been constructed.

References

1. T. Torii, K. Maeda and M. Narita, Phys.Rev. **D64** 044007 (2001).
2. K. Shima and M. Tsuda, hep-th/0012235; K. Shima and M. Tsuda, hep-th/0101178.
3. K. Shima, *Z. Phys.* **C18**, 25 (1983); K. Shima, *European. Phys. J.* **C7**, 341 (1999); K. Shima, hep-ph/0012320, *Phys. Lett.* **B501**, 237 (2001).
4. M. Godina, P. Matteucci, and J. A. Vickers, *J. Geom. Phys.*, **39**(4):265 (2001).
5. L. Fatibene, M. Ferraris, M. Francaviglia, and M. Godina, *Gen. Rel. Grav.*, **30**(9):1371 (1998).
6. M. Godina, P. Matteucci, L. Fatibene, and M. Francaviglia, *Gen. Rel. Grav.*, **32**(1):145 (2000).
7. I. Kolář, P. W. Michor, and J. Slovák, *Natural operations in differential geometry* (Springer-Verlag, Berlin, 1993).
8. G. Giachetta, L. Mangiarotti, and G. Sardanashvily, *New Lagrangian and Hamiltonian methods in field theory* (World Scientific, Singapore, 1997).
9. F.I. Mikhail and M.I. Wanas, Proc. Roy. Soc. Lond. **A356**, 471 (1977).
10. M. I. Wanas, Turk. J. Phys. **24**, 473 (2000) .
11. M. Wanas, M. I. and Kahil, M. E. Gen. Rel. Grav. **30**, 1921 (1999); M.I. Wanas Astrophys. Space Sci. **258**, 273 (1998) .
12. M.I. Wanas, M. Melek and M.E. Kahil, Grav. Cosm. **6**, 319 (2000).
13. H.G. Ellis, gr-qc/0107023.
14. H.G. Ellis, gr-qc/0003024
15. Y. Itin, gr-qc/0103017.

B1: RELATIVISTIC ASTROPHYSICS

JOHN L. FRIEDMAN
Department of Physics, University of Wisconsin-Milwaukee
PO Box 413, Milwaukee, WI 53201
Email: friedman@uwm.edu

This review summarizes the parallel session on relativistic astrophysics at GR16. Much of the work reported here involved the structure and stability of neutron stars and the astrophysics of accretion disks around neutron stars and black holes.. A large part of the recent work in relativistic astrophysics is tied to numerical investigations of binary coalescence and gravitational waves, but these topics demanded sessions of their own; gravitational waves in the present session were mentioned in connection with neutron-star instability and in a talk on coupling of gravitational waves to radio waves. Two talks involved relativistic stellar systems and cosmology. Finally, several authors outlined advances involving gravitational collapse, cosmic censorship, and baby universes.

1 Neutron stars

1.1 Neutron star structure, models and observation

Matthias Hanauske spoke of work with several collaborators, including J. Schaffner-Bielich, H. Stöcker, and W. Greiner, on neutron stars with hyperons in their core and on strange quark stars.[1,2] In particular, he emphasized the possibility of a sudden phase transition from a metastable star (with only neutrons) to a stable "twin" with the same number of baryons but with a more strongly bound hyperon core, or to a stable strange quark star. Uncertainty in the value of the strong coupling constant allows a range of values for which hyperons with large strangeness fraction form stable cores above twice nuclear density. Like strange quark stars, stars with with such cores might constitute a third family of stable compact stars, with central densities higher than those of white dwarfs and neutron stars. The sudden transition from a neutron star to a star with a hyperon core (or to a strange quark star) can produce gamma-rays and gravitational waves, and the authors propose the transition as a supernova-like model for gamma-ray bursts.

Neutron stars are widely believed to have a superfluid interior. Although its effect on stellar structure is ordinarily small, the superfluid gives rise to an additional class of stellar oscillations. Associated with a doubling of the fluid degrees of freedom in the two-fluid model for superfluidity, the additional oscillations are due to independent motions of the two fluids – roughly, its neutrons and protons. R. Prix' talk, reporting work, in collaboration with N.

Andersson and G. Comer,[3] exhibited a class of *equilibrium models* in which the two interacting fluids move independently, rotating uniformly about the same axis, but at different rotation rates. It is an intriguing model, dramatizing the possibility of equilibria that depart significantly from configurations of ordinary fluid.

Typical neutron stars have magnetic fields below 10^{13} G, implying magnetic energy less than 10^{-9} times their binding energies. As a result, their magnetic fields have been ignored in constructing equilibrium models. Interest in such models has suddenly grown, with the recent discovery of magnetars, thought to be neutron stars with magnetic fields that can reach 10^{15} G. Kohkichi Konno [4] reported a recent series of papers, partly in collaboration with Y. Kojima, that obtain axisymmetric equilibria with large magnetic fields. They work in a slow-rotation formalism and keep terms of lowest nonvanishing order in the magnetic field and in the coupling of the field to the star's rotation. The slow rotation approximation is suitable for all or nearly all observed neutron stars, and, even for magnetars, the weak field approximation is appropriate. The main remaining idealizations that restrict the accuracy of the models are a restriction to axisymmetry and a lack of microphysics in the description of the magnetic field inside the star.

Work by Mukherjee and Sharma models compact stars, using solutions due to Mukherjee, Paul and Dadhich of a Vaidya-Tikekar system of equations. The resulting equation of state is claimed to be close to that of a quark-diquark mixture and they consider the structure and stability of their equilibrium models.

Finally, Chongming Xu reported joint work with X. Wu and M. Soffel on rotating stars in general relativity, modeled as a perfect fluid coupled to an elastic solid. Using the Carter-Quintana theory of elastic solids, they develop a first-post-Newtonian formalism for stellar perturbations, finding, in particular, junction conditions for a solid-liquid and solid-solid interface. Although the authors have not yet done so, one might usefully apply their formalism to modes, like surface g-modes, whose character is determined by the existence of a crust.

1.2 Instabilities of neutron stars

Three years ago, a series of surprises appeared dramatically to have improved the likelihood that the spin of rapidly rotating, newly formed neutron stars is limited by a nonaxisymmetric instability driven by gravitational waves – and that the emitted waves may be detectable. First was the discovery that the r-modes, rotationally restored modes that have axial parity for spherical models, are unstable in perfect fluid models with arbitrarily slow rotation. Studies of the viscous and radiative timescales associated with the r-modes revealed a second surprising result: The growth time of r-modes driven by current-multipole gravitational radiation was significantly shorter than had been expected, short enough that the instability appeared easily to survive viscous damping in hot, newly formed neutron stars. (See reviews by Andersson and Kokkotas [5], Friedman and Lockitch [21], and Lindblom [9].) At nearly the same time as the theoretical surprises was an observation by Marshall et al. of a fast pulsar in the supernova remnant N157B was the first clear example of a class of neutron stars that are rapidly rotating at birth and whose rotation might be limited by the r-mode instability.

The question then was whether the instability would survive a detailed incorporation of neutron-star physics. Possible ways to kill the instability include disappearance of the modes due to a continuous spectrum in GR and damping of the modes by a number of possible mechanisms: nonlinear couplings, magnetic-field wind-up, the formation of a crust, an unexpectedly high bulk viscosity, and an unexpectedly rapid cooling of a young star. Four of the speakers at this session addressed the issue, with the present author briefly reviewing work by investigators who could not come to GR16.

Masayasu Hosonuma and Kostas Kokkotas described work on the first question, whether r-modes found in the Newtonian limit persist in the exact theory. Earlier work by Kojima, Beyer and Kokkotas had shown that the equation describing nonbarotropic axial perturbations in a slow rotation approximation could become singular inside the star, and, when it did, there was no discrete spectrum. Kokkotas and J. Ruoff [6] examined the time evolution of initial data that approximates an r-mode, again for slow rotation, but including a radiation-reaction term to drive any instability. For a singularity deep in the star, they see no r-mode, while a singularity near the surface still allows an approximate r-mode. It remains to be seen whether these preliminary results will persist when the approximation is extended to include coupling of axial and polar modes. Hosonuma described what seemed to be similar results; a formalism developed with Kojima was used to find the time evolution of relativistic axial perturbations. The oscillations were found gradually to become incoherent with phase lag caused by the relativistic frame dragging. Hosonuma did not report seeing a true mode or an effective mode of the kind seen in the evolutions of Kokkotas and Ruoff. Andersson et al.[8,7] find r-modes for particular barotropic and nonbarotropic models; but the question of whether some models have no r-modes remains open.

Full nonlinear evolutions by Stergioulas and Font [10] (in GR with a Cowling approximation) and by Lindblom, Tohline and Vallisneri (Newtonian, with an artificially large radiation-reaction term) [11] showed no nonlinear damping until the r-mode amplitude is of order unity; shocks – breaking of surface waves – then damp the mode. Most recently, several workers have developed a second-order perturbation theory, [12,13] and preliminary results (not yet in preprint form at the end of '01) seem to show that a coupling of the r-mode to shorter wavelength g-modes leads to a saturation amplitude much smaller than unity. If true, it means that the fully nonlinear evolutions have grid spacing too large to see the relevant couplings. An old star spun up by accretion might then produce a detectable signal, but one would not expect to see the instability in nascent star.

Work by Jones,[14] by Lindblom and Owen,[15] and by Haensel et al.[16] shows that a hyperon core will greatly increase the bulk viscosity and the cooling rate of a neutron star. As noted in Hanauske's talk, hyperons are likely to be present above 2-4 times nuclear density; and the careful studies appear to leave to leave little doubt that, when present, a hyperon core will not allow an r-mode instability of young neutron stars. Whether hyperons are present is less certain; in particular, if rapidly rotating neutron stars initially have mass between 1.3 and 1.4 $M_\odot$, they may avoid a hyperon core; and a somewhat lower initial mass is suggested by the accretion-disk model of Watts and Andersson, described below.

Finally, the second-order r-mode is differentially rotating, leading several workers to suggest that a magnetic field windup can damp the instability. Rezzola et al.[17,18] examine the windup, and Friedman, Lockitch and Sa [19] find the differential rotation for slowly rotating polytropes and rapidly rotating uniform-density stars. But a consistent second-order calculation that includes the magnetic field has not been completed, and it remains to be seen whether a background magnetic field might simply alter the mode to avoid a secularly increasing differential rotation.

In my view, none of the work on these several ways to kill an r-mode has ruled out the instability; but their combined weight has made it much less likely.

Neutron stars that rotate rapidly at birth are likely to be surrounded by an accretion disk. Anna Watts reported work with N. Andersson [20] on the evolution of a young accreting neutron star, finding that the fallback accretion prolongs the time during which gravitational waves from an r-mode instability can be detected from days to months – possibly to years, if the accretion postpones the formation of a crust.

Karino Shigeyuki presented preliminary results on numerical determination of dynamical instability to a nonaxisymmetric bar mode, done in collaboration with S. Yoshida and Y. Eriguchi. Although the the instability has been investigated in numerical evolutions of the nonlinear fluid equations, the authors study the linearized equations, and that emphasis may well lead to a more accurate determination of instability points and a more comprehensive view than the expensive numerical evolutions can provide. In particular, Karino reports a systematic study of model sequences with varying compressibility, rotation law, and azimuthal wave number. For a highly differential rotation law, they find an unstable bar mode at remarkably small values of rotation: At the onset of instability, the ratio $T/|W|$ of rotational kinetic energy to gravitational binding energy is 0.1.

In the final talk of the neutron-star stability session, Horst Beyer presented a mathematical framework for the rigorous treatment of stellar perturbations and stability.[22] With the fluid perturbation described by a lagrangian displacement vector ξ, the equations governing fluid perturbations have the form

$$\partial_t^2 \xi + B\partial_t \xi + C\xi = 0,$$

with C symmetric, B antisymmetric. The presence of B has made it difficult numerically to locate points of dynamical instability and difficult to find proofs of stability. Beyer rewrites the equations as a system first order in time and uses the domain of the operators to define a suitable Hilbert space, with respect to which time evolution is described by strongly continuous group of operators. He then obtains a sufficient condition for stability in terms of an operator constructed from B and C.

1.3 Astrophysics of accretion disks and black holes.

Three talks in the session touched on parts of this vast field. John Miller [23] reported results of a numerical study(in collaboration with Ewa Szuszkiewicz) of nonstationary accretion. Miller described a program in which the approximations of a slim-disk approach are successively relaxed. In particular, a more physical description of viscosity is used, for which the ϕr component of the viscous stress tensor is proportional to the radial gradient of the angular velocity. The slim-disk approximation involves a vertical integral of the fluid equations, particularly inaccurate for the vertical component; and, in the present work, the integrated vertical component of the hydrostatic equilibrium equation is replaced by a dynamical treatment of vertical acceleration. Present work includes the treatment of two-temperature plasma and investigation of the possible transition to an advection dominated accretion flow (ADAF).

A. R. Prasanna reported work with B. Mukhopadhyay on advection dominated accretion flows around black holes.[24] ADAF models commonly involve self-similar solutions, and earlier work by Bhatt and Prasanna had shown that these solutions are singular for some values of the adiabatic index and the viscosity. In the present work, inclusion of the Coriolis force is found to extend the parameter space for nonsingular self-similar solutions, allowing γ to assume any values in the $4/3 - 5/3$ range used in ADAF models.

A black hole signature might be found in the emission lines from its accretion disk, a possibility dramatized by the observation of X-ray iron lines from Seyfert galaxies. In a phenomenological study, Jun Ogura, N. Ohuo, and Y. Kojima [25] delineate observable properties of emission lines from Keplerian disks around Kerr black holes. Several authors had looked at line profiles from disks about spherical black holes; and Chen and Eardley had considered polarization properties in this case. The present authors have extended the analyses of both line profiles and polarization properties to the observationally more relevant case of rotating black holes.

1.4 Gravitational waves

Peter Dunsby spoke on work with G. Brodin and M. Marklund,[26] that considers radio waves produced by the interaction of a weak gravitational wave with the electromagnetic field of a thin plasma. Because gravitational and electromagnetic waves satisfy the same dispersion relation, electromagnetic waves can be effectively generated as a result of this interaction. Estimates are given for this effect in the case of a binary pulsar and a cold electron plasma, and in each case it is found that the emitted radiation will lie in the radio frequency band.

1.5 Relativistic stellar systems

Robert Mann presented results of a study with P. Chak [27] of the statistical mechanics of a simple and explicit general relativistic self-gravitating system. In 1+1 dimensions they examine a system of N relativistic particles. They compute the partition function and one-particle distribution functions to leading order in $1/c$, recovering known results for the non-relativistic one-dimensional self-gravitating system in the large c limit. They find that curved spacetime effects generally cause both position and momentum distribution functions to become more sharply peaked, and that the temperature of a relativistic gas is smaller than its non-relativistic counterpart at the same fixed energy.

William Stoeger outlined a method developed with M.E. Araújo and S. R. M. M. Roveda for finding spherically symmetric dust spacetimes, associated with galaxy redshifts, observer area distances, and number counts. In particular, when given data for a dust-filled FLRW spacetime, the method reproduces that spacetime.The coordinate freedom of the system and conditions on the observer's worldline greatly simplify the treatment.

D. Khokhlov discussed implications of a time-dependent mass scale. Scaling all masses linearly with time is claimed to meet constraints from the current age of the universe, from the high-redshift supernovae data, and from primordial nucleosynthesis.

1.6 Gravitational collapse and singularities

Megan Govender spoke on work with S. D. Maharaj and Roy Maartens, dealing with the collapse of a spherically symmetric imperfect fluid.[28] They are primarily concerned with the evolution temperature in a collapsing star, and they use the Israel-Stewart formalism for imperfect fluids. Comparing their results to those found from the Eckart (causality-violating) theory, they find relaxation effects introduced by the causal theory can have a significant and, in principle, observable impact on the temperature, rate of collapse and other properties.

Naresh Dadhich presented joint work with P. Joshi and Roy Maartens on naked singularities arising in the collapse of inhomogeneous dust. (See Joshi [29] for a review). The authors emphasize the fact that shear arising from inhomogeneity in the initial data distorts the apparent horizon and delays its formation. The delayed formation of a horizon is then what allows singularities to be seen. I would add to this picture the fact that the formation of singularities from dust is also due to the unphysical nature of dust itself: as Rendall has shown, collisionless matter, with a smooth distribution function exhibits no naked singularities, while dust does.

S. Ghosh, in collaboration with Dadhich, extended the investigation of naked singularities to higher dimensional Vaidya spacetimes. Singularities arising in 3-dimensions can be suppressed in higher dimensions. The author notes that the increase in dimensions strengthens both gravity and the inhomogeneity. Although greater inhomogeneity delays the formation of a trapped surface and so favors a naked singularity, the strengthened gravity seems to have the upper hand; but naked singularities are completely covered only in the limit as the spacetime dimension becomes infinite.

Irina Dymnikova reported work with A. Dobosz, M.L. Fil'chenkov, and A. Gromov on the quantum production of baby universes inside a vacuum black hole.[30] Their calculation is similar to an earlier calculation by Guth and Fahri; they consider a transition from black hole to white hole, allowing a collapsing geometry to rexpand to form a baby universe. In their model a universe arising inside a Schwarzschild-deSitter black hole via quantum tunnelling, starts from a deSitter stage and ends as a Kasner-like geometry.

Acknowledgments

Work on this summary was supported in part by NSF Grant PHY00-71044.

References

1. J. Schaffner-Bielich, M. Hanauske, H. Stöcker, and W. Greiner, astro-ph/0005490.
2. M. Hanauske, D. Zschiesche, S. Pal, S. Schramm, H. Stöcker, and W. Greiner, Ap.J. **537**, 50320 (.)
3. R. Prix, G.L. Comer, N. Andersson, astro-ph/0107176 (2001).
4. K. Konno, Astron. & Ap. **372**, 594 (2001), and references therein.
5. K. D. Kokkotas and N. Andersson, Proceedings of SIGRAV XIV, Genoa 2000, Spinger-Verlag (2001), gr-qc/0109054.
6. J. Ruoff and K. D. Kokkotas, gr-qc/0106073 (2001); MNRAS **328**, 678 (2001).
7. K.H. Lockitch and N. Andersson, gr-qc/0106088 (2001).
8. K.H. Lockitch, N. Andersson, and J. L. Friedman, PRD **63**, 024019 (2001).
9. L. Lindblom, in "Gravitational Waves: A Challenge to Theoretical Astrophysics," ed. by V. Ferrari, J.C. Miller, and L. Rezzolla (ICTP, Lecture Notes Series); astro-ph/0101136.
10. N. Stergioulas and J. A. Font, PRL **86**, 1148 (2001).
11. L. Lindblom, J.E. Tohline, and M. Vallisneri, PRL **86**, 1152 (2001).
12. A. K. Schenk, P. Arras, E.E. Flanagan, S.A. Teukolsky, PRD **65**, 2001 (024001).
13. S. M. Morsink, "Nonlinear coupling between r-modes of Rotating Neutron Stars," preprint (2001).
14. P.B. Jones, PRL **86**, 1384 (2001); PRD **64**, 084003 (2001).
15. L. Lindblom and B.J. Owen, astro-ph/0110558 (2001); submitted to PRD.
16. P. Haensel, K.P. Levenfish, and D.G. Yakovlev, astro-ph/0110575 (2001); Astron. & Ap., in press.
17. L. Rezzolla, F.L. Lamb, D. Markovic, and S. L. Shapiro, PRD **64**, 104013 (2001)

18. L. Rezzolla, F.L. Lamb, and S. L. Shapiro, Ap.J. **531**, L141 (2000).
19. J.L. Friedman, K. Lockitch and P.S. de Sa, in preparation.
20. A.L. Watts and N. Andersson, astro-ph/0110573 (2001), submitted to MNRAS.
21. J. L. Friedman and K. H. Lockitch, "Implications of the r-mode instability of rotating relativistic stars," gr-qc/0102114 (2001); Proceedings of the 9th Marcel Grossman Meeting, World Scientific, ed. V. Gurzadyan, R. Jantzen, R. Ruffini.
22. H. Beyer, astro-ph/0007342.
23. E. Szuskiewicz and J.C. Miller, astro-ph/0107257 (2001), MNRAS, in press.
24. A.R. Prasanna and Bukhopadhyay, astro-ph/0107119.
25. J. Ogura, N. Ohuo, and Y. Kojima, Pub. Astron. Soc. Japan **52**, (2000).
26. M. Marklund, G. Brodin and P. Dunsby, Ap.J. **536**, 875 (2000).
27. R.B. Mann and P. Chak, gr-qc/0101106.
28. M. Govender, R. Maartens, and S. Maharaj, MNRAS **310**, 557 (1999).
29. P.S. Joshi, Pramana **55**, 529 (2000).
30. I. Dymnikova, A. Dobosz, M. L. Fil'chenkov, A. Gromov, gr-qc/0102032 (2001), to appear in Phys. Lett. B.

B2II: STRUCTURE FORMATION

WILLIAM M. LESAME
Department of Mathematics, Applied Mathematics and Astronomy,
University of South Africa
PO Box 329, Pretoria 0003, RSA
Email: lesamwm@unisa.ac.za

This review summarizes the parallel session B2ii on structure formation. Most of the presentations centered on recent data from cosmic microwave background, large-scale structure and supernovae and the two large surveys *viz.* the Sloan digital sky survey and the 2-degree-Field galaxy redshift survey.

1 Large surveys in cosmology

Multi-fibre technology now allows measurements of redshifts of millions of galaxies. The two major redshifts surveys reported their latest results at GR16.

1.1 The SDSS group

The US Sloan Digital Sky Survey (SDSS)[a] aims to image 10,000 square degrees of the North Galactic Cap in five passbands, with an r-band limiting magnitude of 22.5, to obtain spectroscopic redshifts of 10^6 galaxies and 10^5 quasers, and to obtain similar data for three approximately 200 square degree stripes in the South Galactic Cap.[1] Such studies of galaxy clustering and redshift-space distortions are useful to test cosmological models and their parameters and to infer the relation between galaxy and dark matter distributions. Galaxies are currently understood to form in the dark matter haloes that are built up hierarchically (galaxies, clusters of galaxies, etc.) through mergers or accretion of small objects. The clustering patterns of galaxies are therefore determined by the spatial distribution of dark matter haloes.

Joshua Frieman[2] reported on the first SDSS measurements on angular and redshift space clustering based on a sample of 29,300 galaxies. This sample spans about 7% of the total projected survey area of SDSS and the analysis of the SDSS team focused mainly on the small-scale clustering. For the full flux-limited galaxy, the team have

• measured the angled-averaged redshift-space correlation function $\xi(s)$ and the two-dimensional correlation function $\xi(r_p, \pi)$,

[a] http://www.sdss.org/

• measured the angular moments to quantify the anisotropy induced by peculiar motions and

• modeled the small-scale anisotropy to infer the galaxy pairwise velocity dispersion $\sigma_{12}(r)$.

The results for the full galaxy sample are in good agreement with those obtained from clustering analysis of the LCRS [3,4] which has similar selection, geometry and size. The large sample and high quality of the SDSS imaging data allowed for a detailed examination of the dependence of real-space correlations and redshift-space distortions on galaxy photometric properties, for example, the red galaxies exhibit a higher amplitude and steeper real-space correlation function and much stronger finger-of-God distortions than the blue galaxies.

1.2 The 2dFGRS group

The Anglo-Australian 2-degree-Field Galaxy Redshift Survey (2dFGRS)[b] is designed to measure redshifts for approximately 250,000 galaxies selected from the revised and extended version of the APM galaxy catalogue. The main survey regions are two declination stripes, one in the southern Galactic hemisphere spanning $80^\circ \times 15^\circ$ around the SGP, and the other in the northern Galactic hemisphere spanning $75^\circ \times 10^\circ$ along the celestial equator; in addition there are 99 fields spread over the South Galactic Cap.

Ofer Lahav spoke on new results from 2dFGRS. He reported, firstly, on the principal component analysis (PCA)[6,7] used to classify galaxy spectra. This analysis makes use of spectral information in the rest frame wavelength range 3700Å to 6650Å. He stressed that the advantages of the PCA approach is that it is completely objective and reproducible and can be applied over the full magnitude range of the 2dFGRS whenever spectra are of sufficient signal to noise. Using PCA, spectral types are assigned into two broad classes: (a) galaxies with strong emissions lines ('late-types') and (b) more quiescent galaxies ('early types'). A study of luminosity function and clustering per spectral types supports the view that luminosity, and not spectral type, is the dominant factor in determining how clustering strength of the galaxy population varies with luminosity. The second part of the talk focused on the notion that galaxy motions relative to the Hubble flow arise from gravitation forces due to mass-density fluctuations and hence reflect the underlying distribution of both dark and luminous matter. In this manner galaxy motions provide constraints on the cosmological density parameter Ω_m and the

[b] http://www.mso.anu.edu.au/2dFGRS/

fluctuation amplitude parameter σ_8. Lahav, in collaboration with Bridle *et. al.*, [8] gave a contrast of the 2dFGRS data set with that of the two CMB experiments viz, BOOMERANG and MAXIMA-I. The comparative analysis is restricted to the scale-invariant flat ΛCDM model under the additional nucleosynthesis constraint of $\Omega_b h^2 \sim 0.019$. [14] They find that all three data sets agree well, overlapping significantly at the 2σ level. A joint analysis of the three data sets gives a joint best point fit and 95% confidence limits of $\Omega_m = 0.28(0.17, 0.39), h = 0.74(0.64, 0.86)$, and $\sigma_8 = 1.17(0.98, 1.37)$. Also for the best fit point, $Q_{rms-ps} = 19, 7\mu K$ and the age of the universe is 13.2 Gyr.

2 Light cone effects

The two-point correlation function and the power spectrum are the most used tools to quantify the clustering of cosmological objects. These statistical quantities are well defined on the constant-time hypersurface. Cosmological observations, however, are carried out on a light cone and hence light cone effects need to be incorporated in the statistical analysis. Hiroaki Noshioka gave a discussion of the effectiveness of the multipole moments of the redshift-space two-point correlation function on a light cone as a probe of the cosmological parameters and models for bias. The appropriate two-point correlation function, which incorporates light cone effects, is derived and applied to galaxy and quasar samples which roughly match 2dF and SDSS samples. The dependence of multipole moments of the two-point correlation function on cosmological parameters and models for bias are then examined.

3 Primordial perturbations

3.1 Fluids with negative pressure

Fluids of negative pressure are being increasingly studied in cosmology, in particular, such fluid were first considered in the context of the inflationary scenario in which the early universe experiences a very short period of accelerated expansion. Such fluids are however plagued with instabilities in density fluctuations in the small wavelegth limit.[9] Fabris *et. al.* reported on a field representation that involves coupling gravity with a scalar field. For the field representation, with the same behaviour for the scale factor as the corresponding hydrodynamical representation, the instabilities are absent. This mainly to the fact that a relativistic Euler type equation for the hydrodynamic representation is replaced by a Klein-Gordon type equation for the field representation. This indicates that instabilities detected in fluids of negative pressure could be due to the hydrodynamic representation. For large scale perturbations, that is, long wavelength limit, both approaches lead to the same result.

The common element to inflationary models is an exponential expansion phase of the size of the universe prior to the beginning of the standard scenario that describes the adiabatic era. For the moment there is no consensus about how the inflationary phase ends. Two types of transition to an adiabatic era are possible: a first order phase transition which deals with an effective cosmological constant given by the energy of a scalar field and a second order phase transition based on a variable cosmological constant. In their second talk Fabris *et. al.* discussed the evolution of scalar perturbations in the inflationary model that joins the radiative regime by second order phase transition. Their analysis is restricted to $\Lambda_\infty = \sigma_0 = 0$ and $k = 0$. The obtained value of the spectral index is equal to 2, a result that depends very weakly on the various parameters of the model and on the initial conditions used.

Kavilan Moodley, in collaboration with Bucher M. and Turok N, [10] make the point that perhaps more interesting and of significance, to discovering how the perturbations from homogeneity and isotropy were generated in the early universe, is to determine the fundamental character of the primordial perturbations. The talk focused on prospects of testing the adiabaticity of the primordial perturbations using MAP and PLANCK experiments, with and without polarization information, and with just the adiabatic mode and three isocurvature modes and their correlations. The term adiabatic here means that the primordial stress-energy of the universe was governed by a single, spatially uniform equation of state, that is, on surfaces of constant temperature the densities of the various components (eg, baryons, neutrinos, etc.) are uniform and that these components share a common velocity field. Moodely reported that with just the adiabatic mode allowed, the errors on the cosmological parameters are as indicated elsewhere in the literature. With just one isocurvature and its correlation with the adiabatic mode added, these errors increase modestly by less than a factor of two. When three isocurvature modes and their correlations are allowed, the uncertainties are enormous implying a breakdown of the quadratic approximation. Their results demonstrate that polarization information plays a crucial role in testing adiabaticity of primordial perturbations.

Takayuki Tatekawa *et. al.* presented a perturbative approximation theory, based on the Lagrangian description, by extending the method of Adler and Buchert [11]. Analytic approaches usually treat matter in the universe as a self-gravitating fluid and attempt to solve the hydrodynamic equations for the fluid. Including pressure effect of the fluid, Tatekewa *et. al.*, derived and solved the Lagrangian perturbation equations up to second order, assuming a polytropic equation of state. The power spectra of density fluctuations contains large amplitude at small scale and they interpret this as due to the fact that the scale of the inhomogeneity is shortened as inhomogeneity grows because of gravitational instability. Furthermore, in the computation of the power spectra of density perturbations the shell crossing singularities occur even in the presence of the pressure effects. However, the epoch of the occurrence of shell-crossing is later than that in the Zel'dovich approximation.

4 Gravitational entropy

Manfred Leubner discussed structure formation using the quantized gravitational entropy formalism. This formalism is of the view that increasing inhomogeneity due to gravitational clumping reflects increasing gravitational entropy in a time evolving universe. Leubner first makes the assertion that for an ensemble of uniformly distributed particles the ensuing gravitational clustering is subject to a specific quantization rule for the amount of increase of the gravitational entropy. The ground state is identified as quintessence matter background. The gain of gravitational entropy at each higher order merging process is subject to an extremal condition yielding a discrete spectrum of nested bound structure scales of specific mass and radius. Consequently a master arrow of time is expected to point in the direction of increasing gravitational entropy evolution as a manifestation of irreversibility.

5 The fractal debate

The main issue in the fractal debate is that if the distribution of galaxies follows a scaling, or fractal, pattern up to the limits of the presently available observations, this would contradict the standard Friedmann-Robertson-Walker universe, which in turn would lead to the dismissal of the cosmological principle.[12] Ribeiro [13] first posed the question of whether or not a perturbed model could turn the density decay at increasing redshift depths into a power law type. If this happens, then galaxy distribution of this cosmological model would then have a scaling behaviour compatible with the power-law behaviour of density-distance correlation observed in the galaxy catalogues. In an attempt to resolve the fractal debate Ribeiro[?] formulated a conjecture, called *the apparent fractal conjecture*, which states: *the observed fractality of the large-scale distribution of galaxies should appear when observational relations necessary for fractal charaterizations are calculated along the past light cone in a perturbed metric of standard cosmology.* Elcio Abdalla *et. al.* presented a talk on work done to check the validity of the apparent fractal conjecture by deriving observational relations in a perturbed Einstein-de Sitter cosmology by means of a perturbation scheme where the scale factor is expanded in a power series to yield perturbative terms. The results show that the Einstein-de Sitter cosmology can be approximately described by a decaying power-law like average density profile and, in this context, the apparent fractal conjecture is correct.

6 Acceleration in the late universe

A number of high quality, but indirect, experiments are suggesting that the universe is currently accelerating. The luminosity distance estimates from type Ia supernova [15] favour recent acceleration while the recent CMB data[16,17,18,19] suggest the universe has almost zero spatial curvature. The best studied explanation for this uses quintessence; a very light scalar field, whose effective potential leads to acceleration in the late universe. However quintessence requires extreme fine-tuning since not only must the cosmological constant be zero but the quintessence field must dominant at late times only. Bruce Bassett in collabortion with Kunz, Silk and Ungarelli presented a possible explanation whereby quantum effects become important at low redshifts and stimulate the universe to begin accelerating. They considered a phenomenogical model (vacuum metamorphosis) in which the dark energy of the universe is described by a scalar field Q whose equation of state w undergoes a sudden transition from $w_0 = 0$ (dust) to $w_f < -0.3$ at a specific redshift. They then used the current CMB, large-scale structure (LSS) and supernovae (SN1a) data to constrain the phenomenological parameter space variables (Ω_Q, z_t, w_f) and found that

- The global best-fit to the current data occurs for $z_t = 1.5, w_f = -1.0$ and $\Omega_Q = 0.73$ while the marginalised 1d likelihood for z_t peaks at $z_t = 2.0$. The best-fit model is consistent with the data and is a marginally better fit than the best ΛCDM model.
- The CMB provides the best constraint on parameters, especially on z_t due to the integrated Sachs-Wolfe effect.

7 Galaxy interactions

Jorge Cervantes-Cota *et. al.* presented simulations within the framework of scalar-tensor theories, in the Newtonian limit, to investigate the influence of massive scalar fields on the dynamics of the collision of two equal spherical clouds. They employed a SPH code modified to include the scalar field to simulate two initially non-rotating (proto)galaxies that approach each other, and as a result of the tidal interaction, intrinsic angular momentum is generated. They obtain sufficient large values of J/M to suggest that intrinsic angular momentum can be the result of tidal interactions.

Acknowledgments

The workshop was arranged by Joshua Frieman.

References

1. D. G. York *et. al.*, Astronom. J., **120** 1576 (2000).
2. I. Zehavi *et. al.*, astro-ph/0106476.
3. D. L. Tucker, MNRAS, **285** L5(1997).
4. Y. P. Jing, Atsophys. J., **494,** 1 (1998).
5. M. Colles, MNRAS, **328** 1039 (2001).
6. P. Norberg *et. al.*, MNRAS, **328,** 67 (2001).
7. P. Norberg *et. al.*, MNRAS, submitted (2001).
8. S. L. Bridle, MNRAS, **312**, 333 (2001).
9. J. C. Fabris and J. Martin, Phys. Rev. **D55**, 5205 (1997).
10. M. Bucher, K. Moodley, and N. Turok, astro-ph/0011025.
11. S. Adler and T. Buchert, Astronomy and Astrophysics, **343**, 317 (1999).
12. P. H. Coleman and L. Pietronero, Phys. Rep., **213**, 311 (1992).
13. M. B. Ribeiro, astro-ph/0104180.
14. D. Tytler, astro-ph/0001318.
15. A. G. Riess et al Astronom. J., **116** 1009 (1998).
16. C. B. Netterfield *et. al.*, astro-ph/0104460.
17. A. T. Lee et. al. Astrophys. J., **561** L1 (2001).
18. N. W. Halverson *et. al.*, astro-ph/0104489.
19. S. T. Padin *et. al.*, Astrophys. J., **549** L1 (2001).

B2III: THE COSMOLOGICAL CONSTANT

BRIAN P. SCHMIDT
via Cotter Rd., Weston Creek, ACT 2611 Australia
E-mail: brian@mso.anu.edu.au

Observations of type Ia supernovae, large scale Structure, and anisotropies in the cosmic microwave background have provided compelling evidence that the energy density of the Universe is dominated by something like a cosmological constant. Although there is increasing amounts of data providing corroborating evidence in favor of a cosmological constant, no clear theoretical understanding of this majority component of the Universe has emerged. In this session of GR16, we explore several observational and theoretical approaches to better understand the Cosmological Constant.

1 Introduction

Since the 1920s astronomers have been actively trying to understand the dynamical evolution and geometry of the Universe. However, it is only after seven decades of work by hundreds of individuals and groups that we reached the point where meaningful conclusions can be drawn from data.

Over the past 5 years two teams have successfully used Type Ia supernovae (SN Ia) to trace the expansion of the Universe to look-back times exceeding 8 billion years. [1,2] Both group's observations show that supernovae are, on average, fainter than what would be expected for even an empty Universe. This is an unexpected result: the Universe is accelerating. The easiest way to explain the observed acceleration is to include an additional component of matter with an equation-of-state parameter more negative than $w = p/\rho < -1/3$; the most familiar being the Cosmological constant ($w = -1$).

Ultimately, because any experiment is subject to unknown systematic effects, it is sensible to look for consistency between completely independent experiments, rather than to only push one method forever. Unfortunately, there is not a plethora of methods that can provide believable results on cosmology by themselves. However, the cosmic microwave background (CMB) anisotropies, with recent experimental measurements from Boomerang [3] and Maxima [4], provide strong constraints on the total matter density, and the 2dF and Sloan redshift surveys are beginning to provide a greater handle on the total density in normal matter [5]. These two experiments along with the SN Ia observations overlap in a region around $\Omega_M = 0.25$, $\Omega_\Lambda = 0.75$. To **not** take this overlap region seriously, we must assume that not one, but two methods have a significant systematic error. If we assume that the CMB measurements are correct, then the Universe cannot have a low value for Ω_M, and both the SN Ia and large scale structure observations must be wrong at the $>> 5\sigma$ level. Alternatively, we can assume that the SN measurements are incorrect at the 3σ level, the large scale structure measurements are correct, and the CMB measurements are completely wrong. The position of the first doppler peak, which provides most of strength of the current CMB derived constraints on the total matter density, cannot be moved to small enough angles to accommodate small values of Ω_M. This conjecture then requires one to reject the CMB theory, a theory which pre-predicted all of features seen in the Cosmic Microwave Background observations. By far the easiest solution is to accept the consistency of the three methods, and assume they all are correct: The Universe is dominated by something like the cosmological constant.

In the following sections participants in GR16:B2iii will describe their efforts to better understand the cosmological constant from both an observational and theoretical viewpoint.

2 Cosmological Parameters, Gravitational Lensing and Galaxy Evolution: N. Panchapakesan, V.B. Bhatia, S. Mahajan, Deepak Jain, and Abha Dev

A distant object, like a quasar, is gravitationally lensed by a galaxy in the intervening space between the quasar and us. The probability for lensing depends on the travel distance (or time) from quasar to us. This in turn is a function of cosmological parameters specially the nature and amount of matter or energy in the universe; like the cosmological constant and matter both baryonic and otherwise. The statistical analysis of the observed number of lensed objects, based on the gravitational theory of lensing, serves to determine (or at least constrain) the cosmological parameters. All calculations are consistent with present values of cosmological constant (of 0.7), and also predict the equation of state for the dark energy if it is not a constant.

However the predicted number depends also on the assumed distribution of galaxies as a function of their luminosity or mass. Galaxy luminosities are related to the observed velocity dispersion of matter (through an exponent γ called the Faber-Jackson parameter). The predictions also depend on the models used for the evolution of galaxies.

We (the group at Delhi university) were the first to point out that $\gamma = 4$ makes the predictions independent of the nature of evolution for most models of galaxy evolution involving mergers, and so values of γ other than four (as are being suggested now) must be used.

The predictions depend also on the slope of the distribution function, α, in the Schechter distribution function for the galaxies. As elliptical galaxies dominate lensing, due to their larger exponents in the relation between luminosity and dispersion, it is their slope that is needed accurately. Unfortunately this is not yet well known. In addition, the normalisation of the distribution, related to the number density of galaxies is also not wekk constrained. These uncertainties lead to wide variations in the constraints on the cosmological parameters.

Reduction of uncertainties in the galaxy parameters (like γ, ϕ_*, α) in the coming years should help in more precise prediction of the cosmological parameters. Ongoing galaxy surveys (like Sloan and 2dF) will be increase the total galaxy survey size by more than an order of magnitude, and provide more data useful for the physical characterisation of the galaxies. These observations should then allow a much more precise predictions of the cosmological parameters from this method[6,7].

3 Quantum Gravitational Back-Reaction: N. C. Tsamis and R. P. Woodard

Quantum gravitational back-reaction offers an attractive model of cosmology which also resolves the problem of the cosmological constant. The idea is that there is no fine tuning of the cosmological constant, Λ, or of scalar potentials [8]. In fact there need not be any scalars. Inflation begins in the early universe because Λ is positive and not unnaturally small. Inflation eventually ends due to the accumulation of gravitational attraction between long wavelength virtual gravitons which are pulled apart by the rapid expansion of spacetime. Inflation persists for many e-foldings because gravity is a weak interaction, even at typical inflationary scales, and it requires an enormous accumulation of gravitational potential to overcome this. Because the model has only a single free parameter — $G\Lambda$, where G is Newton's constant — it can be used to make unique and testable predictions. When Λ is adjusted to make the COBE 4-year RMS quadrupole come out right, the model predicts the following CMB anisotropy parameters [9]:

$$r \approx .0017 \quad , \quad n_s \approx .97 \quad , \quad n_t \approx -.00028 \,. \tag{1}$$

The infrared character of the physical mechanism means that it can be studied reliably using perturbative quantum general relativity. The expectation value of the gauge-fixed metric was computed [10] on the manifold $T^3 \times R$

and in the presence of a state which is free graviton vacuum at $t = 0$,

$$\left\langle \Omega \middle| g_{\mu\nu}(t, \vec{x}) dx^\mu dx^\nu \middle| \Omega \right\rangle = -dt^2 + e^{2b(t)} d\vec{x} \cdot d\vec{x} \ . \tag{2}$$

The resulting Hubble constant is,

$$\dot{b}(t) = H \left\{ 1 - \left(\frac{G\Lambda}{3\pi} \right)^2 \left(\frac{1}{6} (Ht)^2 + O(Ht) \right) + O(G^3) \right\} \ . \tag{3}$$

The correct interpretation of this result is that quantum gravitational back-reaction slows inflation as long as perturbation theory remains valid, and that the effect must eventually grow nonperturbatively strong.

The great unsolved problem is how to treat the period beyond the end of inflation. That progress is possible derives from the fact that this is not a strongly coupled theory like QCD; the dimensionless coupling constant is $G\Lambda \approx 5 \times 10^{-11}$. It is not that any particular particle creation event matters much but rather that an enormous number of them add coherently over the ever-increasing volume of the past light cone. It therefore seems possible that the incremental screening effect derived from each newly visible element of the past light cone can computed perturbatively provided the accumulated effect is subsumed into a stochastic background [11] and provided one can obtain the propagators in the back-reacting geometry.

This past summer progress has been made in developing a new infrared expansion which may give the most important part of the propagator in an arbitrary homogeneous and isotropic background. It also seems inevitable that inflation must end suddenly with a cataclysmic decay into deflation. As modes re-enter the horizon which left during inflation, they become part of a hot, dense plasma. This plasma forms a thermal barrier by scattering the infrared particles which must carry quantum correlations forward from the era of inflationary particle production in order to keep the bare cosmological constant screened. When the thermal barrier becomes too opague, the bare cosmological constant reasserts itself and the universe begins to expand again. Subsequent evolution seems to be governed by the balance between the growth of the inflationary past light cone — which increases screening — and the continued buildup of the thermal barrier — which inhibits screening. It has not been shown yet whether this results in a plausible cosmology. More details can be found elsewhere [12].

4 Global Defect and its Cosmological Implications: Michiyasu Nagasawa

It is a natural idea that the cosmological constant may consist of some scalar fields. On the other hand, cosmological defects are also considered to be formed by a certain configuration of fields. Particularly the global defect is made up only with scalar fields.

It is possible that the late time phase transition of a certain mass scale $\sim m_\Lambda$ can supply a small non-vanishing cosmological constant[13]. When this mass scale satisfies the condition as

$$m_\Lambda \sim 10^{-3} \text{ eV} , \tag{4}$$

then the energy scale of the false vacuum $\sim m_\Lambda^4$ becomes comparable with the critical energy density of the present universe. That is, the observed value of the cosmological vacuum energy might be explained.

This late time phase transition is accompanied by the production of topological defects. These topological defects can explain some of the cosmological problems and may provide interesting observational effects. Particularly such a phase transition can be realized in the two-family see-saw model[14], in which global strings are expected to form. It is worth to argue what kind of phenomenon would be predicted if such a string exists.

In this presentation, the interaction of particles with the defect is considered. Here we concentrate on the global string case. Note that more complicated but interesting effect would be observed in the Alice string like scenario or the case some amount of CP violation exists.

We employ a simple toy model and estimate whether the global string detection would be possible. Two examples of the interactions of the global string with several particles will be investigated. One is the coupling with photons and the rotation angle difference of the polarization is dependent on the coupling strength. The other is the coupling with fermions and the scattering amplitude is dependent on the coupling constant.

If the Nambu-Goldstone boson, ϕ which corresponds to the phase of the defect constituting scalar field, has an interaction with the electromagnetic field such as

$$\mathcal{L}_p = -g_p \frac{\phi}{\eta} \mathbf{E} \cdot \mathbf{B} , \tag{5}$$

where $\mathbf{E}$ is an electric field and $\mathbf{B}$ is a magnetic field, then the polarization of the electromagnetic waves will be converted. The variations of polarization angle of the light rays between whose paths there is a global string have the same magnitude but the opposite sign[15].

Moreover, we consider the derivative coupling of the ϕ particle with the fermions. Then the interaction Lagrangian is expressed by

$$\mathcal{L}_f = g_f \frac{\partial_\mu \phi}{\eta} \overline{\psi} \gamma^\mu \psi \ , \tag{6}$$

where ψ is a fermion field. The scattering problem in the background of the global string can be solved and the resulting cross section shows how the fermion trajectory should be modified[16].

One of the possible applications of the above consideration might be found in the brane world model and another example can be seen in the topological inflation scenario. Both of them can be related with many unresolved problems in particle physics and cosmology. These branes and inflatons may be made up with scalar fields and the detection mechanism presented here could be promising.

Although the description here is only a preliminary indication, the current constraints on the model by various experiments and observations and the possibilities of the global defect detection using the future observations should be studied further.

5 Self-Tuning Branes: Sean Carroll

There are two fine-tuning problems associated with the cosmological constant: why it is so much smaller than its natural value ($\rho_\Lambda/M_{\rm pl}^4 \approx 10^{-120}$, where $M_{\rm pl} = 10^{18}$ GeV is the Planck mass), and why it is of the same order as the matter density today. Despite the success of the "best-fit cosmology" with $\Omega_\Lambda = 0.7$, these naturalness issues cause us to question whether the framework of general-relativistic cosmology might somehow be incorrect; specifically, whether the Friedmann equation

$$H^2 = \left(\frac{\dot{a}}{a}\right)^2 = \frac{8\pi G}{3}\rho - \frac{k}{a^2} \tag{7}$$

could be altered.

How might we alter the Friedmann equation to avoid the cosmological constant problem? One distinguishing feature of vacuum energy is its large negative pressure, $p_\Lambda = -\rho_\Lambda$. This raises the possibility of modifying gravity in such a way that the effective Friedmann equation takes the form

$$H^2 \propto \rho + p \ . \tag{8}$$

Such a relation would lead to conventional behavior in the matter-dominated era, while making the geometry completely insensitive to the vacuum energy.

A specific example of this kind modification occurs in the context of brane-world scenarios. The idea of "self-tuning branes" [17] imagines that we live on a $(3+1)$-dimensional brane embedded in a $(4+1)$-dimensional bulk, in which there lives a scalar field ϕ with very specific couplings to matter on the brane. The original authors showed that the induced geometry of the brane could be flat (Minkowski space), regardless of the value of the vacuum energy on the brane. Such a situation necessitates an alteration in the Friedmann equation, since this geometry has $H=0$ and $k=0$ even though $\rho \neq 0$.

Carroll and Mersini [18] have investigated how cosmological evolution proceeds in the self-tuning scenario. (For related work see [19].) They find that the Hubble parameter obeys

$$H^2 = \frac{4\pi G}{9} a^{-4} \int a^3(\rho + p)da \ . \tag{9}$$

This equation assumes a flat spatial geometry ($k=0$), and neglects evolution of the scalar field and quadratic terms in the energy density and pressure. We see that the combination $(\rho+p)$ enters the right-hand-side, implying that the vacuum energy decouples completely from the expansion rate.

Consider a universe dominated by a combination of matter ($\rho_{\rm M} \propto a^{-3}$, $p_{\rm M}=0$) and radiation ($\rho_{\rm R} \propto a^{-4}$, $p_{\rm R} = \rho_{\rm R}/3$). The Hubble parameter then obeys

$$H^2 = \frac{4\pi G}{9}\rho_{\rm M} + \frac{16\pi G}{27}\rho_{\rm R} \ln\left(\frac{a}{a_*}\right) , \tag{10}$$

where a_* is a constant of integration. It is this relation which may be compared to the conventional Friedmann law. In a matter-dominated era, the functional dependence of H on a is conventional, leading to the familiar $a \propto t^{2/3}$ behavior. The coefficient of ρ, however, is different, suggesting a possible empirical test of the model: in this theory, the Hubble parameter during matter domination will be smaller by a factor of $1/\sqrt{6} \approx 0.41$ than the conventional expectation (at fixed ρ). Unfortunately, we do not have very precise empirical information about the expansion rate during the matter-dominated era. Given that our current universe is apparently dominated by a smooth component causing it to accelerate, we cannot directly constrain the coefficient of $\rho_{\rm M}$ by contemporary observations.

In the radiation-dominated era, quantitative constraints on the behavior of the Hubble parameter may be derived from Big-Bang Nucleosynthesis (BBN) [20]. Expanding the logarithm in (10), we have

$$H_{\rm R}^2 = \frac{16\pi G}{27}\rho_{\rm R} \ln a + Ca^{-4} , \tag{11}$$

where C is an integration constant. This constant is not merely a nuisance that can be set to zero, but rather reflects the normalization of a; if we set $a = 1$ today, the first term is always negative in the past, which is clearly unworkable. On the other hand, we can choose C such that the second term is dominant — and with the correct magnitude to be compatible with observation — during BBN. (BBN is compatible with alternative expansion histories, but only with a highly restricted set [21].) This seems like an unlikely bit of fine-tuning, although it cannot be rigorously excluded. Therefore we see no way to definitively state that the self-tuning cosmologies we have considered are incompatible with observation, although it would require a mysterious coincidence to predict the correct light-element abundances. Since these models were invented to solve the fine-tuning problem associated with the cosmological constant, this must be considered as a strike against them.

Another issue with the self-tuning scenario is that it offers no obvious answer to the question of why the universe is accelerating today. If the cosmological constant is invisible to the expansion of the universe, some new phenomenon would be required. However, a conventional "quintessence" scalar field would still be able to make the universe accelerate; indeed, any energy density which was not strictly constant, but redshifted very slowly as the universe expanded, would be a plausible candidate. It is worth exploring whether the combination of dynamical dark energy plus an alternative theory of gravity could ameliorate the fine-tuning problems which plague the conventional best-fit cosmology.

6 Vanishing of cosmological constant in non-factorizable geometry: S. Shankaranarayanan

6.1 Introduction and Motivation

Any realistic theory of gravity should be able to reproduce r^{-1} behavior of the gravitational potential in the Newtonian limit. Generically, the potential falls off like r^{-d+3}, where d is the number of extra dimensions with infinite extend. Thus, to obtain r^{-1} behavior, higher dimensional theories of gravity — like Kaluza-Klein, Supergravity (also String theory) — have been assuming compactness ($\approx$ Planck length) of the extra dimensions which is commonly referred as Kaluza-Klein compactification. The scale size of the internal dimensions (in these theories/models) is sufficiently small in relation to the energy scale of excitations in the lower dimension, ensuring that the internal dimensions are unobservable, and the world will essentially appear to be effectively lower dimensional. If an extra dimension is non-compact, there would be a continuum modes with masses extending down to zero, when seen from the lower-dimensional view-point.

Recently, Randall and Sundrum(RS, hereafter) [22,23] have shown that these extra dimensions in 5-D space-times need not be compact. They have shown that for 5-D, gravity can be localized on a single 3-brane (where the standard model particles are confined) even when the fifth dimension is infinite. The non-compact localization arises via the exponential warp/conformal factor in the nonfactorizable metric:

$$ds^2 = \exp(-2k|y|)\left[dt^2 - d\mathbf{x}^2\right] - dy^2. \tag{12}$$

For $y \neq 0$, this metric satisfies the five dimensional Einstein's equation with negative five dimensional cosmological constant, $\Lambda \approx -k^2$. The brane is located at $y = 0$, and the induced metric on the brane is a Minkowski metric. The bulk is a five dimensional anti-de Sitter metric, with $y = 0$ as boundary, so that $y < 0$ is identified with $y > 0$, reflecting the Z_2 symmetry with the brane as fixed point, that arises in the string theory.

Perturbations of the metric (12) shows that the Newtonian potential on the brane is recovered at lowest order:

$$V(r) = \frac{GM}{r}\left(1 + \frac{2}{3k^2r^2}\right). \tag{13}$$

Thus, the four dimensional gravity is recovered at high energies, with a first-order correction that is constrained by current sub-millimeter experiments. The zero mode produces the standard $1/r$ gravitational potential along the brane, and the Kaluza-Klein modes give rise to corrections of order $1/r^3$. In the case of RS, the 3-brane is flat Minkowski. Here we show that the results of Randall and Sundrum can be generalized to a wider class of four-dimensional space-times including the four-dimensional Schwarzschild background and de Sitter universe [24].

6.2 Results

Consider the generalization of metric (12), i.e.,

$$ds^2 = g_{ab}dx^a dx^b = \exp[-2a(y)]\left[g^{(4)}_{\mu\nu}dx^\mu dx^\nu\right] - dy^2, \tag{14}$$

with the condition that it satisfies the full five dimensional Einstein's equation with the five dimensional cosmological constant. [The lowercase Latin letters are for the full five dimensions and the lowercase Greek letters for four dimensions.] Denoting the perturbed metric by $\tilde{g}_{ab} = g_{ab} + h_{ab}$ and using the gauge $h_{55} = h_{5\mu} = 0, \nabla^\mu h_{\mu\nu} = 0, h^\mu_\mu = 0$, it is easy to show that $h_{\mu\nu}$ can be written as plane wave gravitons, *i.e.* $h_{\mu\nu} = e_{\mu\nu}\Phi$. The solution to the wave equation for the above line element can be found for the case $m = 0$ [24] and is given by

$$Z(y) = \exp[-2a(y)]. \tag{15}$$

where $Z(y)$ is the y dependent part of Φ. This shows that the existence of a zero mass graviton is a very general result and does not require much of the extra assumptions in RS except that Z should be well behaved and *normalizable* as a function of y, in the relevant range. The stability of the 3-brane is now associated with the normalizability of the graviton wave-function. However, it difficult to show the normalizability of the wave-modes for an arbitrary 4-D metric in Eq. (14). As a special case, let us consider the 4-D metric to be spherically symmetric, i.e.,

$$ds^2 = \exp(-2a(y))\left[A(r)dt^2 - B(r)dr^2 - r^2 d\Omega^2\right] - dy^2, \tag{16}$$

where, $d\Omega^2$ is the angular line element and $a(y)$, $A(r)$ and $B(r)$ need to be determined *via* the five-dimensional Einstein's equations. Solving the 5-D Einstein's equations, we get,

$$A(r) = 1 - \frac{C}{r} - \frac{\lambda}{3}r^2; \quad B(r) = 1/A(r) \tag{17}$$

$$\frac{d^2 a(y)}{dy^2} = \frac{\lambda}{3}\exp(2a(y)), \tag{18}$$

where C and λ are the constants of integration. [The four-dimensional metric is the well known Schwarzschild - de Sitter metric for the choice of $C > 0$, where λ is the four-dimensional cosmological constant.] From the above equations, it is clear that the conformal factor has *only* the λ dependence and independent of C. For $\lambda = 0$, we get $Z = \exp(-2a(y))$ and is normalizable. However, for $\lambda \neq 0$, we get,

$$\exp[-2a(y)] = \exp[2k|y|]\left[\exp[-2k|y|] - (\lambda/12k^2)\right]^2 \tag{19}$$

with k being a constant related to Λ by $\Lambda = -6k^2$. The ground state wave function $Z = \exp(-2a(y))$ is not normalizable for $\lambda \neq 0$ and hence we do not get a massless graviton for $\lambda \neq 0$. [Note that as $\lambda = 0$, the wave function(Z) is normalizable and reduces to the ground state wave function obtained by RS.]

To conclude, we have shown that that the existence of a zero graviton mode is general, i.e. it exists for a wide class of four dimensional metrics in the case of RS model. In particular, the results of RS are valid without modifications for a four dimensional Schwarzschild black hole. However, the presence of non-zero 4-D cosmological constant modifies the results obtained by RS, i. e., the presence of a non-zero cosmological constant does not provide a normalizable ground state wave function corresponding to the zero mass graviton. We thus provide a dynamical reason for the strict vanishing of the cosmological constant *within the context of these models.* The stability of the 3-brane to different classes of matter fields in the context of the general five dimensional metric is under investigation.

7 The oscillating G model: Marcelo Salgado, José A. González, Hernando Quevedo and Daniel Sudarsky

The standard cosmological model is perhaps the simplest and best model for describing the large scale evolution of our universe. Strong evidence like primordial nucleosynthesis, and the cosmic microwave background (CMB) support the underlying hypothesis of the standard cosmology. Yet, there remain some open and fundamental questions concerning some features of our universe that cannot be accounted by the model itself. Namely, the nature of the dark matter and dark energy and the clumpiness properties of galaxies at large scales. Primordial nucleosynthesis constraints the baryonic content of the universe of about $\Omega_{\rm bar} \sim 0.02h^{-2}$ (in units of critical energy density), where h (the present Hubble expansion rate in units of $100 {\rm kms}^{-1}{\rm Mpc}^{-1}$) has likely values in the range $[0.65, 0.75]$. This implies that the baryonic contribution to the total energy density of the universe hardly exceeds 5%. Structure formation simulations and the current models for explaining the CMB anisotropies favor a matter contribution $\Omega_{\rm matt} \sim 0.3$. Therefore, some 25% of the energy-density in our universe appears in a rather unknown form which by interacting only gravitationally has been termed dark matter. On the other hand, the same models for the CMB anisotropies and the observations of high red-shift supernovae (SNIa) favor a cosmological constant contribution (dark energy) of about $\Omega_\Lambda \sim 0.7$ [1,2,3,25]. The different kinds of energy-density contributions add-up to a total energy-density of the universe of $\Omega_{\rm tot} \sim 1$. At this regard, one can comment that the cosmological constant while simply explaining the features of the peaks in the CMB anisotropies and the acceleration of our universe, rises also the "coincidence problem": it is only after 10 billion years (the order of the age of the universe) that the value of the cosmological constant is of the same order of magnitude of the current energy-density of the universe. Such a problem has motivated the search of more natural candidates for the dark energy like quintessential models consisting of massive scalar fields (usually minimally coupled to gravity). Another problem which has raised some controversy is the discovery of a periodicity of $128h^{-1}$Mpc in the galactic count number which extends beyond 2000 Mpc from our galaxy in the directions of the north and south poles of the Milky Way [26]. While some attempts to explain such strange periodicity have been made (some of them reminding the Ptolomeic misconceptions like the spontaneous breaking of the cosmological principle [27] or others assuming ad-hoc matter distributions [28]) it seems that none of them are one hundred per cent consistent with observations. It seems even less clear if there is a close connection between such galactic periodicity and the dark matter-energy problems. At this respect one can also mention models proposing variations of fundamental constants to explain the above galactic periodicity [29]. Among them, the less restrictive seemed to be the oscillating G model consisting in a massive scalar field non-minimally coupled to gravity which oscillates coherently in cosmic time with period $\tau \sim 400h^{-1}$ million years. The model while preserving the spirit of the standard cosmology (homogeneity and isotropy) could explain the galactic periodicity as an illusion produced by the oscillation of the red-shift of light within our past light cone (indeed within the past light cone of any comoving observer as well). A true oscillation in cosmic time of the effective gravitational "constant" would be then reflected in an oscillating galactic count number and mistakenly interpreted as an actual spatial periodicity with period $\lambda \sim c\tau \sim 128h^{-1}$Mpc. Clearly this view assumes that the periodicity is isotropic around every comoving observer and that the scalar field affects all the possible light frequencies. Therefore the hypothesis of the oscillating G model will be validated only if the observations show that periodicity is present in all directions and light frequencies and not only along the galactic poles and within the visible spectrum. We have analyzed the specific model

$$\mathcal{L} = \left(\frac{1}{16\pi G_0} + \xi\phi^2\right)\sqrt{-g}R - \sqrt{-g}\left[\frac{1}{2}(\nabla\phi)^2 + m^2\phi^2\right] + \mathcal{L}_{\rm matt} \, , \qquad (20)$$

in several contexts (see Ref.[30] and references therein). In the framework of the standard cosmology and the standard inflationary scenario $\Omega_{\rm tot} = 1$ we showed that the model can account for the galactic periodicity while preserving constraints such like the primordial nucleosynthesis, the age of the universe, and the current expansion rate. It seems also that the model can generically explain the SNIa luminosity-distance curves within the same values of parameters. The bonus is that no cosmological constant is needed, the dark energy being represented by the oscillating scalar field whose effects are manifested by an apparent galactic periodicity and by the acceleration of the universe. Unfortunately the model (20) breaks down when one analyses solar system effects within the range of parameters satisfying the cosmological observations [30]. For instance, the model fails to reproduce the required value of the effective Brans-Dicke parameter $\omega_{\rm BD}^{\rm eff} > 3000$. Alternatively, if one imposes the constraint for $\omega_{\rm BD}^{\rm eff}$ then the galactic periodicity remains unexplained or the constraint $|\dot{G}/(GH)|_{\rm today} \leq 0.3h^{-1}$ is violated. The main conclusion is that the model (20) is tightly constrained to explain simultaneously all the current observations both local (solar system and binary pulsar) and global (cosmological). However, if we were to drop the $128h^{-1}$Mpc constrained, the model could still be competitive as dark energy model. This is not unreasonable considering this observation is considered spurious in most analyses.

References

1. Riess, A. G., et al. 1998, *AJ*, **116**, 1009
2. Perlmutter, S., et al. 1999, *ApJ*, **517**, 565
3. de Bernardis, P. et al. 2000, *Nature*, **404**, 955
4. Balbi, A.. et al. 2000, *ApJ*, **545**, 1
5. Peacock et al. 2001, *Nature*, **410**, 169
6. Panchapakesan, N. et al 1998 Int.J.Mod. Phys. A13, 4227
7. Panchapakesan, N. et al 1999 Int.J.Mod. Phys. D8, 507
8. N. C. Tsamis and R. P. Woodard, Nucl. Phys. **B474** (1996) 235, *hep-ph/9602315.*
9. L. R. Abramo, R. P. Woodard and N. C. Tsamis, Fort. d. Physik **47** (1999) 389, *astro-ph/9803172.*
10. N. C. Tsamis and R. P. Woodard, Ann. Phys. **253** (1997) 1, *hep-ph/9602316.*
11. A. Linde, D. Linde and A. Mezhlumian, Phys. Rev. **D49** (1994) 1783.
12. R. P. Woodard, "Effective Field Equations of the Quantum Gravitational Back-Reaction on Inflation," talk presented at Cosmo-01, Rovaniemi, Finland, August 29 – September 4, 2001, *astro-ph/0111462.*
13. A. Singh, Phys. Rev. **D52**(1995)6700.
14. R. Holman and A. Singh, Phys. Rev. **D47**(1993)421.
15. L. Masperi and S. Savaglio, Astropart. Phys. **3**(1995)209.
16. J. March-Russell, J. Preskill and F. Wilczek, Phys. Rev. Lett. **68**(1992)2567.; M. Nagasawa, Prog. of Theor. Phys. **98**(1997)851.
17. N. Arkani-Hamed, S. Dimopoulos, N. Kaloper and R. Sundrum, Phys. Lett. **B480**, 193 (2000) [hep-th/0001197]; S. Kachru, M. Schulz and E. Silverstein, Phys. Rev. **D62**, 045021 (2000) [hep-th/0001206].
18. S. M. Carroll and L. Mersini, Phys. Rev. D **64**, 124008 (2001) [hep-th/0105007].
19. A. Mennim and R. A. Battye, Class. Quant. Grav. **18**, 2171 (2001) [hep-

th/0008192].
20. K. A. Olive, G. Steigman and T. P. Walker, Phys. Rept. **333-334**, 389 (2000) [astro-ph/9905320]; S. Burles, K. M. Nollett, J. N. Truran and M. S. Turner, Phys. Rev. Lett. **82**, 4176 (1999) [astro-ph/9901157]; D. Tytler, J.M. O'Meara, N. Suzuki and D. Lubin, astro-ph/0001318.
21. S. M. Carroll and M. Kaplinghat, Phys. Rev. D, in press (2002) [astro-ph/0108002].
22. L. Randall and R. Sundrum, Phys. Rev. Lett. **83**, 3370 (1999).
23. L. Randall and R. Sundrum, Phys. Rev. Lett. **83**, 4690 (1999).
24. T. Padmanabhan and S. Shankaranarayanan, Phys. Rev. D. **63**, 105021 (2001).
25. S. Hanany *et al.*, Astrophys. J. **545**, L5 (2000);
26. T. Broadhurst, R. Ellis, D. Koo, and A. Szalay, Nature **343**, 726 (1990); A. Szalay, R. Ellis, D. Koo, and T. Broadhurst in *Primordial Nucleosynthesis and Evolution of the Universe* (ed. K. Sato, & J. Audouze) (Dordrecht: Kluwer Academic, 1991) p.435; A. Szalay, T.J. Broadhurst, N. Ellman, D.C. Koo, and R.S. Ellis, Proc. Natl. Acad. Sci. USA **90**, 4853 (1993)
27. P. Bundich, P. Nurowski, R. Raczka, and M. Ramella, Astrophys. J. **451**, 10 (1995)
28. R. Van de Weygaert, MNRAS **249**, 159 (1991); M.U. Subba Rao and A.S. Szalay, Astrophys. J. **391**, 483 (1992)
29. M. Morikawa, Astrophys. J. **362**, L37 (1990); C.T. Hill, P.J. Steinhardt, and M.S. Turner, Phys. Lett. B **252**, 343 (1990); R.G. Crittenden and P.J. Steinhardt, Astrophys. J **395**, 360 (1992); P.D. Sisterna and H. Vucetich, Phys. Rev. Lett. **72**, 454 (1994)
30. J.A. Gonzalez, H. Quevedo, M. Salgado, and D. Sudarsky, Phys. Rev. D **64**, 047504 (2001).

B3: PHYSICS OF THE EARLY UNIVERSE

KEI-ICHI MAEDA
Waseda University, Department of Physics
Okubo 3-4-1, Shinjuku, Tokyo 169-8555, JAPAN
E-mail: maeda@gravity.phys.waseda.ac.jp

There has been a big progress in research field of *the Early Universe* after the last GRG conference. The higher-dimensional spacetime became more fashionable and general relativity plays a very important role in cosmology.

1 Recent Progress in Research for the Early Universe

After a great success of big bang cosmology, particle physics came into a major stream of cosmology in 1980's. Many interesting and important ideas have been proposed. Baryogenesis, cosmic phase transition and its relics such as a topological defect, inflationary scenario, creation of the universe, quantum cosmology and so on. After the fluctuation of the cosmic microwave background radiation was found by COBE, however, observational cosmology has also joined into the main steam of cosmology. Now, many big projects in observation have been proposed, and then one may expect that we will understand the formation of the large scale structures in the universe in near future. In new century, what we have to do next is to find some link between the early stage of the universe and observation in order to understand our universe more deeply. The origin of density fluctuation is one of the main targets. Inflationary scenario provides one natural explanation for fluctuations beyond the horizon scale. The inflationary potential to explain the observation has been explored. In this session, Tsujikawa[1] and Wands[2] presented some progress in the study of density perturbations. Tsujikawa's discussion was based on the generalized Einstein gravity, such as Jordan-Brans-Dicke theory, induced gravity, Kaluza-Klein theory, and a higher-curvature theory, while Wands introduced new idea of brane world, which will be explained later. Very exotic inflationary model with variable space dimension was also shown by Nasseri[3]. Moreschi[4] claimed that there is still some fundamental problem in thermodynamical treatment of radiation, which changes the equation of state and may resolve the horizon problem.

Some authors have discussed non-potential type inflationary models based on a superstring theory such as k-inflation. Among them, one of the most important cosmological models would be a pre-big bang scenario. This scenario was discussed in the last meetings, but not in this meeting. However, recent cosmological scenarios based on the brane world such as the ekpyrotic scenario, which could be an alternative to the inflationary paradigm, is closely related to this model. Therefore those studies including density perturbations will become very important again.

Another important consequences of the early universe are the problems of a cosmological constant and an initial singularity, which may not be resolved by any inflationary scenario. We may invoke quantum cosmology, or quantum gravity, but we might also have a chance to find a nonsingular evolution of the universe in some unified theory such as a string theory. In fact, Fabris[5] discussed such a possibility based on the effective 4-dimensional action derived from a string theory, while Moniz[6] presented cosmological solutions of the bosonic NS-NS sector of the low-energy M-theory effective action. As for a cosmological "constant", from the latest observational results of the angular power spectrum of the cosmic microwave background (CMB) and the measurements of high-redshift supernovae, we are forced to recognize the existence of the cosmological constant, or some kinds of the energy with negative pressure (dark energy), with almost the same order of magnitude as the present mass density of the universe. Since there is no plausible theoretical background to support the constant dark energy, it is often treated as a dynamical, slowly evolving component, (mainly a scalar field ϕ rolling down its potential $V(\phi)$) so-called quintessence. Huey[7] showed robustness of the quintessence tracking scenario in generic cosmological models, while Maeda[8] pointed out some advantage of the brane scenario also in a quintessence scenario.

We also have a great progress in particle physics, in particular in a string theory. A string theory predicts a new type of nonlinear structure, which is called a *brane*, a nomenclature created artificially from "membrane". It is a boundary layer on which edges of open strings stand. This also suggests a new perspective in cosmology; we are living in a brane world, which is a three-dimensional hypersurface in a higher-dimensional spacetime. In contrast to the original higher-dimensional world picture based on the Kaluza-Klein theory, our world view appears to be changed completely. Since 4-dimensional brane world in higher-dimensional spacetime is highly inhomogeneous because of localization of the brane in extra dimensions, we need advanced technic developed in general relativity and geometry. In the brane world scenario, we then find a close connection between general relativity and cosmology. In this workshop, we found that more than half number of talks are related to the brane world (see the list below). This means that general relativity has became one of the most important research fields in cosmology. Hence, in the following, we are going to briefly review it with the report.

2 Invitation to Brane World

According to the recent progress in superstring theory, different string theories are connected with each other via dualities, making them to be unified to the *M theory* in 11 dimensions. Among string theories, the 10-dimensional $E_8 \times E_8$ heterotic string theory is a strong candidate for our real world because the theory may contain the standard model. Hořava and Witten showed that this heterotic string model is equivalent to an 11-dimensional realization of M theory compactified on the orbifold $R^{10} \times S^1/Z_2$. Each gauge field on E_8 is confined to the 10-dimensional boundary brane of S^1/Z_2. The 10-dimensional spacetime is compactified to $M^4 \times CY^6$, where M^4 and CY^6 are 4-dimensional Minkowski spacetime and Calabi-Yau space, respectively. Particles in the standard model are expected to be confined to this 4-dimensional Minkowski spacetime, whereas the gravitons propagate in the entire bulk spacetime.

Based on this brane world picture, a new type of Kaluza-Klein cosmology was proposed by Arkani-Hamed *et al.* Ordinary matter fields are confined on the brane which is infinitesimally "thin", though it may be thick physically, probably of the order of the Planck length. Compared with this thickness, on the other hand, the extra dimensions could be larger. How large it is can be known only by gravitational experiments. Since any experiments to confirm Newtonian gravity have been done so far above the 1 mm scale, the laws of gravity might be different below this scale.

Suppose that the fundamental theory of gravity is given by the Einstein-Hilbert action in D $(= 4 + n)$ dimensions. The 4-dimensional Planck mass is related to the D-dimensional one as

$$(M_P)^2 = \left({}^{(D)}M_P\right)^{(n+2)} \times V_n. \tag{1}$$

where V_n is the volume of extra dimensions. In the old Kaluza-Klein idea, the size of extra dimensions should be much smaller than 10^{-17} cm ($\sim$TeV^{-1}) because matter fields can also propagate into extra dimensions. If we believe, however, that ordinary matter fields are confined on a brane world, the extra dimensions are not necessarily required to be so small. To remove the so-called "hierarchy problem", we may set the mass scale ${}^{(D)}M_P$ in D-dimensional spacetime at the fundamental level as low as $\sim$ TeV, nearly the same as the electroweak mass scale. Then a typical size r_0 of the extra dimensions is estimated as:

$$r_0 = (V_n)^{1/n} \sim 10^{(30/n)-17}\text{cm}. \tag{2}$$

For $n = 1$, we expect $r_0 \sim 10^{13}$ cm ($\sim$ 1 astronomical unit), which is to be excluded. If n =2, however, we find $r_0 \sim$ 1 mm, precisely the shortest distance of the present experiments. Hence it could be tested in our near-future experiments. Fundamental problem about radiation reaction in various dimensions was also discussed by Gal'tsov[9].

Another alternative scenario based on the brane world has been proposed by Randall and Sundrum. The Hořava-Witten model can be simplified to a 5-dimensional effective theory in which matter fields are confined to 4-dimensional spacetime while gravity acts in 5 dimensions, because the 6-dimensional Calabi-Yau space is at least by an order of magnitude smaller than the remaining 5-dimensional spacetime. Then Randall and Sundrum simplified the model further by assuming that our world is described by 4-dimensional Minkowski spacetime in 5-dimensional anti-de Sitter (AdS) space with a negative cosmological constant ${}^{(5)}\Lambda(= -6/\ell^2)$. The 5-dimensional spacetime is described by the metric, which is not factorizable:

$$ds^2 = e^{-2|y|/\ell}\eta_{\mu\nu}dx^\mu dx^\nu + dy^2. \tag{3}$$

To find the Minkowski space on the brane, a tension of the brane ($\sigma = \pm 6 \left({}^{(5)}M_P\right)^3/\ell$) is required. The "warp" factor $e^{-2|y|/\ell}$ which is rapidly changing in the extra dimension plays a very important role.

They discussed two models (RS I & RS II). In their first model (RS I), they proposed a mechanism to solve the hierarchy problem by a small extra dimension bounded by two boundary branes, with positive and negative tension, located at $y = 0$ and $y = d$, respectively. By assuming that we are living in a negative tension brane ($y = d$), we find a natural resolution of the hierarchy problem, if $d \sim 35\ell$. Because the effective mass scale (or the vacuum expectation value of a Higgs field) is reduced exponentially on the brane as $m_{\text{eff}} = e^{-d/\ell}m_0$, where m_0 is a bare mass ($\sim {}^{(5)}M_P$).

In their second model (RS II), on the other hand, we live in a single positive-tension brane surrounded by AdS. Although the hierarchy problem is still left unsolved, four-dimensional Newtonian gravity is recovered at low energy scale even if the extra dimension is not compact. This is easily shown by a perturbation equation around the above solution (3) with a positive-tension Minkowski brane at $y = 0$. Consider perturbation to the 4-dimensional components; ${}^{(5)}g_{\mu\nu} = e^{-2|y|/\ell}\eta_{\mu\nu} + h_{\mu\nu}$. By setting $h(x, y) = \hat{\psi}(z)e^{-|y|/2\ell}e^{ipx}$ with $z = \ell(e^{|y|/\ell} - 1)$ and $p^2 = -m^2$, we find the perturbation equations of graviton as

$$\left[-\frac{1}{2}\partial_z^2 + U(z)\right]\hat{\psi} = m^2\hat{\psi}, \tag{4}$$

where

$$U(z) = \frac{15}{8\ell^2(|z|/\ell + 1)} - \frac{3}{2\ell}\delta(z) \tag{5}$$

is a volcano-shaped potential. The volcano-shaped potential confines a massless mode on the brane. As a result, even if the 5-th dimension is not compact, Newtonian gravitational potential is recovered in low energy limit as

$$V(r) \sim -G\frac{m_1 m_2}{r}\left(1 + \frac{\ell^2}{r^2}\right), \tag{6}$$

where m_1 and m_2 are masses of two particles on the brane.

This "compactification" is completely different from the Kaluza-Klein type compactification. We recover 4-dimensional gravity in non-compact 5-dimensional spacetime rather than 5-dimensional gravity. Therefore, in strong gravity region, we expect some deviation from the Einstein's general relativity. To see such difference, we first discuss the RS II model (one-brane scenario). Since the massless graviton is confined on the brane $\mathcal{B}$, gravity in our world can be described by the intrinsic metric $g_{AB} = {}^{(5)}g_{AB} - n_A n_B$ with the vector n_A orthogonal $\mathcal{B}$. To find the basic equation for this intrinsic metric, we use the Gauss equation,

$$R^{\mu}{}_{\nu\rho\sigma} = {}^{(5)}R^{A}{}_{BCD}\; g^{\mu}_{A} g_{\nu}{}^{B} g_{\rho}{}^{C} g_{\sigma}{}^{D} + K^{\mu}{}_{\rho} K_{\nu\sigma} - K^{\mu}{}_{\sigma} K_{\nu\rho}\,, \tag{7}$$

and the Codacci equation,

$$D_\nu K_{\mu}{}^{\nu} - D_\mu K = {}^{(5)}R_{AB} n^A g_{\mu}{}^{B}\,, \tag{8}$$

where the extrinsic curvature on $\mathcal{B}$ is defined by $K_{\mu\nu} = g_{\mu}{}^{A} g_{\nu}{}^{B} \nabla_A n_B$, $K = K_{\mu}{}^{\mu}$ is its trace, and D_μ is the covariant derivative with respect to $g_{\mu\nu}$.

We assume the 5-dimensional vacuum Einstein's equations with a negative cosmological constant ${}^{(5)}\Lambda(= -6/\ell^2)$. To treat a singular hypersurface, we adopt the Israel's junction condition,

$$[g_{\mu\nu}] = 0\,, \qquad [K_{\mu\nu}] = -\left({}^{(5)}M_P\right)^{-3}\left(S_{\mu\nu} - \frac{1}{3} g_{\mu\nu} S\right), \tag{9}$$

where the discontinuity of the extrinsic curvature $S_{\mu\nu}$ is related to the energy and momentum densities $T_{\mu\nu}$ of matter fields in the brane as $S_{\mu\nu} = -\sigma g_{\mu\nu} + T_{\mu\nu}$. Assuming also the Z_2-symmetry of the brane, $K^{+}_{\mu\nu} = -K^{-}_{\mu\nu}$, from the Gauss equation (7) with the above ansatz, we obtain the gravitational equations on the 3-brane:

$$G_{\mu\nu} = -\Lambda g_{\mu\nu} + M_P^{-2} T_{\mu\nu} + \left({}^{(5)}M_P\right)^{-6} \Pi_{\mu\nu} - E_{\mu\nu}\,, \tag{10}$$

where

$$\Lambda = \frac{1}{2}\left[-\frac{6}{\ell^2} + \frac{\sigma^2}{6}\left({}^{(5)}M_P\right)^{-6}\right], \tag{11}$$

$$8\pi G = M_P^{-2} = \frac{\sigma}{6}\left({}^{(5)}M_P\right)^{-6}, \tag{12}$$

$$\Pi_{\mu\nu} = -\frac{1}{4}T_{\mu\alpha}T_{\nu}{}^{\alpha} + \frac{1}{12}TT_{\mu\nu} + \frac{1}{8}g_{\mu\nu}T_{\alpha\beta}T^{\alpha\beta} - \frac{1}{24}g_{\mu\nu}T^2, \tag{13}$$

and

$$E_{\mu\nu} \equiv {}^{(5)}C^{A}{}_{BCD}\, n_A\, n^C g_{\mu}{}^{B} g_{\nu}{}^{D}. \tag{14}$$

is some components of the 5-dimensional Weyl tensor ${}^{(5)}C^{A}{}_{BCD}$. The conservation law for the matter ($D_\nu T_\mu{}^\nu = 0$) is derived from Eq. (8).

The above equation (10) looks quite similar to the Einstein equations. In fact, in low energy limit, we recover the conventional Einstein equations. The 4-dimensional cosmological constant Λ must vanish to find the Minkowski brane, which fixes the strength of the tension σ. There are two new terms ($\Pi_{\mu\nu}$ and $E_{\mu\nu}$) which become important in strong gravity region. The quadratic term $\Pi_{\mu\nu}$ comes from a correction due to a localization of the brane. This term is local on the brane and is easily included in the analysis of the brane world. On the other hand, $E_{\mu\nu}$ is the 5-dimensional object and then it provides non-local contribution on the brane. This makes the problem difficult to solve except for some special cases such as an isotropic and homogeneous brane. Hence we have to solve the inhomogeneous 5-dimensional Einstein equations because of existence of the brane. We need advanced technics developed in general relativity.

If we assume, however, the Friedmann-Robertson-Walker (FRW) spacetime in our brane world, we find the Friedmann equations from (10):

$$H^2 + \frac{k}{a^2} = \frac{\rho}{3M_P^2} + \frac{\rho^2}{36({}^{(5)}M_P)^6} + \frac{\mathcal{C}}{a^4} \tag{15}$$

where ρ is the total energy density of matter fields, while $\mathcal{C}$ is a constant, which describes "dark" radiation coming from $E_{\mu\nu}$. The quadratic term of energy density and "dark" radiation appear. Those effects on the early history of the universe have been discussed widely. The quadratic term changes expansion rate of the universe in the early stage, which will affect some scenario of the early universe. A natural quintessence model is discussed including this term by Maeda[8]. Wands[2] discussed density perturbations in this brane model, and showed some difference from the conventional one. $E_{\mu\nu}$ term plays an important role in some situation. FRW spacetime on the brane is extended to Bianchi type I model by Toporensky[10]. Dadhich[11] showed that one-brane model with non-conformally flat bulk is unstable.

For the case of inhomogeneous brane, although a subject becomes very difficult to solve because it is at least two-dimensional problem, it is very important to discuss it. This is because it includes the topics of gravitational collapse and formation of a black hole or a relativistic star, which is another strong gravity region. Such topics were discussed by Ghosh[12] and Germani[13]. Studying the Vaidya type solution on the brane, Ghosh showed that the back-reaction of the bulk is supposed to strengthen gravity on the brane, and then it would favour formation of a black hole rather than a naked singularity. Germani proved that the outside of a collapsing sphere is always time-dependent when we consider Oppenheimer-Snyder-like dust collapse. This is due to the non-local effect $E_{\mu\nu}$. Although no explicit solution has been found, a black hole is different from Schwarzschild solution also because of existence of $E_{\mu\nu}$. The study of such compact objects may reveal an essential role of the brane.

For RSI model (two-brane scenario), since gravity is not confined in the visible brane, non-local effect from bulk ($E_{\mu\nu}$) becomes essential. The property of gravity changes completely. Garriga and Tanaka showed that the effective gravity theory in 4 dimensions is in fact the prototype Brans-Dicke model with the parameter $\omega = \frac{3}{2}\left(e^{\pm d/\ell} - 1\right)$,where $\pm$ corresponds to the sign of the tension of branes. This is because there appears new type of scalar field which corresponds to a "distance" between two branes, often called *radion*.

In the context of string theory or M-theory, one would also expect additional fields in the bulk as well as gravity e.g. moduli fields in the gravitational sector. Lukas, Ovrut and Waldram in fact derived an effective five-dimensional action by dimensional reduction from 11-dimensional M-theory. Some bosonic part of the effective action is given by

$$S = \int d^5x\sqrt{-{}^{(5)}g}\left[\frac{1}{2}{}^{(5)}R - \frac{1}{2}(\nabla\phi)^2 - \frac{\alpha^2}{6}\,e^{-2\sqrt{2}\phi}\right] + \sqrt{2}\alpha\int_{\mathcal{B}^{(1)}} d^4x\sqrt{-g}\,e^{-\sqrt{2}\phi} - \sqrt{2}\alpha\int_{\mathcal{B}^{(2)}} d^4x\sqrt{-g}\,e^{-\sqrt{2}\phi}, \tag{16}$$

where ϕ is a dilaton corresponding to the volume of the Calabi-Yau space [$V_6 = \exp(-\sqrt{2}\phi)$], α is a constant, and $\mathcal{B}^{(1)}$ and $\mathcal{B}^{(2)}$ represent two 4-dimensional boundary branes with negative and positive tensions, respectively. They found two cosmological solutions. Kodama[14] proved, however, that if we take into account the back reaction of perturbations, one of them is unstable.

The study of the brane cosmology has just started. It may provide us a new paradigm for the early universe with help of advanced research of general relativity and gravitation.

3 List of oral presentation

speaker	*title of talk*
O. Moreschi[4]	Change of the photon equations of state at the early Universe
D.V. Gal'tsov[9]	Gravitational Radiation in the Brane Universe
N. Dadhich[11]	Non-conformally flat bulk spacetime and the 3-brane world
A.V.Toporensky[10]	The shear dynamics in Bianchi I cosmological model on the brane
K. Maeda[8]	A scalar field on the brane
S.G. Ghosh[12]	Gravitational collapse of null fluid on the brane
C. Germani[13]	Gravitational collapse in the brane
D. Wands[2]	Cosmological perturbations in a brane-world
H. Kodama[14]	Backreaction problem in braneworld models
G. Huey[7]	Robustness of the quintessence tracking scenario in cosmology
P. V. Moniz[6]	FRW Cosmological Solutions in M-theory
J.C. Fabris[5]	Has there been a quantum universe ?
S. Tsujikawa[1]	Cosmological perturbations from multi-field inflation in generalized Einstein theories
F. Nasseri[3]	An upper limit for the space dimension at the Planck epoch: based on a model Universe with variable space dimensions

References

1. A.A. Starobinsky, S. Tsujikawa, J.Yokoyama, Nucl. Phys. **B610** (2001) 383.
2. H.A. Bridgman, K.A. Malik, D. Wands, Phys. Rev. **D65** (2002) 043502.
3. F. Nasseri, S. Rahvar, Int. J. Mod. Phys. D. [gr-qc/0008044].
4. O. Moreschi, preprint (1999) [gr-qc/9911105].
5. J.C. Fabris, in the "Mario Novello Festschrift" [gr-qc/0112050].
6. M. Cavaglia, P.V. Moniz, the proceedings of the Ninth Marcel Grossmann Meeting (World Scientific, 2002) [gr-qc/0011098].
7. G. Huey, R. Tavakol, Phys. Rev. **D65** (2002) 043504.
8. K. Maeda, Phys. Rev. **D64**(2001) 123525.
9. D.V. Gal'tsov, preprint (2001) [hep-th/0112110].
10. A.V.Toporensky, Class. Quantum Grav. **18** (2001) 2311.
11. P. Singh, N. Dadhich, Phys. Lett. **B511** (2001) 291.
12. N. Dadhich , S. G. Ghosh, Phys. Lett. **B518** (2001) 1.
13. M. Bruni, C. Germani, R. Maartens, Phys. Rev. Lett. **87** (2001) 231302.
14. O. Seto, H. Kodama, Phys. Rev. **D63** (2001) 123506.

B4: GRAVITATIONAL WAVE SOURCES

MANOJ S MAHARAJ
University of Durban-Westville, Durban, South Africa
E-mail: mmaharaj@pixie.udw.ac.za

A very brief overview of gravitational wave sources[4] is covered and then highlights of the main ideas contained in the talks in this session are presented.

1 Introduction

Within General Relativity the theoretical framework underlying gravitational waves is well understood and very good indirect verification thereof has been obtained through observations (e.g. the Hulse-Taylor binary pulsar system [1,2]) With the advent of sophisticated detectors it may be now possible to obtain direct evidence of gravitational radiation.

The very nature of gravitational wave sources makes it very difficult to obtain independent confirmation of its characteristics. This is because gravitational waves are emitted by sources that either do not emit electromagnetic radiation (black holes), are obscured from view (gravitational collapse of a stellar core) or are so distant that the photons reaching us cannot be distinguished from other photons.

Gravitational wave sources may be classified into three categories:

1. burst sources

   - Compact Binary inspiral: This system consists of two compact stellar mass objects, either neutron stars or black holes or black hole - neutron star systems. These systems are strong gravitational radiation sources. The radiation carries away orbital energy and angular momentum which in turn leads to faster and more compact orbits until coelescence. The radiation from this inspiral is an important source of gravitational radiation for the detectors under construction.
   - Compact Binary Coalescence: Another significant source of gravitational radiation would be that arising from the final stages of inspiral to coalescence. Within this realm it is necessary to adopt numerical techniques to solve the full Einstein Field equations.
   - Black Hole Formation: Before a black hole settles down into its final axisymmetric state it must radiate away its deformations. This

radiation, in the form of gravitational waves, could be observed by suitable detectors.

- Stellar Core Collapse: Gravitational Radiation from supernova events within our own galactic neighborhood may yield gravitational radiation of sufficient intensity to be observed by present detectors.

2. Periodic Signals. Rapidly rotating neutron stars are a source of periodic gravitational waves. The strength of the radiation depends upon the degree of asymmetry.

   - Non-axisymmetric rotators: Neutron stars have a solid, crystalline crust, covered by a fluid surface and a fluid interior. As the star cools or spins down the fluid surface and the stars interior remain axisymmetric and do not contribute to any gravitational radiation. However, the solid crust cannot deform continuously and the stresses that are created result in fractures. The final crust shape is likely to be non-axisymmetric and will be responsible for gravitational radiation as the star rotates.
   - Orbital Precession: If the angular momentum of a neutron star is not aligned with the symmetry axis the star will precess. This precession will give rise to gravitational radiation.
   - R-mode instability: Axial fluid modes in rotating stars (r-modes) are generically unstable to the emission of gravitational waves. The r-mode instability can slow down a rapidly rotating neutron star to Crab-like rotation rates[3]. Present understanding of neutron stars suggest that for about one year after its birth the r-mode may be unstable and the star may be a strong nearly periodic source of gravitational radiation. The modes of interest for rotating stars are[3]:
     - Polar fluid modes, these are slowly damped modes analogous to the Newtonian fluid pulsations
       * f(undamental)-mode: surface mode due to the interface between the star and its surroundings
       * p(ressure)-modes: nearly radial
       * g(ravity modes): nearly tangential, only exist for finite temperature stars
     - Axial fluid modes
       * r(otation) modes: degenerate at zero-frequency for nonrotating stars. In a rotating star, generically unstable. Fre-

quencies from zero to kHz, growth times inversely proportional to a high power of the star's angular velocity.

  - Polar and axial spacetime modes
    * w(ave)-modes: Analogous to the quasi-normal modes of a black hole. High frequency, strongly damped modes

- Stochastic Signals. By their nature, these signals are not generated by isolated sources, are not incident upon the detector from a single direction and have no characteristic waveform. Such a signal may be treated as just another source of detector noise. The only way to detect stochastic signals is to look for correlations between two or more detectors that are dependent on their relative separations, relative orientations and the stochastic signals spectrum. Such signals are not purely primordial in origin, that is radiation arising during the inflationary era, from the decay of cosmic strings, or from the phase transitions in the early universe. Stochastic signals are also generated by discreet but unresolved sources. An interesting theorem by Phinney[5], shows that there is a simple relationship between the spectrum of the gravitational wave background produced by the cosmological distribution of such sources and the present-day comoving number density of remnants. This discussion suggests that there is a stochastic Cosmic Gravitational-wave Background (CGB) produced during the formation of the Universe, not unlike the CMBR. Cornish and Larson[6] have suggested a GABI (GrAvitational Background Interferometer) mission to detect the CGB would be successful.

2 The Workshop B4

In this section we provide a very brief overview of each of the talks presented at this workshop. Note: the speaker is listed in italics when there are multiple contributors.

The r-mode instability in relativistic stars

Contributors: *Keith Lockitch*, Nils Andersson, John Friedman
Here the r-modes and rotational "hybrid" modes of slowly rotating, fully relativistic stars were examined. BY assuming that the star and its perturbations obey the same one-parameter equation of state as barotropic stars, it was shown that there exist no pure r-modes at all - no modes whose limit, for a star with zero angular velocity, is an axial-parity perturbation. Instead, it

was shown that the generic rotationally restored mode is a hybrid mixture of r- and g-modes, i.e., of axial and polar parity terms.

The rotational mode frequencies and eigenfunctions for polytropes and uniform density stars in full general relativity were obtained numerically and timescale for the growth rate of the gravitational-wave driven instability of these modes were computed.

Detecting a relic background of scalar waves with LIGO

Contributors: Carlo Ungarelli

The possible detection of a stochastic background of massive, non-relativistic scalar particles, through the cross correlation of the two LIGO interferometers in the initial, enhanced and advanced configurations was discussed. It was concluded that if the frequency corresponding to the mass of the scalar field lies in the detector sensitivity band, and the non-relativistic branch of the spectrum contributes significantly to the energy density required to close the Universe, then the scalar background can induce a non-negligible signal, in competition with a possible signal produced by a stochastic background of gravitational radiation.

Dynamical Instability in Rotating Supermassive Stars to the Onset of Collapse

Contributors: *Motoyuki Saijo*, Masaru Shibata, Thomas W. Baumgarte, Stuart L. Shapiro

The gravitational collapse of rotating supermassive stars to the onset of collapse in the first post-Newtonian approximation of general relativity was discussed. It was shown that:

- Most components of the star follow the collapse of the central core, i.e. total mass was still well-conserved even when the outside components of the "comoving" grid were excised.
- Although T/W exceeds the critical value ($T/W = 0.27$ for uniformly rotating incompressible Newtonian stars) for the onset of dynamical instability there was no bar formation prior to the formation of a black hole.

Hence it was concluded that supermassive collapse was not an exciting candidate for gravitational wave sources.

How a dense neutron star pretends to be a black hole

Contributors: Hans-Peter Nollert

When the w-modes of neutron stars were discovered, the question arose as to whether their frequencies will approach the quasinormal frequencies of a black hole if the star is made extremely compact. It turned out that this is not the case. Intuitively, however, one expects that a non-rotating neutron star would become very similar to a Schwarzschild black hole as far as its reaction to gravitational radiation is concerned, especially since the metric outside the star is identical to the Schwarzschild metric. It was shown that this was the case, although not on the level of its characteristic modes. It was shown that if a gravitational wave pulse was to hit a very dense neutron star, the response will show two bursts of damped sinusoidal radiation. The first burst has a characteristic frequency corresponding to the fundamental quasinormal frequency of a black hole of the same mass; this frequency is not part of the spectrum of the star itself. Only the second burst exhibited a frequency which appears in the characteristic mode spectrum of the star. In this sense, the star pretends to be a black hole for a while after having been hit by a pulse of gravitational radiation.

Are rotating strange stars good sources of gravitational waves?

Contributors: Dorota Gondek-Rosinska

By calculating the ratio of the rotational kinetic energy to the absolute value of the gravitational potential energy T/W for rapidly rotating compact stars it was shown that the T/W for strange stars is significantly higher than for an ordinary neutron stars. The effects of general relativity upon the non-axisymmetric "bar" mode secular instability of rapidly rotating strange stars was calculated and it was shown that the viscosity driven instability could be a promising source of gravitational radiation in the case of rapidly rotating strange stars.

Quasiequilibrium sequences of synchronized and irrotational binary neutron stars in general relativity

Contributors: *Keisuke Taniguchi*, Eric Gourgoulhon, and Silvano Bonazzola

The quasiequilibrium sequences of close binary neutron stars with a polytropic equation of state in general relativity was discussed and calculated using a multi-domain spectral method. The computations were performed for both cases of synchronized and irrotational binary systems with adiabatic indices $\gamma = 3,\ 2.5,\ 2.25,\ 2$ and 1.8. The difference of last stable orbit between the synchronized case and the irrotational case was discussed.

Gravitational waves from wobbling neutron stars

Contributors: Nils Andersson

In this talk Nils Anderson summarized recent work on gravitational waves from freely precessing neutron stars. The relevant model, suitably parameterized, and the problem of calculating the gravitational radiation reaction on a star modelled as an elastic shell containing a fluid core was discussed. The likely gravitational wave amplitudes of precessing neutron stars in our Galaxy was assessed.

First law of thermodynamics for binary neutron star and determination of the ISCO

Contributors: *Koji Uryu*, John Friedman and Masaru Shibata

The quasiequilibrium sequence of a binary neutron star was constructed numerically and the turning point method was applied to locate the ISCO (Innermost Stable Circular Orbit). It was shown that the ISCO was dependent upon the compactness of neutron star. This further suggested that analytical methods using point particles in calculating the ISCO could not be applied to a binary neutron star system. Other properties of binary neutron near the ISCO were also discussed.

Physics and Astrophysics from Distorted Pulsars

Contributors: *Stephen Fairhurst*, Lee Samuel Finn, Joshua L. Willis and Jacek Wiśniewski

It is known that distorted neutron stars are good candidate sources for gravitational waves which may be detectable in the near future. Working within the framework of the Brans-Dicke generalization of GR, the form of such radiation and the response of an interferometric detector was calculated. Also, by searching for radiation from ***known*** pulsars, it was demonstrated how to obtain information about the distortion of the star and the orientation of its axis of rotation. Finally, numerical simulations were performed in order to model possible results from LIGO II.

Limits On The Time Variations Of The Intensity Of The Primordial Gravitational Waves

Contributors: Melek M.

Limits on the cosmic time variations of the intensity of the Primordial Gravitational Waves (**PGW**), were calculated in the spatially perturbed Friedman- Robertson- Walker (**FRW**) expanding universe

Low frequency gravitational waves from the Galactic halo

Contributors: William A. Hiscock

The MACHO project has detected 13-17 gravitational microlensing events in the direction of the Large Magellanic Cloud, while the EROS collaboration has detected two. One result of the analysis of the observations is that, independent of the halo model assumed, there are of order 2×10^{11} MACHOs of mean mass $0.5 M_{\odot}$ in the Galactic halo. It is noteworthy that this number is similar in magnitude to the number of stars in the Galaxy. The possibility that Galactic halo MACHOs are white dwarfs has recently attracted much attention. Using the known properties of white dwarf binaries in the Galactic disk as a model, the possible contribution of halo white dwarf binaries to the low-frequency (10^{-5} Hz $< f <$ 10^{-1} Hz) gravitational wave background was estimated. Assuming the fraction of white dwarfs in binaries is the same in the halo as in the disk, it was shown that the confusion background from halo white dwarf binaries could be five times stronger than the expected contribution from Galactic disk binaries, dominating the response of space based detectors such as LISA. it was concluded that low-frequency gravitational wave observations will be the key to discovering the nature of the dark MACHO binary population.

Numerical mode analysis of differentially and rapidly rotating stars in general relativity

Contributors: Shin'ichirou Yoshida

Oscillation modes of differentially and rapidly rotating stars in general relativity were numerically investigated using a method based upon the Cowling approximation. Sequences of several oscillation modes (f-, p- and quasi-radial modes), for stellar models with different degrees of differential rotation were obtained.

For the counter-rotating f-mode whose phase velocity propagates in the retrograde direction with respect to the stellar rotation, it was found that the neutral stability points of the secular instability driven by gravitational radiation (Chandrasekhar-Friedman-Schutz instability), appear at faster rotation rates than those of uniformly rotating stars. It was concluded that the differential rotation makes this instability less important in the context of stability of newly-born neutron stars.

Gravitational Wave Damping of Neutron Star Wobble

Contributors: *C. Cutler* and I. Jones

The effect of gravitational wave (gw) back-reaction on realistic neutron stars (NS's) undergoing torque-free precession was calculated. It was shown that gw's damp NS wobble on a timescale $\tau_\theta \sim 2 \times 10^6$ yr $[10^{-7}/(\Delta I_d/I_0)]^2(\text{kHz}/\nu_s)^4$, where ν_s is the spin frequency and $\Delta I_d/I_0$ is the fractional distortion in the NS's moment of inertia (compared to a fluid body) due to shear stresses in the crust.

Deformations of Accreting Neutron Star Crusts and Gravitational Wave Emission

Contributors: G. Ushomirsky, *C. Cutler*, and L. Bildsten

The extent to which the composition and temperature variations in the crust of an accreting NS can generate "mountains" in the NS crust was investigated. It was shown that modest lateral variations in T and μ_e of order 1% were sufficient to halt the spin-up of accreting NS's at 300 Hz, due to gravitational wave backreaction. A formula for the maximum quadrupole moment sustainable by a NS crust, as limited by its breaking strain was also presented.

Recent Progress in 3D Numerical Simulations of Binary BH Merger

Contributors: Ed Seidel

Recent developments in new formulations of Einstein's equations, excision techniques, and gauge conditions for 3D numerical relativity have enabled dramatic improvements in our ability to carry out full scale, nonlinear evolutions of binary black hole mergers. These techniques were briefly described and recent results on binary mergers of black holes, showing waveforms, convergence tests, as well as current limitations and outlook for the future were discussed.

The Lazarus Project

Contributors: Manuela Campanelli

New developments in numerical relativity that bring the field for the first time close to providing useful astrophysical information about the orbital merger of binary black holes relevant for interferometric gravitational wave detectors were presented.

Gravitational radiation from the coalescence of an equal-mass binary system including complete waveforms, gravitational energy and angular momentum radiated for the dynamics beginning from the innermost stable circular orbit (ISCO) were calculated.

New constraints on the scalar coupling of gravitation through the analysis of gravity wave emission in a recently discovered neutron star/white dwarf binary pulsar

Contributors: Yves Wiaux

Eventhough scalar-tensor theories are the simplest generalizations of Einstein's theory, the presence of a scalar component of gravitation implies major conceptual differences with general relativity. Probably the most fundamental of these is the violation of the strong equivalence principle. At the classical level, the Nordtvedt effect is probably the best known among the phenomenological consequences of this violation. However, to post-Keplerian order, this also affects the structure of gravitational waves. In particular, the energy lost in compact binary stars by gravity wave emission may be considerably enhanced by the presence of a strong dipolar scalar component. At least, this will be the case if the stars have different enough internal structures. The reason for a revived interest for this problem is the recent discovery (May 2000) of a neutron star/white dwarf binary pulsar, for which the prospects for the relativistic orbital parameters determination are excellent. The measurement of the orbital period damping rate would in any case be conclusive. It will either constitute an evidence for a scalar gravitational interaction (if the observed value deviates from the predictions of general relativity) or put further constraints on the scalar coupling to gravitation (in case the observed value is in agreement with the predictions of general relativity).

Acknowledgments

Workshop B4 was organized and chaired by Patrick Brady.

References

1. J H Taylor and J M Weisberg. Astrophys. J. **253**, 908 (1982)
2. J H Taylor and J M Weisberg. Astrophys. J. **345**, 434 (1989)
3. Nikolaos Stergioulas,Living Reviews in Relativity , (2001)
4. Lee Samuel Finn, Gravitational Radiation Sources and Signatures , (http://gravity.phys.psu.edu/ lsf/Publications/finn99d.pdf)
5. E.S.Phinney,Mon. Not. R. Astron. Soc. **000**, 1 (2001)
6. Neil J. Cornish and Shane L. Larson, arXiv:gr-qc/0103075v2 , (2001)

B5: MATHEMATICAL COSMOLOGY

PETER DUNSBY
Department of Mathematics and Applied Mathematics University of Cape Town, Rondebosch 7701, Cape Town, South Africa.

1 Introduction

In the past decade there have been substantial advances in theoretical cosmology coupled with a revolution in observational cosmology notably the measurement of the first and second peaks of the Cosmic Microwave Background spectrum, rapidly growing galaxy redshift surveys and observations of type 1A supernovae. Mathematical Cosmology has played an important role in these developments providing many of the theoretical tools needed for studying both the dynamics of the universe and obtaining clear predictions which can be tested by both current and future observations.

Over forty papers were submitted to this workshop, from which sixteen were selected for oral presentation. I will now briefly review these presentations, adding references to publications where available.

2 The Early Universe

Winfried Zimdahl: "Inflation in a self - interacting gas universe"

Zimdahl presented a general thermodynamic scheme for cosmic fluids with internal self - interactions and discussed equilibrium and non - equilibrium aspects of such systems in connection with (generalized) symmetry properties of the dynamics of such systems.

As an example he constructed an exactly solvable gas dynamical model of a "deflationary" transition from an initial de Sitter phase to a subsequent Friedmann - Lemaître - Robertson - Walker (FLRW) period. Finally he demonstrated that this dynamics represents a manifestation of a conformal symmetry of an "optical" metric, characterized by a specific effective refraction index of the cosmic medium. More details of this work are given in [1].

Roy Maartens: "Brane - World Cosmology"

Maartens discussed how recent developments in string theory have led to 5-dimensional warped spacetime models in which standard-model fields are confined to a 3-brane (representing the observed universe), while gravity can propagate in the fifth dimension. Gravity is localized near the brane at low energies, even if the extra dimension is non-compact. He demonstrated how the field equations on the brane modify the general relativity equations in two ways: local 5-D effects are imprinted on the brane as a result of its embedding, and are significant at high energies; non-local effects arise from the 5-D Weyl tensor. Local high-energy effects can modify the dynamics of inflation, and increase the amplitude of scalar and tensor perturbations generated by inflation. He showed how non-local effects introduce new features in cosmological perturbations theory: a non-adiabatic mode in scalar perturbations, and massive modes in vector and tensor perturbations. These may have important implications for testing brane models against CMB and galactic observations [2].

William Lesame: "Bulk null geodesics in a FLRW brane - world"

Lesame continued with is theme by considering null bulk geodesics that intersect a cosmological FLRW brane. He showed that it is possible for a null geodesic to leave the brane and then return to the brane at a later time, leading to causality implications. In the final part of his talk he showed that the turning point of a returning geodesic is related to the brane tension and to the amount of matter on the brane via the brane energy density.

3 Cosmological Perturbations and Structure Formation

Carlos Sopuerta: "Extending the scope of the Long Wavelength Approximation Scheme"

The Long Wavelength Approximation Scheme (LWAS) is an iterative method to solve Einstein equations which is based on the assumption that the scale of spatial variation is much smaller than the scale of time variation. Sopuerta demonstrated how one can use this scheme to find solutions in the form of a series in spatial gradients. He then showed that when the LWAS is applied in cosmology, the first-order solution corresponds to inhomogeneities larger than the Hubble scale, and higher-order terms then correspond to corrections which make the solution suitable for describing smaller scales.

A major part of the talk focused on a new formulation of the LWAS in terms of the covariant fluid approach and its application to inhomogeneous irrotational dust cosmological models. Using this formulation he analysed the behaviour of the kinematic and gravitational variables both near the initial singularity and for late times.

In particular, it was shown the magnetic part of the Weyl tensor plays a central role in the LWAS and can be used to assess the rank of validity of the approximation. On the other hand, using tetrad methods in combination with the covariant fluid approach it is possible to generalize previous work on the LWAS in cosmology by applying it to the connection variables instead of the metric components. This allows one to incorporate non-linear dynamical effects due to the spatial curvature which is well known to be very important in the evolution of homogeneous cosmological models.

David Matravers: "Stress Effects in Warm Dark Matter in Structure Formation"

Simulations using cold dark matter models show that they fail to match the observed core density profiles in galaxies and clusters. Also the simulations predict more small scale structure than is observed in our local group. However, there is recent evidence from simulations, that including slight velocity dispersion (warm dark matter) improves the fit between the models and observations. Matravers presented an analytic description of the implications of velocity dispersion in the cosmological matter distribution a relativistic framework in two ways. Firstly he considered a self-consistent treatment of the Einstein-Liouville problem using a truncation of the BBGKY hierarchy. Then he considered the problem using linear perturbation theory involving a first order matter pressure term and taking the discrete nature of cosmological matter seriously. These approaches lead to new perturbation modes which decay more slowly than in the standard gravitational instability picture. This work provides new insight into the effects of slight velocity dispersion and the Zeldovich approximation [3].

Filipe Mena: "Cosmic no-Hair: Second Order Perturbations of de Sitter Universe"

An analysis of the asymptotic behaviour of second order perturbations in a flat FLRW universe with dust plus a cosmological constant was presented by Mena [7].

He showed that as in the case of linear perturbations, the non-linear perturbations also tend to constants, asymptotically in time, showing that the earlier results concerning the asymptotic behaviour of linear perturbations is stable to non-linear (second order) perturbations. It was argued that this demonstrates the validity of cosmic no-hair conjecture in a non-linear inhomogeneous settings.

Mattias Marklund: "Generation of Electromagnetic Waves in the Early Universe"

Magnetic fields play an important role in our Universe, appearing on all scales from the solar system, through interstellar and extra-galactic scales, to intra-cluster scales of several Mpc. The effect magnetic fields have on density perturbations has been studied extensively by a number of authors, both in the context of Newtonian and Relativistic Cosmology, but as yet there have been no studies of their effect on gravitational wave perturbations. It is well known that electromagnetic waves can scatter off gravitational fields, and because gravitational and electromagnetic waves satisfy the same dispersion relation in vacuum, it is to be expected that the interaction between the two types of fields could be very efficient.

In his talk Marklund considered the dynamics of electromagnetic fields in an almost-FLRW universe using the covariant and gauge-invariant approach of Ellis and Bruni [5]. He focusing on the situation where deviations from the background model are generated by tensor perturbations only and demonstrate that the coupling between gravitational waves and a weak magnetic test field (which propagates on the background) can generate electromagnetic waves. He showed that this coupling leads to an initial pulse of electromagnetic waves whose width and amplitude is determined by the wavelengths of the magnetic field and gravitational waves. A number of implications for cosmology were discussed, in particular he presented an upper bound of the magnitude of this effect using limits on the quadrapole anisotropy of the Cosmic Microwave Background.

4 The Einstein-Straus Model

Marc Mars: "On the uniqueness of the Einstein-Straus model"

Mars showed that the Einstein-Straus model does not give a robust answer to the problem of the influence of the cosmic expansion on the local physics by finding the most general static region embeddable in a FLRW expanding cosmology and proving that the model must be almost spherically symmetric.

More precisely, he showed that the boundary of the static region must be a two-sphere at each instant of cosmic time. The motion of this two-sphere in spacetime is as follows: its "would-be" center (if there was no static region) moves along a path whose projection on any 3-space of constant cosmic time is a geodesic with respect to the induced 3-metric. The velocity of this geodesic is determined from the matching. In particular, this center must be at rest with respect to the cosmological flow (thus giving a spherically symmetric model) when any of the standard energy-momentum tensors inside the static region is imposed.

5 Exact Cosmological Solutions

Henk van Elst: "Scale - invariant dynamics for Abelian G_2 perfect fluid cosmologies"

In his talk van Elst first remarked how the use of scale - invariant variables has proved to be of great benefit in the analysis of spatially homogeneous perfect fluid cosmologies, both as regards qualitative analysis and numerical simulations.

He argued therefore that it is thus natural to use a similar approach to study spatially inhomogeneous cosmologies, in particular, the so - called Abelian G_2 perfect fluid cosmologies, which permit one spatial degree of freedom. It was pointed out that earlier investigations have shown that a direct generalisation of the procedure for spatially homogeneous cosmologies, i.e., scaling with the rate of expansion scalar, does not lead to an analytic system of evolution equations [6] and therefore proposed a new type of scaling which overcomes this difficulty, and at the same time leads to autonomous evolution equations which have first - order symmetric hyperbolic form.

This approach allows for a transparent isolation of (i) the physical degrees of freedom in both the gravitational and the matter source fields and (ii) the gauge degrees of freedom associated with the time slicing. In addition, the self - similar (asymptotic) states can be determined systematically. Known exact solutions fit well into this framework, which offers the possibility of viewing self - similar as well as other solutions in a broader perspective [7].

Jose Senovilla: "New Family of Inhomogeneous γ-law Cosmologies: Example of Gravitational Waves in a Homogeneous $p = \varrho/3$ Background"

Senovilla presented an explicit three - parameter class of $p = \gamma\varrho$, $(-1/3 \leq \gamma < 1)$, cosmological models admitting a two - dimensional group G_2 of isometries acting on spacelike surfaces. The family is self - similar in the sense that it has a further homothetic vector field and it contains sub-families of both (previously unknown) tilted and non - tilted Bianchi models with that equation of state. This is the first algebraically general class of solutions of this kind including dust inhomogeneous solutions. The whole class presents a universal spacelike big-bang singularity in the finite past. More interestingly, the case $p = \varrho/3$ constitutes a new two - parameter inhomogeneous sub-family which can be viewed as a Bianchi V background with a gravitational wave traveling orthogonal to the surfaces of transitivity of the G_2 group. This wave generates the *inhomogeneity* of the spacetime and is related to the sound waves *tilting* the perfect fluid. It seems to be the first explicit exact example of a gravitational wave traveling along a homogeneous background that has a realistic equation of state $p = \varrho/3$ [8].

Bernard Carr: "Spherically symmetric self-similar cosmological solutions and critical phenomena"

Carr presented a survey of spherically symmetric perfect-fluid solutions of Einstein's equations which have equation of state $p = \alpha\mu$ and are self-similar in the sense that all dimensionless variables depend only upon $z \equiv r/t$. He pointed out that the analysis is complete subject to the restrictions that α lies in the range 0 to 1 and that the solutions are everywhere physical and shock-free. The presentation combined the comoving approach of Carr and Coley [9] and the homothetic spate space approach of Goliath, Nilsson and Uggla [10] leading to a deeper physical understanding of these models. The comoving approach demonstrates that all self-similar models are asymptotic to solutions which depend on either powers of z at large and small $|z|$ or powers of $\ln|z|$ at finite z. There are only two self-similar solutions which have an *exact* power-law dependence on z: the flat Friedmann solution and a static solution. The asymptotically Friedmann solutions are inhomogeneous generalizations of the Friedmann model itself in that they are asymptotically Friedmann in their past but evolve fluctuations in the energy density at later times. The so-called asymptotically "quasi-static" solutions include models that undergo self-similar gravitational collapse and are important for studying the formation of naked singularities. For $\alpha > 1/5$, there are also two new families of solutions which are asymptotically Minkowski at large distances from the origin.

Finally it was shown that for different values of α these various solutions incorporate the self-similar solutions associated with the critical behaviour observed in gravitational collapse calculations.

6 Cosmic Topology and Observations

Hideo Kodama "Dynamical degrees of freedom in spatially compact Bianchi models"

Motivated by the recent increasing interests in cosmological models with compact space in the observational cosmology and in the quantum gravity, Kodama demonstrated how one calculates the dynamical degrees of freedom for spatially compact Bianchi models using the canonical formalism for such systems. The calculation was done for all possible space topologies for vacuum models and models with perfect fluid. He showed that compactification significantly alters the parameter count, in particular, that the type V and VIIh systems must be locally isotropic if their space is compact. This result is in a sharp contrast to the fact that the type VIIh system is the most generic system in the spatially open case. He also showed that the type I system is more generic than the type VII0 system for the vacuum case, while the latter is more generic than the former for the perfect fluid case and that the dynamical degrees of freedom increase without bound for the type VIII and type III systems as the space topology becomes more and more complicated due to the contribution of the moduli freedom [11,12].

Reza Tavakol: "Constraints on detectability of cosmic topology"

Recent constraints on the cosmological density parameters from large scale structure information, high redshift supernova observations, BOOMERANG and MAXIMA experiments indicate that we live in a almost-flat universe ($\Omega_0 \sim 1$). Tavakol discussed the question of whether it is possible to observationally detect non-trivial topologies of nearly flat FLRW universes with dust and cosmological constant. Using various topological indicators he showed that as $\Omega_0 \to 1$, increasing families of possible topologies become undetectable observationally. He also showed how, given a non-zero lower bound on $|\Omega_0 - 1|$, one can rule out families of topologies as possible candidates for the shape of the universe [13].

Volker Perlick: "Cassification of image distortions in terms of Petrov types"

An observer surrounded by sufficiently small spherical light sources at a fixed distance will see a pattern of elliptical images distributed over the sky, owing to the distortion effect (shearing effect) of the spacetime geometry upon light bundles. In a classical paper by Kristian and Sachs [14] it was shown that, in lowest non-trivial order with respect to the distance, this pattern is completely determined by the conformal curvature tensor (Weyl tensor) at the observation event. Perlick presented new formulae that allow one to calculate these distortion patterns in terms of the Newman-Penrose formalism. He then showed how to represent the distortion patterns graphically for all Petrov types, and discussed their dependence on the velocity of the observer [15].

7 Inhomogeneous Cosmologies

Roberto Sussman: "Mass and Volume Measurements in Inhomogeneous Specetimes Characterised by the Lemaître - Tolman - Bondi (LTB) Metric"

Sussman examined the issue of coordinate independent determination of the mass of a localized object with a spherical boundary. The relevant spacetimes are the LTB dust solutions and a generalization of these solutions whose source is an imperfect fluid that can be interpreted as a viscous fluid or as collisionless matter with an anisotropic distribution function. He compared different mass measurements from the Schwarzschild and Vaidya exteriors with their Newtonian estimates assuming reasonable initial density and curvature configurations in the interiors. For positive curvature interiors with a symmetry center, the obtained masses can be up to three times larger than that of a Newtonian estimate. Since both LTB dust solutions and their generalization admit regular configurations without symmetry centers, and these configurations can have a much larger proper volume than a Newtonian sphere, the mass measurements (for suitable initial conditions) can be as high as various hundreds of times larger than the Newtonian estimates. Since interior models without centers have particle horizons, number counts performed by the observers in the Schwarzschild or Vaidya "exteriors" would not be able to explain the excess of mass with respect to the Newtonian estimate. This case provides a simple toy model illustrating how the need for dark matter could be eliminated by considering simple curvature effects absent in Newtonian gravity.

References

1. W. Zimdahl and A. B. Balakin *Phys. Rev.* D **63** 023507 (2001).
2. R. Maartens, gr-qc/0101059, and refs. therein.
3. R. Maartens, J. Triginer and D. Matravers, *Phys. Rev.* D **60** 103503 (1999).
4. M. Bruni, F. C. Mena and R. Tavakol, gr-qc/0107069.
5. M. Marklund, P. K. S. Dunsby and G. Brodin, *Phys. Rev.* D **62**, 101501 (R) (2000).
6. C. G. Hewitt and J. Wainwright, Orthogonally transitive G_2 cosmologies *Class. Quantum Grav.* **7** 2295 (1990).
7. H. van Elst, C. Uggla and J. Wainwright, *Class. Quantum Grav.* **19** 51

(2002).
8. J. M. M. Senovilla and R. Vera, *Phys. Rev.* D **63** 084008 (2001).
9. B. J. Carr, A. A. Coley, *Class. Quantum Grav.* **16** R31-R71 (1999).
10. M. Goliath, U. S. Nilsson and C. Uggla, *Class. Quantum Grav.* **15** 167, (1998).
11. H. Kodama, *Prog. Theor. Phys.* **99**, 173 (1998).
12. J. D. Barrow and H. Kodama, gr-qc/0012075 (to be published in Class. Quantum Grav.).
13. G. I. Gomero, M. J. Reboucas and R. Tavakol, *Class. Quantum Grav.* **18** 4461 (2001).
14. J. Kristian and R. Sachs, *Ap. J.* **143** 379 (1966).
15. T. Chrobok and V. Perlick, *Class. Quantum Grav.* **18** 3059 (2001).

C1I+II: LAND BASED DETECTORS OF GRAVITATIONAL WAVES

WILLIAM O. HAMILTON

Department of Physics and Astronomy, Louisiana State University, Baton Rouge, LA 70803 USA

E-mail: hamilton@phys.lsu.edu

This session consisted of only two papers. Aguiar reported on the Brazilian effort to build a spherical detector. Bassan reported on a new transducer for the Nautilus and Auriga detectors. There was also a brief discussion by the author on the collaborative results of the IGEC. Most experimental papers may be found in the proceedings of the 4th Edoardo Amaldi Conference.

1 Introduction

The size of this section was an unfortunate consequence of scheduling. GR 16 began only two days after the close of the Fourth Edoardo Amaldi meeting in Perth, Western Australia. Since the Amaldi meeting is specifically devoted to experimental issues, many of the experimentalists were able to attend only the Australian meeting, leaving very few talks and papers from the active experimental groups at this conference. This is a very active time experimentally with 4 interferometers (TAMA, GEO 600, LIGO, and VIRGO) in the final stages of testing and with the resonant detectors increasing their activity. Thus, for a complete description of the experimental activity in the field of gravitational wave detection, the reader is directed to the conference proceedings of the Fourth Edoardo Amaldi Conference.

There were, however, three presentations that touched on experimental activity. A new spherical detector is being constructed in São Paulo, Brazil and a successful test of a new transducer was reported by the Rome group. In addition a report by the International Gravitational Wave Event Collaboration (IGEC) was given.

2 New Results

2.1 The Brazilian Spherical Detector

Odylio Aguiar reported on the Mario SCHENBERG gravitational wave detector at the Physics Institute of the University of São Paulo. This is a new experimental effort which has just received its initial funding and which is intended to build a spherical detector of gravitational radiation. The initial spherical antenna is planned to feature a sensitivity better than $h = 10^{-21}/\sqrt{\mathrm{Hz}}$ in the 3.0-3.4 kHz bandwidth, and to work not only as a detector, but also as a test bed for the development of new technologies. It is a collaboration between 6 Brazilian institutes and the University of Leiden in the Netherlands. It will use a copper-aluminum alloy for the spherical mass and the investigators expect to obtain a quality factor of 10^7 at ultra-cryogenic temperatures. They anticipate operating this detector at a temperature of $.015 - .020$ Kelvin. The high sensitivity is expected because of the ultra-low temperature and the high density of the copper nickel alloy. Aguiar showed pictures of the constructed facilities and of the first test spherical mass. First cryogenic tests are expected at the end of 2001.

Aguiar expects to operate this omnidirectional detector in coincidence with the detectors being constructed by the mini-GRAIL project in the Netherlands and the SFERA project at the Università di Roma and, ultimately, the various interferometers.

2.2 The New Italian Transducer

Massimo Bassan submitted results of initial tests of a heavy mass transducer on which he has been working for several years. This transducer features a clever cantilever spring arrangement which keeps the test mass in alignment with its sensing elements (either plates for a capacitive transducer or coils for one using an inductive readout). The stability of this arrangement allows much closer spacing and hence higher coupling. This, in turn, leads to a wider bandwidth for the operating resonant gravity wave detector. The transducer has been tested and has both a high mechanical and high electrical Q. The coupled Q is greater than 10^6.

The plan is to incorporate this transducer as a part of the Nautilus detector in its next runs. Its design is such that it is suitable for inclusion on any of the existing resonant bar detectors.

2.3 IGEC Upper Limits for Kilohertz Bursts

The IGEC consists of a collaboration of the following resonant detector groups: ALLEGRO (Baton Rouge), AURIGA (Legnaro), EXPLORER (CERN), Nautilus (Frascati), and NIOBÉ (Perth). All of these detectors have operational experience measured in years. The IGEC has published upper limits obtained by the analysis of two years of coincident data. [1]

More recently, the collaboration analyzed an additional two years of data and, with added experience, reanalyzed the earlier data. The upper limits were improved. This work was reported in Perth by Prodi[2] and by Heng.

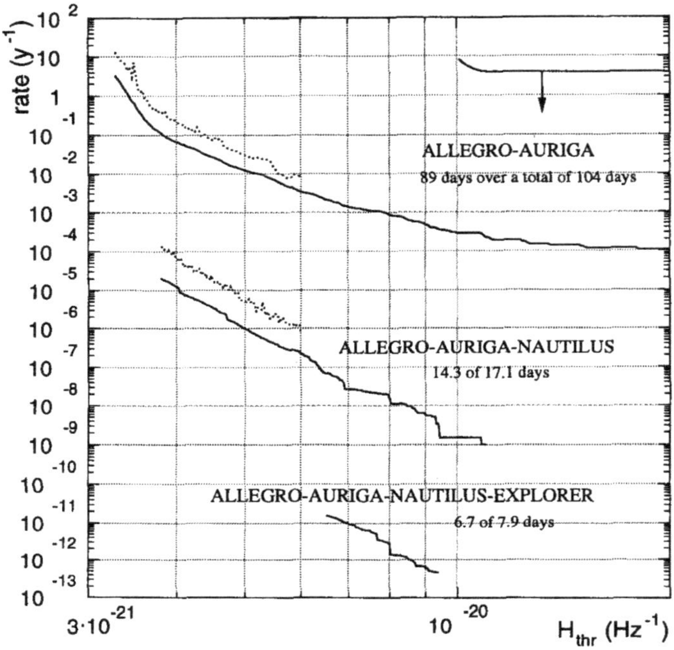

Figure 1. Upper limits to gravitational wave flux established from data taken during 1999 and 2000. The lines show the influence of multiple detectors.

A limited dataset is shown in Figure 1. The newly published data show an improvement in the upper limit by approximately a factor of two from that shown in the figure.

Acknowledgments

The work at Louisiana State University is funded by the National Science Foundation and LSU.

References

1. Z.A. Allen *et al*, *Phys. Rev. Lett.* **85**, 5046 (2000).
2. Z.A. Allen *et al*, Proceedings of the Fourth Edoardo Amaldi Conference, Perth 2001.

C1III: SPACE BASED INTERFEROMETRIC DETECTORS

ALICIA M. SINTES
Max Planck Institut für Gravitationsphysik. Albert Einstein Institut, Am Mühlenberg 1, D-14476 Golm. Germany

Departament de Física, Universitat de les Illes Balears, E-07071 Palma de Mallorca, Spain
E-mail: sintes@aei-potsdam.mpg.de

1 Overview

The session was very well attended and started with an invited review by B.F. Schutz, titled *"LISA science update"*. This complemented a very nice plenary talk, by the same speaker, on *"LISA- Pioneering Space Gravitational Wave Detection"* and set the stage for the presentations done in the workshop. The session continued with Neil Cornish talking about *"measuring and mapping the gravitational wave background"* and ended with B.S. Sathyaprakash with a talk entitled *"digging signals out of LISA confusion noise"*.

2 Summary of The Talks

B. F. Schutz started reporting the latest news on the LISA project. In 2001, ESA and NASA exchanged letters of agreement to share the project equally, and ESA invited NASA to contribute to a technology demonstration mission called SMART2, due for launch in 2006. The two agencies formed a joint LISA International Science Team (LIST), that will organize the community. It has two chairs, T. Prince (JPL, Caltech) and K. Danzmann (AEI). Theorists and astronomers who want to contribute to the science required before LISA's launch in 2011 are welcomed to join in the projects encouraged by the *LIST's Sources and Sensitivities Working Group*, jointly chaired by S. Phinney and B.F. Schutz. There are still many open questions about sources of gravitational waves that have the potential to affect LISA's final design. Schutz encouraged further research in critical areas of relativity and astrophysics, such as studies of the equations of motion of small bodies near black holes, numerical studies of black-hole mergers, studies of binary evolution, and estimates of astrophysical and cosmological sources of background gravitational radiation. Schutz reviewed the different contributions to these topics and how LISA data analysis is envisaged. Unlike the ground-based detectors,

LISA will observe many of its sources with extremely high signal-to-noise ratio. LISA could have a mission lifetime of up to 10 years, and is likely to be the first of a sequence of space-based detectors over the next few decades.

Next, **N. Cornish** discussed strategies for measuring and mapping the gravitational wave background with LISA and LISA follow-on missions. The gravitational wave sky, like the microwave sky, is expect to consist of isolated bright sources superimposed on a diffuse gravitational wave background. The background radiation has two components: a confusion limited background from unresolved astrophysical sources; and a cosmological component formed during the birth of the universe. A great deal could be learned by measuring this cosmological gravitational wave background (CGB), but detecting the CGB presents a significant technological challenge. The signal strength is expected to be extremely weak, and there will be competition from unresolved astrophysical foregrounds such as white dwarf binaries. Cornish presented two different strategies to solve this problem. The first is to cross-correlate the output of two independent interferometers. The second is an ingenious scheme for monitoring the instrument noise by operating LISA as a Sagnac interferometer. He derived the optimal orbital alignment for cross-correlating a pair of LISA detectors, and provided the first analytic derivation of the Sagnac sensitivity curve.

Ending the session, **B. S. Sathyaprakash** talked about how to extract information from the "confusion noise background". In the case of sources for which the waveform emitted is known accurately one could dig deeper into the background and extract useful signals. Sathyaprakash is studying the extent to which techniques such as matched filtering and time-frequency analysis could be useful in digging the signals out of confusion noise.

Space based interferometric detectors will allow exciting exploration of many exotic and interesting astrophysical systems. Whether they manage to see a cosmological background is one of the most unpredictable outcomes of gravitational wave astronomy, but LISA is well situated to at least put stringent limits on many early universe scenarios.

Acknowledgments

I am most grateful to the organizers of GR16 for setting up such an excellent meeting. Preparation of this summary was supported in part by DGICYT Research Project BFM 2001-0988. This session was organized by Robin Stebbins and chaired by Bernard F. Schutz.

C1IV: GRAVITATIONAL WAVE DATA ANALYSIS

B.S. SATHYAPRAKASH
Department of Physics and Astronomy, Cardiff University, Cardiff, UK
E-mail: B.Sathyaprakash@astro.cf.ac.uk

Resonant bar detectors are routinely searching for astronomical sources of gravitational waves and to setting upper limits on event rates. Interferometric detectors are beginning to operate at sensitivity levels good enough to set meaningful upper limits and begin astrophysical searches. With the long baseline interferometers scheduled to take data at unprecedented sensitivity levels the next few years will be a very exciting period for gravitational waves. In session C1iv there were talks focusing on gravitational wave searches (Krolak and Sintes), setting upper limits on astrophysical signals (Brady and Whelan), theoretical developments in modelling binary black holes (Iyer), testing general relativity with gravitational wave data (Will) and tools for gravitational wave data analysis (Schutz). There was also a one-hour round-table discussion on setting upper limits chaired by Andersson

1 Introduction

Gravitational wave (GW) data analysis is a very challenging task. This is because (1) GW antennas are broad-band detectors with near full sky coverage, (2) most GW signals are tracked in phase thereby greatly enhancing their visibility but at the cost of high computational resources, (3) GW antennas operate round the clock taking data at rates up to several megabytes per second placing severe demands on data analysis systems to keep up with the data rates, and (4) the expected signals (in initial ground-based detectors) are pretty weak making it a daunting task to extract signals in what is necessarily a non-Gaussian and non-stationary background. However, GW data analysis can truly improve the effective sensitivity of an antenna since optimal signal processing, based on matched filtering of the detector output with a template that is essentially a copy of the expected signal, can enhance the signal visibility by several factors to many orders of magnitude.

The session C1iv consisted of talks that focussed on various data analysis efforts as well as on theoretical modelling and tools for gravitational wave data analysis. This is a report on a subset of those talks for which the Session Chair had inputs from the speakers.

2 Gravitational Waves from Inspiralling and Merging Binaries

A theoretical challenge faced in searching GW data for signals of known shape, such as binaries consisting of black holes (BH) and neutron stars (NS), is the construction of an efficient and reliable set of templates that are to be used in match filtering the expected signal. The demands placed by the matched filtering technique on signal's phase is quite stringent: A mismatch between a signal, that may be buried in the detector data, and a template, used in digging it out, of about *one* cycle in the entire evolution of the wave in the detector band could result in completely losing the signal. In the case of binary black holes this means that the phase of the signal should be matched by a template to an accuracy of better than one part in a thousand and in the case of NS to better than one part in 10,000. Indeed, in the case of continuous GW (CW) from spinning NS one integrates close to ten billion cycles during an observation period of several months implying a precision of 10^{-10} in the measurement of the signal's phase. Such a requirement has two distinct implications: (1) the phase evolution must be known to a great accuracy (*precise theoretical modelling*) and (2) the data must be passed through a large number of templates corresponding to different plausible values of the signal's unknown parameters (*great computational burden*).

The LIGO-VIRGO-GEO interferometer network has its best sensitivity in the frequency interval 100-300 Hz. In the case of a binary consisting of two black holes each of mass 10 $M_\odot$ the coalescence of the holes is an event whose frequency $1/[6^{3/2}\pi(m_1+m_2)] \simeq 220$ Hz happens to lie right at the heart of the network's sensitivity band. In other words black hole collisions are among the best possible candidate events for initial interferometers[1,2,3]. Consequently, it is important to have an accurate knowledge of the evolution of the binary close to the coalescence as well as the range of values taken by the parameters on which the evolution depends. Such a knowledge is also important in setting upper limits on rates of binary coalescences in the Galaxy and the Universe.

2.1 Improved Waveforms for Black-Hole Collisions

B.R. Iyer (in collaboration with *T. Damour and B.S. Sathyaprakash*) reported on the recent progress made on theoretical modelling of black-hole binary mergers. The last stable orbit (LSO) of a particle in circular orbit around a black hole is at a frequency $f_{\rm LSO} = 220(M/20M_\odot)^{-1}$ Hz, while for comparable masses it is most likely to occur at a frequency[4] higher than the test-mass value. Time-domain template waveforms used in detecting the inspiral phase of a merging BH binary is truncated at the LSO. This is because until recently we did not have a reliable knowledge of the waveform beyond the last stable orbit. This time truncation, harmless as it may sound, does mean a specific modelling of the frequency content of the merger signal which can influence the extraction of the true GW signal depending on how the true signal behaves. Damour et al. showed that if the merger signal terminates quickly, say over less than half an orbital time-scale, then there is no harm in using the time-truncated waveforms, or their Fourier-domain counterparts[2].

Recent theoretical progress in the Hamiltonian description of[5] and the effective one-body approach to solving the dynamics of binary black holes has shed some light on how a binary may end its life after crossing the LSO[6]. Buonanno and Damour[6] introduced a new approach which is no longer limited to the adiabatic approximation and expected to describe, rather accurately, the transition between the inspiral and the plunge, and to give also an estimate of the ensuing plunge signal. The approach of Buonanno and Damour[6] is essentially a re-summation technique[4,2] for the poorly convergent post-Newtonian quantities. It consists of two main ingredients: (i) the "conservative", that is non-dissipative, part of the dynamics is re-summed by an effective one-body dynamics that replaces the two-body dynamics, and (ii) the dissipative part of the dynamics is constructed by borrowing the Padé re-summation technique introduced in Damour et al.[4].

Thus, one has now a reasonably good picture of the last several milli-seconds in the life of a black-hole binary when the two associated Schwarzschild black holes merge leading to a single spinning Kerr hole. It has been suggested that a good strategy to detect compact binaries would be to use as search templates the waveforms based on the effective one-body approach. These waveforms comprise the inspiral and merger phases followed by the quasi-normal mode ringing of the newly formed spinning black hole. Such a template bank should be augmented by an extended set of search templates that result from a slight variation of the effective one-body waveforms. Such a strategy may be good enough to take into account the uncertainty that lies in modelling the late stages in the evolution of these systems. Now the range of masses where one can improve the signal visibility by taking into account the merger part of the signal is roughly $[10, 40]M_\odot$. Fortunately, the number of templates needed to search for binaries in this range of masses is only about 10. Hence we will not increase the computational cost, nor will we worsen the statistics (false alarm and false dismissal rates) by enhancing the number of templates in this range of masses.

2.2 *Setting Upper Limits on the Galactic Rate of Neutron Star Mergers*

The sensitivity of gravitational wave detectors are currently reaching levels that are not yet good enough to conclusively detect events frequently but enough to set meaningful upper limits on astrophysical events. The LIGO and GEO ran in coincidence for a little over two weeks during Dec 28, 2001 and Jan 14, 2002. The data collected over this period will be used in setting upper limits on different kinds of astrophysical events. An event of great astrophysical interest is the inspiral and coalescence of double neutron stars since in this case the signal shape is known to a very high accuracy. This is because the (initial) interferometers would only see the adiabatic inspiral of the NS-NS coalescence and not the merger of the system. The inspiral phase is not as strongly non-linear as the merger phase and therefore high-order post-Newtonian theory is adequate to compute the waveforms to a good accuracy.

P. Brady reported on the plans of the LIGO-GEO collaboration to place upper limits on the Galactic rate of double NS coalescence by using data from the GEO and three LIGO detectors [7]. This will be the first ever effort in which data from a network of gravitational wave detectors will be used in placing upper limits on astrophysical events. It is therefore not only scientifically interesting but it will be an opportunity to test, validate and verify the data analysis software developed by the LIGO Scientific Collaboration and GEO. In essence, the upper limit on double NS coalescence will be calculated in the following way [8]: The sensitivity of an interferometer is characterized by the noise spectral density $S_h(f)$ which is nothing but the Fourier transform of the detector autocovariance function, assuming that the the noise is a stationary stochastic process. The match filtered signal-to-noise ratio (SNR) for the inspiral part of the gravitational wave emitted by a binary of total mass m, symmetric mass ratio η and located at a distance r is given by [10]

$$\rho = \frac{m^{5/6}}{r\pi^{2/3}} \left(\frac{5\eta}{6}\right)^{1/2} \left[\int_{f_{\rm low}}^{f_{\rm high}} f^{-7/3} \frac{df}{S_h(f)}\right]^{1/2}, \tag{1}$$

where $[f_{\rm low}, f_{\rm high}]$ is the frequency range in which more than 99% of the SNR is concentrated. The coalescence rate limit R is based on the maximum SNR $\rho_{\rm max}$ observed coincidentally in the four LIGO-GEO detectors, within the coincidence window, during the entire observation period T. If the actual rate is larger than R then we would observed a larger SNR event. One expects the LIGO-GEO network to eventually place upper limits that are better than previous limits based on the Caltech 40 m detector [8] as well as close to the astrophysically based limits [9].

3 Continuous Gravitational Waves

Our Galaxy is expected to have at least 10^8 spinning neutron stars that form roughly at the rate of one every 30 years. Neutron stars with a non-axisymmetric deformation are good candidate sources of GW. There are several plausible mechanisms that may lead to the deformation of the star or to precession of its rotation axis, and hence to the emission of continuous signals. There are a number of ways in which a spinning neutron star could radiate GW: (1) Neutron stars normally spin at high rates (several Hz to several hundred Hz) and this must induce some equatorial bulge and flattening of the poles. The presence of a magnetic field may cause the star to spin about an axis that is different from the symmetry axis leading to a time-varying quadrupole moment. (2) The star may have some density inhomogeneities in the core/crust set up during its formation and/or subsequent convectively unstable motions of the core. (3) The presence of an accretion disc, with its angular momentum not necessarily aligned with that of the neutron star, can potentially alter axisymmetry. That and electromagnetic radiation reaction torques can induce and sustain wobble. (4) The normal modes of the neutron star fluid (radial and other oscillations) can extract rotational energy and re-emit in the form of GW. (5) There are certain relativistic instabilities in the neutron star fluid which may cause the star to radiate away energy at increasingly higher amplitudes until the instability condition is violated. The best example of this is the Chandrasekhar-Friedman-Schutz instability in which normal modes of a star become unstable in the presence of rotation.

A simple model for the non-axisymmetric deformation on a neutron star emitting GW is the following: If I_{zz} is the moment of inertia about the spin axis of a neutron star emitting GW at a frequency f then the wave amplitude at a distance r is:

$$h = 3 \times 10^{-26} \left(\frac{10\ \text{kpc}}{r}\right) \left(\frac{I_{zz}}{10^{45}\ g\ cm^2}\right) \left(\frac{f}{200\ \text{Hz}}\right)^2 \left(\frac{\epsilon}{10^{-5}}\right), \qquad (2)$$

where ϵ is a dimensionless number that gives a measure of deviation from axisymmetry which is possibly no greater than 10^{-5}. The instantaneous amplitude of the signal is pretty small compared to the broadband amplitude noise spectral density of 10^{-21} expected in initial interferometers. The signal-to-noise ratio (SNR) is enhanced by coherently integrating the signal for a long period – typically several months to a year. But there are several problems to cope with: Firstly, coherent integration requires a precise knowledge of the signal's parameters – the direction to the source, the gravitational wave frequency, the spin down rate of the neutron star, etc. Secondly, the signal is monochromatic only in a frame attached to the solar-system barycenter. In a frame attached to the antenna one has to take into account the motion of the detector relative to the source which, over an integration time of several months, causes a Doppler modulation of the signal's amplitude and phase. This has the implication that the signal will be smeared out in a large number of frequency bins around the true frequency of the signal. Finally, the source might lie in a binary causing additional modulations that must be taken into account. In the past four to five years there have been efforts to design and invent algorithms that are capable of optimally detecting a CW source.

3.1 The Hough Transform Algorithm to Search for Continuous Gravitational Waves

A. Sintes (in collaboration with *M.A. Papa and B.F. Schutz*) reviewed the hierarchical *Hough Transform* algorithm which is a very powerful tool, and the best so far, to integrate a phase-modulated CW signal [11]. The algorithm works by first selecting candidate sources using a sub-optimal filtering technique but one that is computationally inexpensive. Candidate sources selected in the first stage are followed up in the second stage using a fully optimal search algorithm. However, the entire data set is never searched with full sensitivity.

In detail, there are three key steps: The first step consists of breaking a long data set (of three to four months duration) into many shorter segments and constructing the demodulated power spectra of the shorter segments using a coherent (that is, phase-sensitive) method. Since the data segments are short the number of points to be investigated in the parameter space is quite small compared to a coherent search of the full data segment. A CW signal would appear as an *unsmeared* peak in the power spectrum of a demodulated data segment. In consecutive data segments, however, the peak is likely to have shifted to a different position based on the parameters of the signal. In the second step, therefore, one employs an incoherent method (that is, a phase-insensitive search) and tracks the frequency evolution of the peaks which will be known in advance for each chosen parameter value. This is known as the *incoherent Hough map*. The second step, being a phase-insensitive search, is also computationally inexpensive. At the end of the first two steps one makes a list of candidate events or the parameter values where a signal could lie. In the third step one performs a fully coherent search, *the coherent Hough map,* but only for those values of the parameters that passed the first two steps.

The Hough transform algorithm is at an advanced stage of development at the Albert-Einstein Institute, Golm, Germany. It will be the principal search algorithm to used in searching for CW sources in LIGO and GEO interferometers.

3.2 All-sky search for CW sources in EXPLORER

A. Krolak (in collaboration with *P. Astone, K. Borkowski and P. Jaranowski*) reviewed the data analysis tools to perform search for continuous gravitational-wave sources in the data from a resonant bar detector. He presented results of the analysis of the data from the EXPLORER. The analysis involved a random search of the parameter space consisting of all sky positions and a range of the spindown parameter. It was found that EXPLORER detector data in the bandwidth analysed were to a good accuracy Gaussian. The number of threshold crossings agreed very well with the hypothesis that data were white Gaussian noise.

4 Triana – A Quick-Look Data Analysis Tool

Gravitational waves that are likely to be observed by the first generation of interferometers will only have a low SNR of about 5-10. Therefore, GW detectors record and analyse a number of different instrumental and environmental monitors, such as laser frequency, intensity, and beam modes, seismometers, magnetometers, microphone signals, etc., to either firm up candidate events as true GW events, or to veto them out as spuria arising as a consequence of background noise. This is especially important during the first few years of observation though in the course of time lessons learned from earlier runs could help us in reducing the number of different environmental channels that are recorded. This piece of activity of understanding detector response and using that information in improving detector performance and characterising data is termed *detector* or *data characterisation* (DC) and it is an essentially step in making a firm GW detection.

Triana is a graphical programming environment for DC and quick-look analysis whose development began in 1989 at Cardiff University under the leadership *B.F. Schutz* when he and Justin Shuttleworth started putting together the basic ideas in C++. When the SUN Microsystem released Java, Cardiff recognized its versatility, flexibility and portability and chose it as the language for their data analysis system. The first Java version of Triana was written by Schutz and Taylor in 1996 using the Java Development Kit (JDK) version 1.0x. This consisted of about 40,000 lines of source code. In January 1997, the emergence of the JDK 1.1 (and therefore the different event-handling API) led them to re-write the GUI side of the application and at the same time they redesigned the structure of the application.

Triana is particularly suited for prototyping complicated algorithms aimed at understanding the operation of an interferometer. Triana allows users to create complex programs out of basic building blocks supplied with the system, just by dragging the blocks, dropping them into a workspace, and connecting them together. Triana is one of the most powerful applications ever written in Java. It was initially developed at Cardiff to help analyse massive data sets from GEO600 and other GW detectors. Triana has been extended to work on diverse kinds of data: numerical data, images, audio data, and even text. It comes with many toolkits and modules and with wizards that help users to create new tools. Programs created using Triana may be run from within its graphical environment or in the background, using Triana's scripting language called *TrianaOCL*. Triana can also write into remotely located databases, send complex queries and receive results in textual form and display them graphically. Such features are being routinely used by GEO600 scientists.

Perhaps the most exciting feature of Triana is the client-server application: Triana can distribute a network of units to a heterogeneous set of platforms, manage communication amongst different processors, create logs, etc. This feature has been demonstrated this on a cluster consisting of a PC running windows, a Unix workstation and University's high performance computer system. GRID can therefore adopt Triana as a software tool to dynamically distribute programs visually within a convenient and interactive graphical user interface. This will allow the utilisation of the computational resources of the entire "super network" consisting of all the machines connected to GRID that run Triana. In other words, Triana can optimally and efficiently use the resources of the network.

Triana is being supported for development as a Grid programming tool by the EU-funded GridLab project (www.gridlab.org) and the PPARC-funded GridOneD project (www.gridoned.org). These projects will help build custom interfaces between Triana and Globus so as to make use of the resource discovery and management utilities in Grid, offer safe compute-serving, remote execution, program control and supervision from any platform, including handheld PDAs and develop interfaces with Cactus for data and instruction exchange. Triana will become a work-flow tool providing graphical control and distribution of programs over computer networks. It will also include wrappers to pre-existing C, C++ and Fortran libraries thereby making it very convenient for users to effortlessly export massive computational resources into a single working environment. It is also aimed to transfer and store data using XML-complaint formats.

Acknowledgments

I would like to thank all the contributors to the Session for their exciting talks and in particular to P. Brady, B.R. Iyer, A. Krolak, B.F. Schutz and A. Sintes for their input to this contribution and to Nils Andersson for coordinating the round-table on setting upper limits. This report was written at the Albert-Einstein Institute and I am indebted to Bernard Schutz for hospitality.

References

1. V.M. Lipunov, K.A. Postnov, and M.E. prokhorov, New Astron. **2,** 43 (1997);
2. T. Damour, B.R. Iyer, and B.S. Sathyaprakash, Phys. Rev. D **62**, 084036 (2000).
3. T. Damour, B.R. Iyer and B.S. Sathyaprakash, Phys. Rev. D **63,** 044023 (2001).
4. T. Damour, B.R. Iyer, and B.S. Sathyaprakash, Phys. Rev. D **57**, 885 (1998).
5. T. Damour, P. Jaranowski and G. Schäfer, Phys. Rev. D **62**, 021501; *ibid,* **62**, 044024; *ibid,* **62**, 084011 (2000).
6. A. Buonanno and T. Damour, Phys. Rev. D **62,** 064015 (2000).
7. The activities of the *Inspiral Upper Limits* group can be found at `http://www.lsc-group.phys.uwm.edu/iulgroup/`
8. B. Allen, et al., Phys. Rev. Lett. **83,** 1498 (1999).
9. E.S. Phinney, Astrophys. J., **380** L17 (1991).
10. L.P. Grishchuk, V.M. Lipunov, K. Postnov, M.E. Prokhorov, and B.S. Sathyaprakash, Physics-Uspekhi **71**, 3 (2001)
11. M.A. Papa, B.F. Schutz and A.M. Sintes, in Gravitational Waves: A challenge to theoretical astrophysics, V. Ferrari, J.C. Miller and L. Rezzolla (eds) (ICTP Lecture Notes Series, Volume III, ISBN 92-95003-0505, 2001), pp 431-442.

C2: TESTS OF SPECIAL AND GENERAL RELATIVITY

AROONKUMAR BEESHAM
Department of Mathematical Sciences,
University of Zululand, P. Bag X1001, Kwa-Dlangezwa 3886, South Africa

We begin with the oral presentations. D. L. Khokhlov considered an approach to special relativity based on a transformation law for mass which results in the energy of a moving particle being the same in a laboratory frame as well as in its rest frame, i.e., the description of phenomena in the laboratory frame coincides with that in its rest frame. It is claimed that this approach can be tested through the decay of energetic pions. The ratio of the probabilities of the decay of a pion into an electron to that of a pion into a muon is estimated. Results of the Super-Kamiokande and other experiments which detect fluxes of energetic neutrinos from decayed atmospheric pions seem to favour the approach presented.

L. Iorio discussed the possibility of accurately measuring a combination of the PPN parameters β and γ through detecting the general relativistic perigee shift of the SLR satellite LAGEOS-2 in the field of the earth. The proposed observable is a weighted sum of the orbital residuals of, *inter alia*, the perigee of LAGEOS-2 and the nodes of LAGEOS and LAGEOS-2. The proposed measurement is conceptually similar to the Lense-Thirring one. The major source of uncertainty is the effect of higher degree even zonal harmonics of the geopotential. By dealing with other systematic errors it should be possible to reach a higher level of accuracy, especially with improved data for the geopotential due to be released. The accurate results expected for the combination of these PPN parameters could be used with those for the Nordvedt parameter η to get new and independent measurements of β and γ.

V. Iafolla gave an update on the development of a high sensitive Italian spring accelerator that is to be part of some experiments in the proposed Bepicolombo Esa Cornerstone mission to Mercury. The inertial acceleration acting on the orbiter of mercury is to be measured. A prototype of the instrument has been developed and tested in the laboratory, and a report was given on its description, sensitivity and thermal stabilisation. It is hoped that the instrument will develop into a cyrogenic differential detector that will be sensitive enough to improve the test accuracy of the weak equivalence principle (WEP) in free fall conditions.

The above was not the only proposed test of the WEP. A. M. Nobili described the development of the satellite, Galileo Galilei, to test the WEP to about 10^{-17}, roughly four orders of magnitude higher than the most accurate ground-based experiments. Apart from this Italian project, there are two others also planned, viz., the French μSCOPE and the American STEP. All these experiments are designed to test the universality of free fall. Laboratory tests are presently being undertaken on a prototype, and data are being analysed. The main features of the experiment were pointed out and compared to the other two proposals, and an update given on the mission study and error budget.

N. Ashby spoke about placing an advanced Cesium clock in the International Space Station as part of the PARCS project, in which several groups are participating. The microgravity space environment allows laser-cooled Cesium clocks to spend longer times in space, resulting in improved clock performance. Significant improvements are expected in several fundamental special and general relativity experiments, e.g., gravitational frequency shift and local position invariance. World-wide timing, time transfer and navigation will also benefit. Ashby briefly described the clock, principal limitations on performance of relativity experiments, scientific objectives and benefits, and projected outcomes.

Another benefit arising from the proposed Esa mission to mercury is to possibly tighten the limits on the PPN parameters by one order of magnitude, as discussed by A. Milani. Analysis of the orbital data can determine γ at the 10^{-5} level. Through spatial polarization of mercury's heliocentric orbit in the gravitational field of Jupiter, one can constrain β at the 10^{-5} level. The Shapiro effect can provide an uncorrelated value for β. These expected results are deduced from a full cycle data processing simulation, taking into account not only random measurement noise, but also correlated noise and systematic effects both in the tracking and in the onboard accelerometer data. Thus results should indicate realistic rather than formal accuracies.

H. J. Paik discussed the design of a cryogenic null test for Newton's $1/r^2$ law with expected resolution of 10^{-3} at $50\mu m$. Achieving the desired sensitivity requires not only extremely low intrinsic noise but careful attention to metrology errors and extremely low coupling to seismic, gravitational, and magnetic disturbances. A null source is be employed to minimize Newtonian errors. A description of the experimental design was given. Most metrology and density errors will average out due to the cylindrical symmetry in the design. Superconducting magnetic levitation will allow soft, low-loss suspension of the test masses. The low intrinsic noise of the detector, combined with reduced metrology errors, should lead to high resolution of the experiment.

A. Tartaglia discussed a scheme of an interferometric measurement similar to the classical Michelson-Morley experiment in order to spot possible anisotropies due to the angular momentum density of the Earth. Considering non geodesic (circular) trajectories for the light beams (using wave guides) it was shown that time of flight differences between an East-West and a South-North round trip would be proportional to the angular momentum density. With geodesic rays (free light beams) the effect is much lower and out of any hope of detectability. In the first instance we should be in the range of the sensitivity required for detecting gravitational waves even with a laboratory size apparatus. It is hoped that the gravitomagnetic fringe shift will be seen slowly rotating the experimental set-up in a horizontal plane at the equator. Other ground based experimental configurations such as circularly counter-rotating light beams (in light pipes) around a massive spinning sphere were discussed.

Poster presentations were as follows. M. J. Ermamatov proposed a general relativistic kinetic theory of electrical conductivity in the approximation of the relaxation time. Some consequences of the theory, for example, for rotating conductors, have been investigated.

It has been recently claimed that cosmologies with a time dependent speed of light might solve some of the problems of the standard cosmological scenario, as well as inflationary scenarios. H. Vucetich has shown that most of these models, when analyzed in a consistent way, lead to large violations of charge conservation. Thus, they are severely constrained by experiment, including those where c is a power of the scale factor and those whose source term is the trace of the energy-momentum tensor. In addition, early Universe scenarios with a sudden change of c related to baryogenesis are discarded.

N. Charmoun and H. Vucetich have computed the weak interaction contribution to the proton neutron mass difference both from simple models of the nucleon (the Bag model and the constituent quark model as two extreme relativistic and non–relativistic representations of the quarks inside the nucleon) and from a model independent sum rule. When this contribution is included in the analysis of the Eötvös experiment, they reduce the bound of a possible weak interaction violation to the equivalence principle by one order of magnitude from 10^{-2} to 10^{-3}.

C. S. Unnikrishnan attempted to answer the important question of whether or not leptons generate gravity. Certain experiments aimed at re-determining the Newtonian gravitational constant, G, have reached the level of sensitivity required to provide the first meaningful result on this important issue. Stronger constraints on the hypothesis that leptons do not generate gravity could be more decisively provided, to accuracies better than 30%, if sensitivities can be maintained in existing experiments when source masses of different materials are used. Any more improvement in the near future will require new experiments improving on the Kreuzer-type experiment, or those measuring self acceleration of composite objects.

The analogy between general relativity and electromagnetism suggests that there is a new galvano-gravitomagnetic effect, which is the gravitational analog of the Hall effect and takes place when a current carrying conductor is placed in a gravitomagnetic field. In connection with this galvano- gravitomagnetic effect, B. J. Ahmedov explored the possibility of using current carrying conductors for detecting the gravitomagnetic field of the Earth.

It has been shown that the interplay between magnetic and gravitomagnetic fields leads to the general relativistic effect of charge redistribution inside conductors. The electromagnetic measurements of gravitomagnetic effects which can be performed within a conductor in the presence of the gravitational field of both a slowly rotating metric source and an applied magnetic field both were proposed by B. J. Ahmedov and M. Karim.

Theories based on a metric-affine geometry of spacetime predict that in a gravitational field a pair of orthogonal linear polarisation states of light propagate with different phase velocities. This gravity-induced birefringence could in principle be measured in local test experiments and, so, violates the Einstein equivalence principle. O. Preuss has used polarization measurements in solar spectral lines and of white dwarfs to constrain the essential coupling constants for this effect predicted by metric-affine gravity and other prototypes of nonmetric theories.

S. Pireaux discussed the predictions regarding light deflection of a scalar-tensor theory and the present/future technological developments allowing one to test this effect.

Acknowledgments

This session was organized by H. Vucetich and chaired by C.M. Will.

D1I: QUANTUM GENERAL RELATIVITY

CARLO ROVELLI
Physics Department, University of Pittsburgh, PA-15260, USA
Centre de Physique Theorique, Luminy, Marseille, F-13288, EU
E-mail: rovelli@cpt.univ-mrs.fr

1 Highlights

Current research in quantum gravity is very lively. The research is actively developing both towards the construction of the fundamental theory and towards applications.

A large part of the efforts to develop the general theory of the quantum gravitational field is presently concentrated in *loop quantum gravity,* in its various forms. The general theory of loop quantum gravity was reviewed at this conference in a plenary talk by Abhay Ashtekar. In the session, several specific aspects of the theory were discussed. The current efforts to understand the low energy behavior of loop quantum gravity were reviewed by Abhay Ashtekar and Jerzy Lewandowski. The covariant spacetime version of the theory, denoted *spinfoam* theory, and some notable finiteness results due also to Alejandro Perez and Louis Crane were illustrated by myself. Possible experimental tests of the quantum discreteness of geometry were discussed by Seth Major. Unfortunately, Martin Bojowald had to cancel his participation to the conference at the last minute, and his scheduled communication on his exciting recent results on the application of loop quantum gravity to the initial singularity had to be cancelled. The simplicial approach, whose recent developments were reviewed by Ruth Williams, has also recently converged with loop quantum gravity. In fact, on the one hand the spinfoam formalism include the Barrett Crane simplicial model, on the other hand it can be viewed as the covariant version of (canonical) loop quantum gravity. Altogether, loop quantum gravity appears to be in a very exciting phase of intense development and novel results.

Other approaches to quantum gravity were considered in the session as well. In particular: non commutative geometry (Kamesh Wali), the string motivated AdS/CFT theory (Robert Mann), lattice and functional formalisms (Renate Loll).

The most studied applications of quantum gravity are in black hole physics, gravitational collapse, and cosmological models. The application of quantum gravity to gravitational collapse is a long studied problem. A new and rather surprising results presented by Peter Hájíček, obtained via a careful hamiltonian analysis of the problem came therefore as a surprise: a shell bounces and re-expands, and the singularity disappears. The collapse problem was discussed also by Mike Ryan, Cenalo Vaz and Valentin Gladush.

The most characteristic physical system for quantum gravity is certainly the quantum black hole. There are now two well-known derivations of the Bekenstein-Hawking black hole entropy from a complete quantum gravity theory, obtained respectively within string theory and within loop quantum gravity. But a complete physical understanding of the quantum black hole thermal behavior is not yet available, and the problem keeps raising interest. The problem was discussed in the session by Steven Carlip, Gabor Kunstatter and Louis Witten. Finally, a cosmological model with a perfect fluid was discussed by Julio Fabris.

I'll now illustrate all the contributions to the session in more detail.

2 General theory: loop quantum gravity

Abhay Ashtekar: "Semiclassical issues in nonperturbative QG" Ashtekar discussed the use of a detailed mathematical theory of quantum geometry, in order to analyze the fundamental discreteness of spacetime predicted by loop quantum gravity. This discreteness makes it difficult to relate the underlying 'polymer excitations' of quantum geometry to the Fock states normally used in low energy physics. To bridge this gap, using some recent results of Varadarajan, Ashtekar and Lewandowski have located, analyzed and used Minkowskian Fock states in the background independent framework. The fundamental discreteness is lost in the standard semi-classical description because Fock states constitute only a small subset of all non-perturbative states. While the quantum geometry operators have purely discrete eigenvalues in the full theory, none of these eigenvectors belong to the Fock space. This framework is well-suited to address two key questions: Can the background-independent, non-perturbative theory reproduce the familiar low-energy physics on, say, suitable coarse graining? and, Can one pin-point where and why perturbation theory fails? Ashtekar described work in progress on both these fronts.

Carlo Rovelli: "Spin foam models" A spin foam model can be seen as the covariant, or Feynman integral, version of a loop quantum gravity theory. A spin foam represents a four-geometry and the model is formulated as a sum over these four-geometries. Remarkably, there is a duality between these models and certain particular field theories defined on group manifolds (Group Field Theory, or GFT). The sum over spin foam can be generated as the Feynman perturbative expansion of the GFT. Each spacetime appears therefore as the Feynman graph of the auxiliary GFT theory. This phenomenon is a 4d analog of a phenomenon first noticed in 2d quantum gravity, or "'zero dimensional string theory", where the spacetime could be obtained as Feynman graphs of an auxiliary matrix model. The main result presented is that (up to certain degenerate terms) the theory is *finite* order by order in the perturbation expansion. This allows finite three-geometry to three-geometry transition amplitudes to be perturbatively computed explicitly in the physical, lorentzian, 4d theory.

Seth Major: "Tests of quantum geometry" Using the framework of Alfaro et al, Major described several tests of quantum geometry. These include modified thresholds for the GZK cutoff and Cerenkov radiation and shifts in atomic energy levels. The modified thresholds already set limits on parameters in the Alfaro et al framework. The shifts, including the Lamb shift, are currently unobservable.

Ruth Williams: "Progress in discrete quantum gravity"
Approaches to discrete quantum gravity include sums over histories using the simplicial analogue of the Einstein action formulated by Regge, and state sum models, such as the Ponzano-Regge and Turaev-Viro models in three dimensions and the Barrett-Crane model in four dimensions. There has been a remarkable convergence with the loop quantum gravity approach, in recent years. Williams described progress with simplicial and state sum models, including a discrete version of the calculation of the ground state wave function for linearized gravity, and an investigation of perturbations of three-dimensional state sum models.

3 General theory: others

Kamesh Wali: "Action functionals on a two-sheeted spacetime"
Wali proposes a model of left- and right- chiral fields living on the two-sheeted space-time along with two distinct gauge fields, within the framework of non-commutative geometry. The mathematics presented generalizes the fundamental concepts of Riemannian geometry, such as differential forms, connection and curvature. The model has a rich and complex structure with new interaction terms to be explored. One

of them is a parity violating interaction arising from the gravitational sector of their generalized theory.

Robert Mann: "Conserved quantities in AdS/CFT" The talk reported on work concerning the problem of defining conserved charges in gravitational theory. This is a long-standing problem in general relativity – the equivalence principle implies that locally gravity cannot be distinguished from acceleration, and the problem of defining a local measure of gravitational energy and angular momentum has remained elusive. A reasonable measure of success has been attained in recent years using the Brown/York quasilocal formalism, which proposes a definition of conserved quantities enclosed by a $(d-2)$ surface in d dimensions. A generic problem with the quasilocal formalism is that the conserved quantities typically diverge as the mean radial size of the enclosed surface becomes infinite. A recent proposal inspired by the AdS/CFT correspondence offers a potential resolution to this difficulty. This conjectured correspondence states that a quantum theory of gravity whose boundary conditions are the same as those of asymptotically anti de Sitter (AdS) spacetime is in 1-1 correspondence with a quantum conformal field theory (CFT) defined on the boundary of this spacetime. Quantum field theories generically have counterterms, and the correspondence suggests that these counterterms should be geometric invariants of the induced metric on the boundary at infinity. Employing this proposal, Mann computed the energy, angular momentum and entropy of spacetimes containing rotating black holes with NUT charge (ie in the general Kerr-NUT-anti de Sitter class). He found in all cases that the conserved quantities were finite. No reference spacetime was required at any stage of the calculation. He then showed how to extend this prescription to asymptotically flat spacetimes, and obtained corresponding results that are finite. The interpretation of these quantities in spacetimes with NUT charge is still obscure.

Renate Loll: "The conformal factor problem" Under certain assumption on the behavior of the partition function under renormalization, one finds the interesting result that the well known divergence due to the conformal modes is cancelled by a Faddeev-Popov determinant. This result confirms that absence of the conformal sickness noticed by Loll in the lorentzian lattice models.

4 Gravitational collapse

Peter Hájíček: "Gravitational collapse" Hájíček studies the quantum dynamics of a self-gravitating null shell for the spherically symmetric case, using only gauge invariant quantities (Dirac observables). The definition of evolution is based on an asymptotically non-trivial symmetry of the model (time translation element of the BMS group). Hájíček proves in particular, that a version of quantum theory exists, in which the shell bounces and re-expands, and in which the singularity disappears. The quantum metric outside the shell contains a time-dependent mixture of black and white holes. These remarkable and surprising results raised much interest and a lively discussion, in the session.

Michael Ryan: "Quantum collapse of dust shell: solutions" A hamiltonian formulation of the problem of the collapse of a dust shell given by Hájíček and Kijowski and Kuchař allows one to study the the minisuperspace problem of the quantum collapse of a spherically symmetric dust shell, albeit with a very complicated Hamiltonian operator. Ryan presented both analytic approximate solutions and exact numerical solutions to the problem. Since the solutions involve only geometrical variables defined on the shell, it is necessary to reconstruct the quantum spacetime geometry around the shell (possible in this special spherically symmetric case) in order to discuss the interpretation of the solutions in terms of spacetime geometry, especially the position of the shell with respect to horizons. This allowed Ryan to consider the quantum evolution of the shell in spacetime.

Cenalo Vaz: "LeMaitre-Tolman-Bondi collapse" Vaz applied the Kuchar transformation to LeMaitre-Tolman-Bondi models of inhomogeneous dust collapse. Dirac's constraint quantization leads to a simple Wheeler-DeWitt equation with a general mass function. A solution was presented over the entire class of models considered. These solutions cover collapse into both black holes and naked singularities (up to the Cauchy Horizon) and can be employed in the study of quantum gravitational effects in the final stages of collapse with different initial conditions as well as, it is hoped, in the study of inhomogeneous cosmologies.

Valentin Gladush: "Quasiclassical spherical configuration" Gladush has considered a quasi-classical model of self-gravitating spherical dust shell, and studied the energy spectrum using quasi-classical quantization rules, and the stability of the system.

5 Black holes

Steven Carlip: "Boundary conditions, constraints, BH entropy" Recent work by a number of authors has suggested that black hole entropy may have a microscopic description in terms of a boundary conformal field theory, living either near the horizon or at spatial infinity. Carlip described a new approach to boundary conditions near such a boundary, in which the fall-off requirements are introduced as a set of second class constraints that modify the algebra of diffeomorphisms. For Liouville theory in a dynamical background, this approach leads to the standard central charge; for dilaton gravity, the result is a Virasoro algebra with a classical central charge of the form required to explain the Bekenstein-Hawking entropy.

Gabor Kunstatter: "Quantum Mechanics of Charged BHs" Basic properties of black hole thermodynamics were used to quantize generic charged black holes in the Euclidean sector. The analysis led to a discrete spectrum for the horizon area that is qualitatively consistent with the results of Bekenstein, Mukhanov and others, but different from the spectrum predicted by loop quantum gravity. In particular, the area spectrum was shown generically to be equally spaced. Quantization of the charge sector gave the condition $Q = me$, for integer m, as expected, but a consistency condition constrained the fine structure constant $e^2/\hbar$ to be a rational number. Near extremal black holes were shown to be highly quantum objects. Moreover, with a standard choice of factor ordering, extremal black holes did not appear in the physical spectrum at all.

Louis Witten: "BH Quantization in Canonical Gravity" Witten started from the observation that the Schwarzschild black hole can be viewed as a special case of the marginally bound LeMaitre-Tolman-Bondi models of dust collapse corresponding to a constant mass function. Using a midi-superspace quantization of the collapse for an arbitrary mass function illustrated in this same session by C Vaz, Witten showed that the solution leads both to Bekensteinss area spectrum for black holes as well as to the black hole entropy. In this context, the entropy is naturally interpreted as the loss of information of the original matter distribution within the collapsing cloud.

6 Finite dimensional models

Julio Fabris: "Quantum cosmological perfect fluid" Fabris quantized the gravity-perfect fluid system, employing the Schutz's formalism. The matter degrees of freedom allow to recover the notion of time. Wave-functions are obtained in closed form, in the mini-superspace, and the expectation value for the scale factor is computed, yielding a singularity-free, bouncing Universe. He showed that these results are recovered by a classical system where, besides the ordinary matter employed in the quantization procedure, a universal repulsive fluid with a stiff matter equation of state is added.

D1II: STRING THEORY AND SUPERGRAVITY

DONALD MAROLF
Physics Department, Syracuse University
Syracuse, NY 13244-1130, USA

The strings and supergravity session featured a small but varied collection of talks ranging from studies of exact solutions and solitons to supersymmetry, cosmology, and talks related to the gauge-theory/gravity dualities of[1,2]. Given the breadth of topics and the rather liberal amount of space made available in the proceedings, it seemed best to allow each speaker to describe their talks in some length. What follows is therefore a description of each talk in the order that they were presented. Each contribution was written by the speaker and only slightly edited by myself as session chair. For full details of the works, please refer to the references provided below.

Reduced $D = 10$ $\mathcal{N} = 4$ Yang-Mills Theories

Matthias Staudacher

Staudacher discussed various aspects of the dimensional reductions of the maximally supersymmetric gauge theory, namely $\mathcal{N} = 1$ in $D = 10$ to lower dimensions, and in particular the cases $D = 10 \to d = 0, 1, 2, 4$. These reductions are relevant since supersymmetry survives the reduction process, allowing one to obtain a number of exact and analytic results. This is particularly important since there exist manifold, largely conjectural relationships to string theory, supermembrane quantization and supergravity models.

The first half of the talk focused on the reductions to $d = 0, 1, 2$. All three cases have been used in different proposals for non-perturbative definitions of string theory and M-theory, going, respectively, by the names IKKT model, deWit-Hoppe-Nicolai or BFSS model, DVV model or matrix string theory. A report was given on a number of recent results concerning the d=0 reduction[3,4]. This reduction is of interest in its own right and is also relevant to the bound state problem of the $d = 1$ reduction. This work demonstrates that the current techniques for counting the number of ground states are not yet consistent and in fact quite incomplete. It is also relevant to the issue of computing exact partition functions in the $d = 2$ reduction.

The second half of the talk discussed an ongoing investigation of Maldacena-Wilson loops in the $d = 4$ gauge field theory: Using the AdS/CFT correspondence between classical supergravity and the strong coupling limit of the field theory, a number of exact results for these loops have been proposed in the literature. An important problem consists in relating these results to weakly coupled gauge theory. The first-ever two-loop perturbative calculation[5] was discussed. The first main result is that the Maldacena-Wilson loop operator is completely two-loop finite, suggesting finiteness to all orders. The second chief result is that vertex diagrams contribute to the two-loop static potential. Previous lower order calculations in the literature found only (trivial) ladder diagrams to contribute. This indicates that, if AdS/CFT is correct, it indeed solves, at strong coupling, 't Hooft's longstanding planar diagram summation problem for this field theory.

Rotating black holes in higher dimensions

Roberto Emparan

There are several motivations for studying General Relativity in dimensions higher than 4. In addition to allowing rich dynamics and new qualitative behavior, higher dimensions are required by most unification schemes, such as string theory. Moreover, in scenarios with large extra dimensions and a low fundamental scale there is the possibility that black holes will be produced and detected in future colliders[6]. Such black holes will generically be rotating.

Solutions describing neutral, rotating black holes in higher dimensions were found in[7]. For $D \geq 6$ they present new qualitative features which, perhaps surprisingly, have so far attracted little attention. This is the subject of this contribution, which is based on work in progress with Rob Myers.

In $D \geq 5$ dimensions, rotation can take place in more than one plane. However, the talk considered the case where the black hole spins in a single plane, with rotation parameter $a \propto J/M$. As is well known, the amount of rotation that a 4D Kerr black hole can support is limited: if the bound $M \geq a$ is violated, a naked singularity results. As shown in[7], a similar bound appears for $D = 5$. However, for $D \geq 6$ the horizon is present for arbitrary values of a. There is no extremal limit and no bound on the spin. Hence, these ultra-spinning black holes are distinctive of higher dimensions.

What is the shape of such an ultra-spinning black hole? An analysis of the proper size of the horizon reveals that the black hole is highly flattened along the directions parallel to the plane of rotation—a 'pancaked' horizon. Moreover, if the rotation parameter is sent to infinity (with the mass per unit area kept finite), the geometry that results is that of a black membrane. The latter is known to be classically unstable[8], so it is natural to ask whether the highly pancaked, ultra-spinning black holes will not become unstable before reaching this limit, i.e., already at finite values of a.

The local stability of 4D black holes has been established in classic work[9]. Global stability has been addressed using the area theorems (and their thermodynamic interpretation), which state that the sum of the areas of the future horizons can not decrease. As for black branes, both local and global arguments lead to the conclusion that they are unstable[8].

Both approaches should be applied to ultra-spinning black holes. While the local analysis of linear perturbations is still in progress, global thermodynamics points clearly to an instability of black holes in $D \geq 6$ for sufficiently large values of a/M. To see this, the possible decay modes of the black hole were identified and the area of the final products were compared to that of the initial black hole for the same mass and spin. Several decay modes were studied, such as emission of waves (modeled by a gas of null particles, or using the radiation formulas for near-Newtonian sources), or the black hole breaking apart into several smaller black holes (a classically forbidden process). In all cases the instability sets in at $a/M \approx$ a few. Consistently, this happens only for $D \geq 6$; for $D = 4, 5$ this approach predicts no instability.

Black hole entropy calculations based on symmetries

Jacek Wiśniewski, with Olaf Dreyer and Amit Ghosh

A microscopic derivation of black hole entropy is one of the greatest challenges to candidate quantum theories of gravity. As an alternative to the existing quantized models, a set of very attractive ideas was recently suggested by Andrew Strominger[10] and Stephen Carlip[11]. Their symmetry based approaches are very general and mostly classical. The starting point is an observation that symmetry generating vector fields of black hole space-time form a Diff(S^1) algebra which, on the level of the algebra of conserved charges associated with these vector fields, becomes a centrally extended Virasoro algebra[12]. The black hole space-time turns out to be a representative state of the algebra with a fixed conformal weight. Then, solely using representation theory, one can count the degeneracy of such a state and this gives the entropy of the black hole : $S = 2\pi\sqrt{cH_0/6}$. Here, c is the central extension and H_0 is the conformal weight of the black hole or, more precisely, the eigenvalue of the Hamiltonian (same as the zero-mode of the Virasoro algebra) for the black hole space-time as the eigenstate.

These calculations, however, face some conceptual and technical problems. Strominger's calculations are based on *asymptotic* symmetries. It is not apparent how these symmetries capture the essence of the black hole space-time. In fact, the results are equally applicable to a star having similar asymptotic behavior. Subsequently, Carlip improved on this idea by making the symmetry analysis in the near-horizon region. Conceptually this approach is much more satisfactory in that the black hole geometry is now at the forefront. Ref.[11] also, however, faces some technical problems which corrected in[13] and some interesting results emerge :

a) The Lie brackets of the vector fields form a Diff(S^1) algebra both on and near the horizon. b) The vector fields do not admit a well-defined limit to the horizon (horizon penetrating coordinates were used to make this explicit). c) It is essential that the entire calculation of the Poisson bracket of charges (i.e., Hamiltonians of the vector fields) is performed at a distance ϵ away from the horizon and the limit $\epsilon \to 0$ is taken at the end. The lack of limit of the vector fields forbids a clear interpretation of the symmetries in the classical gravity. Presumably, this feature is an indication that the true origin of the central charge is quantum mechanical where ϵ is to be regarded as a regulator.

There is in fact a whole one-parameter family of such vector fields forming Diff(S^1), as above. As a result both c and H_0 are modified, but in such a way that the entropy is always reproduced (up to a multiplicative factor of $\sqrt{2}$). Only for a specific choice of this parameter, motivated by some arguments in the Euclidean signature, one can shift H_0 by some 'ground state' value to get rid of $\sqrt{2}$. However, the meaning of this choice is unclear.

Finally, one can perform a symmetry analysis completely on the horizon within the framework of *isolated horizons* (see other talks). The framework is naturally suited to address the question of symmetries of black holes in equilibrium (very weak assumptions are made). It turns out that no central extension appears in this case. Summarizing, the current viewpoint of the authors is that probably there is some truth in the symmetry based approaches, but it is hard to avoid details of quantum theory, especially in order to understand the origin of the central charge. This could be the case because one is attempting to give an essentially classical argument for a phenomenon that is inherently quantum mechanical.

Exact Super Black Hole Solutions

R.B. Mann, with J. Kamnitzer and M.E. Knutt

Mann reported on a project which considered the problem of finding exact solutions to supergravity coupled to matter. Very few exact classical solutions to supergravity theories are known that have non-trivial fermionic content. A superparticle (which, if massive, is a D0-brane), a cosmological constant and a super-Liouville field were included as matter sources, each minimally coupled to $(1+1)$-dimensional supergravity.

One of the pitfalls of finding exact solutions is in ensuring that they cannot be reduced by infinitesimal local supersymmetry transformations to purely bosonic solutions. Working in superspace offers a straightforward means of avoiding this difficulty, since a superspace supergravity solution – one which satisfies the constraints – has nonzero torsion beyond that of flat superspace.

The torsion is a supercovariant quantity, and as such its value remains unchanged under a gauge transformation. Hence any exact superspace solution with non-zero torsion must necessarily be non-trivial in this sense.

Mann reported on several exact solutions obtained for each of the supermatter sources mentioned above. The exact compensator superfield that describes the supergravity can be used to construct models of two-dimensional supersymmetric black holes with non-trivial curvature. To our knowledge these are the first solutions found using this technique. The superparticle and cosmological solutions had locally constant supercurvature[14], but the super-Liouville solution had locally non-constant curvature[15]. In the latter case possibility that a gravitini condensate formed was considered and examined the implications for the resultant spacetime structure. All such condensate solutions were found to have a condensate and/or naked curvature singularity.

Inverse dualisation and non-local dualities between gravity and supergravity theories

Dmitri Gal'tsov, with Chiang-Mei Chen and Sergei A. Sharakin

Gal'tsov's talk was based on the preprint hep-th/0109151 which finds classical dualities of a new type between Einstein vacuum gravity in certain dimensions and ten and eleven-dimensional supergravities. The main idea is that Kaluza-Klein two-forms arising in toroidal compactification of vacuum gravity can be dualized in dimensions $D \geq 5$ to higher rank antisymmetric forms and these forms may be identified as matter fields belonging to bosonic sectors of supergravities. While it is perhaps not surprising that the Maxwell equations and the Bianchi identities for the KK fields translate into similar equations for dual higher rank forms, a non-trivial test is whether the dilatonic exponents in the reduced actions are the same. Several cases of such dualities are described. The most interesting is the correspondence between $2 + 3 + 6$ dimensional reduction of the eleven-dimensional supergravity and eight-dimensional Einstein gravity with two commuting Killing vectors. A related duality holds between both (suitably compactified) IIA and IIB ten-dimensional supergravities and eight-dimensional Einstein gravity with three commuting Killing vectors. Another case is the correspondence between the ten-dimensional Einstein gravity and a suitably compactified IIB theory. It is worth noting that all dualities of this sort are non-local in the sense that variables of one theory are related to variables of the dual theory not algebraically, but via solving differential equations.

A remarkable fact is that the $11D$-supergravity/$8D$-gravity duality holds not only in the bosonic sector, but also extends to Killing spinor equations exhibiting unbroken supersymmetries of the $11D$ theory. Namely, the existence of Killing spinors in the supergravity framework is equivalent to the existence of covariantly constant spinors in the dual Einstein gravity. It would be interesting to check whether this correspondence found at the linearized level extends non-linearly, i.e. holds for suitably supersymmetrized $8D$ gravity. A more challenging question is whether classical dualities found here have something to do with quantum theories. Although an answer was not presented in the talk, the results concerning the ten-dimensional supergravities look promising in this direction.

New supersymmetry algebra on gravitational interaction of Nambu-Goldstone fermion

Motomu Tsuda, with Kazunari Shima

A supersymmetric composite unified model for spacetime and matter, superon-graviton model (SGM) based upon SO(10) super-Poincaré algebra, is proposed in the papers[16,17]. In SGM, the fundamental entities of nature are the graviton with spin-2 and a quintet of superons with spin-1/2. The fundamental action which is the analogue of Einstein-Hilbert (E-H) action of general relativity (GR) describes the gravitational interaction of the spin 1/2 N-G fermions in Volkov-Akulov (V-A) model[18] of a nonlinear realization of supersymmetry (NL SUSY) regarded as the fundamental objects (superon-quintet) for matter.

Tsuda's talk performed the similar geometrical arguments to GR in the SGM spacetime, where the tangent Minkowski spacetime is specified by the coset space SL(2,C) coordinates (corresponding to N-G fermion) of NL SUSY of V-A model[18] in addition to the ordinary Lorentz SO(3,1) coordinates, and discussed the structure of the fundamental SGM action[17,19,20]. The overall factor of SGM action is fixed to $\frac{-c^3}{16\pi G}$, which reproduces E-H action of GR in the absence of superons (matter). Also in the Riemann-flat space-time, i.e. the vierbein $e_a{}^\mu(x) \to \delta_a{}^\mu$, it reproduce V-A action of NL SUSY[18] with $\kappa^{-1}{}_{V-A} = \frac{c^3}{16\pi G}\Lambda$ in the first order derivative terms of the superon. Therefore our model (SGM) predicts a (small) non-zero cosmological constant, provided $\kappa_{V-A} \sim O(1)$, and possesses two mass scales. Furthermore it fixes the coupling constant of superon (N-G fermion) with the vacuum to $(\frac{c^3}{16\pi G}\Lambda)^{\frac{1}{2}}$ (from the low energy theorem viewpoint), which may be relevant to the birth (of the matter and Riemann space-time) of the universe. The (spacetime) symmetry of our SGM action was also demonstrated. In particular, the commutators of the new NL SUSY transformations on gravitational interaction of N-G fermion with spin-1/2[17,19] and -3/2[19,21] form a closed algebra, which reveals N-G (NL SUSY) nature of fermions and the invariances at least under a generalized general coordinate and a generalized local Lorentz transformations. In order to linearize the SGM action, the linearization of $N = 2$ V-A model which is now under investigation is extremely important from the physical point of view, for it gives a new mechanism generating a (U(1)) gauge field of the linearized (effective) theory[22].

Quantum Cosmology from D-Branes

P. Vargas Moniz, with A. Yu. Kamenshchik

Recent developments in string theory suggest that, in a Planck length regime, the quantum fluctuations are very large so that string coupling increases and consequently the string degrees of freedom would not be the relevant ones. Instead, solitonic degrees of freedom such as D-p-branes would become more important. Hence, what would be the effect of those new physical degrees of freedom on, say, the the very early universe and in particular from a quantum mechanical point of view?

In this work, one has initiated an investigation on D-p-brane induced quantum cosmology. It could be pointed that it may not be justified to quantize an effective theory (arising from a fundamental quantum theory). However, in so far as new fundamental fields and effects arise from the fundamental theory, a quantization of the effective action could capture significant and relevant novel features.

The starting point is the result (obtained by Duff, Khai and Lu) that the natural metric that couples to a p-brane is the Einstein metric multiplied by the dilaton. Employing an (adequately) modified Brans-Dicke action with a deformation parameter p-dependent, different quantum cosmological scenarios were analyzed. In particular, several early universe scenarios were identified similar to quantum Pre-Big-Bang and Universe-anti-Universe creation. The possible quantum mechanical transition amplitudes were also studied with a view towards determining the effect of quantum cosmological solitons in the very early Universe. Finally, it was found that the solutions of Wheeler-DeWitt equation allowed for sub-class with $N = 2$ SUSY. Other consequences regarding "stringy" cosmology were investigated, namely possible duality transformation in the effective action and their relation to Pre-Post-Big-Bang scenarios. Possible developments of this work include considering D-brane realistic actions and SUSY extensions.

Polarization of the D0 ground state in quantum mechanics and supergravity

Donald Marolf, with Pedro Silva

Marolf's talk addressed a quantum version of the dielectric effect described by Myers in[24]. In[24], the application of a Ramond-Ramond background field to a D0-brane system induces a classical dielectric effect and causes the D0-branes to deform into a non-commutative D2-brane. In contrast,[23] places D0-branes in the background generated by a stack of D4-branes. While no classical dielectric effect results, the four-branes modify the potential that shapes the non-abelian character of the quantum D0 bound state. As a result, the bound state is deformed, or polarized.

Two aspects of the deformation were studied and compared with the corresponding supergravity system. Fundamental to this comparison is the connection described by Polchinski[25] relating the size of the matrix theory bound state to the size of the bubble of space that is well-described by classical supergravity in the near D0-brane spacetime. The near D0-brane spacetime is obtained by taking a particular limit in which open strings decouple from closed strings. The result is a ten-dimensional spacetime with small curvature and small string coupling when one is reasonably close (though not too close) to the D0-branes. However, beyond some critical distance r_c the curvature reaches the string scale. As a result, the system beyond r_c is not adequately described by the massless fields of classical supergravity. The goal was thus to compare deformations of the non-abelian D0-brane bound state with the deformations of this bubble of 'normal' space.

While the detailed effects were beyond the scope of the work presented, the deformations of the quantum mechanics ground state and the supergravity bubble were shown to have corresponding scaling properties. This supports the idea that the gravity/gauge theory duality associated with D0-branes can be extended to include couplings to nontrivial backgrounds such as those discussed in[26,27,24]. A part of this was the analysis in an appendix of infrared issues associated with 't Hooft scaling in 0+1 dimensions. This in turn strengthens the argument that Polchinski's upper bound[25] on the size of the D0-brane bound state in fact gives the full scaling with N. Corresponding arguments can in fact be made for all Dp/D(p+4)-systems for $p \leq 2$.

References

1. J. Maldacena, *Adv. Theor. Math. Phys.* **2** (1998) 231, hep-th/9711200.
2. N. Itzhaki, J. M. Maldacena, J. Sonnenschein, S. Yankielowicz, Phys.Rev. D58 (1998) 046004, hep-th/9802042.
3. W. Krauth and M. Staudacher, Nucl. Phys. B **584**, 641 (2000) [hep-th/0004076].
4. M. Staudacher, Phys. Lett. B **488**, 194 (2000) [hep-th/0006234].
5. J. Plefka and M. Staudacher, JHEP **0109**, 031 (2001) [hep-th/0108182].

6. P. C. Argyres, S. Dimopoulos and J. March-Russell, Phys. Lett. B **441** (1998) 96 [hep-th/9808138];
7. R. C. Myers and M. J. Perry, Annals Phys. **172** (1986) 304.
8. R. Gregory and R. Laflamme, Phys. Rev. Lett. **70** (1993) 2837 [hep-th/9301052];
9. S. Chandrasekhar, *The Mathematical Theory Of Black Holes*, (Oxford University Press, Oxford U, 1985).
10. A. Strominger, JHEP 9802 (1998) 009
11. S. Carlip, Class. Quantum Grav. 16 (1999) 3327
12. J. D. Brown and M. Henneaux, Comm. Math. Phys. 104 (1986) 207
13. O. Dreyer, A. Ghosh and J. Wiśniewski, Class. Quantum Grav. 18 (2001) 1929
14. M.E. Knutt and R.B. Mann, Class.Quant.Grav. **16** (1999) 937; Phys.Lett. **B435** (1998) 25.
15. R.B. Mann and J. Kamnitzer, Nucl. Phys. **B** (to be published).
16. K. Shima, *Z. Phys.* **C18**, 25 (1983);
K. Shima, *European. Phys. J.* **C7**, 341(1999).
17. K. Shima, hep-ph/0012320, *Phys. Lett.* **B501**, 237 (2001).
18. D.V. Volkov and V.P. Akulov, *Phys. Lett.* **B46**, 109(1973).
19. K. Shima and M. Tsuda, hep-th/0101178, *Phys. Lett.* **B507**, 260 (2001).
20. K. Shima and M. Tsuda, hep-th/0109042.
21. K. Shima and M. Tsuda, hep-th/0012235, *Phys. Lett.* **B** in press.
22. K. Shima, Plenary talk at the Fourth International Conference on Symmetry in Nonlinear Mathematical Physics, July 7-14, 2001, Kiev, Ukraine. To appear in the Proceeding.
23. D. Marolf and P. J. Silva, JHEP **0108**, 043 (2001) [hep-th/0105298].
24. R.C. Myers, JHEP 9912 (1999) 022, hep-th/9910053.
25. J. Polchinski, Prog.Theor.Phys.Suppl. 134 (1999) 158-170, hep-th/9903165.
26. W. Taylor, M. Van Raamsdonk, Nucl.Phys. B558 (1999) 63-95, hep-th/9904095.
27. W. Taylor, M. Van Raamsdonk, Nucl.Phys. B573 (2000) 703-734, hep-th/9910052.

D2: CONCEPTUAL AND FOUNDATIONAL ISSUES IN QUANTUM GRAVITY

H. F. DOWKER
Department of Physics,
Queen Mary, University of London,
London E1 4NS.

1 The talks

In her talk, "Doing it the Lorentzian way: a fresh look at path integrals" Renate Loll (Spinoza Instituut, Utrecht) described her work on the attempt to discretize the sum-over-histories for Lorentzian spacetimes. Using a new non-perturbative path-integral formulation for Lorentzian space-times [1,2], allows a new look at some long standing problems in constructing a quantum theory of gravity. Its main idea is to obtain the quantum theory as the continuum limit of a state sum over piece-wise flat geometries. This method is related in spirit to lattice formulations of QCD, but conceptually different in that it is compatible with diffeomorphism invariance and does not distinguish any particular background metric. Unlike in other discrete regularizations of the gravitational path integral, the causal structure of the Lorentzian geometry is taken into account from the outset.

Loll described the construction of the regularized state sum, given in terms of 'dynamically triangulated' Lorentzian geometries. After applying a generalized Wick rotation to the set of simplicial space-time geometries, one obtains a well-defined convergent path integral whose associated transfer matrix (with respect to proper time) is bounded and positive for finite lattice volume.

Loll described a number of interesting properties of the underlying continuum theories in d=2,3 [3]. These include the discovery of a new universality class of 2d quantum gravity models and the emergence of a 'ground state' of well-behaved geometry in three space-time dimensions. The construction provides an explicit example of a non-perturbative Wick rotation in quantum gravity which cannot simply be reduced to an analytic continuation in 'time'. It also shows how a causality condition on quantum states can effectively act as a regulator for the quantum geometry. These features make the Lorentzian path integral an interesting new approach to the search for a quantum theory of gravity.

Charis Anastopoulos (Spinoza Instituut, Utrecht) talked on "Quantum phases and the structure of quantum theory"[4,5]. The talk focused on the approach to quantum mechanics in which quantum theory is viewed as a generalization of classical stochastic processes, like for example a random walk. On this view, the basic entities of the theory are the histories that the system can follow. On this "sample space" of histories, in the classical case we have a classical probability measure. In the quantum case, there is quantum interference between histories in general and this phase information is exhibited by the off-diagonal entries of the so-called decoherence functional. In the consistent histories approach, only partitions of the sample space of histories for which there is no quantum interference are allowed to be considered as "physical," though one can consider the structure of the sample space and the decoherence functional whether or not this consistency condition is imposed.

In the algebraic approach to consistent histories developed by Isham and others, a history Hilbert space can be constructed and a partition is given by a decomposition of the unit operator into projections. These projections commute and Anastopolous stated that in this way the Kochen-Specker theorem can be sidestepped. He also made the claim that such a history based quantum theory is not subject to Bell's theorem because it is not a probability theory. Whether or not this is the case, it is certainly true that the issues of contextuality and non-locality need to be readdressed in a history based interpretation.

Paulo Moniz (Universidade da Beira Interior, Portugal) gave a talk on "Quantum string cosmology from a de Broglie Bohm perspective" in which he reported on work done with J. Marto. In the talk he argued that the de Broglie-Bohm interpretation of quantum mechanics can fruitfully be applied to superstring inspired cosmology, a certain model of scalar-tensor gravity. The analysis was restricted to a flat FRW geometry, with an homogeneous dilaton field and a cosmological constant in the gravitational sector of the theory.

Employing a suitable superposition of solutions of the Wheeler-DeWitt equation, the basic feature of a de Broglie-Bohm framework was derived: the presence of a quantum potential which can be different from a standard classical potential. The cosmological properties of the resulting solutions were analysed and a wider set of solutions in minisuperspace were found. In addition, the regime where effectively the dilaton kinetic energy dominates over Λ is larger and determined a slower cosmological evolution in time than in the usual model. This effect becomes more manifest as ω is closer to $-4/3$. Some cosmological problems (e.g., the horizon problem) which remain with the standard Pre-Big-Bang inflationary cosmology may have to be reconsidered. Moniz claimed that when ω is closer to $-4/3$ their model does not require an early Pre-Big-Bang phase where the Universe would have had to be infinitely huge.

Herbert Balasin (TU-Wien) gave the talk, "Quantum light from beyond the event horizon" in which he began by stressing that one of the major obstacles in the unification of quantum theory and general relativity arises from the fact that quantum mechanically the causal structure of spacetime is presumably subject to the quantum fluctuations. In particular, time itself should become a quantum concept whereas time plays the distinguished role of an evolution parameter in the usual non-relativistic quantum theory. In this talk, Balasin reports on investigations into the quantum analogue of the motion of a point particle in an arbitrary gravitational field. Classically this is at the heart of general relativity since it probes the geodesic structure of the underlying spacetime manifold. Balasin argued that due to the extended nature of the quantum particle, its motion in a Newtonian gravitational field is no longer independent of its mass. Taking this Newtonian result as inspiration, Balasin aimed to quantize the geodesic deviation equation in order to describe the motion of the quantum particle relative to a freely falling observer. This was hoped to provide the resulting quantum theory with a relativistically covariant notion of time namely that of an inertial observer.

He showed that the resulting formalism can be couched in terms of the tangent bundle of the underlying spacetime. The usual structures like the action functional and the corresponding Hamiltonian formulation fit into this picture. As an application of this formalism, he considered the transition amplitude for a particle localized initially behind the event horizon of a Schwarzschild black hole. He also considered a similar situation in flat space, namely the transition amplitude through the lightcone. He claimed that there is a non-zero amplitude for finding a particle, initially localized inside the horizon, on the outside. It wasn't clear if this constitutes a claim of a violation of relativistic causality and the conclusion must be clouded by the known problems with relativistic single particle quantum mechanics. Indeed we know the the correct relativistic theory is quantum field theory and that is believed to be relativistically causal.

References

1. J. Ambjorn and R. Loll, *Nucl. Phys.* B **536**, 407 (1998).
2. J. Ambjorn, J. Jurkiewicz and R. Loll, *Phys. Rev. Lett.* **85**, 924 (2000).
3. J. Ambjorn, J. Jurkiewicz and R. Loll, in *M-Theory and Quantum Geometry*, eds. L. Thorlacius and T. Jonsson, NATO Science Series (Kluwer Academic Publishers, 2000).

4. C. Anastopoulos, *Found. Phys.* **31**1545(2001).
5. C. Anastopoulos and K. Savvidou, "Quantum mechanical histories and the Berry phase", quant-ph/0007093.

D3: QUANTUM FIELD THEORY IN CURVED SPACETIME

L.H. FORD
Institute of Cosmology, Department of Physics and Astronomy, Tufts University, Medford, MA 02155 USA
E-mail: ford@cosmos.phy.tufts.edu

A total of thirteen talks, representing a wide cross section of this field, were presented. The content of these talks and their relation to the larger field are briefly summarized in this report.

Within the general area of quantum gravity, quantum field theory in curved spacetime occupies a position on the transition from classical to quantum gravity. Its conceptual foundations are better understood than those of other quantum aspects of gravity, and it should appear as a limit of any more complete theory. It encompasses a rather wide variety of physical phenomena, and this variety was exhibited in the various talks presented in this session.

The Hawking effect remains perhaps the greatest triumph of quantum field theory in curved spacetime. Thus it is not surprizing that more talks were devoted to various aspects of the Hawking effect than to any other single topic. V. Frolov discussed work with D. Furaev on the mining of energy from a black hole by cosmic strings[1]. The basic idea is to consider a black hole with one or more cosmic strings attached to it. The vibrational modes of these strings provide extra degrees of freedom which can become excited and serve as channels for the radiation of energy from the black hole. The net effect is to enhance the rate of black hole evaporation by a factor that can be as large as 10^{31} for electroweak strings. Backreaction and information loss are two aspects of the Hawking effect that are still poorly understood, and which were addressed in two of the papers presented. A. Fabbri discussed an exactly soluable 2D model for the backreaction in the case of near extremal charged black holes, and also addressed the information loss problem within the context of this model. R. Casadio discussed a model of a collapsing shell made of quantized matter, and discussed the backreaction problem for this model[2]. One of the subtle aspects of quantum field theory in curved spacetime involves the definition of quantum states. This issue was discussed in the context of the Hawking effect by S. Shankaranarayanan, who discussed the method of complex paths for the definition of states in Schwarzschild spacetime.

The Hawking effect has inspired some very interesting work in flat spacetime quantum field theory, including the Fulling-Davies-Unruh effect, the thermal radiation seen by an accelerated detector. This was the subject of a talk by G.E.A. Matsas, in collaboration with D.A.T. Vanzella[4]. Here the decay of accelerated protons was discussed. Vanzella and Matsas show how this process can be viewed from both the inertial and the accelerating frames, with the same answer for the rate in both cases. They also present some detailed calculations for the decay rate. Another effect inspired by the Hawking effect is the creation of phonons by "dumb holes", first proposed by Unruh. In the session, Matt Visser presented a variant of this idea, the creation of photons by black hole analogs in which an effective event horizon is created by the physical motion of a material at speeds greater than the speed of light in the material. The specific model discussed by Visser used a Bose-Einstein condensate, in which the speed of light at certain frequencies can be many orders of magnitude smaller than in vacuum. Visser discussed the intriguing possibility that one might actually be able to create such black hole analogs and observe the resulting quantum radiation[5].

Another topic of long standing interest related to quantum effects in strong gravitational fields is that of singularity avoidance. It is well-known that quantum effects in principle can evade the classical singularity theorems by violation of energy conditions. One way in which quantum effects might manifest themselves in avoiding singularities is through large quantum particle creation and the subsequent backreaction. This idea has been discussed by numerous authors over the years, but it still leads to several unanswered questions. In this session, this topic was addressed by T.P. Singh, reporting on work in collaboration with T. Harada, H. Iguchi, K. Nakao, T.Tanaka and C. Vaz[3]. These authors examine the limits of semiclassical analysis and discuss a model in which the backreaction is not significant before the analysis breaks down at the Planck scale. A somewhat different singularity issue was discussed by D.A. Konkowski, in collaboration with T.M. Helliwell. These authors discuss a notion of singularity which arises when wave operators are not essentially self-adjoint. They apply this to the Klein-Gordon, Maxwell and Dirac operators in quasiregular spacetimes, spacetimes with mild singularities such as conical or dislocation singularities. They find that these spacetimes are quantum mechanically singular in the sense that the wave operators fail to be essentially self-adjoint.

One of the important applications of quantum fields in curved space is to cosmology. The strong gravitational fields in the early universe provide a likely place where quantum effects may play a significant role. In this session there was one talk devoted to cosmological issues, given by Paul Anderson on work with B.L. Hu and C. Molina-Paris. These authors treat a model containing N scalar fields on a cosmological background, and examine the backreaction question with the aim of applying the results to chaotic inflationary models. A $1/N$ expansion was presented, and the effects of the leading terms discussed.

The Casimir effect, the effect of boundaries on quantum fields, has long been a useful way to model the effects of spacetime geometry on quantum fields. In the session, there was one talk relating to the Casimir effect by M. Karim. This talk dealt with the effects of a weak gravitational field on the original parallel plate configuration of Casimir. The results claimed are controversial, with some arguing that the magnitude of the effect is much smaller than that suggested by Karim. In any case, this configuration illustrates an interesting effect that should exist on some level.

There are still a number of more formal issues of quantization in curved spacetime that are active areas of research. One of these has to do with the ambiguity in the definition of vacuum states. (As already noted above, this was discussed by Shankaranarayanan for Schwarzschild spacetime.) Tevian Dray also addressed it, in a report on work with J. Fisher[6]. These authors reexamined and generalized the Ashtekar-Magnon condition, which is an energy condition for selecting a unique quantization.

The study of quantum matter fields in a classical spacetime is part of the larger subject of quantum gravity, and the boundaries are often not well defined. This session contained two talks which lay somewhat outside of quantum field theory in curved spacetime, but on related issues. These were both given by R.B. Mann, in collaboration with I.S. Booth. The first was on pair creation of rotating black holes in cosmological backgrounds[7,8]. The second was on quasilocal energy and naked black holes[9].

In summary, this session contained a diverse set of contributions which illustrated the variety of interesting mathematical and physical questions which are being actively pursued to better understand quantum processes in the presence of gravity.

References

1. V. Frolov and D. Fursaev, Phys. Rev. **D63**, 124010 (2001), hep-th/0012260.
2. G.L. Alberghi, R.Casadio, G.P. Vacca and G. Venturi, Phys. Rev. **D64**, 104012 (2001), gr-qc/0102014.
3. T. Harada, H. Iguchi, T.P. Singh, K. Nakao, T.Tanaka and C. Vaz, Phys. Rev. **D64**, 041501 (2001), gr-qc/0010101.
4. D. A.T. Vanzella and G. E.A. Matsas, Phys. Rev. Lett. **87**, 151301 (2001), gr-qc/0104030.
5. C. Barcelo, S. Liberati and M. Visser, gr-qc/0110036.

6. J. Fischer and T. Dray, Gen. Rel. Grav. **31**, 511 (1999), gr-qc/9807028.
7. I.S. Booth and R.B. Mann, Nucl.Phys. **B539**, 267 (1999).
8. I.S. Booth and R.B. Mann, Phys. Rev. Lett. **81**, 5052 (1998).
9. I.S. Booth and R.B. Mann, Phys. Rev. **D60**, 124009 (1999); Phys. Rev. **D59**, 064021 (1999).

www.ingramcontent.com/pod-product-compliance
Ingram Content Group UK Ltd.
Pitfield, Milton Keynes, MK11 3LW, UK
UKHW020404150926
8624IPUK00004B/338

9 789812 381712